Component-based Product Line Engineering with UML

The Addison-Wesley Object Technology Series

Grady Booch, Ivar Jacobson, and James Rumbaugh, Series Editors
For more information check out the series website [http://www.aw.com/cseng/otseries/]

Armour/Miller, *Advanced use-case Modeling, Volume 1*

Binder, *Testing Object-Oriented Systems: Models, Patterns, and Tools*

Blakely, *CORBA Security: An Introduction to Safe Computing with Objects*

Booch, *Object Solutions: Managing the Object-Oriented Project*

Booch, *Object-Oriented Analysis and Design with Applications, Second Edition*

Booch/Rumbaugh/Jacobson, *The Unified Modeling Language User Guide*

Box, *Essential COM*

Box/Brown/Ewald/Sells, *Effective COM: 50 Ways to Improve Your COM and MTS-based Applications*

Cockburn, *Surviving Object-Oriented Projects: A Manager's Guide*

Collins, *Designing Object-Oriented User Interfaces*

Conallen, *Building Web Applications with UML*

D'Souza/Wills, *Objects, Components, and Frameworks with UML: The Catalysis Approach*

Douglass, *Doing Hard Time: Developing Real-Time Systems with UML, Objects, Frameworks, and Patterns*

Douglass, *Real-Time UML, Second Edition: Developing Efficient Objects for Embedded Systems*

Fowler, *Analysis Patterns: Reusable Object Models*

Fowler/Beck/Brant/Opdyke/Roberts, *Refactoring: Improving the Design of Existing Code*

Fowler/Scott. UML *Distilled, Second Edition: A Brief Guide to the Standard Object Modeling Language*

Gomaa, *Designing Concurrent, Distributed, and Real-Time Applications with UML*

Gorton, *Enterprise Transaction Processing Systems: Putting the CORBA OTS, Encina + + and Orbix OTM to Work*

Graham, *Object-Oriented Methods, Third Edition: Principles and Practice*

Heinckiens, *Building Scalable Database Applications: Object-Oriented Design, Architectures, and Implementations*

Hofmeister/Nord/Dilip, *Applied Software Architecture*

Jacobson/Booch/Rumbaugh, *The Unified Software Development Process*

Jacobson/Christerson/Jonsson/Overgaard, *Object-Oriented Software Engineering: A use-case Driven Approach*

Jacobson/Ericsson/Jacobson, *The Object Advantage: Business Process Reengineering with Object Technology*

Jacobson/Griss/Jonsson, *Software Reuse: Architecture, Process and Organization for Business Success*

Jordan, *C + + Object Databases: Programming with the ODMG Standard*

Kruchten, *The Rational Unified Process, An Introduction, Second Edition*

Lau, *The Art of Objects: Object-Oriented Design and Architecture*

Leffingwell/Widrig, *Managing Software Requirements: A Unified Approach*

Marshall, *Enterprise Modeling with UML: Designing Successful Software through Business Analysis*

Mowbray/Ruh, *Inside CORBA: Distributed Object Standards and Applications*

Oestereich, *Developing Software with UML: Object-Oreinted Analysis and Design in Practice*

Page-Jones, *Fundamentals of Object-Oriented Design in UML*

Pohl, *Object-Oriented Programming Using C + +, Second Edition*

Pooley/Stevens, *Using UML: Software Engineering with Objects and Components*

Quatrani, *Visual Modeling with Rational Rose 200 and UML*

Rector/Sells, *ATL Internals*

Reed, *Developing Applications with Visual Basic and UML*

Rosenburg/Scott, *use-case Driven Object Modeling with UML: A Practical Approach*

Royce, *Software Project Management: A Unified Framework*

Ruh/Herron/Klinker, *IIOP Complete: Understanding CORBA and Middleware Interoperability*

Rumbaugh/Jacobson/Booch, *The Unified Modeling Language Reference Manual*

Schneider/Winters, *Applying use-cases: A Practical Guide*

Shan/Earle, *Enterprise Computing with Objects: From Client/Server Environments to the Internet*

Warmer/Kleppe, *The Object Constraint Language: Precise Modeling with UML*

White, *Software Configuration Management Strategies and Rational Clearcase: A Practical Introduction*

Component Software Series

Clemens Szyperski, Series Editor
For more information check out the series website [http://www.awl.com/cseng/cbdseries/].

Allen, *Realizing eBusiness with Components*

Cheesman/Daniels, *UML Components: A simple process for specifying Component-based Software*

Atkinson et al, *Component-based Product Line Engineering with UML*

Component-based Product Line Engineering with UML

COLIN ATKINSON
JOACHIM BAYER
CHRISTIAN BUNSE
ERIK KAMSTIES
OLIVER LAITENBERGER
ROLAND LAQUA
DIRK MUTHIG
BARBARA PAECH
JÜRGEN WÜST
JÖRG ZETTEL

Addison-Wesley

An imprint of Pearson Education

London · New York · San Francisco · Toronto · Sydney
Tokyo · Singapore · Hong Kong · Cape Town · Madrid
Paris · Milan · Munich · Amsterdam

KOBRA

PEARSON EDUCATION LIMITED

Head Office:
Edinburgh Gate
Harlow CM20 2JE
Tel: +44 (0)1279 623623
Fax: +44 (0)1279 431059

London Office:
128 Long Acre
London WC2E 9AN
Tel: +44 (0)20 7447 2000
Fax: +44 (0)20 7240 5771

Websites: *www.it-minds.com*
www.aw.com/cseng/

First published in Great Britain in 2002

ISBN 0 201 73791 4

British Library Cataloguing in Publication Data
A CIP catalogue record for this book can be obtained from the British Library

Library of Congress Cataloging in Publication Data
Applied for.

10 9 8 7 6 5 4 3 2 1

Designed by Claire Brodmann Book Designs, Lichfield, Staffs
Typeset by Pantek Arts Ltd, Maidstone, Kent
Printed and bound in Great Britain by Biddles Ltd of Guildford and King's Lynn

The Publishers' policy is to use paper manufactured from sustainable forests.

About the authors

In total ten authors from Fraunhofer Institute for Experimental Software Engineering (IESE) contributed towards the writing of this book. Colin Atkinson was the leader of the KobrA method development team at Fraunhofer IESE and is the prime author and editor of the book. He is also primarily responsible for the component model and the UML modeling approaches employed in KobrA. Joachim Bayer is a specialist in product line engineering and techniques for separating concerns, and mainly contributed towards the product line engineering, component reuse, patterns, and method application parts of the book. Christian Bunse, the manager of IESE's "Object-oriented and Component-based Development" competence area, is primarily responsible for the implementation approach. Erik Kamsties, the manager of IESE's "Application Engineering" competence area contributed mainly to the component specification and realization approaches. Oliver Laitenberger, the manager of the "Cost-effective Software Development" competence area at IESE, is the main author of the quality assurance aspects of the method. Roland Laqua, a specialist in configuration management and maintenance, is primarily responsible for the configuration management aspects of the method. Dirk Muthig, an expert in the area of product line engineering, is the chief author of KobrA's product line approach, and along with Joachim Bayer, is also the lead author of the Library Case Study featured in the book. Barbara Paech, the head of the quality software development department at IESE, is the primary author of the context realization part of KobrA. Jürgen Wüst, a specialist in the area of quality modeling and measurement, is primarily responsible for the quality modeling parts of the method. Finally, Jörg Zettel, a specialist in software process modeling, is responsible for the overall KobrA process description.

The Fraunhofer IESE focuses on applied research, development, and technology transfer in the areas of innovative software development approaches, quality and process engineering, software product lines as well as continuous improvement and organizational learning. To prepare industrial software developers and users for current and future information technology challenges, new techniques, methods, processes, and tools are being developed to base software development on sound engineering principles.

Contents

Foreword

Component-based approaches to system design and construction are so intuitively appealing that it seems obvious that component-based systems must be easy to build. Were that it were true! Over the years, we have learnt the hard way that intuition alone does not suffice. Software engineering has been struggling with componentization since the discipline first emerged in the 1960s. During this time we have made much progress and for the last decade(s) we have known how to develop systems from small-grained components called objects.

Today the battleground has shifted to dealing with large-scale, web-based systems built from components of varying granularity. This book represents a significant advance in this area. It addresses the complex issue of how to architect large-scale, component-based systems. It adopts a pragmatic approach built on the best practice of object-oriented methodologies such as OMT, Objectory, Fusion and Catalysis as well as formal approaches like the Cleanroom approach.

The book describes the KobrA methodology for modeling architectures. KobrA is a systematic approach to using UML that is refreshingly simple and straightforward. One of the problems for authors on components is that the term "component" is so heavily overloaded that it is hard to make any statement that won't confuse somebody. To overcome this, KobrA has introduced a new term "Komponent." Komponents are the "logical" components that represent the logical building blocks of a software system. The initially unfamiliar term really pays off and helps KobrA be very clear and concise when it needs to be.

KobrA is conceptually clean because it uses only a few concepts which are applied uniformly at all levels of granularity and abstraction. The method is thus applicable in essentially the same way to small and large systems alike, without disruptive changes in concepts, products or activities. The benefit is that the development activities applicable at a given point in a project are primarily determined by the properties of the artifact under development rather than by its size or location in the architecture.

KobrA is practically useful, because it is prescriptive in nature. The method proposes a logical and recursive process for producing the engineering artifacts. For each artifact it explains what inputs are needed and what consistency checks should be applied. The systematic process does not straitjacket the developer. KobrA does not require the developer to start with a greenfield; indeed KobrA devotes significant attention to software reuse.

KobrA is aimed at product lines as well as single applications and systems. Software product lines are a critical new paradigm for software development. By allowing the cost-effective development of product variants, product lines enable organizations to make impressive gains in time to market and cost reductions. KobrA is ahead of the pack in proposing a systematic methodology for framework engineering.

KobrA is aligned with industry standards. Industry has become increasingly aware of the need to be able to model software architectures in a consistent way independent of application domains and middleware platforms. In spring 2001, the Object Management Group (OMG), which is the leading standards body for distributed systems and component technologies, announced the Model Driven Architecture (MDA) to support this new approach to architecting systems using UML. MDA enables cross-platform interoperability, increased reuse of the design and implementation, and reduced cost of application development and management. MDA is based on UML and thus the KobrA method can claim to be the first comprehensive methodology to support model driven architectures.

With the publication of this book, Colin Atkinson and his colleagues at Fraunhofer IESE have made a very significant contribution to component-based methodologies and architectures. I hope you will be able to learn from the book as I have. If you want to know how to approach developing component-based architectures in a systematic manner, I cannot think of a better place to start than with *Component-based Product Line Engineering with UML*.

Derek Coleman
Chief Technologist
RosettaNet
www.rosettanet.org

July 2001

Preface

Bringing the levels of reuse in software development into line with that in other engineering disciplines has been a long sought after but elusive goal of software engineering. In the last few years, however, new development paradigms have emerged which promise to radically change this situation, and allow software engineering to claim its rightful place in the family of engineering disciplines. Most prominent among them are component-based software development and product line engineering. At one end of the granularity spectrum, components represent reusable software building blocks that can be quickly and easily assembled into new systems, while at the other end of the spectrum product line engineering consolidates all the common parts of an organization's range of products within a single, highly reusable software core.

Unfortunately, the exploitation of these paradigms for enterprise software development is frustrated by the lack of prescriptive, systematic methods for their application. Contemporary software methods that accommodate components or product lines typically do so either very late in the software life-cycle (in the case of components), or very early in the life-cycle (in the case of product lines). Few, if any, existing methods exploit the natural synergy between components and product lines at all stages and levels of development. This book presents a new method, known as KobrA, that aims to address this problem by offering a simple, prescriptive and systematic approach for component-based product line engineering across the full software life-cycle. It also leverages the Unified Modeling Language (UML) to support the two paradigms in a way that is compatible with most mainstream implementation and middleware technologies, including the leading component technologies. It therefore provides a foundation for a Model Driven Architecture (MDA) approach to software engineering.

A major goal of KobrA is to avoid the feature overload found in many other methods by defining as simple and orthogonal a feature set as possible. In particular, both the product and the process of KobrA are based on the recursive application of a minimal set of core principles. Another major emphasis of KobrA is on quality. The very power of components and product lines to promote reuse makes it imperative that the reusable assets are of the highest possible quality. Clearly, an organization that systematically develops and reuses poor-quality artifacts will not fare well in the long run. Well-defined consistency rules, and systematic techniques for verifying them, are therefore integrated into all parts of the method.

The evolution of software development methods is characterized by periods in which there is a rapid proliferation of concepts followed by periods of unification and consolidation. The story of the UML is perhaps the clearest example of this pattern. We hope the ideas in KobrA will represent a step towards the consolidation of component-based and product line development concepts, and help to expose and refine some of the key principles involved. Product line engineering is characterized by the idea of analyzing and explicitly capturing the commonalities and variabilities among different members of a product family. Component-based development is characterized at the analysis and design levels by "localized" development artifacts (e.g. UML diagrams) focussed on the description of individual logical abstractions (i.e. components). These two core ideas form the backbone of the book.

The KobrA project

The method described in this book was developed as part of a project funded by the German Federal Ministry of Education and Research (BMBF) known as the ***Ko**mponenten**b**asierte **A**nwendungsentwicklung*[1] or KobrA project. Four organisations collaborated in the project:

Fraunhofer-Institut für Experimentelles Software Engineering (IESE), Kaiserslautern

GMD-Forschungszentrum Informationstechnik (GMD-FIRST), Berlin

PSIPENTA Software Systems GmbH, Berlin

Softlab GmbH, München (project leader)

with Fraunhofer IESE being primary responsible for the KobrA method. GMD-FIRST and Softlab developed a complementary workbench to support the use of the KobrA method in conjunction with commercial CASE tools, while PSIPENTA provided the test-bed for the approach in the domain of enterprise resource planning.

Target audience

This book is intended primarily for software engineers who are looking for a simple but prescriptive method for developing software, especially those involved in the development of a family of applications who seek leverage from existing software assets. Because of the modular nature of the method, the main technological ingredients can be applied independently of one another.

1. German for "component-based application development."

Developers seeking support with component-based developments can apply the component-oriented ideas independently of product lines. Similarly, developers seeking help with model-driven architectures can apply the specification and realization ideas without worrying about components.

The book will also be of interest to academics and students who wish to gain a greater insight into the key principles of component-based development and/or product line engineering and how they relate to one another.

Readers should ideally have a familiarity with the general concepts of software engineering, the fundamentals of object-oriented software development, and the main features of the UML.

Document structure

The book is organized in five parts. Part 1 provides an introduction to the KobrA method and its background, and gives an overview of its key features. Part 2 focuses on the activities involved in modeling a system of logical components using the UML, without concern for product line issues. Part 3 describes how executable incarnations of these logical components can be attained, either by direct implementation or by the reuse of existing components such as Commercial Off-The-Shelf (COTS) components, and discusses how this fits into an incremental development approach. Part 4 focuses on product line engineering, and shows how support for product lines can be added. Finally, Part 5 focuses on the critical supporting activities, namely maintenance, quality assurance, quality modeling, and the introduction of KobrA into real-life software development settings.

Two appendices follow the main body of the book. Appendix A describes the KobrA metamodel, and includes a complete listing of the consistency rules that govern the structure and relationships of components. Appendix B provides a detailed specification of the KobrA process, and includes a complete listing of the activities that make up the process and the product flow dependencies between them.

The library case study

A single, running case study is used in all but Part 1 of the book to illustrate the application of the KobrA method. The domain of library systems is used for the case study since its basic concepts are familiar to nearly everybody. Systems in this domain help librarians and information professionals perform the tasks involved in running a library. This includes assistance in customer interaction, stock management, and accounting.

The basic library system supports the management of registered users and available items and tracks all performed loans, returns, and reservations. This basic library is selected as the particular member of the library system family used to illustrate the development of single systems in Parts 2 and 3 of the book.

The illustration of the product line concepts in KobrA requires many more details about the library system domain to be useful. For this purpose, three library systems were analyzed and their common and distinct features modeled in a generic framework. The extension of the case study to cover product lines is explained in Part 4.

The complete case study, including both the basic library system and the library system product line is described in [BMG01], which is available on the book's website.

Acknowledgments

The authors are grateful for the support and contributions of numerous individuals to the ideas described in this book. First, we would like to thank our colleagues on the KobrA project and on the KobrA board: Marion Dikel, Robert Hagemann, Gernot Krause, Jörg Lucas, Jens Reinhold and Torsten Sander of PSIPENTA Software Systems GmbH, Martin Dehn, Günther Merbeth, Friedrich Rössler and Uwe Zeithammer of Softlab GmbH, Matthias Anlauf, Marko Fabiunke, Joanna Filipek, Boris Groth, Stefan Jähnichen and Ronald Melster of GMD-FIRST, and Günther Ruhe of Fraunhofer IESE. Second, we would also like to thank our current and former colleagues at IESE: especially Stan Jarzabek, Dieter Rombach, Peter Rösch and Tanya Widen. Third, we would like to extend our thanks to our reviewers for their valuable feedback on drafts of the book, especially Morgan Bjorkander, Mathias Clauss, Mike Fischer, Hans-Gerhard Groß, Bernhard Humm, Heinrich Hussman, Thomas Kühne, Cris Kobryn, Antje von Knethen, and Frank Maurer. Finally, special thanks are due to Brigitte Göpfert for her insights and expertise in the library information domain of the case study, Maud Schlich for her contribution to the testing chapter, Jhair Tocancipa for his help with the EJB and CORBA examples, and to our editor, Alison Birtwell, of Addison-Wesley for her encouragement and support. And last but not least we are grateful to the Fraunhofer IESE and the BMBF for supporting this work.

Website

Visit the book's websites for more information on the topic:
www.iese.fhg.de/KobrA/book
www.kobra-project.org

PART

Introduction

1

Every new method is built upon the shoulders of the methods that have gone before it. The first part of the book sets the scene for the more detailed parts that follow by describing the background to the KobrA approach, the methods that have been most influential in shaping its features, and the goals that drove its development.

Chapter 1 starts by first explaining the significance and background of the paradigms integrated within KobrA, and reviews the existing methods upon which it is based. These can be regarded as the ingredients of KobrA. The chapter then describes the strategic principles by which these ingredients were combined. These can be regarded as the recipe for KobrA.

Chapter 2 then continues with an overview of KobrA and compares its features with those of other leading methods. Finally, the main properties of KobrA are summarized.

Background 1

"A program that is used in a real world environment must change or become progressively less useful in that environment."

Lehman

Lehman's law of software engineering is perhaps even more valid today than when it was first postulated in 1985 [LB85]. Success in today's competitive software industry is increasingly dependent on an enterprise's ability to react rapidly to changing market conditions and user requirements.

There are various ways of addressing the demand for change, but one of the most effective is software reuse. Clearly, the higher the proportion of existing software assets that can be adapted or included in a new product (or product variant), the lower the costs involved, and the greater the competitive position of a company. However, a systematic and reliable approach to software reuse has proved elusive. Numerous approaches have been tried over the years, but truly systematic software reuse is still the exception rather than the rule. One of the core goals of KobrA is to help address this problem and facilitate more cost effective software development through reuse.

The objective of this introductory chapter is to explain the background to the KobrA method, and to set the context for its description in the ensuing chapters. To this end we briefly examine the leading technologies upon which the software industry is currently pinning its hopes for greater reuse, and identify some of the obstacles that currently stand in their way. We then consider the strengths and weaknesses of some of the most influential development methods, and discuss how well they support these reuse technologies. Finally, we describe some of the ideal characteristics that a method should exhibit, and identify the main principles that guided the development of the KobrA method.

1.1 Reuse technologies

Over the years numerous technologies have been suggested as solutions to the perennial software reuse problem. However, only a few of these have succeeded in having a significant impact. Software developers today still generally spend inordinate amounts of time "reinventing the wheel," and any reuse that does take place tends to be more *ad hoc* than planned. In the last few years, however, several interrelated technologies have emerged that have the potential to make a serious impact on the reuse problem, particularly:

- component-based development
- architecture styles and design patterns
- product line engineering.

1.1.1 Component-based development

Many industry observers predicted that object-orientation would be the "silver-bullet" for reuse [Cox84]. Classes and objects, supported by the mechanisms of inheritance, polymorphism and dynamic binding, were supposed to facilitate widespread, institutionalized reuse and encourage a marketplace of reusable COTS (Commercial Off The Shelf) components. Unfortunately, however, these expectations were never fully realized.

One of the key problems was the fact that mainstream object-oriented languages are compiler bound technologies. This means that, at run time, objects are entombed within larger, monolithic artifacts that are not amenable to any of the powerful reuse-oriented features applicable to classes and objects. In C++, for example, the only directly executable unit is the "program"[1] which has very limited reuse properties. Reusing individual classes and objects requires developers to work through the tedious edit/compile/link cycle to embed the reused class/object within a new program.

The component paradigm alleviates this problem by allowing individual objects to be independently executable units. It can therefore be viewed as an extension of the traditional object paradigm that frees objects from the shackles of an encapsulating program. In the guise of components, classes enjoy all the reuse advantages of the object paradigm, but can also be packaged into ready-to-assemble and ready-to-run capsules. Using components in the construction of new applications is therefore more of an assembly than a programming process.

The component paradigm represents a substantial step forward in the quest for greater reuse. However, at present, it lacks adequate methodological support. Most development approaches that accommodate components essentially view them as merely executable modules that only have an important role in the final implementation and deployment phases of development. This is certainly the model of components offered by the current leading component technologies

1. Denoted syntactically by the "main" construct.

(Enterprise Java Beans, COM+ and CORBA). In order to unlock the full potential of components, the component paradigm must be supported in all phases of development, and must be viewed as an integral part of the development process rather than just the end result.

1.1.2 Architectural styles and design patterns

The component-paradigm promotes "reuse in the small," since components represent the smallest coherent packages of software that make sense as stand-alone, reusability entities. Architectural styles and design patterns can be viewed as addressing reuse at the next level of granularity, since they facilitate the reuse of strategies for assembling basic building blocks. They therefore provide a natural complement to the component-oriented approach – components represent the reusable building blocks from which solutions are constructed, and patterns/styles represent the reusable strategies for putting them together.

Architectures are generally assumed to be of a larger scale than designs, and styles are typically viewed as more oriented towards automated application than patterns, but both pairs of concepts overlap. Indeed, what some authors present at the architecture level [BMR96] others present at the design level [GHJ95] (e.g. the Model View Controller Pattern) and what some authors present as a style [GS94] others present as a pattern [BMR96] (e.g. layering). Architectural Styles and design patterns[2] are both essentially based on the underlying idea of defining generic assembly concepts that have proven their efficacy in solving recurring software problems. This idea can also be exploited at higher and lower levels of abstraction in the form of analysis patterns [Fowl97] and idioms [BMR96], respectively, and in a more concrete, encompassing fashion in the form of frameworks.

Architectural styles and design patterns not only typically support reuse at a larger level of granularity than components, but also at a higher level of abstraction. In contemporary technologies, components are represented in a concrete, executable language but architectural styles and design patterns, by their very nature, need to be expressed in a more abstract way typical of an architecture or a design. In other words, they need to "abstract away" from the details of a specific implementation, and capture the essence of the solution concept in the form of a model.

The Unified Modeling Language (UML) [OMG00a] is currently the leading notation for modeling architecture and design level information in a graphical form, and provides explicit mechanisms for depicting architectural styles and design patterns. By comparison, the UML's support for the component concept is somewhat limited and conforms to the prevailing "executable module" notion of contemporary component technologies. It provides a specific component symbol for representing deployable software packets, but intends this to be used primarily in component and deployment diagrams at the very latest stages

2. "Architectural style" and "design pattern" are commonly used terms because they were introduced by seminal publications [GHJ94] [GS94].

of development. This restricted view of components not only limits their role in the earlier phases of development, such as analysis and design, but also complicates the exploitation of associated architectural styles and design patterns. To enable the full potential of architectural styles and design patterns to be used with the component paradigm, methodological support is needed for modeling components at a higher level of abstraction using the full spectrum of UML modeling capabilities.

1.1.3 Product line engineering

Reuse at the largest level of granularity is supported by the product line approach. This exploits the fact that most software companies develop and maintain a family of similar products rather than one single product. A product line is designed to leverage the overlap among members of the family by consolidating the common parts within a single, high-quality core, and by facilitating easy selection and/or tailoring of the variable parts. Developing new applications is thus more an issue of instantiating the reusable asset rather than of programming in the traditional sense.

For many years practical product line engineering was frustrated by the fact that traditional software implementation technologies, such as programming languages, did not really provide the mechanisms needed to support the rapid and cost-effective adaptation of implemented code. As a result, contemporary product line methods have been forced to concentrate on the early life-cycle activities, and thus are often seen as rather "high-level" and somewhat divorced from the "real business" of coding. There is also an erroneous perception that product line development is incompatible with "regular" development of single systems. Many developers therefore feel the adoption of a product line approach would force them to discard their current "single system" practices with which they feel comfortable.

The advent of component based development promises to change this situation by providing mechanisms that enable software elements, right down to the binary level, to be rapidly and efficiently assembled into new applications. This allows the basic tenet of product line development to be applied at all phases and levels of software development, and to work with software in all its different forms, including binary forms. Components, therefore, provide the perfect foundation for the practical application of product line development.

Another critical factor in the success of a product line approach is the nature of the reusable "core." At a minimum this should contain a reference architecture supported by techniques for capturing and selecting the points of variation among family members. In its most general form, the reusable asset takes the form of a framework, which embodies implementation (i.e. code level) artifacts for the common parts of the family, as well as higher-level design and architecture models. Since they embody every possible reusable asset, at all levels of abstraction, frameworks constitute the largest possible reusable artifacts for a particular product family.

Like components, the practical exploitation of frameworks and product lines is also hindered by the lack of modeling support from the UML. Other than the general stereotype mechanism, the UML provides no explicit mechanism for capturing variable parts of a system, or for selecting among them. There is also a lack of methodological support for utilizing product lines and frameworks in conjunction with components.

1.2 Development methods

Technologies of the kind described in the previous section hold the key to increased software reuse in the development of software systems. However, without appropriate methodological support their value is limited. To be effectively applied to practical problems, advanced development concepts must be accompanied by guidelines describing when, where and how they should be applied. This is the role of software development methods.

In the relatively short time that computers have existed, the number of different methods for developing software for them has become quite large, particular of the object-oriented variety [GOW01]. Providing a comprehensive overview of them is consequently a topic for a substantial book in its own right. In the following subsections, we briefly review the methods that have had the most direct influence on the development of the KobrA method. The goal is not to provide a comprehensive description of each method, or to compare their pros and cons in a systematic way, but rather to highlight their primary strengths and weaknesses as a means of motivating the features and characteristics of KobrA, and to help define KobrA's position within today's methods landscape. In the next chapter, after providing an overview of KobrA, we compare and contrast its features with the methods introduced below.

1.2.1 First-generation object-oriented methods

Almost all modern software-development approaches have roots in the first generation of object-oriented methods. The explosion in the number of object-oriented methods in the early 1990s, and the subsequent cross-fertilization of ideas, makes it very difficult to trace the precise influence and contribution of each method. Below we discuss five relatively old object-oriented methods that have had the greatest influence on the KobrA method. Other well-known first-generation methods such as Shlaer/Mellor [SM88], Objectory [Jac92], Syntropy [CD92], Booch [Boo94] and Coad/Yourdon [CY91] are not discussed explicitly because their influence has been less direct. Since they were developed relatively early in the history of object technology, none of these methods directly considers components, design patterns and product lines. Nevertheless, they all embody important ideas that can be helpful in using these technologies.

1.2.1.1 OMT

Probably the single most influential method in the evolution of object-oriented development methods is the OMT (Object Modeling Technique) method developed [RBP91] at General Electric research labs. Not only was it the most widely used object-oriented method in its heyday, but also many of its ideas have been adopted in ensuing generations of methods and notations. Its two most significant legacies are the UML, whose notational concepts are primarily based on the OMT, and its approach to analysis.

The OMT identifies three distinct, but overlapping, models to be generated during the analysis of a system:

- **Object model.** Using a class diagram, the object model identifies the user visible data abstractions (or types) that the system manipulates, and the relationships between them. This part of the method is heavily influenced by classic entity-relationship modeling techniques.
- **Dynamic model.** Using statechart diagrams and sequence diagrams, this model describes the dynamic behavior of the system in terms of states, events and object interactions.[3]
- **Functional model.** Using an enhanced form of data flow diagram, this model shows the computations, or data transformation, performed by the system.

The analysis phase is followed by a design phase in which the system is divided into subsystems and algorithms are designed for the methods of the identified classes. Although it is the strongest part of the method, OMT's analysis approach has one main weakness. The approach used to document the functional model, based on data flow diagrams, proved to be difficult to use consistently with the other analysis models. In practice, therefore, OMT users often ignore this model. The design process is the major shortcoming of OMT as a general purpose method. It is much less well defined than the analysis process, and lacks any kind of support for incremental development.

Despite these shortcomings, OMT's overall approach to system analysis is very sound, and still represents one of the most effective ways of organizing the models that describe a system's required properties.

1.2.1.2 Fusion

Fusion [Col94] is a descendent of OMT that also enjoyed significant success in its heyday. As well as fixing some of the problems with OMT's analysis approach, and significantly enhancing the design process, the Fusion method also pointed the way towards increased rigor and prescriptiveness in object-oriented development.

The Fusion analysis phase adopts the same three basic analysis viewpoints as OMT, but with slightly different models and terminology:

3. Here we use the UML terminology rather than the original OMT name for the diagrams.

- **Object model.** A class diagram which plays the same basic role as the object model in OMT. Its main purpose is to define the externally visible types manipulated by the system.
- **Operation model.** Using textual operation schemata rather than data flow diagrams, this model replaces the functional model of OMT, and provides a declarative definition of each system operation in terms of pre- and post-conditions.
- **Life-cycle model.** Using regular expressions, this model describes a subset of the information typically included in the behavioral model of OMT, namely the allowable sequences of system operation invocations.

One of the main innovations of Fusion's analysis approach was to require that the textual operation schemata describe the effects of system operations in terms of the concepts in the object model. This not only improves the presentation of the functional model (as compared to data flow diagrams), it also gives rise to a substantial set of consistency rules between the functional and object model. This in turn, can significantly improve the quality of the models, and provides one of the most concrete strategies available for determining closure on the content of an object model. Most methods provide only very vague guidelines as to when to stop modeling classes and their relationships.

The other main contribution of the Fusion method was the use of interaction modeling and collaboration diagrams to drive a systematic design process. Before Fusion, design was largely viewed as an extension of analysis, but at a slightly lower level of abstraction. Fusion showed that the design process could be organized in a highly systematic fashion based on the elaboration and documentation of interaction scenarios on an operation-by-operation basis. The basic idea in the Fusion design phase is to create a collaboration diagram to document how each system operation is realized in terms of lower-level interactions. Analyzing these interactions not only leads to the identification of design classes, but also systematically identifies what methods and attributes they have to support.

Unfortunately, Fusion suffers from the same problem as OMT in having a "flat" waterfall-based process model. Later variants of Fusion, such as EVO Fusion [Cot95] explicitly support incremental development, but Fusion itself is essentially based on a strict "waterfall" sequence of activities. However, the method does hint at the possibility of recursively applying the method to "large" objects (e.g. subsystems) that are identified in the first pass through design. The implication is that the same process steps can be reapplied, but at lower levels of granularity.

Although it does not address the most recent advances in object technology, Fusion's basic strategies for the analysis and design of a system and its use of operation schemata to control the quality and information content of models, still offer one of the best foundations for systematic object-oriented development.

1.2.1.3 ROOM

Another influential method published around the same time as Fusion was Real-Time Object-Oriented Modeling (ROOM) [SGW94]. In contrast with OMT and Fusion, which essentially represent a synergy of data modeling and structured development strategies, ROOM's view of software is based more on the concept of interacting processes. The basic building blocks of ROOM, known as actors, are active "logical machines", rather than simple ADTs, and typically encapsulate an active thread or process as well as state information. At the language level they therefore correspond to active constructs such as tasks in Ada or threads in Java.

Actors communicate with each other via ports, each of which defines a set of messages that an actor supplies or requires. In general, an actor can have multiple ports, some of which may be incoming ports that define the messages the actor receives and processes, and some of which may be outgoing ports that define messages the actor sends to others. Bi-directional ports are also supported. The protocols represented by ports are defined independently of actors by means of protocol classes. Thus, in order to arrange for two actors to communicate, a binding has to be established between them in which an outgoing port of one party is bound to a compatible incoming port of the other party. The compatibility of ports can be ensured by defining one to be the "conjugate" of the other.

Although ROOM was published before component-based development became a buzzword and dedicated component technologies such as CORBA, COM and JavaBeans became available, it contains all the basic characteristics of a classic run-time component model. Actors represent self-contained, autonomous components that can be realized as hardware elements as well as software elements, ports represent independently defined interfaces that allow components to be connected together in arbitrary configurations, and bindings represent concrete connectors that link components together to solve particular problems. Moreover, ROOM allows configurations of actors at one level to be regarded as the internal realization of actors at a higher level, thus facilitating the recursive nesting of components to arbitrary depths.

The main problem with ROOM is its lack of integration with the data-modeling arm of object technology epitomized by the object models of OMT and Fusion. Although ROOM defines an advanced and highly systematic way of using state machine diagrams, it makes little use of the core object-modeling concepts such as associations, attributes and multiplicities etc., and gives little indication of how they fit in. This is a symptom of the fact that ROOM is more process- (i.e. thread) oriented than data-oriented. ROOM is also rather vague about the development time organization of software artifacts.

These problems notwithstanding, ROOM still provides one of the cleanest run-time models of a component-based system. Furthermore, it shows how these concepts are naturally recursive, and enable larger components to be constructed from smaller ones to create arbitrarily nested structures.

1.2.1.4 HOOD

The Hierarchical Object-Oriented Design (HOOD) [Rob92] method is much less well known and widely used than the previous three methods, with a user base largely limited to the European Space Agency (HOOD's creator) and its contractors. However, it has powerful and unique concepts not found in any other methods.

HOOD views a system as a community of objects organized in terms of two hierarchies: the seniority hierarchy and the usage hierarchy. The seniority hierarchy reflects the containment of objects within one another, and always yields a tree structure. The usage hierarchy reflects the usage of one object by another, in the sense of a client–server relationship, and is usually a graph rather than a tree.

The key idea in HOOD is to organize the development steps around the containment hierarchy (or seniority hierarchy as it is called). The overall development approach is thus one of recursive, top-down refinement in which progressively smaller objects are identified, modeled and implemented. The heart of the process is the so-called basic design step that is responsible for decomposing an object into its sub-objects and elaborating successively deeper parts of the tree.

Unfortunately, HOOD lacks many of the concepts considered fundamental to the object paradigm, including inheritance, polymorphism and classes. Moreover, its details are very much oriented towards Ada, making it difficult to use with other languages. Nevertheless the idea of organizing the development process around the object containment hierarchy provides a strategy for exploiting the advantages of recursive refinement in connection with object technology.

1.2.1.5 OORAM

OORAM [RWL96], like HOOD, is not among the most widely known object-oriented methods, but introduced some important ideas relating to the way in which systems can be modeled. The key innovation in OORAM is to focus on roles throughout the modeling process as a way of tying different perspectives of a system together. Following the use-case ideas pioneered by Objectory [Jac92], many object-oriented methods incorporate the ideas of roles when describing how external agents interact with the system. However, OORAM extends the idea to the internal parts of the system, allowing the interactions between all kinds of objects to be described in terms of roles.

A role essentially defines how a client object sees a server object, including its operations, behavior and needs. When a system is configured, different objects may play different roles provided that they satisfy the expectations placed upon the role (polymorphism). Different models are thus generated to describe roles from the perspective of different objects in the system. Once the models are stable, object types are defined by comparing these different role models. The role synthesis activity reveals new object types by consolidating the behavior and properties of simpler ones.

Unfortunately, the overall life-cycle process adopted in OORAM is essentially a waterfall model, although there is a high level of iteration within the major phases such as analysis and design. Object containment and hierarchical development play no part in the process, and thus it is essentially a "flat" method like OMT and Fusion. Nevertheless the idea of role modeling, and particularly the idea of basing the information within diagrams on the perspective of particular objects is a powerful one.

1.2.2 Component-oriented methods

In view of the importance of the component paradigm, most modern methods aim to accommodate them in one form or another, but generally components are viewed as just a convenient implementation tool rather than an integral part of the overall software development cycle. In the last few years, however, methods have emerged which orient the whole development process around components, and view them as richer abstractions than just binary code modules. Three of the most prominent are discussed below.

1.2.2.1 Catalysis

Catalysis [DW98] was one of the first methods developed specifically to leverage the UML in connection with component-based development, and embraces many of the other reuse technologies mentioned in Section 1.1, including architectural styles, design patterns and frameworks. The method either introduced or popularized many of the ideas that today are considered natural ingredients of component-based development, and several of these have been explicitly adopted in the UML.

Catalysis is organized in terms of three primary dimensions, called "modeling concepts," "levels of modeling," and "principles." Each of these, in turn is composed of three further basic ideas.

The three main *modeling concepts* emphasized in Catalysis are:

- **collaborations** – a set of abstract interactions between objects that play certain roles with respect to one another. The method popularized the concept of treating collaborations as "first-class citizens" of a methodology with their own dedicated notation.
- **types** – an abstract definition of an object's externally visible behavior. Catalysis adapted the concept of an object model as used in OMT and Fusion to provide the basis for the definition of a component's type model. As in Fusion, the type model forms the foundation for the textual description of a component's operations.
- **refinement** – a relationship between two descriptions of the same thing but at different levels of detail. Catalysis emphasizes the idea of fractal-like concepts in which the same basic structure is evident no matter how far one zooms in or out of a particular part of the system. Refinement is also seen as providing the foundation for recursive development.

To this list of modeling concept is added the notation of a framework – a generic set of reusable models and designs. Any of the previous concepts that are deemed to be potentially reusable are captured in a generic way in the form of a package.

In the second dimension of concern, Catalysis defines *three levels of modeling*:

- **problem domain or business** – the "outside" of a system which describes the concepts of relevance to users of the system; this captures the environment or context in which the system is expected to operate
- **component specification** – the "boundary" of a system or component that captures externally visible behavior
- **component implementation** – the "inside" of a component which describes how it is constructed from smaller parts.

Finally, in the third dimension of concern, Catalysis defines three *principles* to support the development effort:

- **abstraction** – the separation of the relevant aspects of a concept or artifact from irrelevant detail
- **precision** – the development of abstract descriptive artifacts that make concrete, accurate, and testable statements
- **pluggable parts** – adaptable software building blocks that can be connected and used together with minimal effort.

These various concepts and principles contribute towards the products of a software development project. The process aspect is addressed by means of a set of process patterns intended to support different development contexts. These focus on the concepts of component-based development, short cycles, phased development, and variable degree of rigor.

The rich set of features offered by Catalysis is both its strength and its weakness; it is a strength because it equips the method with all the conceptual tools that may be needed to tackle software-development projects. But it is a weakness because it makes the method complex and thus difficult to learn and apply effectively. In particular, the method's reliance on patterns to define the process gives it a feel of being somewhat unfocussed and unprescriptive, and provides potential users with little concrete guidance on how to exploit the rich set of ideas in a systematic way. In short, it can be difficult to "see the wood for the trees."

These problems notwithstanding, Catalysis lays out most of the essential ingredients that a method for component-based development should support, particular the notion of the type model, action refinement, recursive processes, and collaborations as "first-class citizens" of the modeling language.

1.2.2.2 Select Perspective

Another method developed at around the same time as Catalysis to exploit the benefits of the component paradigm is the Select Perspective [AF98]. This method emphasizes the importance of business process modeling as the starting point of development, and follows a clean process that transforms system-independent business processes into implementation-oriented models of the system of interest in a step-by-step way. This includes the explicit identification of components, as well as the potential integration of legacy systems.

A business process model is first developed to gain an understanding of the environment in which the required system is expected to operate and the role it is expected to play. The goal is to identify the core concepts and activities in the "as is" business process, as well as in the "to be" business process. Use-case modeling techniques are then applied to analyze and elaborate upon the precise scenarios by which the system and its users interact.

The structural aspects of a system are described by class diagrams, usually after the business process model and the use-case models have been developed. Class diagrams are used in a global sense to model, among other things, the overall system architecture, but they are also used in a local sense to describe classes related to a particular theme or responsibility in the system. Groups of scoped classes are organized into packages to capture the properties of business-oriented components.

The other two main models developed in the Select Perspective are interaction diagrams and state machine diagrams. Interaction diagrams are primarily used in a subsequent step to use-case analysis as a means of documenting the way in which the user interacts with the system and perceives the interaction among objects. State machine diagrams are then used to describe how business objects and control objects behave according to received events and internal state. The method also describes how legacy assets can be wrapped to facilitate their integration into encapsulating component-based environments.

The Select Perspective is particularly rich in practical recommendations and guidelines which indicate that the approach has been matured in practice. It is also supported by a dedicated professional CASE tool. This combination of practical guidelines with a professional tool strengthens the approach's applicability and acceptance.

The main weakness of the Select Perspective is that it is not always clear which aspects of the underlying business objects are being described by which models. In other words, the distribution of the information describing a business object is somewhat arbitrary. For example, it is not always easy to see which models describe the specification of a business object or component, and which describe the design decisions related to how it works. Also, it is not clear how to apply these models and activities in a recursive way to support a hierarchy of nested components.

Nevertheless, like Catalysis, the Select Perspective defines most of the essential ingredients needed for component-based development in the early stages of the software life-cycle, and provides some useful guidelines for their application. In particular, it reinforces the idea of localizing class diagrams to capture the properties of individual business objects or components. It also makes clear why the properties and advantages of components should be considered as early as the business process modeling stage, and defines practical techniques for addressing them as part of the business process modeling process. Unlike other methods, it also explains the role that component technology can play in integrating legacy systems into new applications, and suggests how this can be achieved.

1.2.2.3 UML components

Cheesman and Daniels describe a method [CD00], strongly influenced by Catalysis, RUP [JBR99] and Syntropy [CD92], that focuses on the specification of components using the UML. The method identifies two main phases (or workflows): the requirements workflow which captures the basic needs that the system must fulfill in terms of use-cases and high-level business classes and the specification workflow which documents the business types, interfaces, and components that have been chosen to satisfy these requirements. Each involves the creation of a number of models:

- Requirements
  - business concept model
  - use-case model
- Specification
  - business type model
  - interface model
  - component specifications
  - component architecture.

The specification workflow is the most interesting from the perspective of component-based development, and consists of three main sub-workflows: component identification, component interaction, and component specification.

During component identification, components and their interfaces are first identified and an initial component architecture is generated. Interfaces play a particularly important role in this phase. They come in two basic varieties. System interfaces are derived from use-cases to support the required interactions, while business interfaces are derived from a business type model to support data abstractions. As an input for the workflows that follow, individual components are identified for each interface, and the initial component architecture consisting of component instances and their connections is defined. The component interaction workflow then refines these initial interface components by developing optimal collaboration patterns.

The component specification workflow documents the interfaces and components identified in the previous workflow in terms of UML models. This is achieved using an enhancement of the type model strategy introduced by Fusion and refined by Catalysis. Basically, every interface has an associated class diagram (type model) and several associated operation specifications which document the effects of the interface operations in terms of pre- and post-conditions. Class diagrams are therefore localized for a specific interface and often overlap in the information they contain. Finally, components are specified using class diagrams which indicate how many instances of each kind of interface a particular component supports.

In essence, UML components offer a subset of Catalysis concepts but with a much simpler, RUP-flavored process. This is both its strength and weakness. On the one hand it packages a core subset of the Catalysis concepts in a more accessible and prescriptive way, but on the other, it loses some of the key ideas of Catalysis, including the nesting of components to arbitrary depths, the recursive application of development concepts, and the use of frameworks to package larger-grained reusable structures than interface and components. Nevertheless, the emphasis on the early definition of component architectures in terms of component instances and their connections, and the enhancement of the idea of focussing diagrams on individual components or interfaces, represent valuable insights.

1.2.3 Product line oriented methods

With the possible exception of Catalysis, the methods described to this point are focussed on the development of single systems. The creation of system variants takes place as part of the maintenance activity, and is generally viewed as a repeated application of the method rather than an integral part of the method itself. With the growing recognition of the value of a product line approach to the software life-cycle, several methods have emerged in recent years that focus on product line oriented software development and maintenance. Catalysis can support such an approach thanks to its advanced framework concepts, but product lines are not its main focus. The product line oriented methods vary in their degree of customizability and the level of abstraction at which they address the variabilities and commonalities in a product family.

1.2.3.1 FODA

Probably the most frequently used product line method is the feature-oriented domain analysis (FODA) method published by the Software Engineering Institute about one decade ago [KCN90].

The key product of FODA is its feature model which captures the essence of domain engineering. The basic idea is that a domain is analyzed to identify the features which a system in this domain must or may provide. These features are hierarchically represented within a feature model. Features are recursively composed of other features, with some features being optional or alternatives to other features. A feature model thus serves as a useful input to the designers of a reference architecture for the domain.

The problem with FODA is that it is not described in sufficient detail to be easily applied without guidance. The lack of a detailed description is also a reason why many organizations often define their own domain-engineering methods, which are mostly variants of FODA but more concrete in those aspects which are relevant to the particular organization.

Nevertheless, the feature model is a useful way of capturing commonality and variability within a system family, and adds value to methods that focus on software reuse and product lines. For example, the Reuse-Driven Software Engineering Business [JGJ97] has been integrated with FODA's feature modeling [GFD98].

1.2.3.2 FAST

Another well-known product line engineering method is Lucent's FAST method [WL99]. The heart of FAST is an activity known as Commonality Analysis which identifies commonalities in a family of systems in terms of general textual statements. The variations among them are also captured in a similar textual form to produce a document, known as the domain model, that characterizes the commonalities and variabilities in the family. This domain model is used as the basis for "implementing the domain," which involves the creation of domain-specific languages, architectures, generators, etc. to facilitate the low-effort creation of new members of the product family.

Once a suitable domain-specific application engineering environment is available there is no doubt that it can lead to significant reductions in the effort involved in creating new family members. However, the development of such an environment itself involves considerable effort, not only in its initial creation but also for its subsequent maintenance. It therefore represents a high-risk "up-front" investment that many organizations cannot or do not want to afford, especially when the additional initial effort to analyze a domain and predict potential changes is taken into account. The success of the investment is also very sensitive to the quality of this domain analysis because a domain-specific environment can be rather expensive to modify when unexpected changes occur.

Another problem with FAST is that it only addresses the product line issues at a very high level of abstraction, akin to the analysis level in conventional development methods. The critical connection to concrete implementation technologies is not directly addressed. Moreover, the guidelines provided for the domain analysis process are vague and unprescriptive.

Despite these problems, FAST clearly sets out the key characteristics that distinguish successful product line approaches from conventional single system methods. Chiefly, these are the ideas of commonality analysis to identify commonalities in, and variabilities between, family members, and the explicit documentation of them as part of the domain description. FAST also promotes the concept of continuously managing commonality and variability while moving from the requirements down to the implementation level.

1.2.3.3 PuLSE

Although not as well known as the previous methods, one of the most mature product line approaches is the PuLSE method, developed at Fraunhofer IESE [BFK99]. PuLSE splits the life-cycle of a software product line into four phases: initialization, product line infrastructure construction, usage, and evolution. The method provides technical components for these different deployment phases which contain the technical know how needed to operationalize the development of the product line. Moreover, these technical components are customizable to the respective context.

The customization of PuLSE to the context in which it will be applied ensures that the process and products optimally match the requirements. In the initialization phase, the other phases and the technical components are tailored. Through this tailoring of the technical components, a customized version of the construction, usage, and evolution phases of PuLSE is created. The principle dimensions of customization are the nature of the application domain, the organizational context, reuse aims and practices, the project structure and available resources.

Unfortunately, PuLSE suffers from the same basic problem as the other product line approaches in that it focuses on the description of family properties at a very high-level of abstraction, and gives no concrete guidance on how the required flexibility should be realized at the implementation level. Like the others it does not explicitly consider the role of components either.

However, PuLSE is one of the most mature and adaptable product line methods and illustrates the importance of customizability in the commonality/variability analysis techniques that characterize these approaches.

1.2.4 Object-oriented method frameworks

Object-oriented methods have come a long way since the early approaches mentioned above. Not only have they become more sophisticated, they have had to embrace a significant set of new concepts, such as components, architectures, frameworks, use-cases and incremental development. One strategy for accommodating all these ideas in a coherent way is to raise the level of abstraction in which a process is described and make it more generic. This leads to methods that are compatible with a large number of specific development strategies, but are not ready to use "out of the box." They must consequently be tailored to the needs of specific projects. Such approaches, of which there are two main examples, are therefore often characterized as method frameworks.

1.2.4.1 Rational Unified Process (RUP)

The Rational Unified Process[4] ([JBR99], [Kru00]) is primarily an amalgamation of OMT [RBP91], Booch [Boo94], Objectory [Jac92] and the Rational Method [Kru00]. The appearance of OMT, Booch and Objectory in this list is no coincidence, since the original authors of the UML played a significant role in the RUP's development. In fact, as its name implies it aims to be for processes what the UML is for modeling notations – a unified standard. It is therefore more accurate to view the RUP as a "process framework" for different kinds of software systems, domains, and organizations, rather than a concrete method ready to use "out of the box."

At the top level of granularity, the overall RUP development life-cycle is divided into four main phases each consisting of a sequence of iterations. The inception phase is responsible for developing a "good idea" into a vision of the end product in terms of an outline of the candidate architecture, a list of identified risks and a "proof-of-concept" prototype. This is followed by the elaboration phase during which the majority of the functional requirements are specified, a stable architecture

4. There are actually three distinct variants of this process framework: the general Unified Process, defined in [JBR99], a more concrete variant known as the Rational Unified Process, defined in [Kru00], and proprietary extensions of the latter that can be purchased from Rational Software. The core ideas of three variants are essentially the same, however.

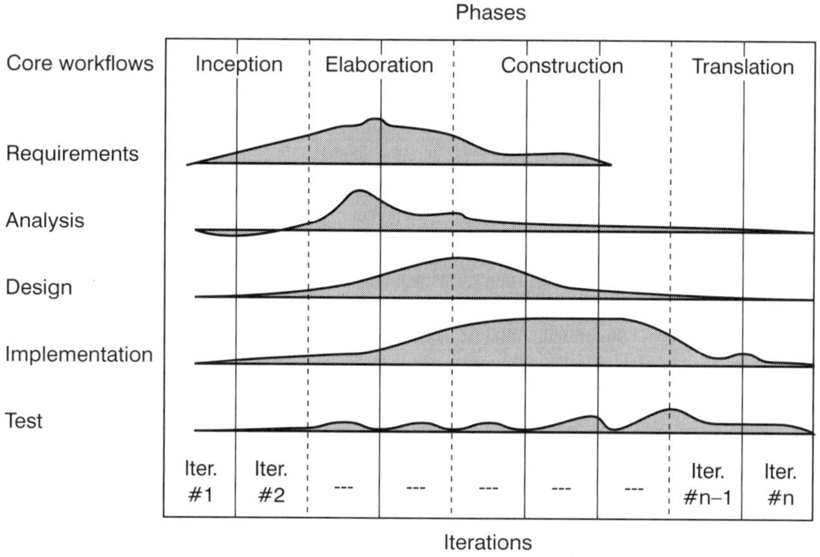

Figure 1.1 Unified Process iterations, workflows, and phases

is designed, and a project plan is established. Most of the implementation and testing work is actually carried out during the subsequent construction phase. Finally, in the transition phase, beta tests are executed, users are trained and the software is given to the end users. At the end of this phase the product is fully released.

Figure 1.1 is the core diagram that illustrates the relationship between activities, phases and iterations in the RUP. Core workflows, similar to the steps in classical software development approaches, are shown on the vertical axis. These include requirements, analysis, design, implementation, and test. The horizontal axis represents the different phases and their iterations. While each iteration includes all described workflow activities, the emphasis of a particular iteration shifts according to the phase. This means that during the inception and elaboration phase most effort is put into the definition of requirements. The construction phase, on the other hand, requires more effort on implementation and testing.

Different artifacts are created at different points in a RUP development project:

- The use-case model represents the system's intended functions and environment. It is used as a means of communication between customers and developers. The use-case representation of the system contains actors, use-cases, and their relationships.
- The analysis model provides an overview of the expected system functionality, but avoids details of how the system actually works. The analysis model consists of classes and interaction diagrams illustrating the realization of use-cases.
- In the design model, the analysis model is adapted to a selected implementation environment, such as CORBA. The design model encompasses the design classes, subsystems, packages, collaborations, and relationships between them.

- The implementation model primarily contains the code for the software system.
- The test model consists of information relevant for performing testing during the various phases. It is a collection of test cases, test procedures, and test components.

Another very important artifact in the RUP is the architectural description which is essentially orthogonal to the models just described. The architectural description provides a view of the system that combines "slices" of the different models most relevant for the architecture. In other words, it contains the architecturally most important use-cases from the use-case model, the architecturally most important classes and subsystems from the design model and so on. Most parts of the architectural description are finalized at the end of the elaboration phase and do not change significantly during subsequent phases. The architectural description therefore captures the structural elements and their interfaces together with their behavior as embodied in their collaboration.

The RUP is clearly an extremely feature-rich method that embraces almost all the generally recognized techniques and artifacts of modern software engineering, including use-cases, architectures, patterns, and components. Moreover, it places great emphasis on elaborating these artifacts in an iterative and incremental manner. In other words, successive iterations gradually add more and more detail to the models until a final executable product is attained.

As with Catalysis, this richness is both a strength and weakness of the RUP. It embraces all the modern best practices of software engineering, but gives little concrete guidelines on how to resolve the tensions that inevitably arise. For example, the process is simultaneously "use-case driven" (i.e. function-oriented) and "architecture centric" (i.e. object-oriented), but it is not always clear which should take precedence when conflicts arise. Also, although the method divides up the work to be performed into small iterations, as illustrated in Figure 1.1, many of the iterations ignore key process steps (or "workflows" as they are termed in the RUP). Detailed implementation activities, such as coding and testing therefore still do not occur until the analysis and design activities have been performed on a system-wide basis. The basic phases of the waterfall model are therefore lurking in the background and still have a strong influence on the organization of products and process steps.

Surprisingly for a method intended to be the standard for UML development, the method is not very prescriptive about the use of the UML. It requires the creation of various models, but is not specific about which diagrams should be used to represent them and what precisely they should contain. In particular, the role of components is not entirely clear. As an artifact, a component is defined as "the physical packaging of model elements," in line with the UML viewpoint. As a workflow, however, "component engineering" deals with all kinds of models at all abstraction levels, suggesting that a component is more of an architectural unit than a physical executable.

In summary, the RUP has many of the right conceptual ingredients but does not really integrate them in a simple, "ready to use" way. Despite these shortcomings, however, the RUP provides a valuable framework for visualizing the many ingredients of modern object-oriented software engineering methods and how they work together.

1.2.4.2 OPEN

Another approach developed with the unification of object-oriented methods in mind is OPEN (Object-Oriented, Process, Environment, and Notation) [GHY97]. OPEN initially encompassed a unified notation, known as the OML (OPEN Modeling Language) [FHG97] as well as a process, but the former has been subsumed by the UML standardization effort. In its current form, therefore, OPEN, like the RUP, can best be characterized as a highly generic process framework oriented towards development with the UML.

As illustrated in Figure 1.2, OPEN defines the basic activities and products involved in object-oriented development at a very high level of abstraction. In order to use the approach it is first necessary to define a concrete instance of these activities and products tailored to the project at hand. Key ideas in the framework include an object-oriented description of the processes, in which activities are viewed as contract-driven objects, and the use of matrices to map tasks at one level of granularity to appropriate solution activities and techniques at the level below. OPEN is therefore often characterized as a contract-driven process, meaning that there are no explicit sequences defined for the

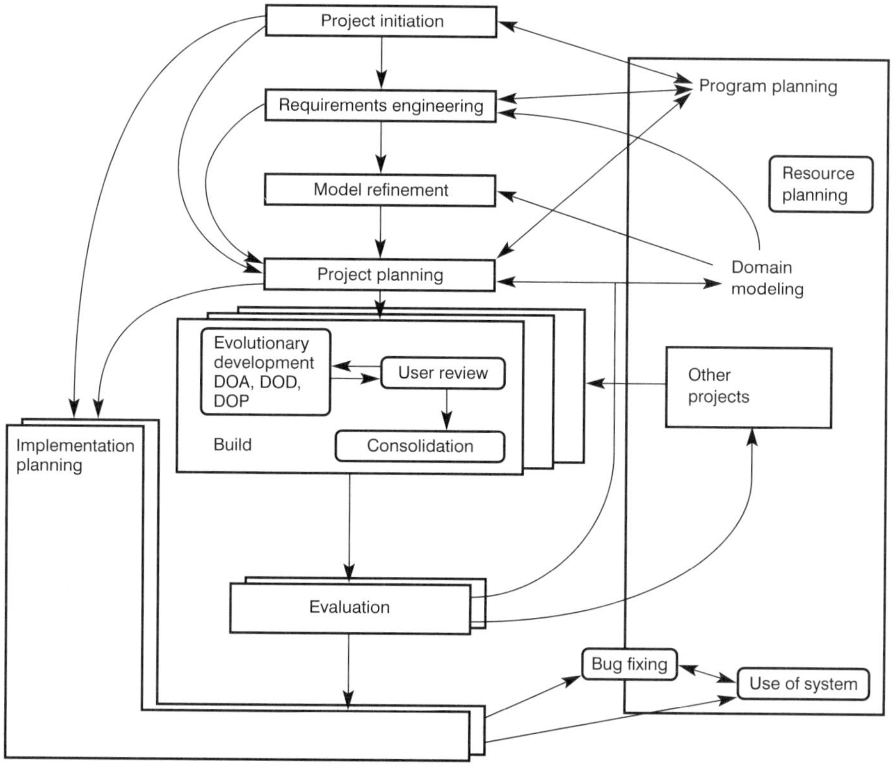

Figure 1.2
OPEN process and products

development activities, only strict dependencies. As a result, the process model can encompass a wide range of distinct enactment policies.

Like the RUP, the generic nature of OPEN is a double-edged sword. On the one hand it means that the ideas of OPEN are applicable, when properly instantiated, to a very wide range of domains. On the other hand, it means that much of the difficulty in using the method is wrapped up in the instantiation process. Poor or incorrect instantiation can easily lead to an incoherent process with very little chance of success. Unfortunately, the instantiation of the OPEN process is one of its least mature parts. Also, in trying to integrate all popular object-oriented techniques, including components, architectural styles and design patterns, OPEN suffers from the same "feature overload" problem as the RUP. Nevertheless, OPEN embraces some important innovations, including the use of contract driven activities to structure the overall development process, and the matrix-based strategy for mapping activities and techniques at different levels of granularity.

1.2.5 Cleanroom techniques

Although no longer as fashionable as the previously mentioned approaches, in its heyday the Cleanroom approach to software development generated as much excitement [MLH87]. Cleanroom still represents one of the most systematic and rigorous approaches to software development available today. Among other things, Cleanroom pioneered the principles of:

- **incremental development.** A system should not be carried through the development phases as one single whole, but should be assembled from individually executable elements (increments). These increments are derived from an analysis of the system specification, and must fulfill various conditions to ensure that they can be merged to form the complete system.
- **recursive development.** A system is developed by the recursive application of common abstraction techniques, rather than by the application of disjoint, monolithic phases.
- **near-zero defect software.** Every software artifact is inspected as soon as it is created to make the software as error free as possible.
- **rigorous specification and design.** Box structures (step-wise refinement) and correctness verification (step-wise abstraction) facilitate the rigorous and semi-formal mapping of high-level requirements to low-level implementation concepts.
- **usage testing.** Software specifications are not limited to functional requirements, but also model expected system usage in terms of externally visible system states, their interdependencies, and probabilities of state transitions. This model is used later to develop test cases and to perform certification.

The core of the Cleanroom development process is the "box-structured method." This identifies three basic views of a software element, or "box," known as a black box, a state box and a clear box (see Figure 1.3). These can be nested hierarchically in a tree-like structure to support a step-wise approach to system elaboration in which individual boxes are expanded and refined within each of the steps.

As illustrated in Figure 1.3, a black box describes a system, or subsystem, by modeling its externally visible behavior. All details about internal structure and operations are omitted. The model captures the behavior in terms of stimuli and responses, the former being input to the system (e.g. a command with parameters), and the latter being externally visible reactions of the system (e.g. printing on a screen). Each response of the system is determined by the system's stimuli history which contains all previously received stimuli.

A black box can be refined into a state box. A state box describes a system's behavior by using an internal state, defined by analyzing required stimulus history and responses. The internal state is used to encapsulate parts of the stimulus history. Each response is determined by stimulus and state histories.

Finally, the state box is refined into a clear box, which provides a procedural description of the system by presenting details about how stimuli are processed

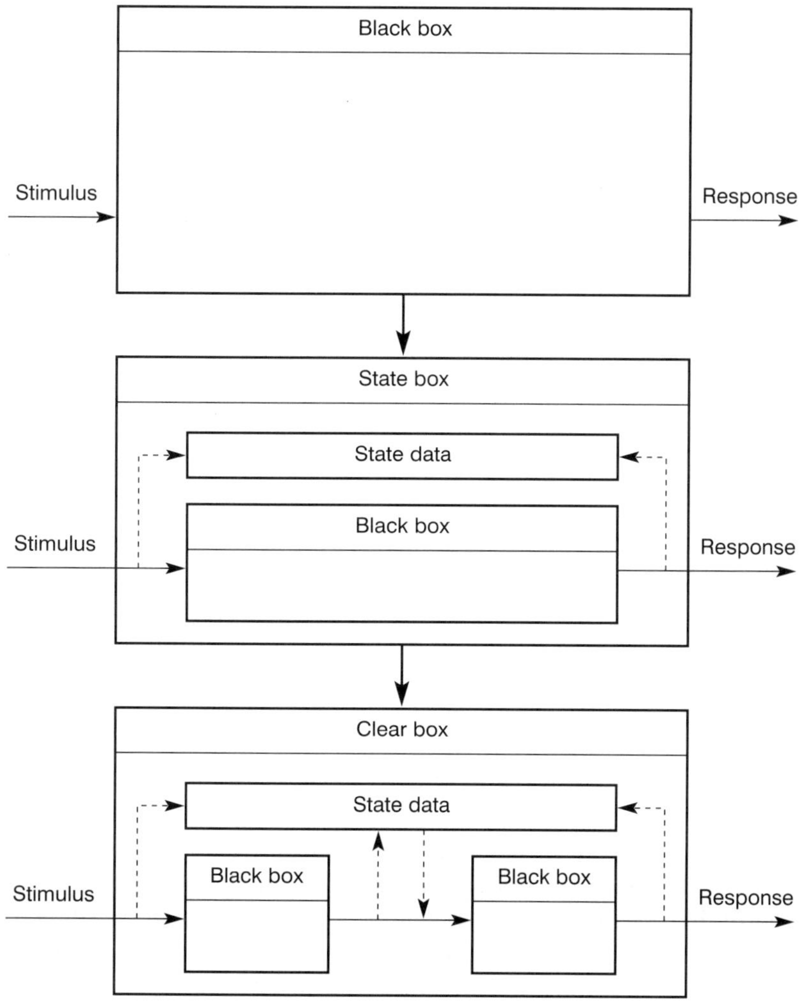

Figure 1.3
Box structured development

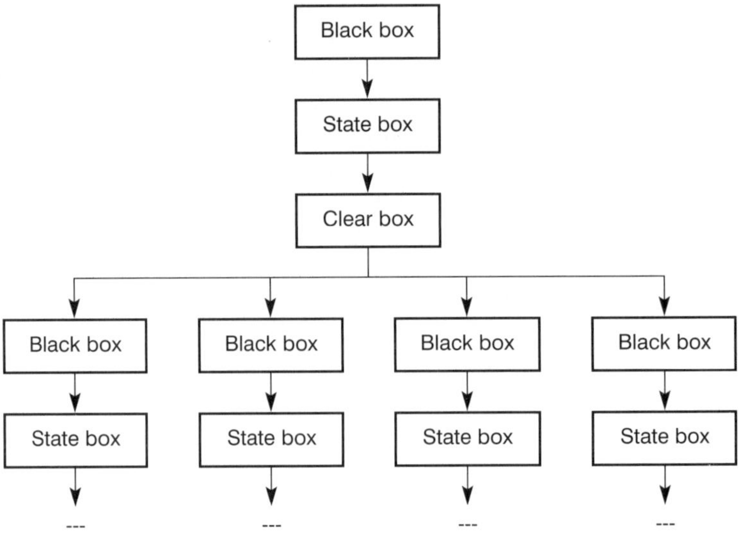

Figure 1.4 Box-structured hierarchy

to determine responses. The black-box function of the state box is replaced by concurrent or sequential usage of new black boxes. These boxes can be expanded at the next level into new state and clear boxes.

The refinement of the three box types produces a hierarchy of box structures, as illustrated in Figure 1.4. These black-box subsystems become well-defined and independent modules in the overall system by further refinement.

Four fundamental principles have been defined to guide system refinement in terms of these three box structures:

1. **Referential transparency.** In the delegation of any system part for design and implementation, all requirements should be specified explicitly and independently, so that no further communication or coordination is logically required to complete the system part.
2. **Transaction closure.** The transactions (i.e. use-cases)[5] of a system or part of the system should be sufficient for the acquisition and preservation of all its state data. Its state data should be sufficient for the completion of all its transactions. In particular, system integrity, as well as user functionality should be considered in achieving transaction closure.
3. **State migration.** System data should be decentralized to the smallest system parts that do not require duplicating data updates. This basically states that, where possible, the implementation of a system part should be decomposed into lower level subparts which encapsulate and protect well-defined data abstractions.
4. **Common service.** System parts with multiple uses should be considered for definition as common services. A corollary principle is to create as many opportunities as possible for reusability within and between system parts.

5. Cleanroom transactions are essentially equivalent to modern-day use-cases.

For all its many strengths, Cleanroom has one overriding problem – its association with structured development, and as a consequence its prime facia incompatibility with object-oriented development. Although numerous authors have pointed out that this is in fact not the case [Dec96], Cleanroom's image as the flagship of structured development has meant that it has never made the transition to object technology, let alone to component technologies. This is unfortunate because, many of its software engineering principles are as relevant to object-oriented and component-based development as they are to traditional structured approaches.

1.3 Basic goals

The development of a method inevitably involves trade-offs. In this section, we describe the basic goals and strategic principles that drove the development of the KobrA method.

1.3.1 Simple

A method should be as simple as possible. Software engineering is already complex enough without unnecessary complexity being added by development methods. However, many of the methods identified in the previous section do just this by introducing many overlapping concepts for a single underlying idea. Such concept redundancy usually complicates, rather than simplifies, the task of the software developer.

In a concrete sense, simplicity means that a method should be as economic as possible with its concepts. This means, in particular, that the features in a method should be as orthogonal as possible. This is closely related to the idea that a method should "separate concerns" to the greatest extent possible.

1.3.2 Systematic

Many methods present potential users with a rich set of concepts, but only provide vague and high-level guidelines on how they should be used. Users of such methods consequently often find themselves stuck, with little idea of what they should be trying to do, or what options they have available. To avoid this problem a method should be as systematic as possible.

This has two distinct aspects. First, it requires that concepts and guidelines defined in the method should (so far as possible) be precise and unambiguous, ideally to the extent of statements in a mathematically formal language. However, this is rarely practicable for industrial scale projects. Second, it requires that a method should (as far as possible) tell developers what they should do, rather than what they may do. In other words, a method should be prescriptive.

The principle of separation of concerns mentioned above is important for systematic development, as well as for simplicity. Without a clear concept of what precisely is to be achieved at a given point in a process, it is very difficult to go about achieving it in a systematic way. A specific and fundamental case of this principle is that the product(s) of a method should be strictly separated from the process.

1.3.3 Scalable

A method should be scalable both with respect to granularity and complexity. Scalability with respect to granularity means that the method should be able to accommodate large-scale and small-scale problems in the same manner using the same basic set of concepts. There should be no quantum changes in concepts or activities when a problem changes only in size. One of the best ways of achieving this is to organize the artifacts and the activities in the method hierarchically (i.e. recursively). This means that both the product and the process have the properties of a fractal – no matter how close you "zoom in" or "zoom out," their basic shape looks the same.

Scalability with respect to complexity means that the concepts of the methods should be applicable incrementally. This is important if a method is to be used cost effectively. Novice users should be able to use the core features of the method effectively without needing to learn the more advanced features. However, they should also be able to introduce these more advanced features seamlessly when they are ready. Strictly separating concerns is one of the key strategies for achieving such incremental introduction of capabilities.

The idea of the "scale" of a method is related to the distinction between "heavyweight" and "lightweight", or agile, methods. Scalability implies that a method has some capacity to adjust its "weight" according to the particular project at hand.

1.3.4 Practical

The three characteristics discussed above are quite similar to one another, and could be perceived as having an academic flavor. A method developed solely on the basis of these three principles is likely to be quite small and theoretically pure – in short, an academic method. However, a mainstream method, capable of supporting enterprise-level software development, must balance the theoretical principles with consideration of practical development and integration issues.

One of the most critical concerns from a practical standpoint is that a method should be compatible with as many commonly used implementation and middleware technologies as possible, particular those that are either *de facto* or *de jure* standards. This is closely related to the Model-Driven Architecture (MDA) approach, which is based on the concept that the essential structure and behavior of a system should be captured in an abstract way so that it can be mapped flexibly to different implementation technologies. This not only allows an architecture to exploit the most appropriate implementation technology available, but also endows it with the capacity to endure inevitable technology evolution.

2 Method overview

This chapter provides an introductory tour of the KobrA method. Like any artifact composed of multiple parts, it can be described from a top-down or a bottom-up perspective. In this chapter we take the top-down approach, describing the big picture of how KobrA can be used for software development. Then, in the following parts of the book we take a bottom-up approach, describing the various parts of the method in detail and showing how they go together to form the overall development approach.

The basic goal of this chapter is to furnish the reader with a mental picture of the method upon which to base his/her navigation through the detailed chapters in the rest of the book. First, we explain the main areas of concern in KobrA, and then describe the artifacts that are developed during a KobrA project and the activities that are responsible for producing them. Once the main concepts of the method have been introduced, we then compare and contrast KobrA to other leading methods, and summarize its main characteristics. The final subsection provides an explicit "road map" to guide readers through the book.

To help explain the method, in this chapter we use a small case study known as the simple banking system. This enables customers to open accounts in some specified currency, and to deposit money into, or withdraw money from, accounts in any currency. To facilitate the necessary currency conversions, the system maintains a set of exchange rates between euros and other currencies. Clerks working for the bank can update a particular exchange rate at any time. Customers can also request for an amount in a foreign currency to be converted to and from euros. Finally, an account can have an associated limit specifying to what extent a customer can overdraw money from the account.

Four different variants of the system are available: one offering all the facilities described above; one with all the facilities except those related to international currencies; one with all the facilities except support for overdrawing to a set limit; and one which has support neither for overdrawing nor for foreign currencies.

2.1 Core concepts

"Separation of concerns" is a time-honored strategy for handling complexity in science and engineering. By dividing a complex problem into parts that can be "conquered" separately, it allows projects to be tackled which would otherwise be beyond the capacity of the human intellect. In the context of a software engineering method, it is the basic means by which a method can be powerful enough to handle enterprise scale problems, yet still remain fairly simple. A core principle of KobrA is therefore the strict and systematic separation of concerns, so that at all times during a development project developers are aware of what they should be attempting to do and what "concern" they are working on.

Perhaps the most obvious manifestation of this principle in KobrA is in the separation of the "product" from the "process." Many methods arbitrarily mix the description of *what* engineers should be trying to produce with the definition of *how* they should produce. However, this makes it difficult to distinguish what is strictly necessary from what is merely optional. In contrast, explicitly separating the definition of the products from the process not only simplifies the presentation of the method, but also allows different approaches to be used to produce the required artifacts provided they ensure that these artifacts exhibit the desired properties. To reinforce the strict separation of product (or artifacts) and process (or activities) in KobrA, this chapter and most of the following chapters present the products and the processes separately.

Figure 2.1
Primary development dimensions

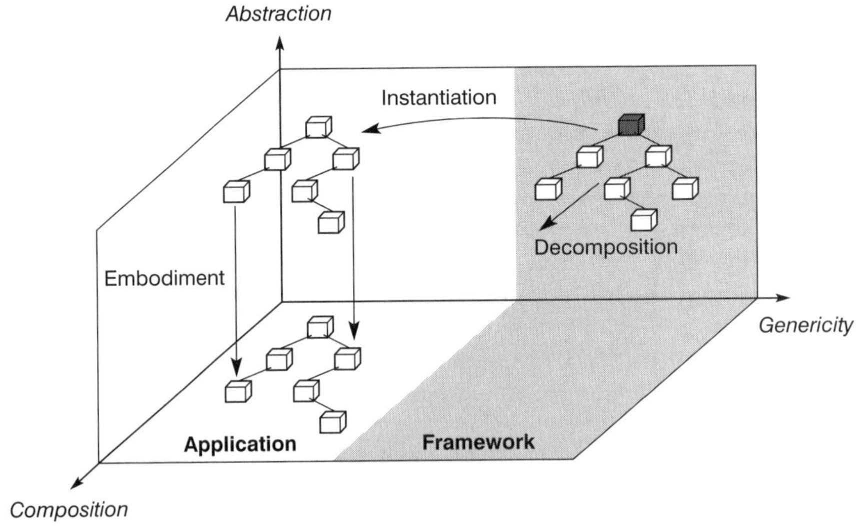

Another fundamental separation of concerns in KobrA is the organization of the method in terms of three orthogonal dimensions of development: one dealing with the level of abstraction; one dealing with the level of genericity; and one dealing with composition. The relationship between these is illustrated in Figure 2.1.

In a typical KobrA project, development begins with the black box in the top right-hand corner of Figure 2.1. This represents a generic, abstract, black-box description of the system to be developed. It is generic because it has possibly variant features (i.e. parameters), it is abstract because it is comprised mainly of UML models at a level akin to analysis and design, and it is a black box because it describes only the externally visible properties of the system when considered as a single whole. In order to create an executable version of this system it is necessary to:

1. Remove the genericity to create an instance of the system that meets the needs of a specific application (instantiation).
2. Decompose the system into finer grained parts to yield a containment tree of nested components (decomposition).
3. Reduce the level of abstraction to create an executable representation of the system and its parts (embodiment).

Performing all three transformations produces the system in the bottom left part of Figure 2.1, which is represented in terms of finer grained components, lower (i.e. zero) genericity, and a concrete (i.e. an executable) representation.

All methods that aim to support component-oriented reuse must support the same basic transformations in one way or another. However, most mix containment, genericity and abstraction issues in arbitrary ways leading to unnecessary complexity and confusion. In KobrA these three dimensions are strictly separated. Development concerns dealing with the genericity dimension come under the umbrella of *product line engineering*, development concerns dealing with the composition dimension come under the umbrella of *component modeling*, and development concerns dealing with abstraction dimensions come under the umbrella of *component embodiment*.

The separation of these concerns does not mean that they are addressed in sequential order. On the contrary, since they are genuinely orthogonal in KobrA, they can be tackled in various orders, and in fact arbitrarily interwoven, depending on the precise set of circumstances found in a particular project. The only requirement for strict separation is that at any point in the development process a developer knows precisely which concern he/she is working on. Figure 2.1 shows only the canonical route for getting from the abstract, generic, and coarse-grained (i.e. user-oriented) view of a system to a concrete, specific, and fine-grained version that can be executed.

The product line engineering aspects of the method are described in detail in Part 4 of the book, the component modeling aspects in Part 2 and the component embodiment aspects in Part 3. Part 5 of the book describes the supporting activities concerned with maintenance and quality assurance.

2.1.1 Product line engineering

At the largest level of granularity, the product line paradigm takes precedence in KobrA. This splits the overall development cycle into two parts, one dealing with the development of a framework – a reusable set of software artifacts whose core is embedded within all products developed by the enterprise; and the other concerned with the development of an application – a concrete instance of the framework, adapted and extended to meet the needs of a specific user or customer. Framework engineering encompasses the activities and artifacts related to the development and maintenance of a framework, while application engineering encompasses the activities and artifacts related to the development and maintenance of an application.

The motivation for product line engineering is to leverage the fact that most enterprises develop a family of similar, but slightly different, systems rather than a single system. Under such circumstances it makes sense to consolidate common parts of the family within a reusable core (i.e. a framework) rather than to develop everything afresh for each specific system. The development of a specific member of the family then becomes more a matter of instantiating the framework than developing everything from scratch. However, for this approach to be cost-effective, the products developed by an enterprise must have a fairly large overlap, and the commonalities and variabilities between these systems must be carefully analyzed and well understood.

A powerful way of capturing both the variability and commonality within the members of a family is by genericity. A generic artifact is an artifact that embraces all possible variants of the family, but provides some means of selecting between them. As illustrated in Figure 2.2, *C* and *L* are variants of some artifact that share a common set of features (represented by the intersection of the circle) but also have some features distinct from one another. The generic representation of the artifact is captured by the thick line. In principle, four groupings of features can be derived from this generic artifact, one with the features embraced by the circle *C*, one with the features embraced by *L*, one with the intersection of the features and another with the union of features. However, they may not all be meaningful for the particular product in hand. In the case of the simple banking system, if *C* represents the features related to currency manipulation, and *L* represents the features related to account overdrawing, all four combinations are meaningful.

Product line engineering addresses the genericity dimension illustrated in Figure 2.1. Framework engineering focuses on the development of the generic artifacts capturing the feature sets which are characteristic of all members of a product line, while application engineering is concerned with instantiating these to create specific members. Since these concerns are orthogonal to the

Figure 2.2
Generic artifacts

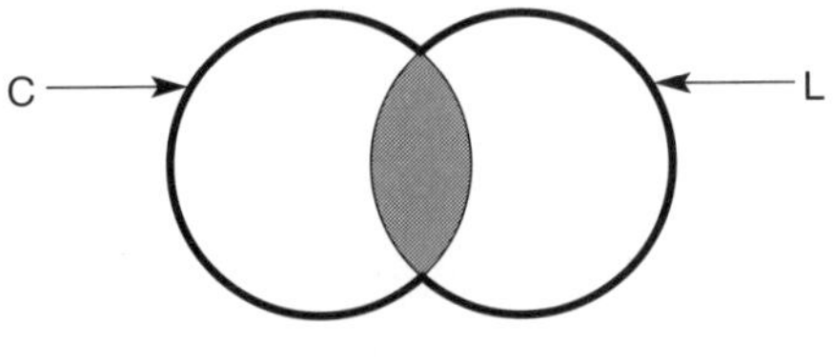

containment and abstraction dimensions, both frameworks and applications can contain artifacts at all levels of granularity and at all levels of abstraction, right down to executable binaries.

2.1.2 Component modeling

At the intermediate level of granularity, KobrA is driven by the component paradigm. KobrA frameworks and applications are all organized in terms of hierarchies of components. However, the components in a KobrA framework or application are not physical components in the sense of contemporary "physical" component technologies (e.g. CORBA, .NET/COM+, J2EE/EJB), but rather "logical" components that represent the logical building blocks of a software system. To help distinguish the logical KobrA components from regular physical components we use the word "Komponent" for the former. A Komponent is thus a "**K**obrA **component**". The term component refers to any kind of component, including Komponents.

2.1.2.1 Komponents versus Komponent instances

Another important separation of concerns in KobrA is the careful distinction between the run time and development time incarnations of Komponents. A running system is regarded as being composed of a collection of Komponent instances that essentially possess the fundamental characteristics of objects in object-oriented programs, including encapsulation, modularity, unique identities, and unified data/functions. Moreover, they communicate using the same basic method invocation mechanisms as objects.

At development time, software engineers are concerned with the description of Komponent types that capture the common properties of their run time Komponents instances. Like the UML, KobrA uses the unqualified term "component" (or Komponent) to mean a component type.

Komponents have both the properties of classes and the properties of modules (i.e. packages in UML terminology). The class-like facet allows a Komponent to define the attributes, operations and behavior possessed by its instances, while the module-like facet enables a Komponent to represent a name space and act as a container for features, classes, and other Komponents. In this respect Komponents are somewhat like subsystems in the UML. Like a subsystem, a Komponent defines a grouping of model elements that represent a behavioral unit of a system. The difference between UML subsystems and Komponents is that a subsystem cannot possess behavior of its own, whereas Komponents most definitely can.

2.1.2.2 Aligning composition and containment trees

The aim of the composition dimension depicted in Figure 2.1 is to capture how coarse-grained components are "made up of" finer-grained components, in a recursive manner, down to the level of small, primitive components. However, the concept of "being made up of" has a different interpretation at run time than at development time. Being precise about the differences and relationships between the two interpretations is essential for systematic component-based development.

At run time, the notion of one Komponent instance being "made up of" other component instances is captured by the concept of composition. KobrA adopts UML's concept of composition, which is viewed as a variant of aggregation in which a part is private to, and created by, the composite. At development time, however, the notion of one Komponent being "made up of" other Komponents is captured by module containment (a variant of UML ownership).

Since composition and containment both define tree structures, a basic principle of KobrA is that the run time and development time organization of Komponent instances and Komponents should be "aligned." In other words, the shape of the development-time containment tree should be driven by the nature of the run time composition tree. The basic requirement for alignment is that the shape of a containment tree, and the location of Komponents within it, should support the required visibility between Komponent instances at run time. Visibility in a containment tree is governed by the usual visibility rules of a module or package hierarchy – namely, a Komponent can see its immediate children, its peers, and any of its ancestors or their peers.[6] Aligning the two trees provides a systematic way of ensuring the high cohesion and low coupling of Komponents.

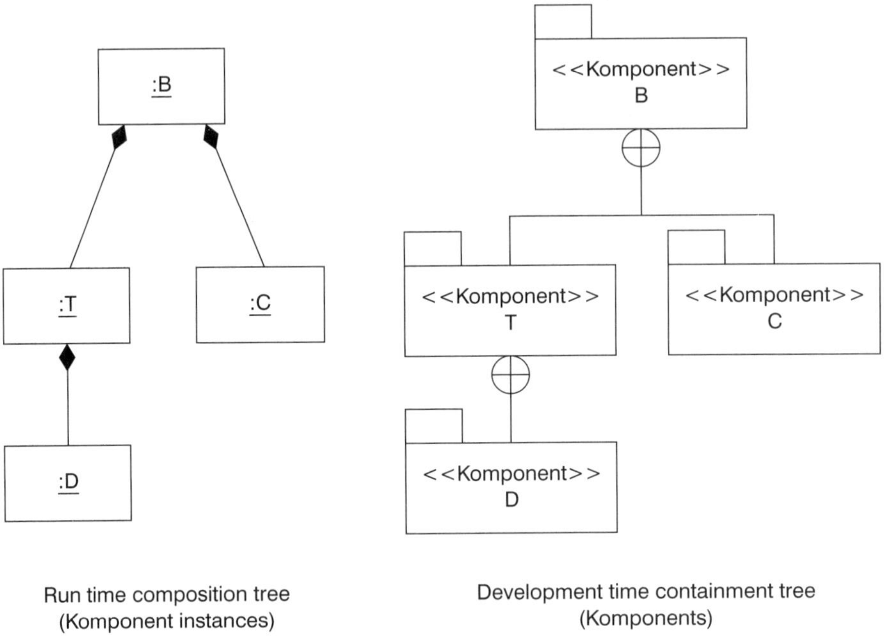

Figure 2.3
Aligned composition and containment trees

6. This is true for the regular "private" forms of composition and containment. KobrA also supports "public" forms of these, which can be used together or separately.

Figure 2.3 shows a simple example of aligned composition and containment trees using the UML notation. In this case the shape and organization of the trees are identical. However, attaining alignment is not always this simple. One problem is that a system of Komponent instances does not always have a single composition tree. Often a system can contain multiple, dispersed composition trees, or may contain none at all. Moreover, the shape of these trees may change over time. For these reasons, a different run time tree is used as the starting point for alignment – the so-called "creation tree." The creation relationship exists between two Komponent instances when one creates the other. Since a Komponent instance must be created by exactly one other Komponent instance, the creation relationship traces out a tree structure, and once established the creation tree is fixed. It also subsumes any composition tree that may exist at start-up time.

The alignment of the creation tree and composition tree is also governed by the "clientship" relationships that exist between Komponent instances at run time. The clientship relationship corresponds to the basic "usage" relationship between Komponent instances in which one instance, the client, invokes operations of another, the server. The alignment of the composition and creation trees has to accommodate the inter-component visibilities needed to support the required network of clientship relationships at run time.

The containment and composition trees in Figure 2.3 actually represent a meaningful architecture for the simple banking system. In this architecture, *B* corresponds to the *Bank* Komponent, *C* corresponds to a *Converter* Komponent which is responsible for handling all exchange rates and currency conversions but knows nothing about accounts, *T* corresponds to a *Teller* Komponent which is responsible for handling accounts but knows nothing about currency conversions, and finally *D* corresponds to a *Dictionary*, which represents a regular collection ADT for storing value/key pairs. These will be discussed in more detail in the following subsections.

2.1.2.3 Modeling principles

A central goal of KobrA is to enable the full expressive power of the UML to be used in the modeling of components. To this end the use of the UML in KobrA is driven by four basic principles:

- **Uniformity.** Every behavior-rich entity is treated as a Komponent, and every Komponent is treated uniformly, regardless of its granularity or location in the containment tree. In particular, the system as a whole is viewed and modeled as a Komponent, albeit a large-grained one. This endows a Komponent containment tree (and the applications created from them) with the property of a fractal,[7] since the products (and the accompanying processes) are identical at all levels of granularity. It also promotes reuse, because any Komponent, anywhere in a containment tree, can be made into a system if it satisfies the needs of some customer.

7. A fractal is a mathematical curve which has the property that its shape appears identical at all levels of granularity (i.e. no matter how far one "zooms in").

- **Encapsulation.** The description of what a software unit does is separated from the description of how it does it. Encapsulating and hiding the details of how a unit works facilitates a "divide and conquer" approach in which a software unit can be developed independently from its users. This allows new versions of a unit to be interchanged with old versions provided that they do the same thing.
- **Locality.** All descriptive artifacts represent the properties of a Komponent from a local perspective rather than a global perspective. This means that there are no diagrams, or other descriptive artifacts, that take a global perspective and cover all aspects of the system. Even the largest Komponent at the root of the containment tree only has a black-box view of its sub-Komponents, and thus its models only describe its local properties. By reducing the coupling between components this promotes reuse.
- **Parsimony.** Every descriptive artifact should have "just enough" information, no more and no less. This means that all models and diagrams should contain the minimum amount of information needed to describe the necessary properties, but no more. Including more model elements than necessary simply obscures the essential information, and complicates the models.

2.1.2.4 Komponent specifications and realizations

KobrA supports the principle of encapsulation by modeling a Komponent in terms of a "specification," which describes what a Komponent does, and a "realization," which describes how it does it. The specification of a Komponent describes the requirements that the Komponent is expected to fulfill, both functional and non-functional, and the expectations that the Komponent places on its environment. The realization, in contrast, describes the design by which the Komponent fulfills these requirements, partly with the help of server Komponents. A Komponent has exactly one specification and at most one realization.[8]

As illustrated in Figure 2.4, the specification and realization of a Komponent are composed of a suite of interrelated UML diagrams and supporting textual models. When highlighting the distinction between the specification and realization, a Komponent can be visualized as a cube whose top surface represents the specification (i.e. externally visible properties) and bottom surface represents the realization (i.e. private design). The exact set of models needed for a particular Komponent depends on the Komponent's properties, but in general they are all modeled in the same way.

Although specifications and realizations are normally used in pairs to describe the properties of a Komponent, it is possible for a framework to contain specifications that have no accompanying realizations. Such specifications play a similar role to abstract classes in object-oriented programming languages – namely, they facilitate the definition of an interface (or contract) without requiring a commitment to a specific implementation of the interface. KobrA provides an associated generalization mechanism, akin to inheritance in object-oriented programming

8. Multiple realizations are viewed as variants of a single generic realization.

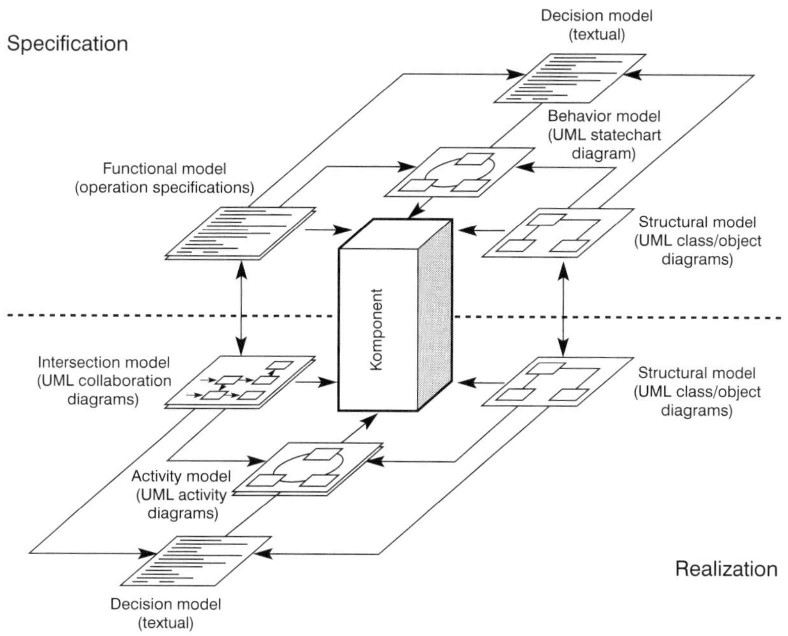

Figure 2.4
KobrA Komponent modeling

languages, which allows one specification to be derived from another, and accompanying conformance rules that define the contexts in which a Komponent can be used to provide a particular service. As in object-oriented programming languages, the combination of these mechanisms allows a Komponent to play different roles during its lifetime (by being conformant to numerous interfaces) and allows a given need for a service to be satisfied by numerous Komponents.[9]

2.1.3 Component embodiment

Embodiment is defined as the "act of giving a concrete form to." This is precisely what is needed in the third dimension of concern in order to produce concrete executable artifacts from the abstract models which are generated during Komponent modeling. Komponent embodiment is therefore concerned with reducing the abstraction level of development artifacts. There are two basic strategies for attaining an executable incarnation of a Komponent – the first is to develop it from scratch and the second is to reuse an existing Komponent that is capable of providing acceptable properties.

2.1.3.1 Reuse

The most cost effective way of attaining a concrete incarnation of a logical Komponent is to reuse an already existing, fully developed component. This need

9. KobrA therefore accommodates the key object-oriented mechanisms of inheritance, abstract classes, polymorphism and dynamic binding, but at the "design" level.

not necessary be a KobrA component, but may for example be a "Commercial Off the Shelf" (COTS) component purchased form a third-party vendor.

The reuse of such a pre-existing component is contingent upon the reusing framework (or application) and the reused component reaching agreement on the nature of the service to be provided. In KobrA this is achieved by developing a UML-based KobrA specification of the reuse candidate, if none already exists, and comparing this to the specification of the desired Komponent. It is rare that a pre-defined component provides exactly the right set of services desired by the reusing Komponents. Therefore, if reuse is to take place a certain degree of adaptation is usually necessary on the part of the reusing Komponents, or the reused components, or both.

2.1.3.2 Implementation

If a suitable pre-defined component cannot be found it will have to be embodied afresh. The first stage of this process is "implementation," which is primarily concerned with developing a tool comprehensible form of a Komponent that can be automatically transformed into an executable form. In many implementation technologies this is known as source code. Since Komponent modeling yields an abstract description of Komponents and their relationships, at analysis and design levels, they can be implemented in virtually all mainstream implementation technologies, including physical component technologies such as Enterprise JavaBeans, CORBA and COM.

This is another aspect of development in which KobrA achieves a strict separation of concerns. As illustrated in Figure 2.5, most methods approach the problem of transforming abstract models into tool-comprehensible source code by defining a set of guidelines specifying how each abstract modeling feature can be mapped into equivalent programming language concepts. Such guidelines therefore essentially aim to add implementation detail and change representation form in one big step, as illustrated by the diagonal line from the realization (i.e. design) to the source code in Figure 2.5. However, this not only increases the complexity of the transformation, potentially leading to reduced quality, but also reduces the opportunities for reuse.

In contrast, KobrA strictly separates the transformation process into distinct refinement and translation steps. As illustrated in Figure 2.6, refinement is concerned with describing model elements at a lower level of detail, but using the same notation, while translation is concerned with mapping model elements into program elements. Separating these concerns not only simplifies the individual transformation steps, but also allows refinement guidelines to be reused with multiple language and implementation vehicles.

Following this approach yields a UML representation of the source code, but in terms of implementation level model elements. Such an "implementation" model complements the source code, and is constructed using a specially defined implementation profile of the UML.

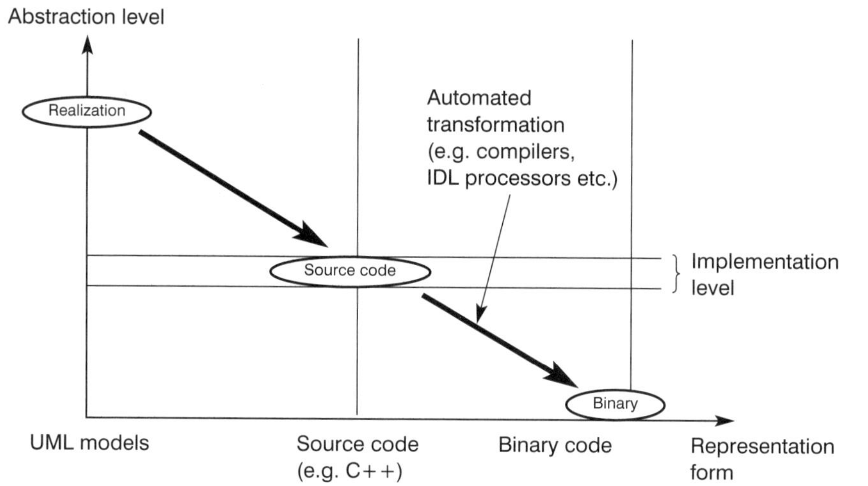

Figure 2.5
Direct transformation of models

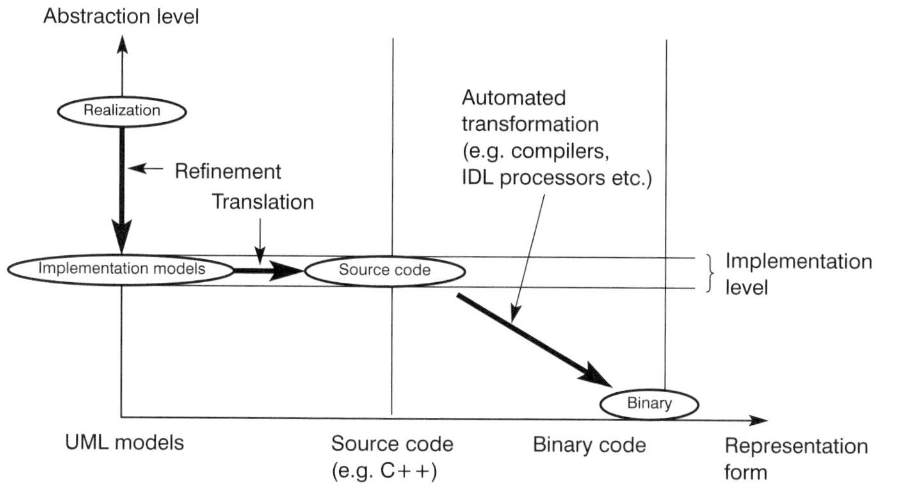

Figure 2.6
Separating refinement and translation

One of the key questions to be resolved during Komponent implementation is how logical components are to be mapped into physical components. A physical component represents an independently deployable and replaceable unit of software. The components of traditional component technologies therefore represent physical components, but in general any separately executable software unit, such as a main program, or an applet etc., is also viewed as a physical component in KobrA. In general, Komponents in a containment tree can be "flattened" into physical components in various ways. At one extreme, the whole tree can be mapped to a single physical component, while at the other extreme every individual Komponent can be mapped to a separate physical component. The choice depends on the nature of the target execution environment, performance requirements, and the desired level of run-time substitutability.

2.1.3.3 Construction, building and releasing

Implementation is only concerned with the creation of a tool-comprehensible representation of a Komponent together with the associated implementation level models. However, to create a deliverable product several other steps are clearly also needed. In general these are highly technology-specific, but KobrA defines three representative activities for completing the transformation to a deliverable form. Construction is concerned with the assembly of the tool-comprehensible artifacts developed during implementation in a form that the associated tool expects and can work with. Building is concerned with creating a directly executable representation of the physical components (i.e. machine or byte code). Finally, releasing is concerned with the configuration of the relevant components into final deliverables. One or more of these activities may also be needed to support software reuse, depending on the precise form in which the reused component is provided.

2.1.4 Project monitoring and control

The three distinct dimensions discussed in the previous sections are concerned with the creation and/or reuse of software artifacts. These are supported by a fourth dimension of concern, referred to as project monitoring and control, which cuts across the other three and affects all the previously mentioned artifacts. The primary issues addressed in project monitoring and control are software quality and maintenance.

2.1.4.1 Quality assurance and measurement

High quality is essential in component-based software engineering. This is because poor-quality components have as much potential to spread poor quality within the systems in which they are reused as high-quality components have the potential to spread high quality. The potential of components to be reused in a wide variety of different systems therefore makes it essential that they are of as high a quality as possible. High quality is of even greater importance in approaches, like KobrA, that facilitate a top-down approach to development based on a tree of artifacts. This is because errors near the top of the tree can have a disproportionate effect, leading to the need to rework much of the subsequent development effort once they are discovered.

For these reasons, quality-improvement and measurement activities are embedded within all parts of the method, in all dimensions of development.

Inspections are the primary quality assurance activities in KobrA. They are applied in all stages and dimensions of development to check that the development artifacts satisfy the large set of rules that control their form and relationships. Figure 2.7 provides a schematic representation of some of the inter-diagram and inter-model consistency rules that relate Komponents and their descriptions. It shows the same containment tree as Figure 2.3, but with

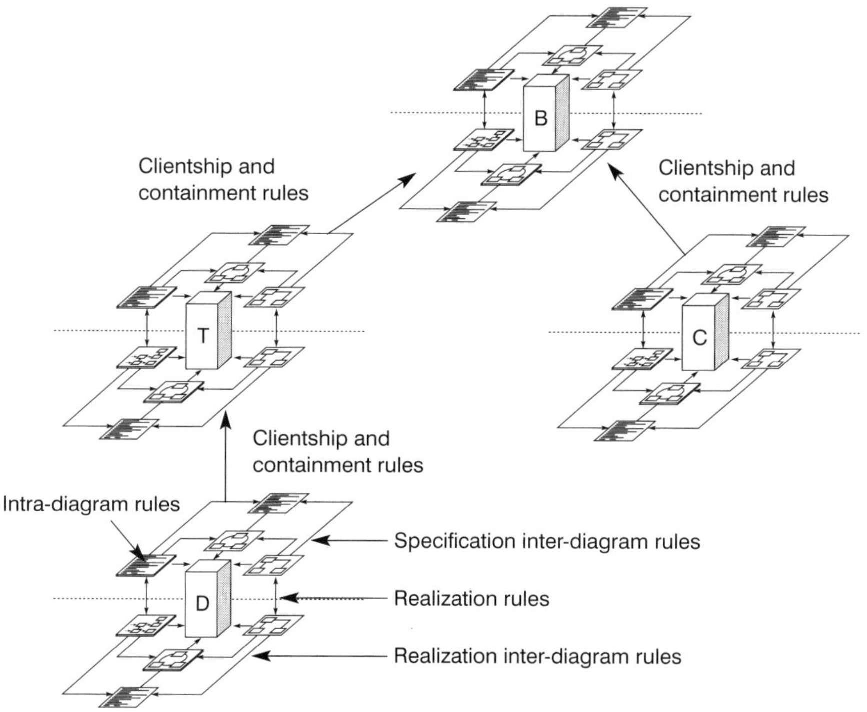

Figure 2.7
Consistency rules

the package representation of Komponents changed to a schematic representation of the models describing the Komponents. Every significant development activity is concluded with a systematic inspection of the relevant consistency rules applying to the constructed artifacts.

KobrA employs a perspective-based and architecture-centric form of inspection, which is optimal for a component-oriented development approach in which concerns are strictly separated and models localized. Inspections are also complemented by testing, with an emphasis on test-case development in tandem with the normal descriptive models. It is recommended that functional and state-based test cases be developed as soon as the specification of a Komponent has been completed, and that structural test cases be developed as soon as the realization of a Komponent has been completed.

The final quality control technique supported by KobrA is quality modeling and measurement. The basic idea is to evaluate certain quality attributes of Komponents by measuring structural properties of their models immediately after they have been created. Such measurement can provide insight into numerous quality factors including design quality, error-proneness etc. Quality models help in analyzing the measured data and can help guide the project management's decision-making process.

2.1.4.2 Maintenance

In a method such as KobrA that emphasizes strict separation of concerns and localized models, the number of development artifacts can quickly become large. When the dimension of time is also taken into account, it is clear that keeping track of the many artifacts and their relationships, and systematically managing their evolution, is essential. This is the role of the maintenance activity of KobrA.

KobrA's maintenance approach is based on the same principle of locality as the modeling activities. This means that information about different incarnations of an artifact is maintained locally by the other artifacts that depend on it. This, in turn, is achieved by making dependencies themselves first-class configuration items subject to configuration control and versioning. Another important aspect of the maintenance activity is to ensure that applications remain consistent with the frameworks from which they were instantiated. When a new version of a framework is generated it may be necessary to update already created applications.

2.2 Artifacts

Having defined the core concepts of KobrA, in this section we focus on the nature of the artifacts that are developed in a typical KobrA project. We first describe the creation of a framework representing the simple banking system product line, including the modeling and embodiment of various Komponents, and then we illustrate how the framework is instantiated to yield a concrete application representing a specific member of the product family.

2.2.1 Frameworks

In order to develop a product line that supports all four variants of the simple banking system mentioned above, emphasis is first placed on the definition of a framework. This captures all the properties of the family in a generic way by explicitly highlighting variant features and indicating to which variants of the system each feature belongs. In general, this includes all Komponents in the framework containment tree.

2.2.1.1. Context realization

As with traditional methods, the development of a new system starts with the analysis and modeling of the environment in which it is to execute. In KobrA this environment is known as the context, and by virtue of the principle of uniformity it is modeled using the same set of artifacts as the realization of Komponents in the body of the containment tree. Thus, the description of the environment is known as the context realization. Despite first appearances, this is actually a highly appropriate name, because the inclusion of a new system into a business process always involves the definition of a new way of performing the process (i.e. of realizing it).

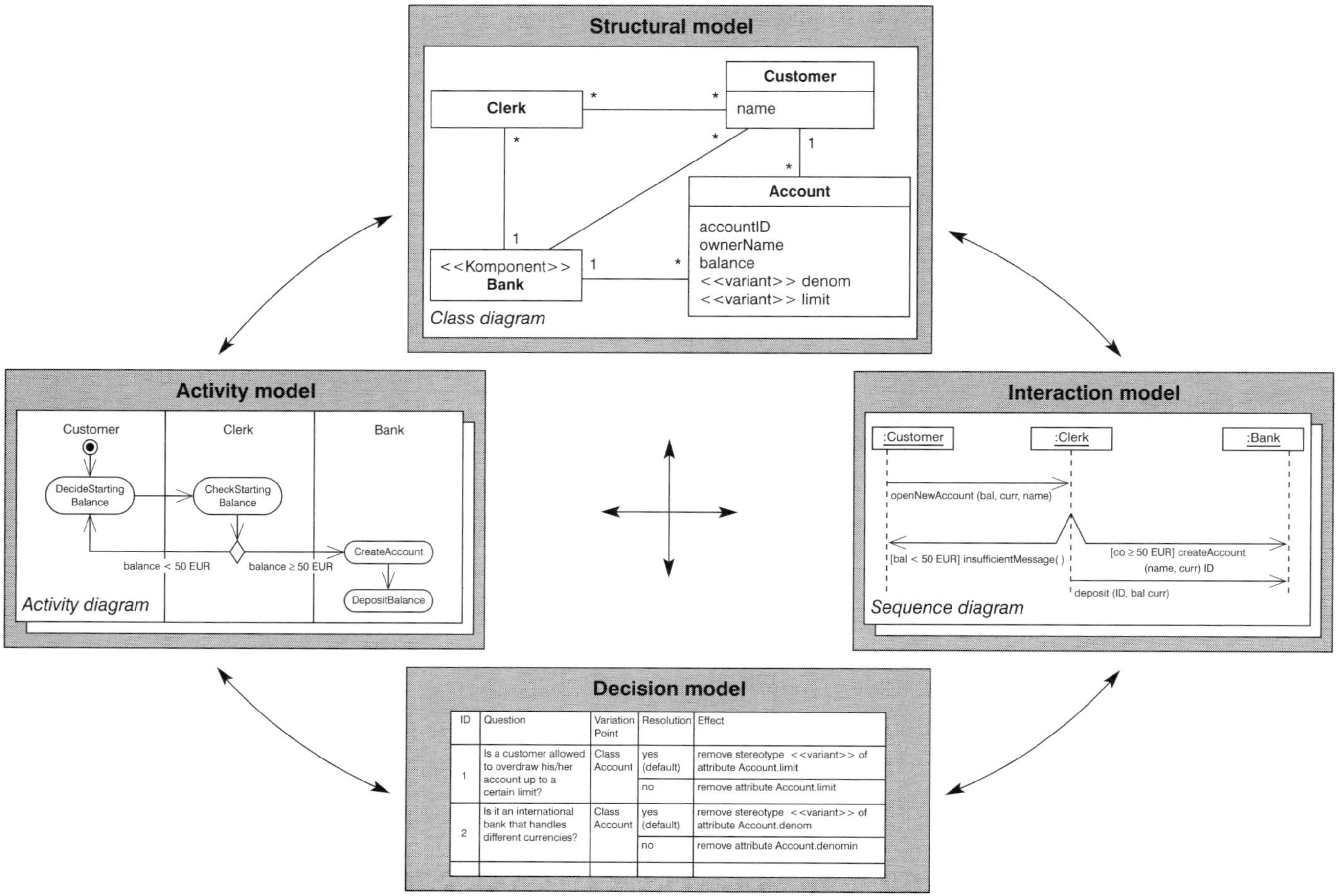

Figure 2.8
Bank context realization

As illustrated in Figure 2.8, the simple banking system's context realization is composed of four closely inter-related models, each providing a distinct view of the overall set of properties of the context. The structural model is composed of a class diagram and conveys information about the actors with which the bank interacts and the externally visible data that it manipulates (in this case only *Accounts*). The interaction model is composed of multiple interaction diagrams (customarily sequence diagrams) that describe the interactions involved in the various system usage scenarios. These are very similar to the sequence diagrams commonly created to describe use-case instances in modern object-oriented methods. The activity model is composed of multiple activity diagrams, which describe the interactions in which the system is involved in a flow chart style rather than an interaction-oriented style.

Because this is a part of a framework, and the context realization encompasses the contexts of all variants of the system, any features that are not common to all variants are labeled with the stereotype <<variant>>. In the models shown, only the class diagram contains variant features, the attributes *denom* and *limit*, but other diagrams not shown in Figure 2.8 will also contain variant concepts. Indeed, sometimes all the information in a diagram may be <<variant>>. The role of the decision model within a generic Komponent is to define how the variant features relate to certain high-level decisions that characterize the distinction between members of the product line.

The models shown in Figure 2.8 only represent a subset of the models that in general can be used to define a context realization. There can also be an enterprise model that captures the high-level concepts and processes that characterize the enterprise, and the activity model may also contain use-case diagrams depicting the high-level usage possibilities. In addition, the UML models and the decision model can be supported by auxiliary artifacts including a data dictionary and various quality artifacts such as test artifacts, quality measures, and inspection artifacts.

There is clearly an overlap between the different models and diagrams making up a context realization, particularly between the interaction model and the activity model. Therefore it is usually not cost effective to develop every possible diagram in a context realization. The choice of which to develop depends on the nature of the Komponent, and on the preferences of the developer or the project manager. However, the diagrams that are created should be consistent with each other, and are thus governed by a substantial list of inter-diagram consistency rules.

2.2.1.2 Komponent specifications

Every Komponent in a containment tree has a specification that defines its externally visible properties, and the expectations it makes on its environment. A Komponent specification can therefore be thought of as defining the requirements, or the interface of the Komponent. For the Komponent representing the entire system at the root of the tree, this environment is described by the context realization, and for all other Komponents this environment is described by their

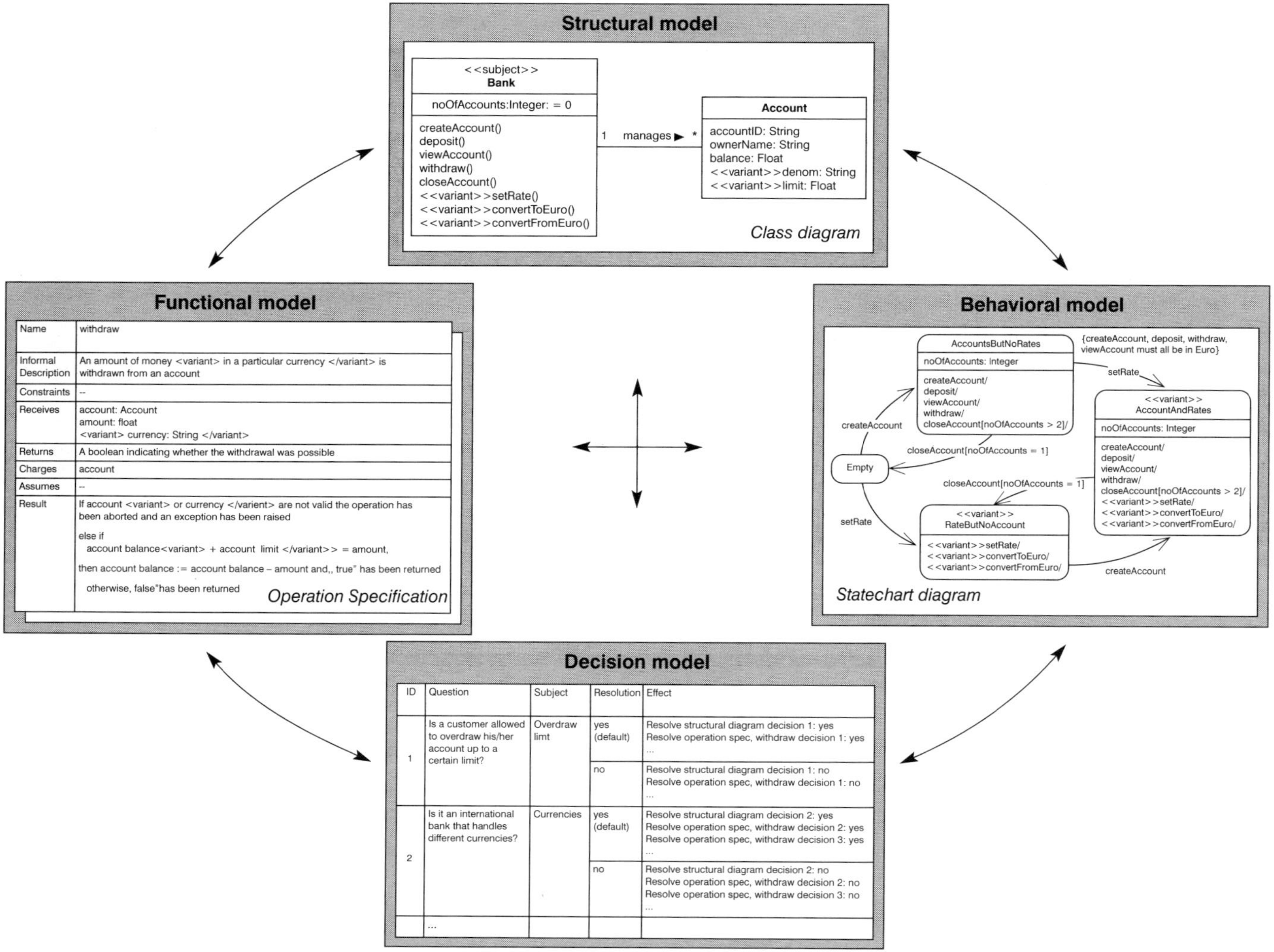

Figure 2.9
Bank specification

parents' realizations. Every Komponent specification is therefore defined with respect to a realization, and must obey a substantial set of related consistency rules.

As illustrated in Figure 2.9 the specification of the *Bank* Komponent (i.e. the system) is also composed of four main models. The structural model defines the information (i.e. the types) that the system utilizes in order to fulfill its contract. As well as data types this can in general also include other Komponents that the Komponent being specified (i.e. the subject) needs to "acquire" at run time. Notice that this model contains some information already contained in the context realization. However, according to the principle of locality, it presents this information from the perspective of the *Bank*, the "subject" of this specification.

The functional model is composed of multiple textual operation specifications, each declaratively describing the effects of one of the Komponent's operations. These effects are captured in terms of the concepts defined in the class diagram and thus provide a concrete way of ensuring that the structural model conforms to the principle of parsimony and does not have unnecessary concepts. The behavioral model is a statechart diagram that captures the externally visible states of the Komponent as they relate to the operations and logical attributes of the Komponent. For highly complex Komponents this can be replaced by a textual statechart table that can summarize certain subsets of the overall behavioral information.

Like the context realization, the *Bank's* specification is generic, and contains variant properties that are not possessed by all variants of the system. As before, therefore, a decision model is needed to describe what sets of features are associated with which variants. Ideally the decisions used in a specification decision model should be the same as those in the encapsulating realization. In this way, decisions have to be resolved only once at or near the top of the containment tree, and are automatically propagated down to all the affected Komponents.

A Komponent specification can have similar auxiliary artifacts to a realization. The only difference is that instead of a data dictionary a Komponent specification can have a non-functional requirements specification that describes how the Komponent should be realized and embodied. As with the context realization, the choice of precisely which of the diagrams to create depends on the properties of the Komponent and the personal preferences of the developer. The only mandatory diagram is the class diagram.

2.2.1.3 Komponent realizations

Most Komponent specifications have an associated realization that defines how the Komponent works. A Komponent's realization therefore describes a Komponent's private design. As illustrated in Figure 2.10, a Komponent realization is composed of the same kinds of models as a context realization. This time however, the diagrams making up the realization must be consistent with the corresponding diagrams in the Komponent's specification.

The class diagram in the structural model constitutes a refinement of the class diagram in the specification. It includes the model elements from the specification class diagram but contains more information related to the Komponent's

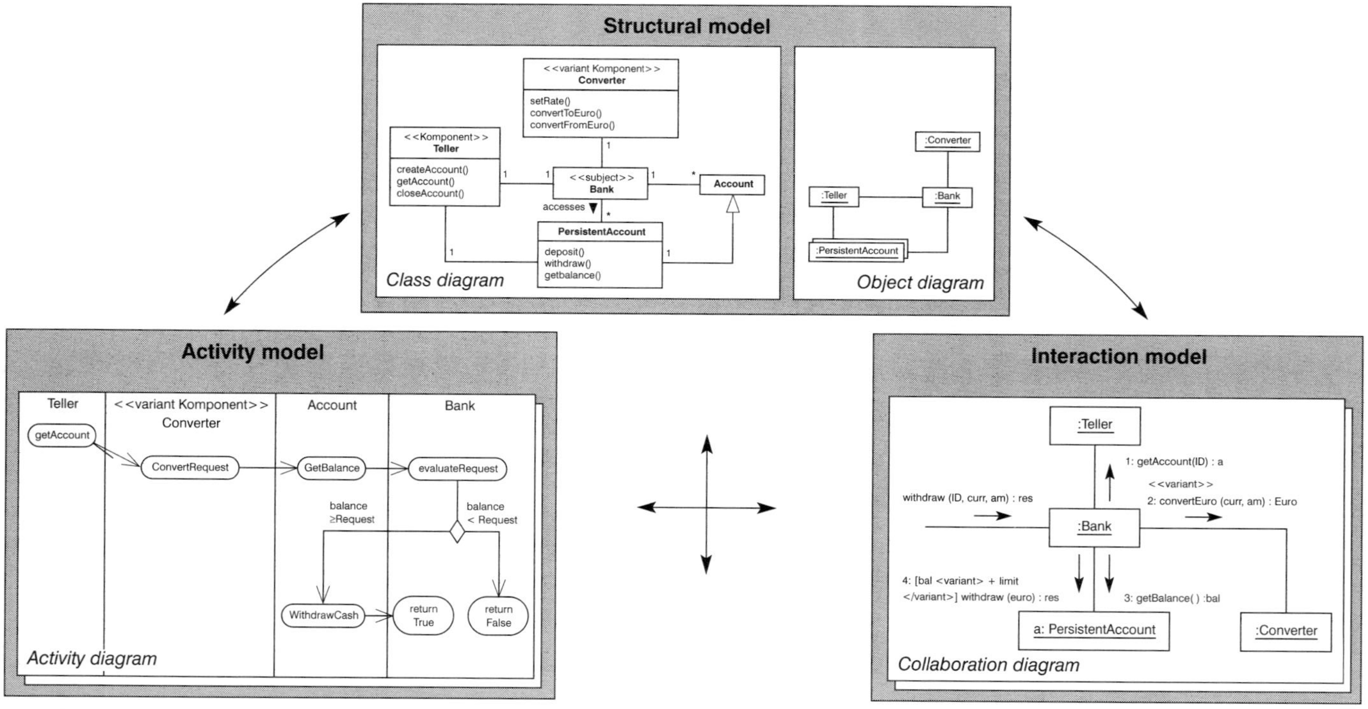

ID		Question	Validation point	Resolution	Effect
Structural diagram	1	Is a customer allowed to overdraw his/her account up to a certain limit?	Class Account	yes (default)	remove stereotype <<variant>> of attribute Account limit
				no	remove attribute Account limit
	2	Is it an international bank that handles different currencies?	Class Account	yes (default)	remove stereotype <<variant>> of attribute Account denom
				no	remove attribute Account denom
....					

Figure 2.10
Bank realization

realization, such as any additional classes or Komponents that are introduced, and the full set of attributes and methods that each model element needs to provide. As can be seen in Figure 2.10, the class diagram can be complemented by an object diagram that illustrates how instances of the supporting classes and Komponents are linked together in the realization of the Komponents. This can be viewed as capturing the essence of the architecture of the Komponent.

As in a context realization, the interaction model is composed of interaction diagrams, each describing how an operation of the Komponent is realized through interactions with other Komponent instances and objects. By convention, collaboration diagrams are used in a Komponent realization rather than sequence diagrams. This is simply because historically collaboration diagrams have tended to be used for design, and sequence diagrams have tended to be used in analysis as a part of the use-case approach. The activity model is composed of a set of activity diagrams, each providing a flowchart-oriented view of the algorithm used to implement an operation, as opposed to an interaction-oriented view. Finally, as before, since this is a generic framework component, a decision model is needed to relate specific features to distinct member of the family, by means of characteristic high-level decisions.

The auxiliary artifacts of a Komponent realization are naturally the same as those of a context realization – namely a data dictionary plus a set of quality artifacts, such as test cases etc.

2.2.1.4 Komponent containment tree

By the recursive development of nested Komponent specifications and realizations the overall Komponent containment tree is gradually developed. The last subactivity of Komponent realization is the identification of sub-Komponents that need to be developed by the repeated application of the basic specification and development activities. Thanks to the principle of uniformity, every Komponent is modeled in the same basic way regardless of its location in the containment tree. As an example, consider the *Teller* Komponent identified during the realization of the *Bank* Komponent. *Bank's* view of this Komponent is documented in its realization models illustrated in Figure 2.10.

Once the realization of *Bank* is complete, its sub-Komponents such as *Teller* can be modeled using the same basic set of artifacts. From a package point of view *Teller* constitutes a sub-package of *Bank* that has ownership of the diagrams and model elements introduced during the specification and realization process. Figure 2.11 shows the specification of the *Teller* Komponent, which is responsible for managing accounts without any knowledge of currency conversion. Nevertheless, *Teller* is also a generic component, since it manipulates *PersistentAccount* which varies according to whether the simple banking system supports foreign currencies and/or overdraft limits. *Teller* therefore need its own decision model.

In the same way that the specification of *Bank* was very closely tied to the context realization, the specification of *Teller* is very closely tied to the realization of *Bank* (its environment). The structural model is derived from the subset of model elements in the *Bank's* structural model that affect *Teller*, and the number and

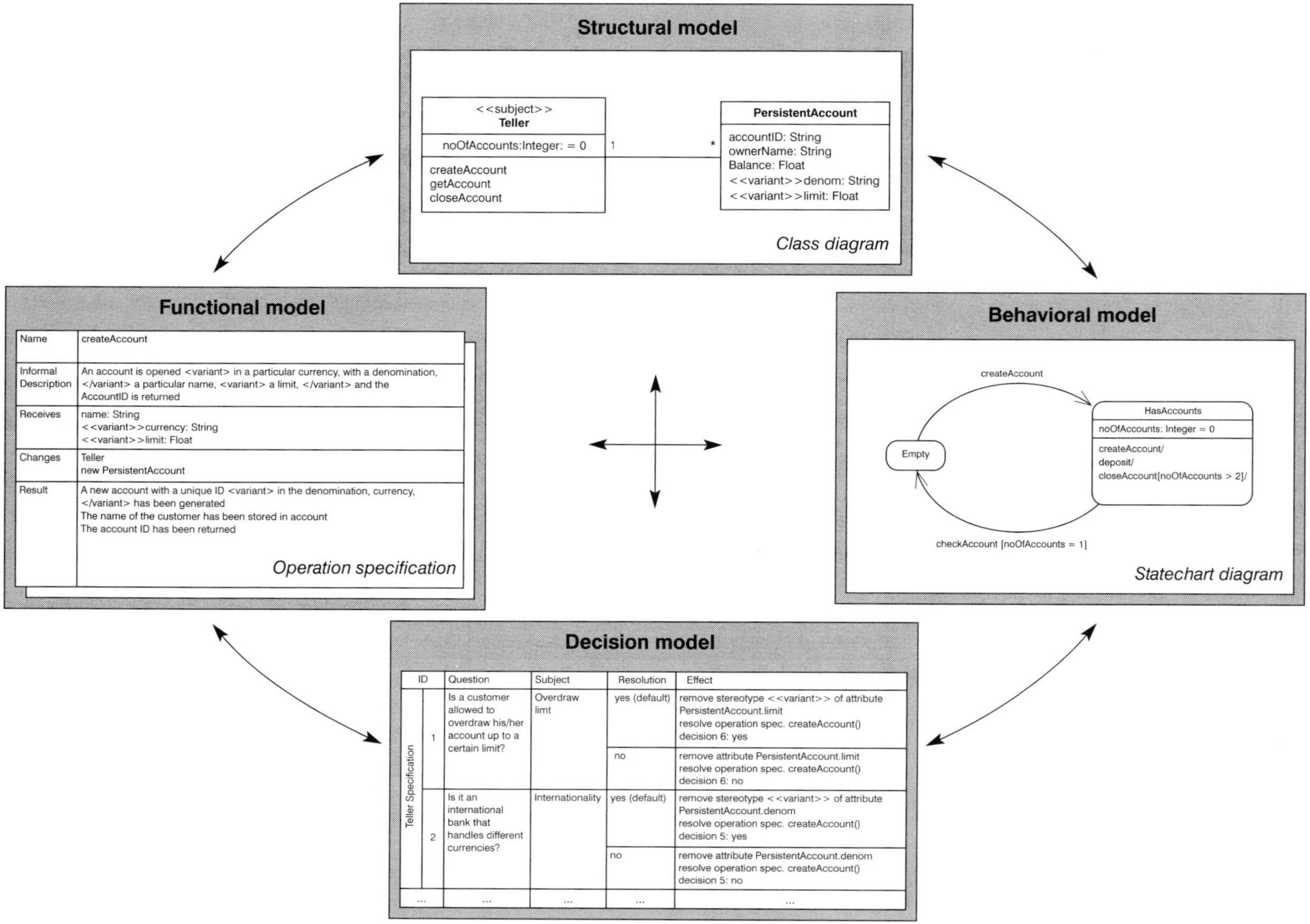

Figure 2.11
Teller specification

form of its operations is defined by *Bank's* interaction model. Basically, every message that is sent to an instance of *Teller* in one of *Bank's* collaboration diagrams defines an operation of *Teller* with the same name and signature.

As a result of the principle of locality, which restricts the extent of the information that can be contained in a Komponent's models, it is rarely the case that an individual Komponent is aware of the nature of the whole containment tree. In general a Komponent only has visibility of (and thus can only import into its model) its immediate children and Komponents at higher levels in the tree. Theoretically, therefore, the containment tree has to be traced out by traversing the containment relationships from the root.

The principle of locality also means that all the diagrams are essentially role models, because they never refer to any global identities. This means that every class or Komponent accessed by a Komponent is subject to substitution according to the rules of polymorphism. KobrA allows any Komponent instance to be supplied as a server of another Komponent instance provided that it satisfies the required set of properties, that is, it is conformant to the expected server. In combination these properties of the containment greatly enhance the ease with which Komponents can be reused in other contexts.

2.2.1.5 Komponent embodiment

The alternating specification and realization of Komponents in a containment tree continues until primitive Komponents are realized whose implementations are "trivial." As part of framework engineering it is also possible to develop some of the artifacts involved in Komponent embodiment. The extent to which it makes sense to do this depends on the nature of the implementation vehicle to be used. If the implementation environment is capable of resolving genericity (i.e. parameterization) during the construction, building and releasing activities, or even at run time, the genericity can be built into the source code. Otherwise, component embodiment during framework engineering can only concern itself with the common parts of the system family, and then only when it is clear how the additional features for specific applications will be added (e.g. inheritance).

2.2.2 Applications

An application constitutes a concrete instance of the framework tailored to meet the needs of a specific member of the system family. Whereas the artifacts in a framework are usually generic, and contain the features for all members of the family, the artifacts in an application are specific, and describe the features of a concrete system. This means that none of the models, anywhere in the application, contains variant features and decision models are replaced by decision resolution models which record the exact set of decisions characterizing the application in hand. Apart from this difference, however, the artifacts making up an application are identical to those making up a framework.

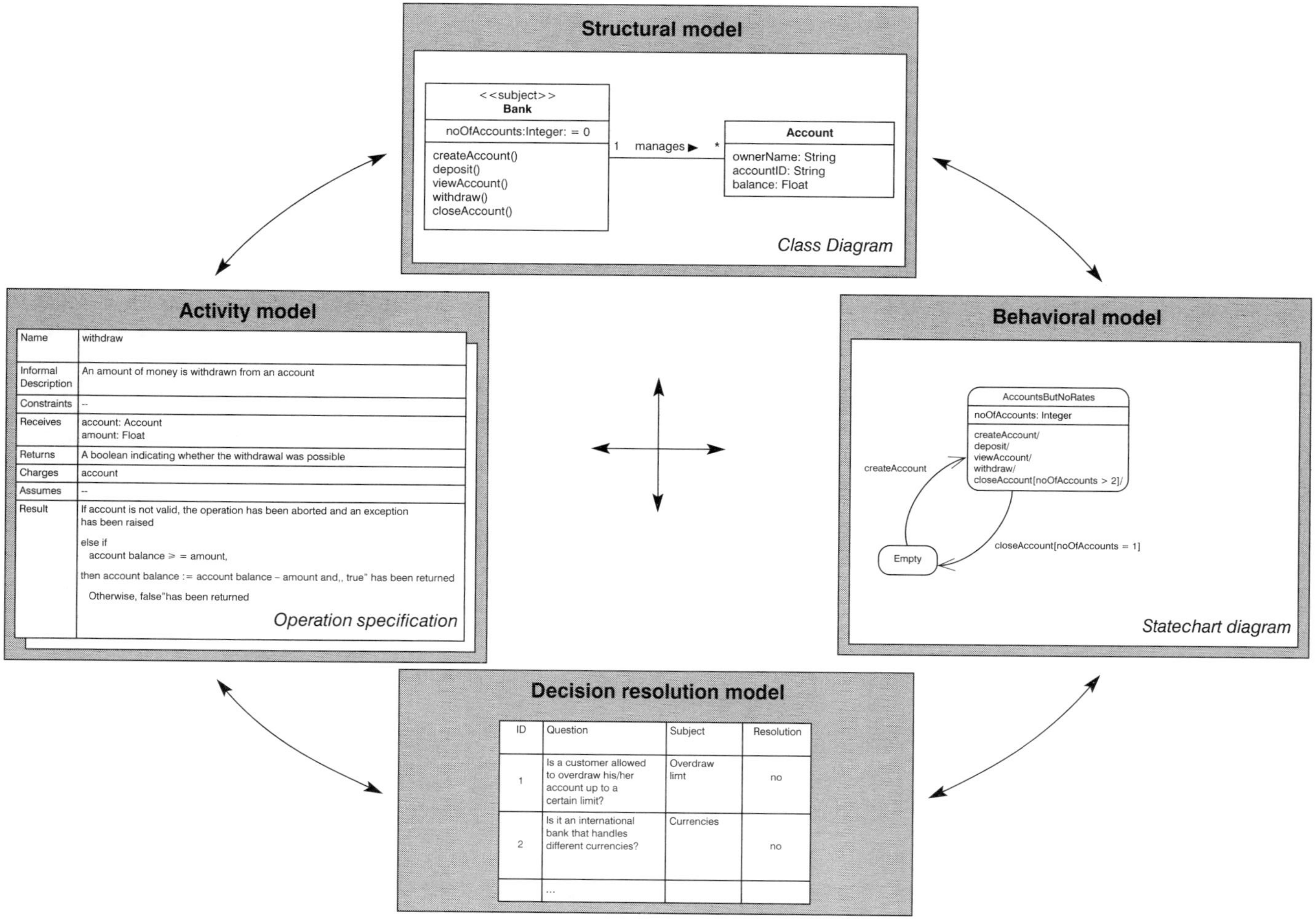

Figure 2.12
Specific Bank specification

2.2.2.1 Specific Komponent specifications and realizations

Figure 2.12 illustrates a specific variant of the *Bank* specification in which all the decisions have been resolved and all non-relevant features have been removed. This variant corresponds to the situation in which no features for overdrawing or for foreign currencies are supported. The models for a different variant would show the features appropriate for that variant.

Figure 2.13 illustrates the specific realization of the *Bank* Komponent for the same variant of the system (i.e. with the same resolution of the decisions). Again, all variant features not possessed by this variant of the system have been removed, and the decision model has been replaced by a decision resolution model which documents the choices embodied within the models.

The decision models in a framework containment tree are therefore related hierarchically, with those lower in the tree being driven by those higher in the tree. This includes auxiliary artifacts such as non-functional requirements specifications, quality artifacts and the data dictionary. User-oriented decisions are resolved at the level of the context realization.

2.2.2.2 Komponent Embodiment

Although Komponent embodiment has been described in association with the framework, typically much of the embodiment effort is performed at the application level rather than the framework level. The optimal balance depends on the nature of the system, and the technology used to implement the Komponents. If the implementation technology supports some form of parameterization, and it is possible to capture variabilities in the framework as parameters in the implementation code, application Komponent embodiment may primarily involve the instantiation of the appropriate parameters, or building the software with the appropriate build options.

On the other hand, if a regular programming language is used, application Komponent embodiment is more likely to involve the definition of subclasses that add the optional features to the core features defined by classes developed during framework-Komponent embodiment. Since there is such a wide range of options, KobrA does not try to be prescriptive about this aspect of development. The documentation of the precise properties of a variant of the system in the form of instantiated UML model provides the foundation needed to apply the implementation technology in the most effective way.

2.3 Process

The previous section has explained the main artifacts that are developed within a KobrA development project. In this section we provide an overview of the related activities. Generally speaking the shape of the process is tightly coupled to that of the product, with many of the activities focussed on the creation of one specific artifact, and usually bearing the same name as that artifact.

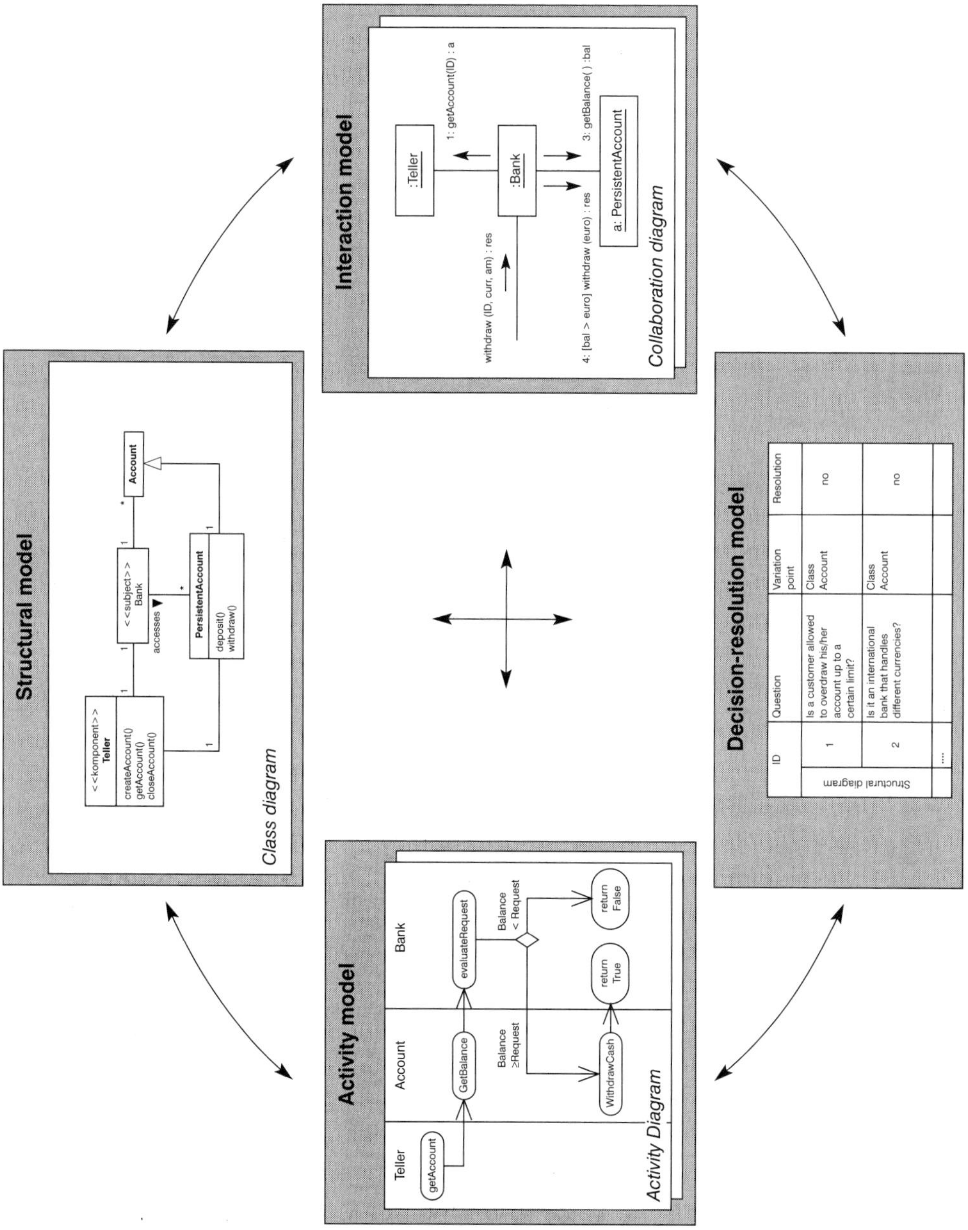

Figure 2.13
Specific Bank realization

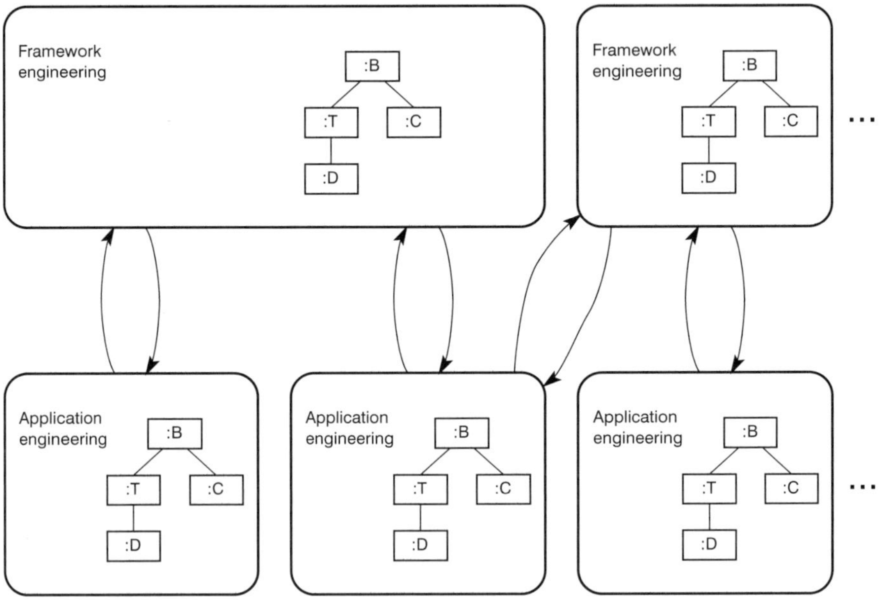

Figure 2.14 Product line life-cycle

As illustrated in Figure 2.14, at the highest level of granularity, the life-cycle of a KobrA project is composed of a sequence of phases in which new versions of the central framework are developed and new applications are instantiated from it to meet the expectations of new customers.

2.3.1 Framework Engineering

Framework engineering is concerned with the construction and maintenance of a framework, and involves two main activities: Komponent modeling, which is responsible for generating the Komponent containment tree; and Komponent embodiment, which is responsible for transforming this abstract containment tree into a deliverable product.

The Komponent modeling activities offer a carefully balanced tradeoff between top-down development and bottom-up development. As illustrated in Figure 2.15, the context realization, Komponent specification and Komponent realization activities essentially assume a top-down strategy for developing the Komponents within a containment tree. However, this is counterbalanced by the component reuse activity, which at any point during the development process allows existing external components (e.g. COTS components) to be inserted into a Komponent containment tree. The overall approach therefore represents a balance between the two.

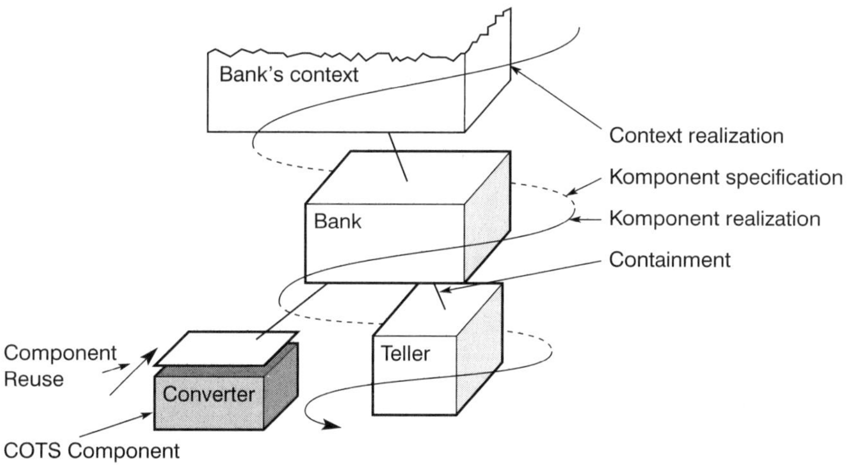

Figure 2.15
Balanced top-down and bottom-up development

2.3.1.1 Context Realization

The goal of the context realization activity is to create the set of artifacts that make up a context realization. It does so by following the basic analysis strategy that has come to be known as use-case analysis, but generalized to accommodate the analysis of variabilities and commonalities in a product line.

Every software system is made up of two basic ingredients, data and functions (or operations) that manipulate this data. In KobrA we use the term "activity" to cover all functional code abstractions (e.g. methods, functions etc.). The manner in which a method combines these two basic ingredients typically characterizes it as being object-oriented, function-oriented, or based on some other paradigm. As illustrated in Figure 2.16, in KobrA the data and activity aspects are first analyzed and modeled separately, and are then gradually brought together to provide a unified object-oriented view of the system.

The early identification of appropriate data concepts and activity concepts is often driven by enterprise modeling (also known as business process modelling) techniques but this does not have to be so. KobrA can work with the output of most well-known enterprise or business process modeling techniques. Once an initial idea of the information and activities that are relevant to a particular

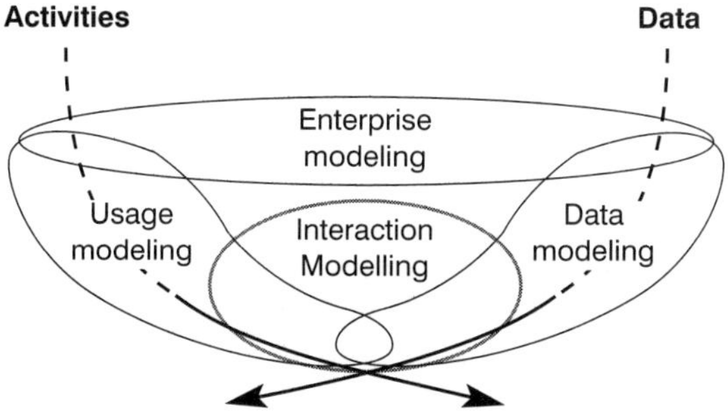

Figure 2.16
Context realization activity

business process have been identified, the use-case modeling techniques can take over. Initially the data models and the activity models are refined separately to lower levels of detail, and then primarily through the vehicle of interaction modeling the two are brought together into a unified picture.

The results of this process are documented using the context realization artifacts described in the previous section. Although KobrA does not make use of the terminology of use-case analysis to conform to the principle of uniformity at all levels, the concepts of use-case analysis are directly subsumed within the context realization process. In particular, it adopts the same transition between view points that characterizes use-case analysis.

2.3.1.2 Komponent specification

The goal of the Komponent specification activity is to create the artifacts which make up a Komponent specification. The encapsulating realization provides much of the information needed to start the process. Essentially, the starting point for the development of the structural model is the subset of the model elements in the encapsulating realization's structural model that have a bearing on the subject of the specification. Because of the principle of locality, none of the other model elements should appear in the Komponent's specification structural model. Similarly, the operations to be supported by the Komponent, and the states that the Komponent should exhibit, can be inferred from the interactions involving instances of the Komponent within the interaction models of its parent and clients.

2.3.1.3 Komponent realization

Komponent realization is responsible for creating the artifacts making up a Komponent realization. As can be seen in Figure 2.17, Komponent realization is based on the same underlying approach as context realization. Data and activities are initially analyzed independently, and are then integrated by focussing on interaction modeling.

The main difference between Komponent realization and context realization is that the former has a much more concrete starting point – the specification of the Komponent. The realization structural model is basically a refinement of the specification structural model, enhanced with the new information derived from the interaction modeling. The activity modeling and interaction modeling activities are also much more directly driven by the operations of the Komponent than in context realization where the starting point is less well defined.

Once the context realization has been completed, the Komponent specification and realization activities can be applied recursively to elaborate the Komponents in the containment tree. In general, therefore, the overall patterns for analyzing data and activities in component modeling is captured by Figure 2.18. This so called DNA (Data aNd Activities) helix illustrates that the modeling process is characterized by cycles in which the data and activities in the system are analyzed separately and then integrated into a coherent object-oriented view. Figure 2.18 illustrates this pattern for the development of one branch of a Komponent tree.

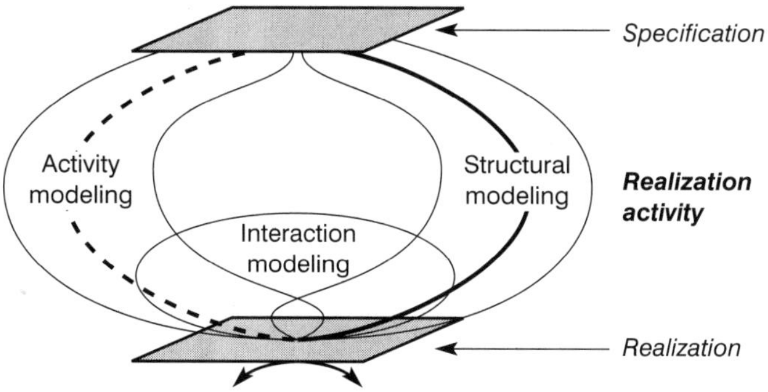

Figure 2.17
Komponent-realization activity

The integrated points of the helix, where the data and activities are merged, correspond to the boundaries between a Komponent's realization and its sub-Komponent's specifications. A Komponent and all its sub-Komponents have to have a unified view of how the data and activities are integrated – that is, what the objects in the system are, and what operations they have.

2.3.1.4 Component reuse

The DNA helix mentioned in the previous section is obviously based on the top-down direction of development in which Komponent realizations are developed from scratch. However, as mentioned previously, this is not the only way of satisfying a Komponent specification. It is also possible, in fact highly desirable, to do so by reusing a pre-existing component. This not only avoids the effort involved in modeling the realization of a Komponent, but usually also obviates much of the Komponent embodiment work as well, since COTS components are usually supplied in an executable form.

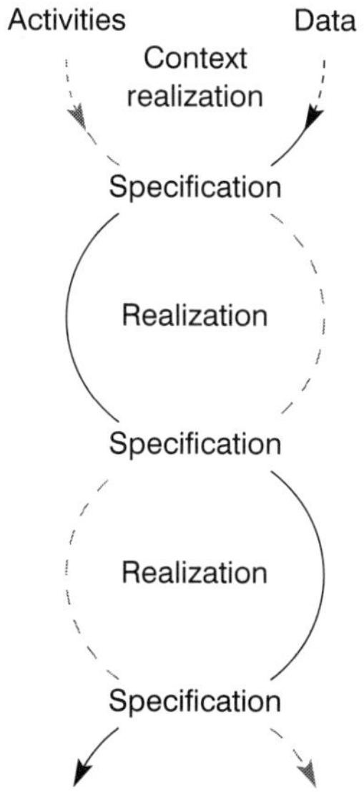

Figure 2.18
DNA helix

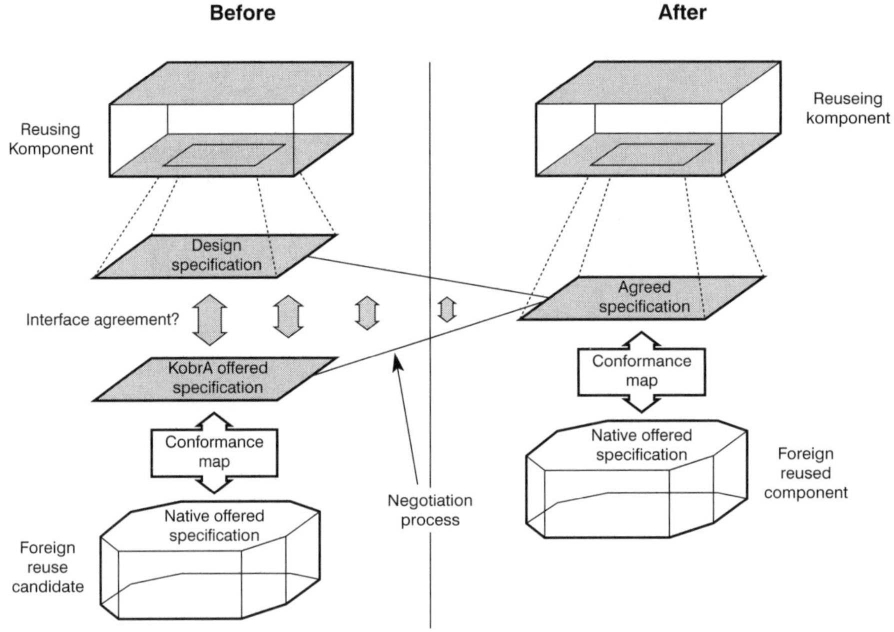

Figure 2.19 Reuse of a foreign component

Figure 2.19 illustrates the basic process involved in reusing a foreign (i.e. non-KobrA) component. The starting point for the process is a specification of the ideal sub-Komponent desired by a Komponent in the tree. This specification can be used to search for a reuse candidate that provides a reasonable match to the desired properties. Once a candidate has been found, if it is not already a Komponent, its properties and services are specified in the form of a Komponent specification. This is known as the offered specification. A conformance map defines how the KobrA form of the specification relates to the native specification, if one exists.

Assuming that the desired specification and offered specification are not already identical, a process of negotiation is entered into with the goal of arriving at a mutually acceptable specification. This is a specification that both the reused component and the reusing Komponent can live with. Being able to arrive at such a specification of course requires some flexibility on behalf of both parties to accommodate a modified specification. Once an agreed specification has been reached, the respective changes are made in both parties and the reused component is logically integrated into the containment hierarchy. Including the executable incarnation of the component in the final deliverables is the concern of the embodiment activities.

2.3.2 Application engineering

Application engineering can involve all the same activities as framework engineering but without the parts dealing with the variabilities and commonalities that characterize the product line. However, instead of these parts, application engineering includes activities dealing with the instantiation of the framework. Two primary instantiation activities complement the regular development activities within application engineering.

2.3.2.1 *Context realization instantiation*

The goal of context realization instantiation is to analyze the problems of a potential user of a system with a view to understanding the extent to which his/her needs overlap with the features supported by the framework. The extent of this overlap is the basis for judging whether it is more cost effective to instantiate the framework or to develop the required system from scratch, possibly using KobrA in "single system" mode.

If a big enough overlap is found to justify the instantiation of the framework, the resolution of the decision models (i.e. the choices) that provide the best match to the required system is established, and the context realization is instantiated by removing all features which do not belong to the specific variant, and by changing the decision models to decision resolution model.

It is rare that the initial instance of a framework perfectly matches the required needs. Therefore, the instantiation process is followed by one or more of the regular modeling activities to adapt the specific context realization models to the exact environment of the users.

2.3.2.2 *Specification and realization instantiation*

Once the context realization has been instantiated, the Komponents within the framework containment tree are instantiated recursively by resolving the decision models and removing the unwanted features. The resolution of the decisions should be clear from the choices made in the instantiation of the context realization. As with context realization instantiation, once the initial versions of the specific models have been generated, the Komponent modeling and embodiment activities may need to be applied again to adapt the models to the precise needs of the new system.

2.3.3 Incremental development

The previous subsection has described the development process from a product line point of view, based on the inter-related enactment of framework and application engineering. However, there is another important, orthogonal dimension concerned with incremental development.

When developing a framework or application, it is of course possible to develop all the models for all the Komponents in a containment tree before generating and testing any executable incarnations of the Komponents. However, this is not always a good approach, because it increases the risk of project failure. Quality problems or errors that are uncovered late in a development project usually require more effort to correct than those found early. Moreover, waiting until late in a project's life-cycle to generate any executable versions of the software decreases the motivation of engineers and management. For these reasons, incremental development is considered an essential part of modern best-practice software development methods.

Figure 2.20
Incremental development

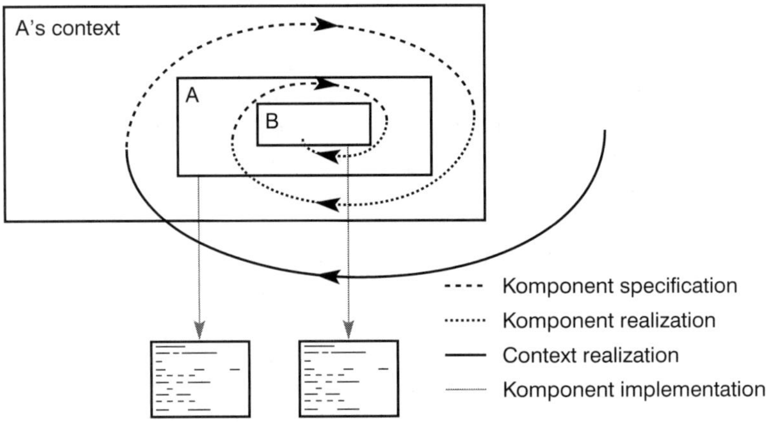

KobrA supports incremental development in a simple but powerful way. Because the Komponent embodiment dimension is orthogonal to the containment dimension, executable incarnations of Komponents can be created as soon as the information necessary to do so is available. As illustrated in Figure 2.20, in KobrA this can occur as soon as the realization of a Komponent, and the specification of its sub-Komponents, have been created. The sub-Komponent specifications are needed in order to support the implementation of sub-Komponent stubs. These in turn are necessary in order to test the executable version of a Komponent once it has been created.

In KobrA, incremental development is closely tied to the component-based structure of a system rather than to the use-cases supported by the system. In other words, in KobrA, the development increments are much more component-oriented than in many other methods.

2.4 Relationship to other methods

KobrA integrates and builds upon many of the best practices and concepts pioneered by the methods described in Chapter 1. Having described the essential features of the KobrA approach, we are now in a position to describe how it relates to other well-known methods.

The product line aspects of KobrA strongly resemble those of the PuLSE approach [BFK99]. In fact, from a product line perspective the KobrA method can be seen as a pre-defined customization of PuLSE. Essentially, the engineering activity of KobrA mirrors PuLSE's product line infrastructure construction phase, the application engineering activity of KobrA mirrors PuLSE's product line infrastructure usage phase, and the maintenance activity of KobrA mirrors PuLSE's product line evolution phase. However, since KobrA is flexible with regard to the high-level techniques used to analyze and identify commonalities and variabilities, it can also be used quite straightforwardly with other mainstream product line methods such as FODA [KCH90] and FAST [WL99]. KobrA essentially offers an alternative way of capturing the results of the domain analysis or commonality analysis in these methods.

The component modeling aspects of KobrA are related to the first-generation object-methods described in Chapter 1 as well as the component-oriented methods. KobrA's run-time model of Komponent instances is very similar to that of ROOM. Although KobrA does not focus on real-time systems, like ROOM's actors, Komponent instances can be viewed as independent "logical machines," and can be nested recursively to arbitrary levels. Therefore, as in ROOM, a Komponent instance is realized in terms of a community of lower-level Komponent instances.

KobrA's development time Komponent model is quite similar to HOOD's development approach [Rob92]. As in HOOD, the description of Komponents is organized in terms of a containment hierarchy (seniority hierarchy in HOOD), but influenced by the required clientship relationships (usage hierarchy in HOOD). KobrA's basic philosophy of Komponent development by the top-down, recursive application of a few fundamental development steps is also very similar to the HOOD approach.

At the level of an individual Komponent, KobrA's use of UML diagrams is strongly influenced by the Fusion method [Col94] as adapted for the UML by FuML [Atk97]. Most notably, the use of textual specifications (operation schemata in Fusion) to specify a Komponent's operations in terms of the content of a class diagram, and the notion of driving the design of a system by the successive elaboration of collaboration diagrams, are all motivated by Fusion. In fact, from the perspective of a Fusion developer, KobrA could be characterized as a recursive version of Fusion.[10] The modeling strategy adopted in Fusion is itself heavily influenced by the OMT method [RBP91], and to a lesser extent the Booch [Boo94] and Objectory methods [Jac92]. KobrA's three fundamental specification models are essentially the basic OMT analysis models.

The KobrA approach also adopts the same basic role modeling philosophy as OORAM [RWL96] by virtue of the principle of locality. Every diagram is focussed on the description of one activity or one Komponent, and only includes information about other external model elements relative to its own perspective. As in OORAM, therefore, role modeling is embedded within all KobrA modeling activities.

KobrA also has a lot in common with the component-oriented methods described in Chapter 1, particularly Catalysis [DW98]. In fact, KobrA incorporates many of the same basic ingredients as Catalysis, such as type models, action refinement, and collaborations, but combines them in a different way. KobrA avoids much of the feature overload of Catalysis, despite adding support for product lines. KobrA also adopts the same basic approach to component specification as the UML components method [CD00]. The externally visible classes utilized by a component are captured in a localized class diagram, while the operations supported by the Komponent are captured in the form of OCL

10. Interestingly, the original developers of Fusion recognized the value of the recursive application of Fusion, but did not elaborate upon the idea.

pre- and post-conditions. Like UML components, KobrA also uses object diagrams to describe localized (i.e. component) architectures. The main difference is that KobrA applies the concepts recursively to enable a tree of nested components to be developed.

The Select Perspective's [AF98] influence on KobrA is primarily related to the "front-end" business process modeling and use-case analysis techniques within the context realization activity. In particular, the overall role of business process modeling and its integration with use-case analysis is influenced by the Select Perspective, as well as the analysis approach of [Paech00]. KobrA's reuse approach has also been influenced by Select Perspective's emphasis on supporting the reuse of legacy software assets.

Although KobrA's product line approach distinguishes it from the RUP [JBR99], [Kru00], the managerial activities that accompany the technical product line engineering activities, such as project management, configuration and change management and quality assurance are quite similar. In fact, KobrA is deliberately much less prescriptive about the management aspects of a software project than it is about the technical aspects. In this sense, therefore, KobrA is compatible with the RUP, and could be used to support a product line approach within the overall project management umbrella of the RUP. Alternatively, the component modeling aspects of KobrA could be applied within the framework of a RUP project to provide a more rigorous and component oriented approach for using the UML.

KobrA's strict decoupling of activities from artifacts, combined with well-defined consistency rules to determine when an artifact is in an acceptable state, essentially give rise to contract-driven activities of the kind emphasized by OPEN [GHY97]. KobrA could therefore be characterized as having a contract driven process in the sprit of OPEN. The product line engineering life-cycle of KobrA also matches OPEN's overall organization of the process in terms of a project- or application-specific part and a domain-oriented part. The KobrA method can therefore be viewed as an instantiation of the OPEN framework.

Finally, although based on a different development paradigm (structural versus object-oriented) the overall philosophy of KobrA has much in common with the Cleanroom approach [MLH87]. A KobrA framework shares the same tree-based structure as a Cleanroom project, and therefore can benefit from many of the organizational strategies perfected in Cleanroom. Also, the principle of fully integrated, systematic, quality assurance activities, especially architecture centric inspections, is motivated by Cleanroom. The elements of the development tree are also analogous. A Komponent specification can be viewed as providing a state-box representation of a software unit, while a Komponent realization can be viewed as providing a clear-box representation.

2.5 Key properties of KobrA

The final section of Chapter 1 defined a set of important characteristics of software engineering methods from the point of view of potential users. In this section we evaluate to what extent KobrA is successful in exhibiting those properties.

The principle of uniformity and the recursive development features built into KobrA serve to significantly reduce the complexity of the method, despite its expressive power. Care has been taken to avoid gratuitous feature and terminology redundancy. Thus, for example, the context and Komponent realization activities have very similar concepts and activities, although their traditional counterparts (requirements analysis and design) usually employ very different concepts and terminology.

These features of KobrA, coupled with strict consistency rules, serve to make the method quite systematic. Developers are provided with concrete guidance as to which diagrams should be developed at each point in a project, and more importantly precisely what information the diagrams should contain. The consistency rules, combined with the principle of parsimony, help developers establish when enough information has been modeled.

Scalability is supported in KobrA primarily through the strict separation of concerns and the recursive (or fractal-like) nature of the development activities (i.e. products). The strict separation of concerns means that developers can apply some aspects of the method (e.g. component modeling) without using others (e.g. product line engineering). The recursive nature of the method means that developers can initially apply the modeling concepts at the system level without worrying about containment issues. In combination, these features allow the capabilities of the method to be exploited in an incremental fashion. The Komponent specification and realization approaches can first be applied in a "flat," "single-system" mode to support systematic UML-based modeling of a system's requirements and design/architecture. Later, the recursive properties of the method can be introduced in "single-system" mode to elaborate upon the design in terms of a hierarchy of sub-Komponents. Finally, product line concerns can be added in the form of variant features and decision models to support a family of systems.

In the sense that KobrA defines concrete rules for structuring development artifacts, and systematic activities for developing them, it can be viewed as a "heavyweight" process. However, its emphasis on the incremental addition of features and the recursive elaboration of artifacts gives it some of the "agility" of so called "lightweight," or agile, methods. The main distinction between KobrA and methods that are typically viewed as being lightweight is its emphasis on modeling.

KobrA's strict separation of concerns makes it compatible with a large number of practical implementation and middleware technologies. Therefore, the approach can be viewed as supporting a Model Driven Architecture (MDA) approach to software development in which the essence of a system's architecture is described independently of platform idiosyncracies. To this end, the concrete guidelines provided in KobrA have been deliberately focussed on the phases of the life-cycle where it can provide the most support to existing implementation technologies. As illustrated in Figure 2.21, the prescriptiveness of the KobrA method goes down in relation to the abstraction level of the artifacts involved in a development project. In other words, KobrA provides a lot of rules and guidelines for abstract models and associated activities involved in Komponent modeling, but provides significantly less rules and advice concerning the source code and related activities. Finally, KobrA has very little to say about the executable artifacts and the associated activities. This allows KobrA to easily dovetail with an enterprise's existing development practices and to provide prescriptive support precisely where it is usually most needed, in the front-end analysis and design-level activities.

Separation of concerns is also a powerful way of improving the tailorability of the method. Although KobrA is ready to use "out-of-the-box," providing concepts and terminology optimally suited for the domain of business software, the modular nature of its process means that activities suited to other domains can easily be substituted. This is particularly valuable for the "front-end" activities which are very sensitive to the concepts and terminology used in a particular domain. The detailed development activities are compatible with the front-end techniques of many domain engineering and requirements engineering techniques.

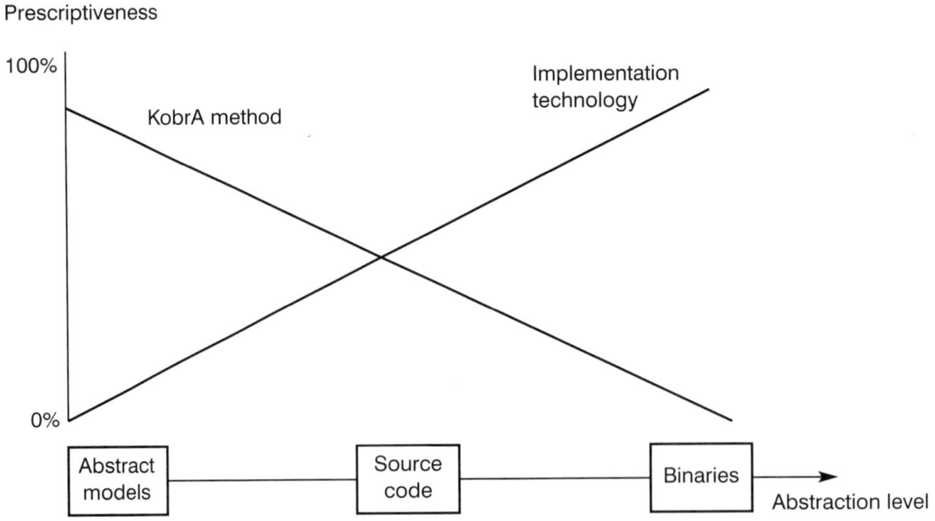

Figure 2.21
Method prescriptiveness

Finally, the modular, incremental nature of the method, coupled with its support for quality control and reuse, make KobrA potentially a highly cost-effective method. With a careful analysis of costs versus benefits, the right amount of effort can be applied to support the right amount of reuse needed to attain an optimal long-term return on investment.

2.6 Road map

The following parts of the book elaborate upon the concepts and techniques introduced in this chapter. In this subsection, we suggest different routes through the material in the book, and explain where details on particular topics can be found.

Part 2 is the core of the book. The material in the other parts will not make sense unless some of the chapters in Part 2 have been understood. In particular Chapters 4, 5, 6 and 7 describe central ideas that should be read first before any other parts of the book are attempted. Chapter 8, 9 and 10 describe more sophisticated ideas that are not essential for a first-time reader of the book. Chapter 3 can also easily be omitted, since it describes the technical details of KobrA's component model and its relationship to the UML. This level of detail is not needed to start using KobrA effectively.

The remaining three parts of the book, Parts 3, 4 and 5, can be read in any order since they are essentially independent. Part 3 describes the concepts and techniques related to the embodiment of Komponents – that is, their transformation into executable forms. Chapter 11 introduces some concepts and activities that can be used to translate the UML models of a Komponent realization into an equivalent tool-comprehensible form. Chapter 12, on the other hand explains how much of the work in the Komponent realization (Chapter 5) and Komponent implementation chapters can be avoided by reusing already existing components. Finally, once all the basic development steps have been introduced, Chapter 13 explains how they can be applied in a manner that supports incremental development.

Part 4 of the book explains the product line engineering aspects of KobrA and can be omitted by readers who are not interested in using these features. Chapter 14 first describes the basic concepts of product line engineering and explains how they are applied in KobrA. In particular it introduces the concepts of framework engineering and application engineering. Chapter 15 then goes on to describe the details of the former by explaining how the models of Komponents can be made generic by the introduction of variability and decisions models. Finally, Chapter 16 explains how the generic Komponents in a framework can be instantiated to provide specific Komponents making up a system for a particular customer.

Part 5 of the book describes the all-important activities that are essential for keeping track of, and controlling the quality of, the artifacts developed during a KobrA project. Chapter 17 describes the details of the KobrA's maintenance approach, particularly the strategy used to manage changes and configurations. Chapters 18 and 19 then discuss approaches for managing the quality of the UML models developed in a KobrA project. Chapter 18 describes how the well-

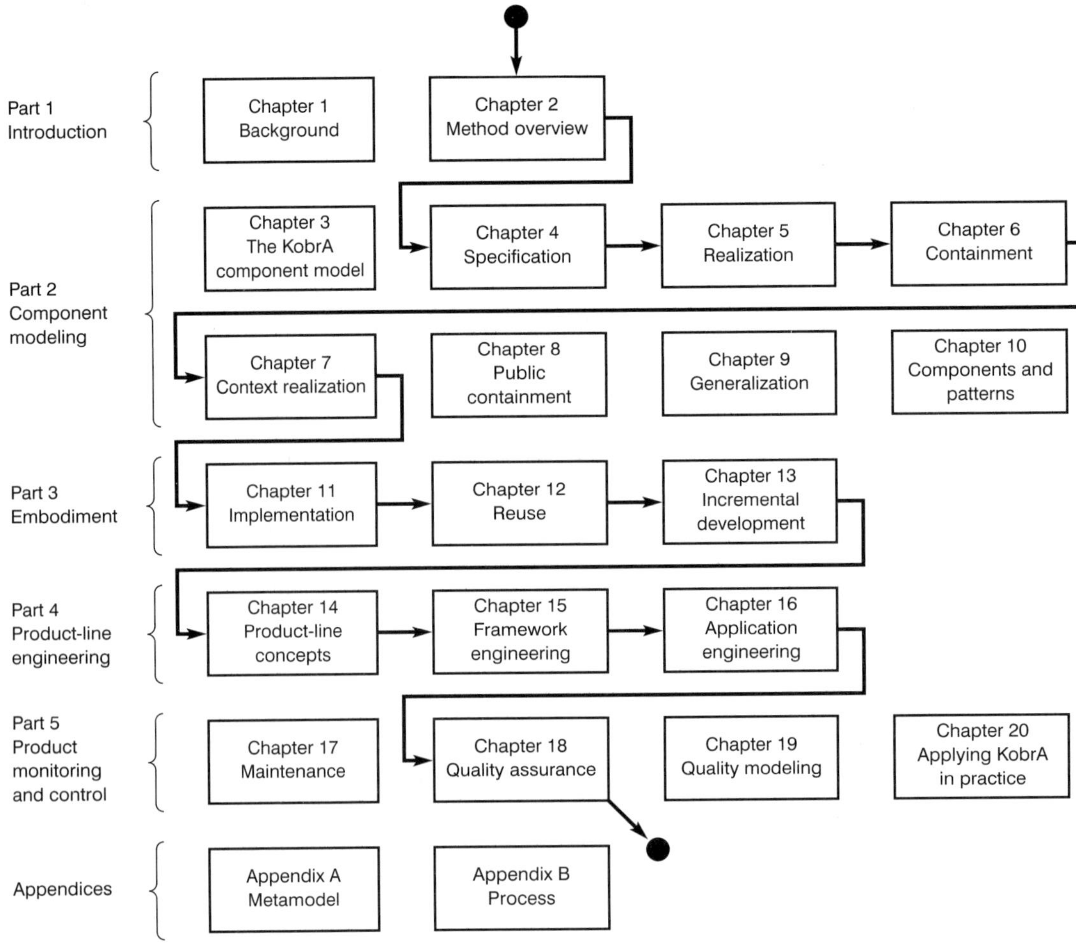

Figure 2.22
Lightweight route

known techniques of testing and inspections can best be utilized in KobrA, while chapter 19 describes how quality can be improved by measuring the structural properties of the UML models. Finally, Chapter 20 provides some guidelines and suggestions for introducing KobrA into real development projects.

The two appendices summarize the details explained in the main parts of the book. Appendix A contains the metamodel, in the form of UML class diagrams, together with descriptions of the abstract syntax and the well-formedness rules which govern how the UML should be applied in KobrA. Appendix B provides a summary of the products and activities making up the KobrA process.

Figure 2.22 provides a map of the book, and illustrates a suggested lightweight route for the first-time reader.

PART 2

Component modeling

The main activities involved in generating a containment tree of KobrA components are the same regardless of whether they are concrete components for a specific system, or generic components within a framework. This part introduces this set of activities, collectively known as component modeling, and describes how they leverage the UML.

Chapter 3 begins by describing the principles underpinning the KobrA component model, and explains the role of trees in KobrA development projects. This is a technically detailed chapter and may be skipped by the first-time reader, since its key concepts are summarized in Chapter 2. Chapter 4 then begins the detailed discussion of the KobrA artifacts with a description of component specifications – UML-based descriptions of the services offered and used by components. This is followed in Chapter 5 by an introduction to component realizations, the corresponding "design" of a component describing how it works. Chapter 6 builds on these two chapters to discuss the notion of containment in KobrA, and explains the nature of the relationships that can connect specifications and realization within a containment tree.

The focus of Chapter 7 is context realization, which is KobrA's "front-end" development activity responsible for describing the environment of the system in the form of a regular component realization. The following two chapters then explain advanced features of the component-engineering approach. Chapter 8 concentrates on a variant of the regular containment relationship, known as public containment, that makes a component's subcomponents externally visible.

Chapter 9, on the other hand, explains the generalization mechanism, which allows one component to inherit the specification properties of another.

Finally, Chapter 10 brings this part of the book to a close with a discussion of the role that architectural and design patterns play in KobrA. After first describing how well-known, published patterns can be applied and supported in KobrA, the chapter explains some patterns that exploit the idiosyncrasies of the KobrA approach.

From Chapter 4 of the book onwards we use the term "Komponent" (as introducted in Chapter 2) as a shorthand for "KobrA component." This not only simplifies the wording, but is also needed for technical reasons to distinguish the UML definition of "component" from the KobrA definition. The unqualified term "component" is used in a general sense to denote any kind of component abstraction (including Komponents). Since Chapter 3 provides a discussion of how KobrA's approach to components relates to other approaches, it uses this more general terminology.

3 The KobrA component model

Although components have been in vogue for some time, there is still no general consensus about precisely what constitutes a component and exactly how they should be put together to build useful systems. In this chapter we define the KobrA component model that explains the KobrA view on these questions.

The KobrA component model has two basic goals. The first is to strike a balance between simplicity and generality, and the second is to support a model of assembly that yields a tree-shaped hierarchy of components. The motivation for the first goal is to capture the important principles of component-based development in a way that avoids much of the complexity and redundancy of contemporary component models. The motivation for the second goal is to provide a software-engineering product that is compatible with recursive development activities, and thus amenable to systematic quality assurance.

We present the KobrA component model by first explaining the basic characteristics of KobrA components and the nature of the key relationships between them. We then describe KobrA's philosophy for modeling components using UML diagrams. Finally, we discuss the basic rules that govern the construction of well-formed component hierarchies.

3.1 Components

What are components? The basic idea of a component is applicable at all levels of abstraction, scales of granularity and stages of development in the software life-cycle. Not surprisingly, therefore, there is a large, and growing, set of definitions from which to choose. One of the best known is the definition formulated at the 1996 European Conference on Object-Oriented Programming [Szy99]:

> A software component is a unit of composition with contractually specified interfaces and context dependencies only. A software component can be deployed independently and is subject to composition by third parties.

This captures the key ideas that a component is a self-contained piece of software, with well-defined interaction points and modes, and is intended to be combined with other components into larger software entities. The same core ideas are also conveyed by the UML's definition [OMG00]:

> A component represents a modular, deployable, and replaceable part of a system that encapsulates implementation and exposes a set of interfaces.

and by Booch's definition [Boo87]

> A reusable software component is a logically cohesive, loosely coupled module that denotes a single abstraction.

From these definitions it is clear that components are built upon the same fundamental foundations as object technology. In fact most authors characterize the component paradigm as an extension of the object paradigm [Szy99]. The principles of encapsulation, modularity, unique identities, and unified functions and data are all basic object-oriented principles that are subsumed by the component paradigm. In addition, the fundamental component-interaction mechanism is also based on that of the object paradigm – namely, procedural operation (or method) invocation.

Although the ingredients of the object paradigm are subsumed with the component paradigm, they are not necessarily integrated in the same way. Indeed some of the confusion that surrounds component technology stems from ambiguity about how certain critical aspects of object-orientation are addressed. Two of the main questions relate to the instance/type dichotomy and the class/module dichotomy.

3.1.1 Instances versus types

The question that probably causes more confusion than any other in connection with components is the role of the instance/type dichotomy. This centers on the distinction between the run-time entities within a deployed system, and the development-time entities that describe them. Because systems often contain multiple copies of a given kind of software entity, these copies are usually instantiated from a type that defines their common properties. A type is thus a blueprint or pattern from which instances can be generated.[11] In object-oriented systems the word "object" is typically used to refer to the instances, and the word "class" is usually used for the types, but these terms have assumed additional meanings in the UML, and so must be used with care.

As an extension of the object-oriented paradigm, the instance/type dichotomy also exists in the realm of components. Components can have a run-time incar-

11. In this section we use the term "type" in the sense of the instance/type dichotomy discussed in Section 3.15 of the UML reference manual, rather than in the more technical sense discussed in Section 3.28.

nation (instance) within deployed systems, and can have a development-time incarnation (type) in the development environment. The run-time incarnations of components are almost always objects, not necessarily in the strict sense of an object-oriented programming language, but in the sense that they expose a set of invocable services and encapsulate a state. The development-time incarnations of components are types, and thus do not themselves contain a state nor offer any invocable services. Component types are therefore not objects [Szy99].

Ideally, the best way to avoid confusion is to always use explicit terms like "component instance" or "component type" to make the distinction between an instance and a type obvious. However, this can quickly become tedious. If the word "component" is used in an unqualified way, the practical question remains as to what should it be taken to mean in the context of the type/instance (development-time/run-time) dichotomy? There are three possible approaches:

1. *Component* means an instance.
2. *Component* means a type.
3. *Component* can mean both.

There are practical examples of each of these options. The prime example of (1) is the COM model from Microsoft. In COM a component is viewed as a binary instance. Approach (2) is used by the UML [OMG00], where component instances are explicitly distinguished from components. Perhaps surprisingly there are also practical examples of (3). In the JavaBeans model, the term "Bean" is used sometimes to refer to component types and sometimes to instances [Szy99]. Although this can lead to confusion, it is less problematic than might at first be expected. This is because components are often singletons, so that only one copy appears in a system. Also, many statements about components apply to component instances and types alike (e.g. components provide a set of services), and thus require no explicit differentiation.

singleton - only one copy appears in a system

In KobrA, approach (2) is adopted. Thus, the term "component", when used without qualification, is taken to mean a component type, at development time. Of course the more specific term "component type" can also be used whenever required to reinforce the idea of typeness. In the following sections, the unqualified term component is used to mean component type.

3.1.2 Classes versus modules

The second major issue is whether components, as development-time artifacts (i.e. types), have class-like properties and/or module-like properties. The difference revolves around whether they themselves present features such as operations and attributes, or whether they only act as containers for elements that do. In the first case they would assume properties usually associated with a class,[12] while in the second they would assume properties usually associated with a module.[13]

12. Strictly speaking, Classifier, is the UML metamodel element that represents entities with the property of possessing features. Class is a subclass of Classifier in the UML metamodel. We use the term "class" in this section since it will be more familiar to most readers.
13. The UML term for "module" is package.

Again, three interpretations are possible:

1 *Component* means a class.
2 *Component* means a module.
3 *Component* means both.

As with the instance/type issue, there are examples of all three choices. CORBA applications are examples of the first approach. The externally visible behavior of a CORBA object is captured in the form of features in the associated IDL class specification. Details of how these are implemented are completely hidden. The UML definition of component, on the other hand, is based on the second choice. Although *Component* is a subclass of *Classifier* in the UML metamodel, the definition of component semantics explicitly states that a component "... does not have its own Features, but instead acts as a container for other classifiers that reside on it." The reason for this view in the UML is that a component is intended to serve only as a representation of a language-level container representing an independently executable unit. The JavaBeans model is also based on the second approach. The behavioral properties of a Bean are described in terms of the regular Java constructs that it contains such as classes and interfaces etc.

Perhaps surprisingly, the best example of the third is also found in the UML, but in the form of subsystems rather than components.

3.1.3 Subsystems

The UML defines a subsystem as a "grouping of model elements that represents a behavioral unit in a physical system." As a subclass of both *Classifier* and *Package* in the UML metamodel, *Subsystem* is an example of a concept that combines both the properties of a class and of a module. A subsystem therefore presents features, but also acts as a container for other entities. UML subsystems may also be instantiatable, and so exhibit the type/instance dichotomy discussed above. The term "subsystem" is used to refer to types, and the term "subsystem instance" is used to refer to instances, just as with components.

In KobrA, components share all these properties of UML subsystems. In addition, KobrA adopts the same basic separation of concerns (specification versus realization) for documenting components, as explained in 3.3.4. They therefore have more in common with the UML subsystem concept than with the UML component concept. Interestingly, the Rational Unified Process [JBR99] also uses the term "component" more in the sense of a subsystem than an implementation-oriented module.

The main conceptual difference between UML subsystems and KobrA components concerns the possession of "behavior." The UML reference manual states that "a subsystem has no behavior of its own." Thus, although a subsystem has behavior associated with it, this is only by virtue of the behavior of its contents rather through its own inherent behavior. The features associated with a subsystem therefore constitute a way of repackaging the features of the subsystem's contents, rather new behavioral units in their own right.

In contrast, a KobrA component definitely can possess its own behavior. KobrA components need not merely repackage the aggregated behavior (or services) of their contents, but can offer new behavior (services) not realized elsewhere. Such new behavior is usually realized *using* the behavior of the component's contents, but is not something associated (in its entirety) with any of the individual contents of the component.

This difference reflects a fundamental difference in philosophy between UML subsystems and components. UML subsystems are intended to serve merely as convenient ways of aggregating and presenting behavior (services) that ultimately has to be implemented by other elements, such as objects etc. Their value is therefore primarily organizational since they themselves make no contribution to the actual implementation of the system. This means that their module facet eclipses their class facet.

KobrA components, on the other hand, most certainly do play a role in the implementation of a system. Moreover, they do so at all levels of granularity from the smallest abstract data types to the system as a whole. Thus, in KobrA the entire system is regarded and manipulated as a component, since it clearly represents a coherent behavioral unit. Moreover, it is quite possible that a behavioral unit that is only a part of one system may be a useful system in its own right for another user.[14] KobrA therefore provides a uniform representation of behavioral units, whatever their granularity, thereby assisting prospects for reuse.

The fact that components can be viewed as classes that contribute towards the realization of a system should not be taken to imply that they are the only classes that do so. On the contrary, although components constitute classes, not all classes constitute components. In addition to components, simple classes also typically play a critical role in the realization of a system's behavior. Generally speaking, simple classes are used as information holders to contain the data manipulated by a system, while components encapsulate the primary behavioral elements in the system. Usually only a few component instances will exist in a system, and their numbers will be relatively constant, while large numbers of objects may exist, and these numbers often vary significantly over time.

Since they possess behavior as well as contents, KobrA components represent a true union of the class and module concepts (i.e. classifier and package concepts in the UML). KobrA components really do sometimes behave as classes (classifiers) and sometimes as modules (packages) depending on the perspective being taken. For this reason, in KobrA components are represented using both the UML class and the UML package symbol, depending on the context in which they are being described. In view of their similarity with subsystems, the UML subsystem symbol can also be used if preferred.

14. A spell checking component, for example.

3.2 Component assembly

How do components go together to make useful systems? This question is perhaps even more fundamental than the question of what a component is, because the truly essential characteristic of a component is that it is a part of something else, not that it is realized in a particular form. In other words, the property of being a component is relative rather than an absolute. As with the previous question, there is an important difference between the run-time relationships between component instances within a deployed system, and the static relationships between the development-time representation of components.

3.2.1 Composition

There are two types of relationships between component instances that are of importance at runtime. The first is *composition*, which captures the idea that one component is a part of another. In the UML, composition is defined as a strong form of aggregation [OMG00]:

> Composite aggregation is a strong form of aggregation which requires that a part instance be included in at most one composite at a time, and that the composite object has sole responsibility for the disposition of its parts.

Key aspects of the composition relationship are:

1. Composite objects are responsible for the creation and destruction of their parts.
2. The parts of a composite object take their identity from their composite object.
3. Composition is transitive, meaning that if *c* is a part of *a* and *e* is a part of *c* then *e* is also a part of *a*.

An important consequence of these properties is that, at any given point in a system's execution, the composition relationships take the form of a tree. Figure 3.1 represents a snapshot of a system of six objects assembled into a single composition hierarchy. In general, the composition assemblies within a system can

Figure 3.1
Composition

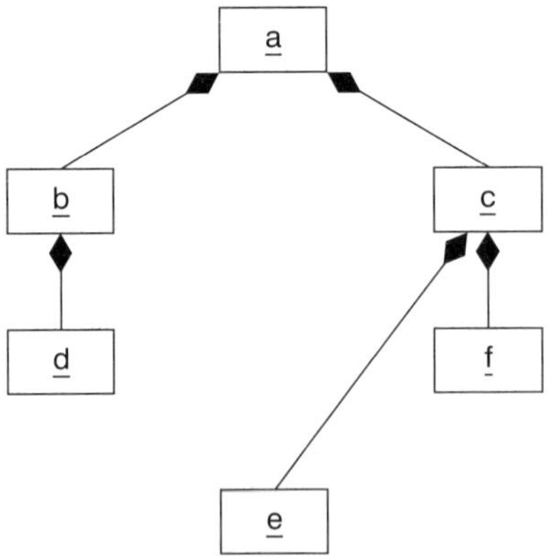

be disjoint, giving rise to a set of trees rather than a single tree. Moreover, the shape of the composition trees may change over time because a part may belong to different composites during its lifetime. However, at any given time the composition relationships must map out trees, since a part must belong to exactly one composite a time.

3.2.2 Clientship

If composition were the only important run-time relationship between components, assembling them into useful systems would be fairly easy, but unfortunately it is not. An even more fundamental run-time relationship, which is essential for building systems of arbitrary complexity, is the clientship or client/server relationship. This is the basic interaction relationship within object-oriented systems, and exists when one component instance, the *client*, invokes the services of another component instance, the *server*. We use the term "clientship" to reinforce the fact that clientship relationships are unidirectional: a client needs to know the identity of a server in order to invoke its services, but a server does not need to know the identity of its clients.

A clientship relationship between two component instances is said to define a *contract* between them. The server promises to deliver services with a particular set of properties provided that the client promises to use them in a particular way. If either of the two parties in the relationship fails to live up to this promise they break the contract into which they have entered.

Clientship is the most fundamental relationship because it subsumes composition. In other words, a composite object is a client of its parts. A composite object almost always creates its component parts in order to interact with them over its lifetime. But even if a composite object never interacts with one of its parts after creating it, it still needs to be a client of it in order to perform the creation. Another important reason why clientship is more fundamental is that clientship links must exist between component instances in order for a deployed system to do anything useful, but no composition links are required. It is possible to build useful systems in which component instances have no composition links.

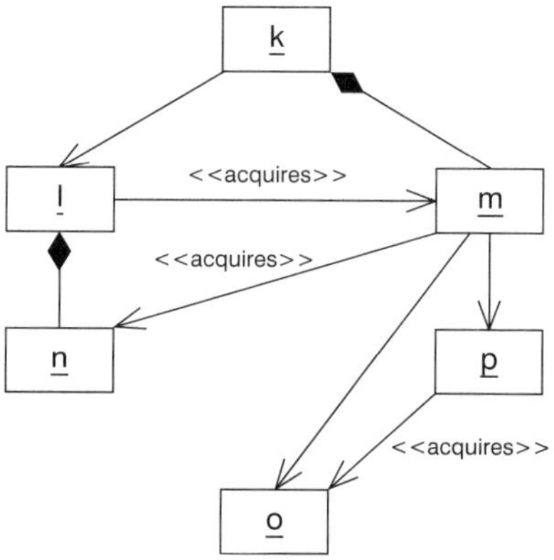

Figure 3.2 Clientship relationships

The run-time clientship relationships within a deployed system rarely, if ever, form a tree structure like composition. On the contrary, the clientship relationships within a system generally form arbitrary graphs. Moreover, the shape of the clientship graph usually undergoes many more changes during the execution of the system than the composition tree does.

Figure 3.2 shows a clientship graph between a set of component instances. Every line in this diagram illustrates a clientship link between two component instances. However, as illustrated in the figure, different kinds of clientship relationships can be identified. Two of the clientship relationships are also composition relationships (those represented by the composition symbol) while three of the clientship links are *acquired* (those labeled by the <<acquired>> stereotype. In contrast with composition links, where clients create their servers (their parts), acquired clientship links are links in which the client does not create the server itself, but gains access to it (i.e. acquires it) during the execution of the system. The unlabeled clientship links are links in which the client creates the server, but the criteria for composition are not satisfied.

Most of the externally visible properties of a component instance, and thus the component type that defines it, are dictated by the clientship relationships in which it is involved. A component instance generally provides a set of services in the form of operations that can be invoked by its clients. This is often called the *supplied interface* of the component. A component instance can also import component instances to act as its servers. These are often called *required* or *used interfaces*.

3.2.3 Ownership

The composition and clientship relationships discussed in the previous sections are links between component instances at run time. When these instances are created from component types, as in the KobrA model, these relationships, as well as all other properties of component instances, have to be described in their respective types. Thus, components that behave as classes, as in KobrA, have some of their development-time relationships dictated by the nature of the run-time links that their instances can participate in. However, in component models like that of KobrA where components also have module-like properties, some of the development-time relationships between components are dictated by their module facet rather than their class facet. Moreover, these two sets of relationships (class-driven and module-driven) need not necessarily be consistent. In this section we describe the nature of these relationships in more detail, and discuss how they can be reconciled.

A module (or a "package" in UML terminology, as noted earlier) defines a name space containing model elements. Any kind of model element can be contained in a package including classes, associations, models and other packages. In the UML a package is said to *own* the elements defined within it, and every element is *owned by* the package in which it is defined. The fact that packages can contain other packages is important, since it allows packages, and hence name spaces, to be nested to arbitrary depths. This not only provides the basis for hierarchical names in the UML, but when the packages are interpreted as components, gives rise to a natural concept of component nesting.

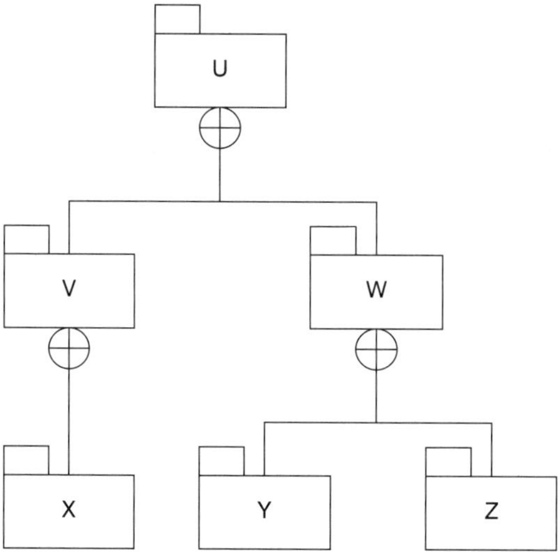

Figure 3.3
Module nesting

Figure 3.3 gives an example of a tree of nested modules. This uses the UML anchor symbol to represent the "scoping" semantics associated with the nesting of packages within one another.

The UML visibility rules are what one would expect for nested packages. In general, the elements owned by a package can have four distinct kinds of visibilities (public, private, protected and package) although private visibility is by far the most common. This means that a model element is not visible to elements outside the package that owns it, but is visible to other elements within that package, or any model elements within sub-packages of that package.

The nesting of one package inside another is viewed as setting up a permission dependency between modules with the stereotype <<imports>>. The nested package is said to have an imports permission dependency on the encapsulated package. This signifies that the public elements of the encapsulating package are added to the namespace of the nested package. This does not change the ownership relationships between the model elements and either of the packages. It just means that public elements in the encapsulating package can be identified in the nested package without using their full path name.

When a module references a model element from another package using its full path name, a permission dependency exists between the two that has stereotype <<accesses>>. The referring package is said to have an accesses permission-dependency on the package defining the model element. When the packages are interpreted as components, this is the type of dependency needed to support acquired clientship between instances of the components. The component of the client has to "access" the component of the server to gain visibility of the required information.

The final form of relationship between packages that has a relevance to the KobrA component model is the generalization relationship. This mimics the semantics of class inheritance in object-oriented languages by making the public and protected elements owned by one package visible in another without the use of the full path names. The package owning the model elements is said to be a generalization of the package gaining visibility of its public and protected model elements.

3.2.4 Containment

The previous subsections have described three fundamental kinds of relationships that have a bearing on components which, like those in KobrA, combine the properties of a class and a module. The composition and clientship relationships, driven by the run-time links between component instances, originate from the class facet of a component, while the ownership relationship comes directly from the module facet. As discussed above, two of these (composition and ownership) yield tree structures, while the third (clientship) generally leads to arbitrary graphs.

There is no *a priori* reason why these sets of relationships should be consistent with each other. For example, it would be easy to nest components in a way that does not support the accesses' dependencies (i.e. the visibilities) needed for the desired clientship relationships. However, this is clearly not acceptable. A coherent component model, which provides a genuine union of the class and module facets of components must find some way of reconciling these distinct views. Furthermore, to support the goal of recursive development activities, the reconciliation of these relationship types should be consistent with a tree-structured hierarchy of components. At first, the existence of arbitrary clientship relationship within a system of objects would seem to be an obstacle to this goal. However, it turns out that there is a way to reconcile clientship with component nesting (at development time) in a way that yields a tree-based hierarchy of components.

In circumstances when *all* the clientship relationships in a deployed system are permanent composition relationships, and *all* the component instances are of different types, there is a clear mapping between the run-time composition tree and the development-time nesting of modules. Basically, the nesting of modules can match, one for one, the nesting of component instances in the composition hierarchy. For example, assume Figure 3.1 shows *all* the clientship relationships within a system (in fact, they happen to all be composition clientship relationships), provided that the component instances have the classes. Then, shown in Figure 3.4, this hierarchy naturally matches the module hierarchy in Figure 3.3.

Consider now, a tree with the same number of instances as in Figure 3.4 above, with the same names and the same links, but with *e* being an instance of component *X* (just like *d*) rather than *Y*. This is illustrated in Figure 3.5

Because of the package visibility rules explained above, the module hierarchy in Figure 3.3 would no longer be compatible with this composition tree, because visibility of the definition of *X* must now be such as to create both *d* and *e*. To rectify the component-module hierarchy the definition of *X* needs to be moved to a higher location in the tree where it is visible to all its clients. This is illustrated in Figure 3.6 below.

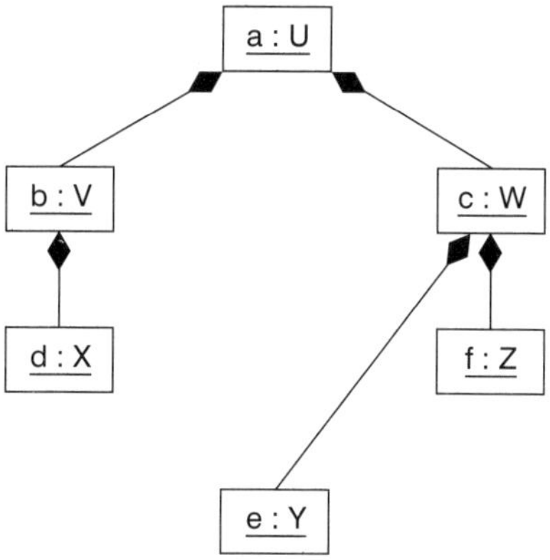

Figure 3.4
Pure composition hierarchy

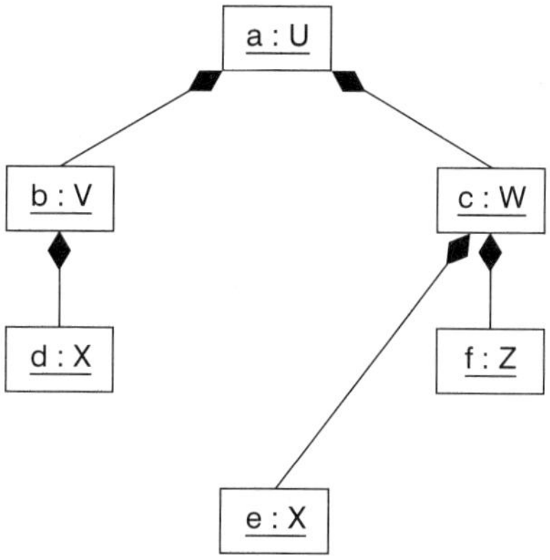

Figure 3.5
Pure composition tree with two instances of the same component

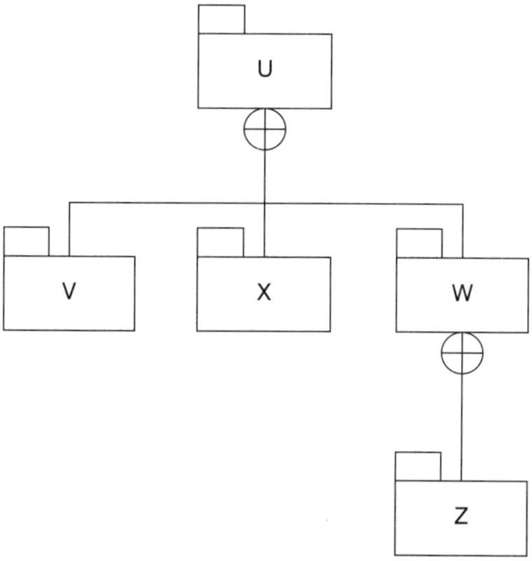

Figure 3.6
Revised module nesting

X is now visible to *V*, *W* and *Z*, whereas before it was only visible to *V*. The reason why the shape of the ownership tree needed to be changed is to facilitate shared visibility of component types. Such shared visibility is needed not only to allow multiple instances of a given component, but also to allow acquired clientship and changes to composition tree over time. Acquired clientship requires the creating and acquiring components to share visibility of the server component type, while changes to the shape of the composition tree require the donating composite and the receiving composite to share visibility of the component defining the exchanged part.[15]

Using the above approach, when a system contains a single, fixed composition tree relating its component instances, it is possible to attain a matching ownership tree relating its component types. Basically the ownership tree has the same structure as the composition tree except that components that need to be mutually visible are moved up to a position where they can be seen by all interested parties.

3.2.4.1 Creation trees

The only problem with this approach is that few if any systems contain a single, permanent composition tree. The number of composition trees in a system can range from 0 through to an unlimited number. Moreover, even if a system has a single tree, this need not be permanent, but can change its shape over time. The above approach therefore does not appear to be applicable to the majority of systems that do not possess a single, fixed composition tree.

All is not lost, however. As mentioned in Section 3.2.1, one of the key properties of the composition relationship between component instances is that the composite creates its parts. In other words, composition implies creation. Fortunately, it turns out that although many systems do not have one fixed composition tree, all systems do have a single, fixed creation tree.[16] In fact, every object-oriented system created to date has a built-in fixed containment tree.[17]

The creation tree has a number of nice properties:

1. When a system does have a single, fixed composition tree, the creation tree is exactly the same shape and structure.
2. When a system has a single but non-fixed composition tree, the creation tree corresponds to the first one.
3. When a system has multiple composition trees, they are all subsumed by the single creation tree.

15. Since composition is one form of clientship, the need to support changes of the composition tree is actually subsumed by the more general need to support the required clientship relationships. Supporting changes to the composition tree is no longer directly mentioned, therefore.
16. We are not concerned here with distributed systems where multiple component instances may be created manually. Even in such cases it is possible to identify a single creation tree, though the root of the tree may lie outside the software components.
17. This statement is based on the fact that every object in a system must be created by exactly one other object in the system, no more and no less.

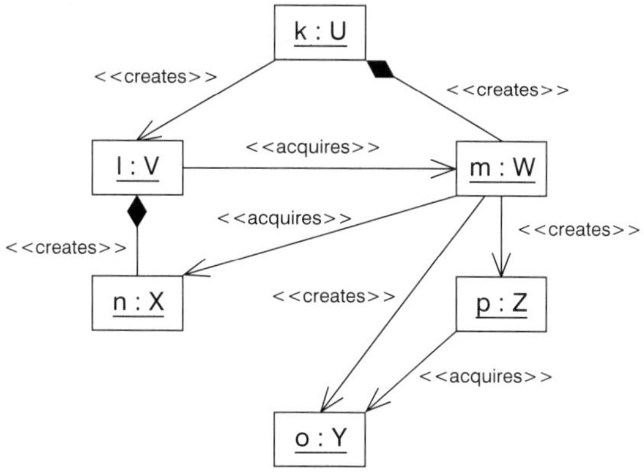

Figure 3.7
Creation tree

The creation tree therefore provides a good foundation for the structure of the development time component-nesting tree. The basic idea is that components are nested in the same way as their instances in the creation tree, but with appropriate changes to support the shared visibility of components needed for clientship[18] and for instantiating multiple instances of the same component type.

Figure 3.7 shows the creation tree within the system of components illustrated in Figure 3.2. The stereotype <<creates>> can be used to identify clientship relationships in which the client creates the server. A client can therefore arrange to have visibility of a server in one of two ways: it can either create it itself, or it can acquire it from some other component instances. Creation and acquisition therefore partition the set of clientship relationship into two disjoint subsets.

3.2.4.2 Containment trees

By basing the structure of the development time component nesting tree on the structure of the run-time creation tree, with appropriate changes to support the mutual visibility needed for the clientship relationships and multiple instances of a given type,[19] the ownership, creation and clientship aspects of a component-based system can be fully reconciled. When this is done, and the development time tree is organized in the way described, we use the term "containment" to refer to the relationship between the components. The ownership relationship is still valid, and the associated terminology can still be used if desired, but in KobrA the term "containment" is used to convey the additional idea of a nesting hierarchy that has been reconciled with the run-time creation and clientship relationships. This terminology is actually consistent with the UML terminology, because although no official containment relationship is defined in the UML, the word "contains" and its derivatives are used extensively to describe the semantics of package nesting.

18. Since composition is a form of clientship, the shared visibility needed for composition changes are handled by the provisions made for clientship.
19. Since creation implies clientship, the need to support multiple instances of the same type is subsumed by the need to support all required clientship relationships.

Figure 3.8
Component containment tree

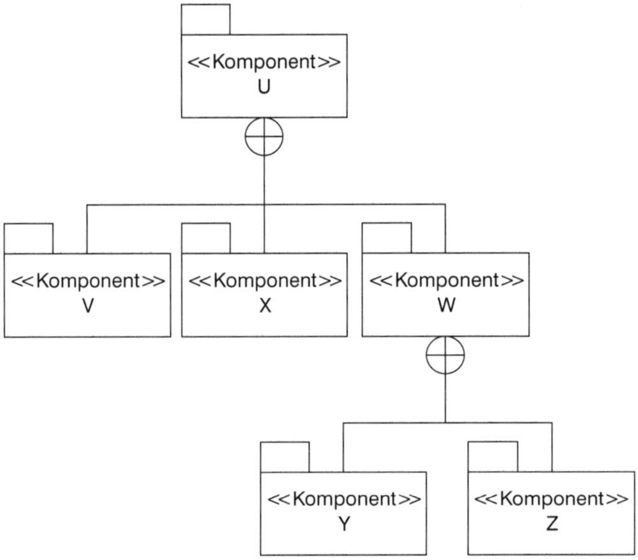

Only when packages are part of a containment tree (rather than just an ownership tree) can they justifiably be called components. For the packages in a containment tree (i.e. a nesting tree reconciled with creation and clientship) we therefore also use the stereotype <<Komponent>>[20] mark them as *bona fide* KobrA components. Figure 3.8 illustrates the component containment tree corresponding to the system of component instances illustrated in Figure 3.7.

For components in a containment tree we use the terms "subcomponent" and "supercomponent" to refer to a component's immediate neighbors in the tree. Basically, if *X* contains *Y*, *X* is a supercomponent of *Y*, and *Y* is a subcomponent of *X*. It is also possible to use other common tree-related terms such as "parent/child" or "ancestor/descendent" to navigate around the tree structure.

3.3 Modeling components

Having described the abstract properties of components and their assembly, we now turn to the question of how they are modeled using the UML. KobrA's use of the UML differs somewhat from traditional object-oriented methods, but is based on a few very simple principles. Once these have been mastered, the use of the UML to describe components is relatively straightforward and systematic. In this section we describe the basic principle by which the UML is applied in KobrA, leaving the details to subsequent chapters.

20. "Komponent" is short for "KobrA component."

3.3.1 The principle of uniformity

Many methods change the way they represent certain abstractions or concepts based on the level of granularity, or phase of development, at which they are being addressed, not because of any inherent change in the concept itself. In other words, they represent and manipulate a given abstraction in distinct ways in different parts of the method. A prime example is the many different ways in which procedural abstractions are handled in modern object-oriented methods. Although the basic ideas are the same, in the early phases of development the concepts, terminology and techniques of use-cases are used, while in later phase of development the concepts, terminology and techniques of collaborations and operation/method invocation are used.

The principle of uniformity requires that, wherever possible, a given fundamental concept is represented and manipulated in the same way in all parts of the method, and in all phases of a project that uses the method. Perhaps the most important effect of this principle in KobrA is that all behavior-rich abstractions or state machines, whatever their complexity and/or granularity, including the system as a whole, are represented as components. The distinction between a system and a part of a system (i.e. a subsystem) is therefore blurred in KobrA. Every component can potentially be a system in its own right, or could be used as a part of a larger component (by composition at run time and containment at development time). The choice is made when a subtree of a containment tree (i.e. a component and its subcomponents) is mapped into an executable form.

The principle of uniformity also requires that all Komponents be modeled in the same basic way regardless of their location in a containment tree. In other words, the nature of the models describing a component should be dictated only by the component's properties. KobrA therefore uses the same basic ideas to describe a system (i.e. a component) and its environment as it does to describe components nested in the containment tree.

3.3.2 The principle of locality

Most of the concepts introduced in the previous sections have been around for some time. In particular, the idea of building systems by connecting together autonomous building blocks, which in turn are built of smaller interconnected building blocks, has appeared in various guises over the years [GW94]. In fact, once one accepts that composition is more fundamental than inheritance, this idea essentially forms the foundation of the object paradigm.

Not surprisingly, therefore, the first generation of object-oriented methods all provide some basic way of expressing building blocks and their nesting within one another to create larger building blocks. In this sense they can all be regarded as supporting the rudiments of component-based development. However, one property that they all share is that their consideration of these component-oriented concepts is limited to their run-time manifestations. Thus, although they provide concepts for describing run-time components and their inter-relationships, the development-time artifacts are not themselves

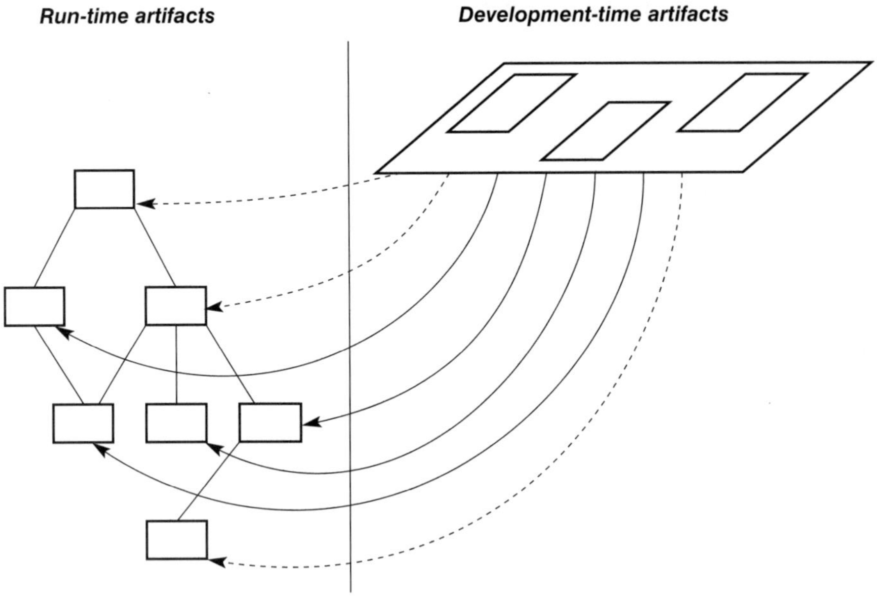

Figure 3.9
"Flat" organization of development-time artifacts

component-oriented. In other words, no attempt is made to organize the development-time artifacts (e.g. UML diagrams), and their contents, in a component-oriented way. In short, although the development-time artifacts *describe* building blocks and their nesting, they are not themselves structured in a way that reflects this structure.

First generation object-oriented methods such as OMT [RBP91], Fusion [Col94], Booch [Boo93], and ROOM [GW94] provide the clearest example of this. As illustrated in Figure 3.9, although they all provide ways of describing building blocks and their nesting, this information is contained in a single class diagram.[21] Thus, the development-time organization of development artifacts (e.g. diagrams) in these methods can best be characterized as "flat," although the run-time situation that they describe can be characterized by nested components. More modern object-oriented approaches, such as RUP [JBR99] and OPEN [GOW01], have a more sophisticated arrangement of diagrams, but this arrangement is more oriented towards abstraction levels than components.

KobrA's approach to modeling is based on the basic idea that the development-time artifacts (e.g. UML diagrams) should be constructed and organized in a way that reflects the run-time component-oriented structure of the system. The previous section explained how this idea known as the principle of locality, is built into KobrA's view of components and their relationships. However, this principle applies to the contents of the development-time artifacts as well as their organization and structure. It requires that every development artifact, including UML models and diagrams, be focussed towards the description of a single run-time

21. The class diagram may evolve in maturity, but at any one time there is only one class diagram.

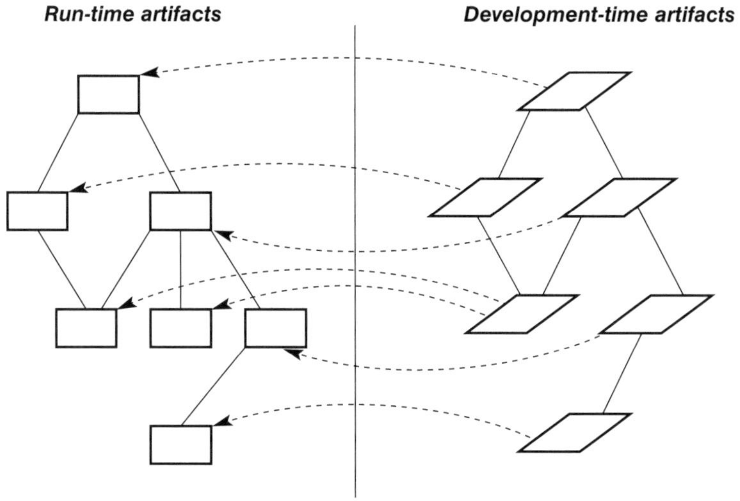

Figure 3.10 Component-oriented organization of Development-Time Artefacts

artifact. In other words, there are no global models or diagrams. Instead, every model or diagram is focussed on the description of a single artifact – that is, it is "local" to that artifact. This view of the organization of the development-time artifacts is depicted in Figure 3.10.

The principle of locality goes beyond simply using subsystems to organize modeling elements. As discussed in Section 1.1.3, there is a connection between the concept of components and subsystems, and organizing modeling elements in terms of subsystems certainly represents a step towards a more component-oriented organization of the development-time artifacts. However, the subsystem concept is only concerned with the ownership and presentation of model elements. It makes no statements about the nature of the diagrams that should accompany a subsystem, let alone about the content of these diagrams. In contrast, KobrA, through the principle of locality, requires that every diagram be local and focussed on the description of a particular entity. Thus, it prescribes what diagrams should be created, and what the contents of these diagrams should be.

UML diagrams may actually be focussed on the description of (i.e. be local to) two kinds of artifacts:

- components
- activities.

In other words, every development-time artifact, including all UML diagrams, is oriented towards the description of a component or an activity. Usually it is clear which entity is the focus of a diagram, but in cases where there is potential ambiguity, the stereotype <<subject>> can be used to identify which entity is being described. This is most useful in class diagrams to indicate which component is the subject of the diagram.

3.3.2.1 Role modeling

Another way of thinking about the principle of locality is in terms of role modeling. Role modeling is an approach to class diagram development that emphasizes the relative nature of classes, objects and their relationships [RWL96]. In other words, a role is a description of an object or a class from the perspective of another object or class. When a run-time system is eventually configured, classes or objects of a different kind may actually be used to satisfy a role, provided they are able to fulfill all the expectations defined for the role. Role modeling is therefore closely related to the idea of polymorphism, namely, the idea that at run time one server object may be replaced by another server object, provided it is able to satisfy the same set of requests.

Roles are needed as a separate concept in the UML because most methods view class diagram development as a mixture of role modeling and absolute modeling. In other words they allow roles (relative classes and/or objects) to be mixed arbitrarily with absolute classes and/or objects. It is therefore necessary to have the concept of a role in order to distinguish the relative use of classes/objects from the absolute use.

In KobrA, however, every diagram can be viewed as a role model, because every diagram is relative. More specifically, every diagram can be viewed as a role model with respect to the component or activity that is its subject (i.e. the focus of the diagram), and almost every element of a diagram can be viewed as a role. The only exceptions are the model elements describing the creation of component instances or objects. When creating an object or a component instance it is necessary to be specific about the precise type (i.e. the class or component) from which the entity is created. This therefore represents a case of absolute modeling. Otherwise, every appearance of a class, object, component or component instance is relative and subject to substitution according to certain conformance rules, which are described in Section 3.4.1.

Since the few instances of absolute modeling are clearly identified (by virtue of their direct involvement in creation) the explicit concept of roles and role modeling is not needed in KobrA. Similar reasoning applies to the interface and type concepts. These serve a similar purpose to roles in the UML, with a few subtle differences in semantics. By the near exclusive use of relative modeling in UML diagrams, and by the exclusive use of components as the system structuring mechanism, KobrA is able to avoid the need for separate interface, role, and type concepts without loss of expressive power. These concepts can be used in certain well-defined places in KobrA, as explained in section 3.3.4.3, but in general they are not needed.

3.3.3 The principle of parsimony

A potential disadvantage of role modeling to the extent used in KobrA is the unnecessary duplication of information. Since separate development artifacts, such as UML diagrams, necessarily overlap to some degree, and each describes a specific piece of information from its own viewpoint, there is a danger of redundancy. This is particularly so for class diagrams which are notoriously difficult to finish, and can in principle be cluttered with an unlimited number of artificial associations. To avoid this problem, KobrA requires every diagram to contain the minimum information needed to convey the required ideas. This principle is supported by concrete guidelines, explained in the coming chapters, for determining what needs to be expressed in a diagram.

3.3.4 The principle of encapsulation

A long-established principle of software engineering is to separate the description of *what* a software unit does from *how* it does it. This facilitates a "divide and conquer" approach in which a software unit can be developed independently of its users. It also allows new versions of a unit to be interchanged with old versions, provided they do the same thing. Various different terms have been used over the years to refer to these two separate pieces. Terms used for the description of what a unit does include "specification," "interface," and "signature," while terms for the description of how a unit works include "realization", "design," "architecture," "implementation," and "body."

This principle also applies naturally to components and therefore appears in many contemporary component models [CD00]. The principle also appears in the KobrA component model, but with one important difference: in KobrA there are *two* descriptions of how a component works rather than one. One description, called the *realization*, describes how the component works in terms of subcomponents and client components, while the other, the implementation, describes how the component works at a level of abstraction that can be understood by automated compilation tools (i.e. at a programming level of abstraction). The realization of a component therefore represents a description of how the component works at the same level of abstraction as the specification, but at a lower level in the containment tree, while the implementation describes how the abstract realization is mapped into a tool-comprehensible form.

In general, therefore, there are three levels of description associated with a component:

- **specification** – describes the externally visible properties of a component's instances and what they need from their surroundings
- **realization** – describes how a component instance realizes its specification by means of interactions with other component instances
- **implementation** – refines a component's realization to a tool-comprehensible form (i.e. source code).

A given version of a component has one specification and at most one realization, but can have multiple implementations. In some situations it is possible for a component specification to have no associated realization or implementations (see 3.4.1.1)

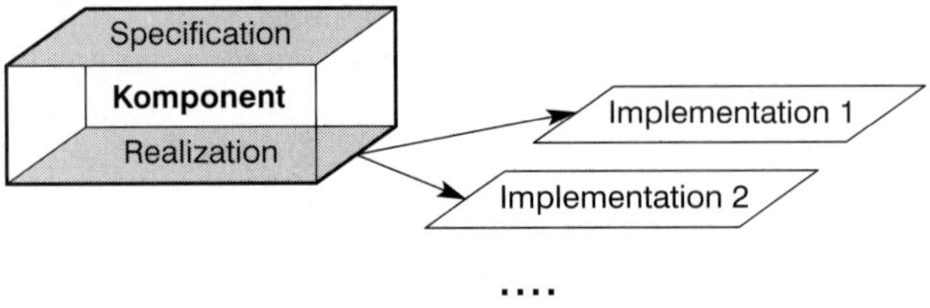

Figure 3.11
Levels of component description

To help reinforce the distinction between component specifications, realizations, and implementations, it is helpful to think of a component as a cube, as illustrated in Figure 3.11. The top surface of the cube can be thought of as the specification that describes what a component does and what it needs. The bottom surface can be thought of as the realization of the component that describes how it realizes the specification using other components. Finally, the implementations can be thought of as separate artifacts, each providing an alternative description of the realization, but at a more concrete level of abstraction akin to a program. The realization and implementation of a component therefore represent elaborations upon how it works but in different dimensions. These distinct levels of description are described in more detail below.

3.3.4.1 Component specifications

The specification of a component describes all the properties of its instances that are externally visible (to other component instances). This includes the set of services that the component instances makes available to others (commonly called the supplied or server interface), and the set of server instances that the component instances need to acquire (commonly called imported, supplied, or used interfaces). The specification also defines the behavior of the component instances over time, including the different states that they may exhibit.

Consider, for example, the system of component instances in Figure 3.12.

Component instance *e* acquires access to component instances *c* and *d*, and is itself acquired by *b*. This means that *e* is a client of *c* and *d*, and a server of *b* and *a*. Because of the asymmetric nature of clientship, however, the fact that *e* is a

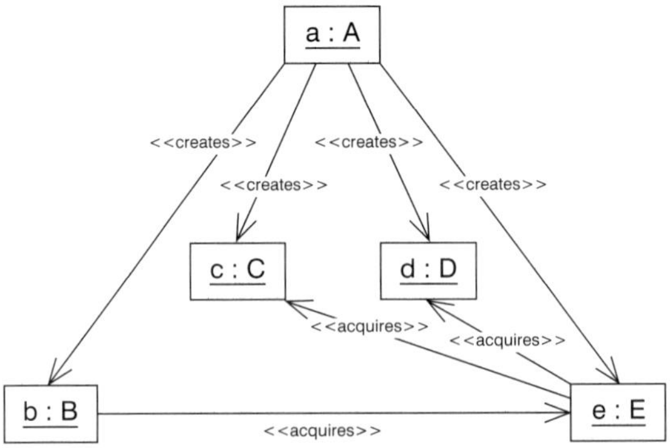

Figure 3.12
Interacting instance

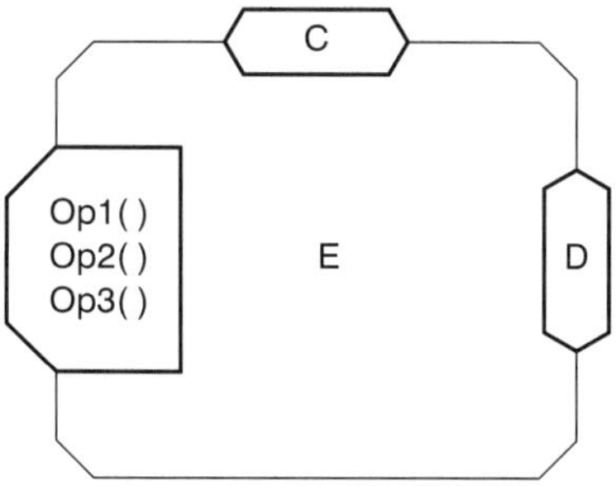

Figure 3.13
Supplied and acquired interfaces

server of *b* does not need to be reflected in the specification of *e*, because servers do not know the identity or type of their clients. However, the fact that *e* is a client of *c* and *d* does need to be reflected in the interface.

Figure 3.13 provides a schematic representation of the services supplied by and acquired by component type *E*. The services supplied by *E* take the form of a set of operations appearing in the polygons on the left-hand side. The services acquired (or required) by *E* are captured in the polygons on the top and right-hand side, and are identified by the name of the component types that supply those services. This specification therefore indicates that an instance of *E* needs to be supplied with (i.e. needs to acquire) instances of components *C* and *D* in order to function.

3.3.4.2 Component realizations

The realization of a component describes the private design by which its instances realize the properties defined in the specification. This includes a description of any server creations performed by the instances, and the identification of any subcomponents. The realization also describes how the various server instances (created and acquired) are linked together within the component's instances. In other words, it describes the structural architecture of the component.

For example, in the system in Figure 3.12, the realization of component *A* would be responsible for creating and configuring the instances of its subcomponents in a way that satisfies all their needs. This would require an instance of component *E* to be connected as a client to instances of *C* and *D* and as a server to an instance of *B*, as indicated in Figure 3.14. In addition, the realization of *A* must declare any subcomponents of *A*. Assuming that none of *B*, *C*, *D*, or *E* are defined outside *A*, they would all be subcomponents of *A* as indicated in Figure 3.15.

If, however, one of the components instantiated by *A* were already defined outside *A*, then of course it would not be a subcomponent of *A*. Instead it would have to be accessed by *A*. This is indicated in Figure 3.16.

As a consequence of this, the specification and the realization of a component may cause it to access other components elsewhere in the containment tree. The specification causes such accesses by acquiring instances of a component, and the realization causes such access by creating instances of a component. In Figure 3.16, for example, component *A* accesses component *D*

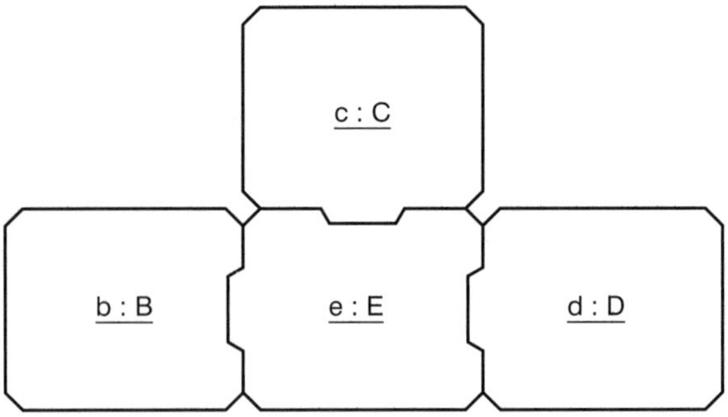

Figure 3.14
Architecture of component *A*

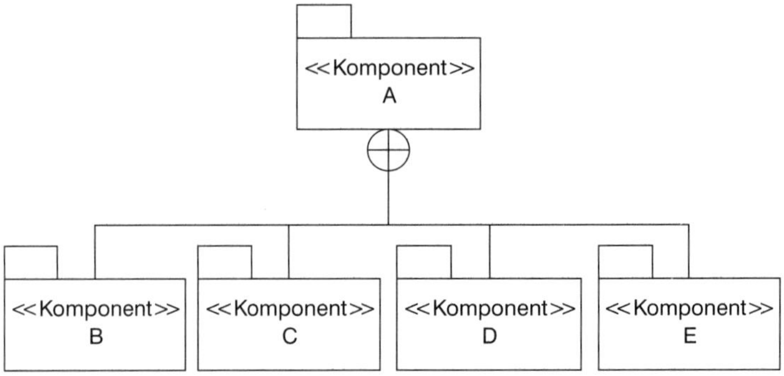

Figure 3.15
Subcomponents of *A*

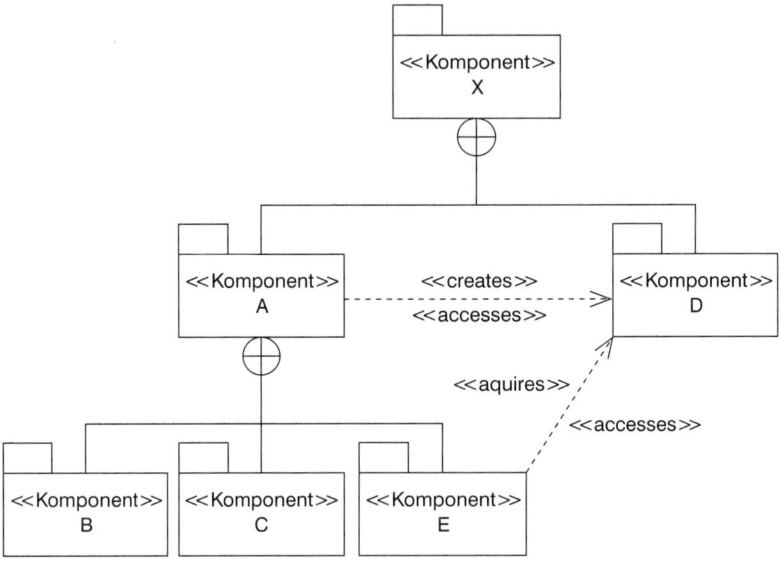

Figure 3.16
Accesses dependencies

because it creates an instance of *D* (defined in its realization), while component *E* accesses component *D* because it acquires an instance of *D* (defined in its specification). Although the predefined permission stereotype <<accesses>> correctly captures the nature of the dependency from a package perspective, the stereotypes <<creates>> and <<acquires>> provide more information since they both imply <<accesses>>, but also indicate why the access permission is required. These two stereotype are therefore generally used in practice.

In terms of UML semantics, the realization of a component has an imports dependency on the corresponding specification.

3.3.4.3 Component implementations

A component realization describes how a component works, but at a level of abstraction akin to design. In order to attain an executable version of a component it is first necessary to create a description of how the component works in a form that can be processed by automated compilation tools. This is the role of the component implementation.

An important decision to be made when implementing components is to decide how the *logical* components in the component tree are to be mapped to physical components that are *deployable* in the final execution environment. A logical component can be mapped individually into a single physical component, or can be grouped together with its children and/or parent into a combined physical component. Instances of physical components represent the smallest granularity of deployment and replacement in the run-time system.

Since the UML (with the exception of a few parts) is not an executable language, a component implementation cannot consist of UML diagrams alone. It must contain some form of program in an executable programming language. However, a component implementation can also contain supplementary UML diagrams to explain how the program works. In principle, any programming language or general-purpose implementation vehicle can be used to implement a component. In practice, however, the implementation process is significantly easier with a programming language that is at least object-oriented such as Java, C++, or Ada. The ideal implementation vehicle for a physical component is an implementation-level component technology such as a JavaBean, a COM object, or a CORBA object. When physical components are implemented using one of these technologies, the advantages of component-based development at the analysis and design levels are combined with the advantage of components at the implementation and deployment levels.

In the description of component concepts in Section 3.1.3, it was pointed out that KobrA components are more like UML subsystems than they are UML components. This naturally raises the question as to whether the UML component concept has any role to play in KobrA. The answer is that UML components provide a natural representation of physical components – the unit of deployment and execution. When implemented using JavaBeans, Enterprise JavaBeans, or COM, a physical component corresponds to a component in one of these technologies, and thus is naturally representable as a UML component.

Although UML models can be used to augment component implementations, there are two important differences between the use of the UML for this purpose and the use of the UML for component specifications and realizations. The first is that the constructs used in component implementation models will, of necessity, be at a much lower level of abstraction than those used within specifications and realizations. This is because the purpose of implementation models is to describe the "as is" implementation of a component, rather than to model its abstract properties. Techniques for creating implementation models are described in detail in Part 3. The second distinction is that, unlike specification and realization diagrams, implementation diagrams are not constructed according to the principle of locality.

3.4 Building Component Trees

Now that the principles by which individual components are modeled in the UML have been described, we turn our attention to how they are put together to build a containment tree representing the desired run-time system. Several fundamental principles guide this process.

3.4.1 Polymorphism

As explained in the previous section, the properties of components are described from a relative perspective in the form of diagrams that represent role models. This raises the question as to where these different perspectives are reconciled, and who is responsible for deciding which component or object plays which role. Primarily it is the component realizations that have this responsibility.

Consider the system illustrated in Figure 3.12 and Figure 3.14 for example. Here the realization of *A* is creating a set of component instances and giving them appropriate visibility of one another so that their needs are satisfied. In Figure 3.14, each component instance was given access to an instance of the precise type that it required as a server, but this does not have to be so. Thanks to polymorphism, a realization can instantiate any type of component, and pass it as a server to any of the other component instances provided that each conforms to the server component type expected. Thus, the assembly of component configurations within component realizations can take full advantage of polymorphism in the usual sense of object-oriented development. The criteria for what component instances and objects can play what roles are determined by the conformance rules identified below.

At run-time, a component instance (or object) can be supplied as a server if its component type (or class) *conforms* to the component type (or class) expected by the client.

A component *X* conforms to a component *Y* if:

1 *X* is a specialization of *Y*, or

2 *X*'s operations are a subset of *Y*'s operations, and have arguments and return parameters that are pair-wise conformant according to the principle of contravariance, or

3 *Y* is a public subcomponent of *X*, and an instance of *X* has an instance of *Y* as a public part (i.e. is related to *Y* by public composition), or

4 *X* has a conformance map to *Y*.

Rules (1) and (2) also apply to simple classes used in KobrA, while the last rule applies only in the context of component reuse, and is explained in Chapter 12.

3.4.1.1 Component specialization

The most familiar conformance rule to object-oriented programmers is likely to be the first rule, since this is the classic object-oriented way to specify conformance. As in object-oriented programming languages, KobrA allows one component to be defined as a specialization of another component. However, unlike most programming languages, KobrA's specialization mechanism only relates the specification of components: component realizations cannot be "inherited." KobrA therefore only supports interface or type inheritance.

Specialization of components serves much the same purposes as specialization in object-oriented programming languages. As well as providing an explicit way of indicating that one component conforms to another, it also allows a given component to assume more than one existing specification (multiple inheritance). This is useful because it allows a component instance to play the role of two distinct servers.

Since component specifications can be specialized as well as realized, it is also possible to have specifications without any realization. Such specifications serve the role of abstracts classes (or interfaces in Java) and provide a way of defining abstract server interfaces without committing to a specific realization. Full details for the component specialization mechanism are defined in Chapter 9.

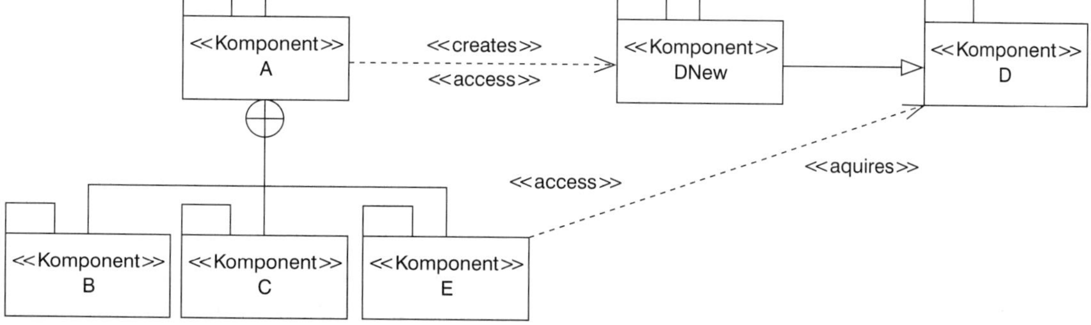

Figure 3.17
Adapted containment hierarchy

Figure 3.18
Adapted architecture of component *A*

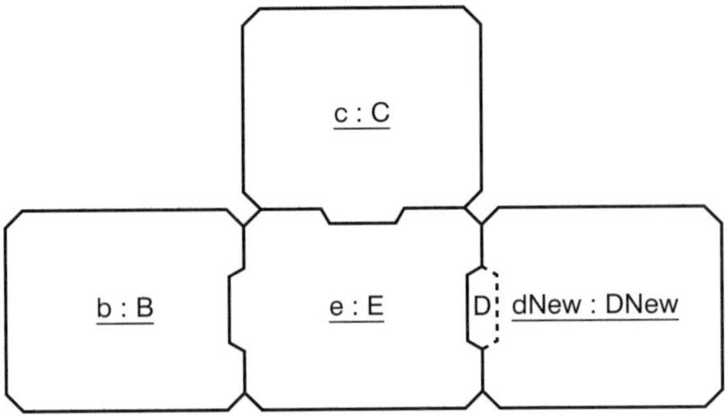

As an example of how specialization can be used to arrange for the polymorphic substitution of instances, consider the following adaptation of the containment hierarchy in Figure 3.17.

In this version of the system, *A* creates an instance of a specialization of component *D* rather than an instance of *D* itself. In the corresponding run-time configuration, illustrated in Figure 3.18, *e* is now connected to an instance of *DNew* rather than *D*, but is unaware of this fact. *d* is playing the role of an instance of *D* from *e*'s perspective.

As might be expected, component specialization implies the generalization dependency from the package perspective. Thus, all the public and protected elements in the generalization component are visible in the specialization component, and can be accessed without using their full path name.

3.4.2 Visibility rules

Another set of rules that has an important impact on how component instances can be assembled within a component realization is the set of the visibility rules. These define what other components in a containment tree a given component can see based on its position in the overall containment hierarchy. KobrA's visibility rules are based on the UML package visibility rules.

A component *X* can only see another component *Y* if:

1. *Y* is *X*'s (immediate) child.
2. *Y* is *X*'s (immediate) parent.
3. *Y* is *X*'s peer (*X* and *Y* share the same parent).
4. *Y* can be seen by *X*'s parent (recursive).

The first rule means that in the containment tree a component can see its immediate children, as one would expect. The next three rules mean that a component can only see (and thus acquire or create) instances of components "up and across" from its location in the tree. It cannot, as it were, go up its branch of the tree and down another branch (except for one level) to gain visibility of a component. The effects of these rules are depicted graphically in Figure 3.19. The components which can be seen (and thus created or acquired) by the black box are those shaded gray. The components represented by white boxes cannot be seen by the black box.

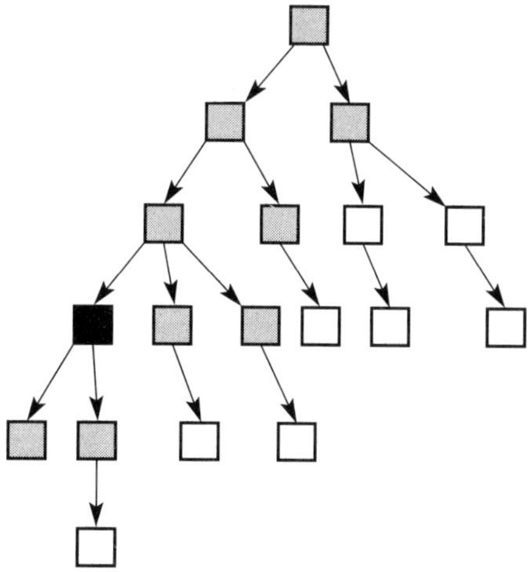

Figure 3.19
Visibility rules

3.4.2.1 Public containment

The above rules are based on the assumption that a component's subcomponents are private, in the sense of the package visibility rules defined by the UML. This is indeed the normal situation, but KobrA also allows subcomponents to be defined as public. This means that the components outside a given component can see its subcomponents, which they would not otherwise be able to do. The changed visibility applies to instances of the public subcomponent as well as the subcomponent type.

With the presence of public containment a fifth visibility rule needs to be added to the previous list:

5 *Y* is a public component of *Z*, and *Z* can be seen by *X*.

Figure 3.20 shows how the presence of public containment changes the visibility situation depicted in Figure 3.19. If the thick arrows in this figure are taken to represent public containment relationships, the set of components visible to the black component changes in the way illustrated.

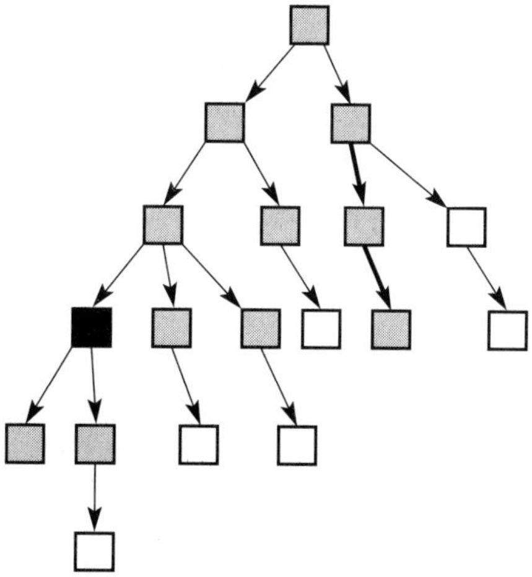

Figure 3.20
Visibility rules with public containment

Public containment has a number of uses in KobrA. First, it provides a much more natural way of modeling scenarios in which parts are clearly visible outside their composites. This is typical of user interfaces, where user interface component instances (e.g. boxes, buttons, panels etc.) are created precisely because they are visible and accessible outside their composite component instance. Second, it simplifies the definition of components that have very little functionality of their own, but which serve primarily as containers for subcomponents that actually provide behavior (very much like subsystems in the UML). Such components can be handled by normal private composition by arranging for all requests to be forwarded to the private parts, but public composition is often much more natural.

Notice the public/private distinction affects the composition relationship between component instances as well as the containment relationship between classes.

3.4.3 Consistency rules

The previous sections have introduced several kinds of rules that play an important role in ensuring that a component containment tree is well formed and consistent. In general there are six distinct sets of rules that must be satisfied in order for a component containment tree to be certified as in a consistent state:

1. **Intra-diagram rules** – ensure that individual diagrams are well formed.
2. **Inter-diagram rules** – ensure that the diagrams within a specification or a realization are consistent with each other.
3. **Realization rules** – ensure that a component's realization represents a correct representation of its specification.
4. **Clientship rules** (contract) – ensure that a client and server both fulfill their contract.
5. **Containment rules** – ensure that a component's relationships with other components are consistent with its location in the containment tree.
6. **Specialization rules** – ensure that a specialized component conforms to the component from which it was specialized.

Figure 3.21 shows the role these various sets of rules play within a component containment tree. This contains the same set of components as Figure 3.17 (with *B* and *C* omitted for simplicity), but with exploded symbols for the components to give a schematic view of the suite of artifacts within the specifications and realizations of the components. The diagram gives an example of the applications of all six types of rules. The role of the intra-diagram, inter-diagram and realization rules are shown only in connection with component *X*, since they are obviously applied the same way to all the other components.

3.4.4 Shaping containment trees

The consistency rules described in the previous section ensure that the components within a containment tree, and the artifacts describing them, are consistent within one another. In other words, they collectively ensure that a containment tree, and the components within it are well formed. However,

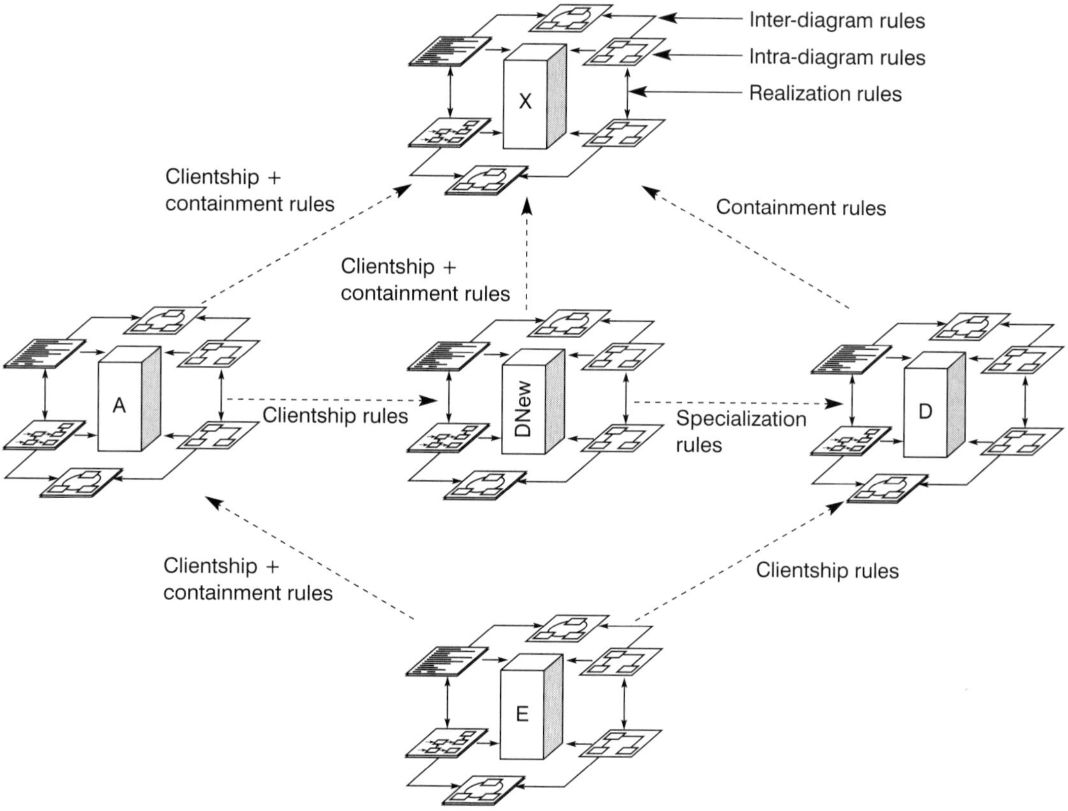

Figure 3.21
Consistency rules

they do not define the shape a tree containing a given set of components should take. Software engineers still have a lot of leeway in choosing the containment tree that best matches the needs of the enterprise and the project. This is the final major question that we consider in this chapter.

The first important point to realize about a containment tree is that in general no individual component knows about the shape of the whole tree. Because of the principles of locality and separation of concerns a component only has visibility of the specifications of its subcomponents, and knows nothing about how its subcomponents are realized (i.e. what subcomponents they in turn may have). The only exception is when the subcomponents are public. The only circumstances under which a single component knows about the whole containment tree is if the tree consists of only two levels (i.e. a root component plus its children) or consists of only public containment relationships.

The shape of a tree represents a balance between the goals of minimizing the replication of information (i.e. the definition of identical components) and maximizing the reusability of components (i.e. the ease with which they can be used in new components). Sharing components, by placing them at locations in the tree where they can be accessed by numerous components, has the advantage that it avoids the replication of the component within all the other components that use it, but has the disadvantage that it creates a dependency between the components and its users. In general the reusability of a component is greater the lower the number of dependencies.

Neither extreme is optimal. One extreme is to maximize shareability by placing all the components within a flat, two-level tree. This allows every component to see (and thus use) every other component in the tree, and thus corresponds to traditional "global visibility" approaches. The resulting tree is as shallow as possible. The other extreme is to have no sharing at all, and to have every component define its server components as subcomponents. This will result in a tree that is as deep as possible, in which all the clientship relationships trace out a tree. Unless there is no logical need for sharing, however (because no two components need the same type of clients), this approach will tend to lead to the replication of component definitions.

Perhaps the best guidelines for shaping a tree and making judgments about sharing are provided by the Cleanroom approach [MLH87] (introduced in Chapter 1). Within Cleanroom a developer creates hierarchies of boxes, akin to KobrA component specifications and realizations, which depend on sound software engineering principles such as:

- **referential transparency** to manage delegation and assignment of responsibility
- **transaction closure** to ensure that a sufficient set of transactions is identified
- **state migration** to ensure that data is decentralized to the smallest system part that does not require duplicating data [MLH87]
- **common services** to create as many reuse opportunities as possible.

Although the Cleanroom approach is fundamentally function-oriented, the basic principles are still valid for the KobrA component tree (see Chapter 6). Consequently, the guidelines for shaping a tree and making judgments about sharing [MLH87, Mil88, Mao93] can be easily transferred to KobrA. A good example is the set of guidelines concerning the principle of common services.

In principle a common service encapsulates functionality with multiple uses without violating the referential transparency principle (i.e., the implementation is just as important as the specification). If a situation should occur in which several boxes work on a common state, a new box structure is added to the hierarchy, which provides access to this state for these boxes. In KobrA this guideline would mean that if several subcomponents access or alter a common part of their super component and if it is not possible to migrate that part to the lower-level components without duplication, it is modeled as a common service. This would involve the creation of a new subcomponent that provides services for altering and accessing this common part and which is invoked or used by the other sub-components. This also permits state migration to be performed in an alternative way since it is only possible to access the migrated common part via the new component. In this sense the common service ensures the integrity of the common part of its super-component.

Specification 4

This chapter describes the nature of the artifacts making up a Komponent specification and the form of the associated development activities. The basic goal of Komponent specification is to create a set of models that collectively describe the externally visible properties of a Komponent. As such the specification can be viewed as defining the interface of a Komponent. However, the term "interface" here is used in a more general sense than that in the UML. In the UML an interface represents a set of operations, but a Komponent specification in addition includes information about the behavior of the Komponent, the logical effects of its offered operations and its expectations on its surrounding Komponents – in fact, everything that is externally "knowable" about the Komponent. The specification therefore defines the *requirements* that the realization of the Komponent must satisfy.

A Komponent's specification can also be thought of as representing the *contract* between the Komponent and its clients and servers. The services supplied by the Komponent in the form of operations define its contract with its clients, and the services of other Komponents used by the Komponent define its contract with its servers. Clientship relationships are thus the main driving force in determining the properties of a Komponent's specification. From the perspective of the Cleanroom approach [MLH87], a Komponent specification is analogous to a state box, since it defines the properties of a Komponent in terms of functional effects and visible states.

An important characteristic of the overall KobrA development process is that it is highly recursive. Every Komponent specification is defined in the context of a Komponent realization and vice versa. This means that a full understanding of Komponent specifications requires an understanding of Komponent realizations and a full understanding of Komponent realizations requires an understanding of specifications. In order to break into this mutual dependency cycle, and present the important concepts in an understandable order, in this chapter we provide only a partial description of Komponent specifications. More specifically, we describe the general form of Komponent specifications, the well-formedness rules governing the specification artifacts (i.e. intra-diagram and inter-diagram constraints), and the related parts of the development process. However, a detailed discussion of the aspects of a Komponent's specification that are driven by its containment and clientship relationships is deferred to Chapter 6, which focusses on Komponent containment.

In KobrA, the term "specification" is used to refer both to the specification artifacts and to the process for creating these artifacts. In the next subsection we describe the artifacts comprising a specification, and show what they look like in a small case study. In the following subsection we describe the associated process for their development.

4.1 Specification artifacts

A Komponent specification may contain up to six distinct artifacts:

1 structural model

2 behavioral model

3 functional model

4 non-functional requirements specification

5 quality documentation

6 decision model.

The first three are referred to as the primary artifacts and collectively capture the functional requirements that the Komponent must satisfy. They roughly correspond to the models used to specify the properties of a system in OMT [RBP91] and Fusion [Col94]. They overlap enough for their consistency to be verifiable, but differ sufficiently to capture distinct aspects of a Komponent's properties and behavior.

The three remaining artifacts, known as the auxiliary artifacts, describe information related to quality and product line engineering, and will be described in Parts 4 and 5. The decision model is required only when developing a product line.

Figure 4.1 provides a schematic illustration of the structural, behavioral, and functional models and their relationships.

4.1.1 Structural model

The purpose of the structural model is to describe the nature of the classes and relationships by which the Komponent interacts with its environment, and also to describe any structure of the Komponent that is visible at the interface. The structural model must contain at least one class diagram, but for large Komponents more class diagrams can be created. The structural model may also contain one or more object diagrams.

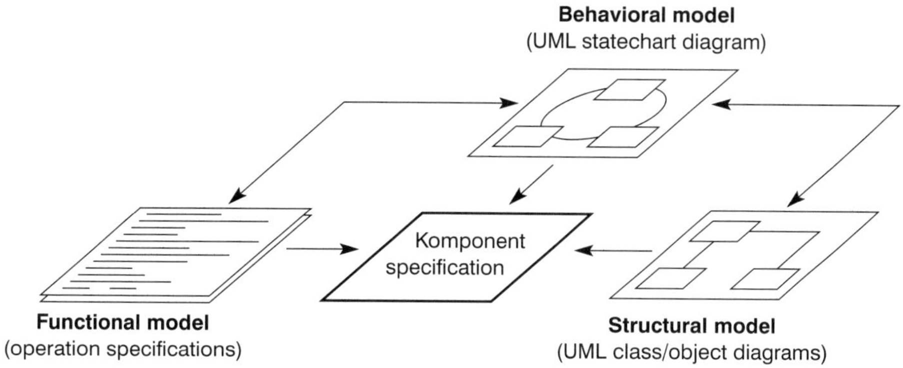

Figure 4.1
Primary Komponent specification models

4.1.1.1 Class diagrams

Class diagrams are the only mandatory diagrams required in the structural model, indeed in the whole Komponent specification. The goal is to define the classes, attributes, and associations that describe the externally visible types characterizing the Komponent's relationship to its environment – that is, the encapsulating realization and clients. The class diagram is therefore focussed exclusively on the description of the Komponent to which it belongs.[22]

Classes appearing in the class diagram may represent *simple classes* (for carrying data), other Komponents, and at the top of the Komponent hierarchy, external *actors* (see Chapter 7). The class diagrams therefore combine the role of a traditional data model (informational) with that of a context diagram (structural). The servers required by the *subject* Komponent (i.e. the Komponent under specification) are represented by other Komponents in the class diagram. Since the principle of parsimony (Section 3.3.3) requires that the class diagram should contain the minimum amount of information necessary to specify the Komponent, it should include server Komponents only if instances of the subject Komponent really need them.

Figure 4.2 illustrates the specification structural model of the *LoanManager* sub-Komponent of the library system which will be used as the running case study in Parts 2 to 5 of this book. The goal of the library system is to support the management of registered users and available items and to track all performed loans, returns, and reservations. The *LoanManager* is specifically responsible for keeping track of the loans. In later chapters of this book this system will be extended with further sub-Komponents and generalized to support a family of similar systems.

The *LoanManager* class in Figure 4.2 represents the Komponent under specification, and is thus referred to as the subject of the diagram. It is labeled with the stereotype <<subject>> accordingly, and its operation compartment lists all the

22. It corresponds roughly to a "system object model" in the sense of Fusion [Col94], or a component "type model" in the sense of Catalysis [DW98].

Figure 4.2
LoanManager specification class diagram

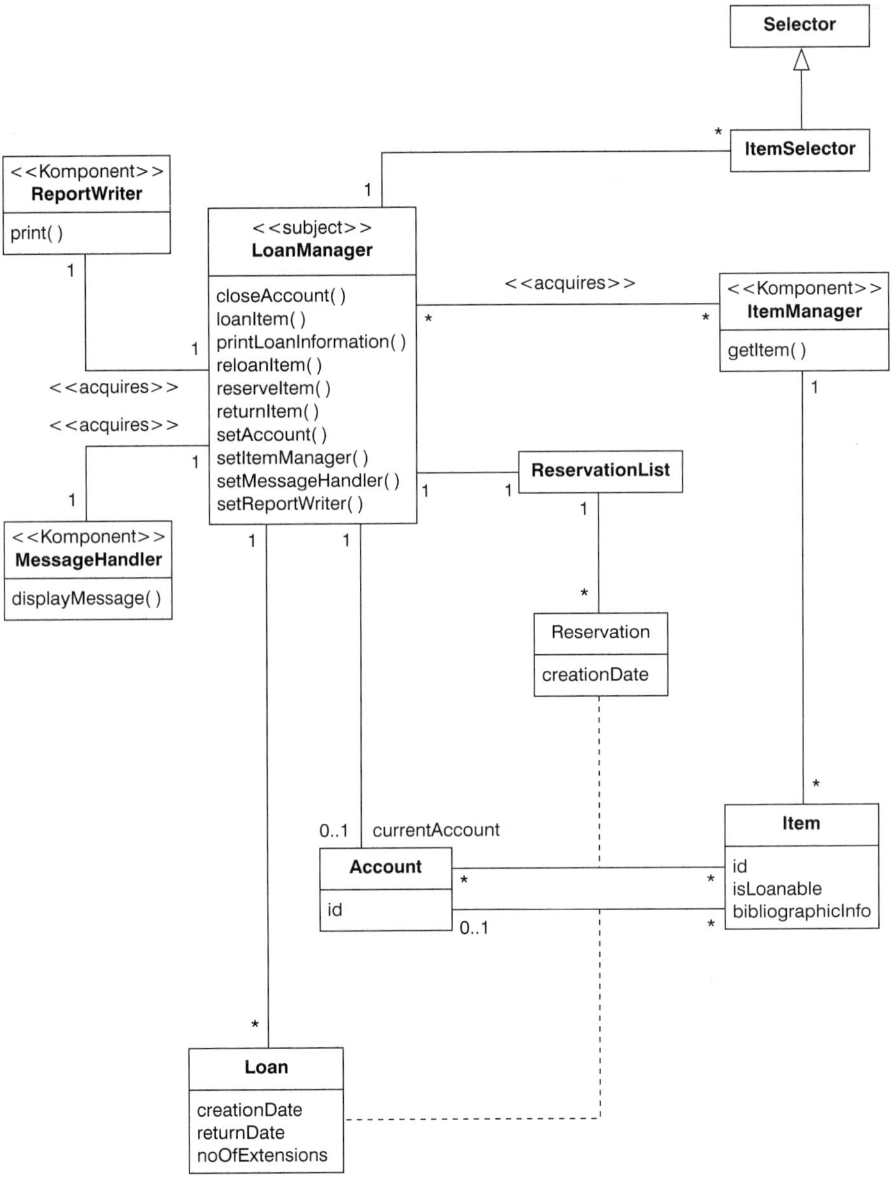

operations offered by the Komponent. The other classes in the diagram either represent Komponents that are acquired by the subject, or data structures that facilitate the Komponent's interaction with its clients. The former are labeled with the stereotype <<Komponent>>. Figure 4.2 indicates that an instance of the *LoanManager* Komponent acquires instances of three other Komponents, *MessageHandler*, *ItemManager* and *ReportWriter*. Acquired Komponents can also have operation compartments, but, due to the principle of locality, they only

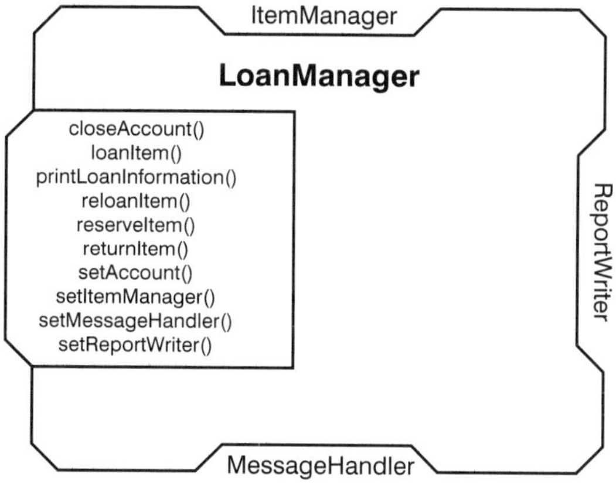

Figure 4.3
Supplied and required interfaces of LoanManager

show the operations invoked by the *LoanManager*. For example, *displayMessage()* is an operation of *MessageHandler* that is invoked by the *LoanManager* Komponent. *MessageHandler* may have more operations, but only those invoked by *LoanManager* are shown in this representation of *MessageHandler*.

The other model elements in the diagram represent the data structures and relationships that characterize how instances of the *LoanManager* interact with, and provide services to, their clients. In particular, all the parameter types and return types of the Komponent's operations must be included (e.g. *Account*, *Reservation*, *Loan*), as well as all logical relationships maintained by instances of the Komponent, such as the fact that a *Reservation* represents an association between an *Account* and an *Item*.[23] Simple classes (as opposed to classes representing Komponents) should not have an operation compartment, but should only show their logical attributes (e.g. *isLoanable* attribute of *Item*). This ensures that issues related to how the Komponent works, rather than what it does, are deferred to the realization models.

Figure 4.3 provides a schematic picture of the *LoanManager's* supplied services (i.e. supplied interface) and acquired servers (i.e. used interfaces), as defined by the class diagram in Figure 4.2.

The operations supplied by the *LoanManager* Komponent are listed in the polygon on the left-hand side of Figure 4.3. This therefore represents the supplied interface that is used by clients. The classes acquired by the *LoanManager* are identified in the indentations on the top, right and bottom edges of the *LoanManager* symbol. These therefore represent the required or used interfaces of *LoanManager*, and define the Komponents of which it is a client. The reciprocal shapes of the indentations and protrusions on this icon are intended to convey the idea of "ports" and "plugs" (or "sockets"). Thus, an instance of *LoanManager* at run-time expects to be "plugged" to Komponent instances that supply the *ItemManager*, *ReportWriter* and *MessageHandler* services (its servers), and to instances that require its own services (its clients). This type of diagram is not really a part of the KobrA notation, but can be helpful in visualizing the interfaces of Komponents in a "plugs" and "sockets" fashion.

23. In UML terms it is an association class.

Figure 4.4
Run-time configuration

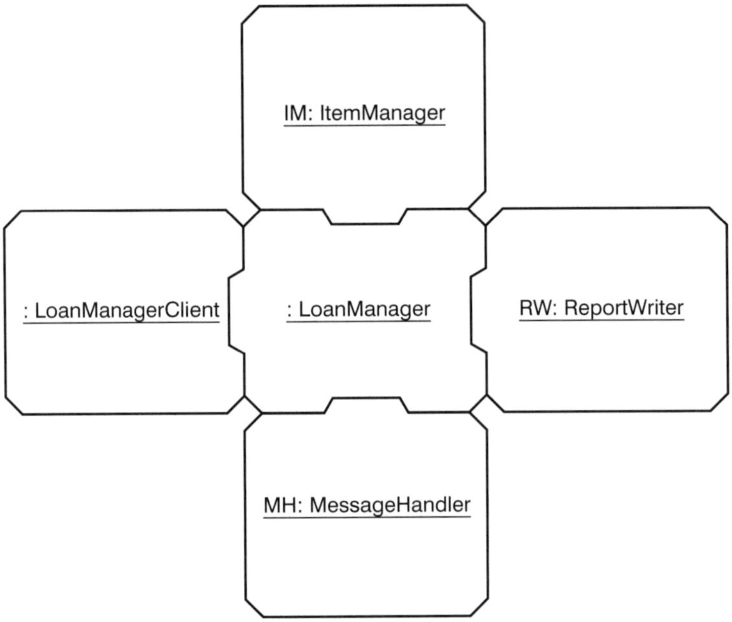

4.1.1.2 Objects diagrams

At run time, an instance of a *LoanManager* Komponent expects to be connected to other Komponents in a manner similar to that illustrated schematically in Figure 4.4. This defines the names by which an instance of the *LoanManager* refers to these Komponents, which is useful in the functional model explained in Section 4.1.2.

In a Komponent specification the information illustrated in Figure 4.4 can be expressed in the form of an object diagram. Object diagrams can thus be used to provide names for acquired Komponents, and to show how many instances of each kind of Komponent the subject Komponent acquires.

The object diagram in Figure 4.5, for example, indicates that the *LoanManager* imports only one instance of the *ItemManager*, *ReportWriter* and *MessageHandler* Komponent, and uses the names *IM*, *MH* and *RW* for these respectively. Notice also that the instance of *LoanManager* is anonymous. This is because the diagram shows the instances and their names from the point of view of any given *LoanManager* instance.

Figure 4.5
LoanManager specification object diagram

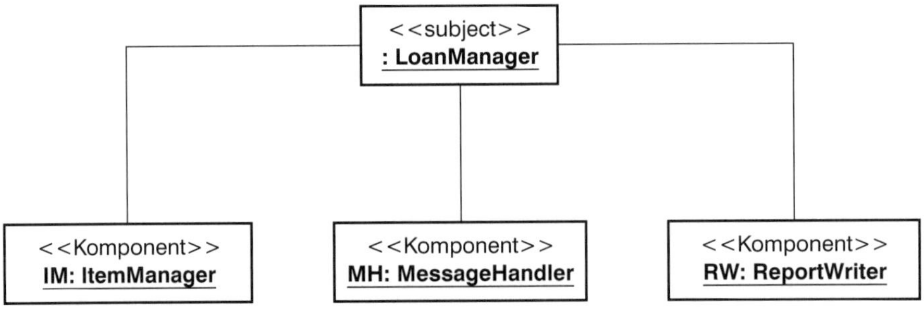

Object diagrams are also important in connection with public composition, described in Chapter 8. In general they are much less frequently used than class diagrams, however.

4.1.2 Functional model

The purpose of the functional model is to describe the externally visible effects of the operations supplied by the Komponent. Based on Fusion [Col94],[24] the model consists of a set of operation specifications, one for each operation, constructed using the template illustrated in Table 4.1.

Table 4.1 Operation specification template

Name	**name of the operation**
Description	identification of the purpose of the operation, followed by an informal description of the normal and exceptional effects
Constraints	properties that constrain the realization and implementation of the Komponent
Receives	information input to the operation by the invoker
Returns	information returned to the invoker by the operation
Sends	signals the operation sends to imported Komponents; can be events or operation invocations
Reads	externally visible information accessed by the operation
Changes	externally visible information changed by the operation
Rules	rules governing the computation of the result
Assumes	weakest pre-condition on the externally visible state of the Komponent and on the inputs (in receives clause) that must be true for the Komponent to guarantee the post-condition (result clause)
Result	strongest post-condition on the externally visible properties of the Komponent and the returned entities (returns clause) that becomes true after execution of the operation with true assumes clause

In general, an operation specification has 11 parts, or clauses. The most important of these are the assumes and result clauses, which represent the pre- and post-condition, respectively. The assumes clause defines what must be true for the operation to be guaranteed to execute correctly, and the result clause describes what becomes true as a result of the operation executing correctly. It is possible to execute an operation when its assumes clause is false, but then the effects of the operation are not certain to meet the post-condition.

The basic goal of the result clause is to provide a declarative description of the operation in terms of its effects. In other words, it describes *what* the oper-

24. Fusion referred to operations specifications as operation schemata.

ation does but not *how*. In general, a result will consist of several subclauses, but an important rule is that the order in which these Boolean expressions appear should not affect the meaning of the clause.

The assumes and result clauses actually contain a full description of the precondition and the effects of the operation, respectively. The other clauses essentially summarize key pieces of information from the assumes and result clauses and thus provide valuable scoping information that facilities a quick understanding of how the operation depends on, and affects, its environment.

As its name implies, the first clause gives the name of the operation. The second identifies its purpose and provides a short informal description of its effects, both for normal and exceptional executions. The next two clauses define the information that is exchanged between the Komponent and the invoker of the operation (an external actor or an instance of another Komponent). The receives clause identifies the information passed to the operation, and is analogous to the arguments of a procedure (or method) in a programming language. The returns clause identifies the information passed back to the invoker, and is analogous to the return value (or "out" parameters) of a procedure (or function) in a programming language. The next three clauses identify the "stored" information that the operation accesses or changes and the messages that it sends to acquired Komponent instances. The reads clause identifies the information which is read but is not changed, the changes clause lists the information that is changed and the sends clause lists the messages which are sent to acquired Komponent instances. The rules clause can be used to capture complex rules governing the computation of the result.

Any convenient notation can be used to define the pre- and post-conditions in an operation specification, ranging from free-style text to fully formal languages. In general, this will depend on the domain and the precise purpose for which the specification is intended. For example, in a safety critical system, a formal notation may be the most appropriate since it facilitates the most precise statements. In other domains, free-style natural language may be more appropriate. In this book we use the basic conventions and syntax of operation schemata in Fusion [Col94]. Structured natural language is used to define pre- and post-conditions, with emphasis on the phrase "has been" in the result clause to indicate that the conditions

Table 4.2 setMessageHandler() operation specification

Name	**setMessageHandler()**
Description	Receives a reference to a MessageHandler Komponent and stores it.
Receives	MH: MessageHandler
Result	MH has been stored

come into effect after the operation has been completed, but may not be true during its execution. Another natural choice for expressing pre- and post-condition is the Object Constraint Language (OCL) defined in the UML standard.

Table 4.2 gives an example of a very simple operation specification. The purpose of the *setMessageHandler()* operation is simply to take a reference to a *MessageHandler* Komponent instance and store it.

The operation specification shown in Table 4.3 is more substantial. This is the specification of the *loanItem()* operation which is responsible for

Table 4.3 loanItem() operation specification

Name	**loanItem()**
Description	The loan of an Item to currentAccount is registered.
Receives	selector: ItemSelector
Sends	MH.displayMessage() Item item = IM.getItem(selector)
Rules	An item is loanable if it is not an item that must always stay in the library (e.g. antique books). An item is currently loanable if it is loanable and not loaned or reserved by another user.
Changes	new Loan
Assumes	Subject is in the state accountIdentified. Selector selects exactly one Item.
Result	Item selected by selector has been obtained from the ItemManager IM by sending the message getItem(selector). If item is currently loanable: a new Loan object, loan, has been created that relates item and currentAccount, and has the attribute values – creationDate = today, – returnDate = today + 4 weeks and – noExtensions = 0 and, loan has been stored. If item is not currently loanable: one of the following messages has been sent to the MessageHandler MH: displayMessage("AlreadyLoaned") or displayMessage("Reserved")

establishing a loan between an account and an item, or reporting back to the user that this is not possible via an instance of the *MessageHandler* Komponent. This interaction with the *MessageHandler* instance, known to the subject as *MH* (from the object diagram), is captured in the sends clause using normal method call syntax.

Another interesting aspect of this operation specification is that is has an assumes clauses which defines what must be true for this operation to be guaranteed to execute correctly. This is an important part of the contract with the client. In this case a reference is made to a particular state called *accountIdentified*, defined in statechart diagram shown in Figure 4.6. Statechart diagrams are one way of providing a behavioral model, as described in Section 4.1.3.

The result clause of the operation specification lists the things that happen whenever *loanItem()* is executed with the constraints in the assumes clause satisfied. First, the *LoanManager* has asked the *ItemManager*, which it knows by the name *IM*, to supply the *Item* identified by the selector that it received. If this *Item* was not currently loanable according to the rule defined in the rules clause, the *LoanManager* has sent a message to the *MessageHandler*, which it knows by the name *MH*, to display an appropriate message. Otherwise, the *LoanManager* has created a new *Loan* object, set its attributes to particular values, and stored it away for safekeeping. Note that although the layout of some of the sub-expressions may resemble "if" statements in a programming language, this is not meant to imply any procedural ordering. On the contrary, the value of the result clause must not depend on the order in which the subexpressions are listed.

Table 4.4 shows another example of an operation specification, this time for the operation *reloanItem()*. This is responsible for extending existing loans. Note that the number of clauses listed for the three operation specifications shown is different. This is because by convention we do not list clauses that are empty, since this is unnecessary. Thus, a clause that is not shown has the default value for that clause. For the assumes clause, the default value is *True*. For the constraints, receives, returns, sends, reads and changes clauses the default value is the empty set. The name, description, and result clause must always have some content.

As mentioned previously, OCL can be used to add more formality to aspects of the pre- and post-conditions in an operation specification. Table 4.5 illustrates an alternative version of the result clause for the *reloanItem()* operation which uses OCL to formalize some aspects of the post-condition. This result clause is intended to have the same meaning as that in Table 4.4. Notice that not all parts of the clause are amenable to expression in the OCL. Describing the effects of operation invocations is not supported by the OCL, but should be describable using the Action Languages currently under development for the UML.

Several important consistency rules must be adhered to when creating operation specifications. First, an operation specification can only refer to classes, attributes and relationships identified in the specification class diagrams, and

Table 4.4 reloanItem() operation specification

Name	**reloanItem()**
Description	An item loaned to an Account is reloaned to the currentAccount.
Receives	selector: ItemSelector
Sends	MH.displayMessage() Item item = IM.getItem(selector)
Changes	loan
Rules	An item is not reloanable if it is reserved or if the number of extensions is greater or equal to MaxExtension (=3).
Assumes	Subject is in the state accountIdentified. The item is loaned to currentAccount.
Result	Item selected by selector has been obtained from the ItemManager IM by sending the message getItem(selector). If item is reloanable: the loan containing item has the attribute values returnDate = today + 4 weeks noExtensions = noExtensions + 1. If item is not reloanable: one of the following messages has been sent to MessageHandler, MH: displayMessage("Reserved") or displayMessage("OverExtension").

to states and events identified in the specification behavioral model. Any references to classes, attributes, relationships, states, or events not defined in these diagrams are erroneous. Conversely, all classes, attributes, and associations in the specification class diagrams must appear in at least one of the operation specifications. Second, received and returned items must match the parameters defined in the corresponding interaction diagrams in client Komponents' realizations. Third, an operation specification can only send messages to imported Komponents, and then only to the operations included in the class diagram.

Table 4.5 reloanItem() result clause using OCL

```
Result   post:
           item selected by selector has been obtained from the
           ItemManager IM by sending the message getItem(selector)
           if item is reloanable
              loan->exists(i: Loan |
                    i. returnDate = today + 4 weeks
                    i. noOfExtensions = i. noOfExtensions@pre + 1)
              xor
           if item is not reloanable
              loan->forAll(i: Loan |
                    i. returnDate = i. returnDate@pre and
                    i. noOfExtensions = i.noOfExtensions@pre)
                 one of the following messages has been sent to MH:
                 displayMessage("Reserved") or
                 displayMessage("OverExtension")
```

These rules not only serve to ensure that the contents of the structural model and the functional model are consistent, but by applying the principle of parsimony they may also be used to judge how much information is actually needed. In particular, the contents of the class diagram can be determined by ensuring that it contains exactly the amount of information needed to specify all the operations – no more and no less. Unused information is obviously unnecessary, and too little information obviously means that the class diagram is incomplete.

4.1.3 Behavioral model

The purpose of the behavioral model is to show how the Komponent behaves in response to external stimuli. The model may take one of two forms: one or more UML statechart diagrams and/or statechart tables.

Three important concepts are related in the behavioral model:

- **events** – indivisible occurrences that have a location in space and time; information can also be carried by an event by means of parameters (e.g. menu item selections etc.)
- **operations** – encapsulated units of functionality associated with Komponents
- **states** – periods of time in between events that determine how the Komponent will respond to the subsequent event.

The next two subsections introduce statechart diagrams and statechart tables.

4.1.3.1 Statechart diagrams

Every external event that is meaningful to a Komponent can be interpreted as a request for the execution of an operation of the Komponent. However, there is not always a one-to-one mapping between events and operations, and an invocation request may not always lead to the execution of the operation. Statechart diagrams can be used to graphically capture the information needed to describe these mappings.

Essentially a UML statechart diagram is a graph in which the nodes represent states and the arcs represent transitions. Each transition can be labeled by the triggering event and the action executed when the transition occurs. When used in the specification of a Komponent, the states must represent client-visible states of the Komponent, the events must correspond to client-visible events, and the actions must correspond to Komponent operations. The clients of a Komponent instance can be external actors or other Komponent instances.

An event in a given state represents a request for the execution of one, and only one operation. However, a given operation can be invoked by more than one event. When there is a one-to-one correspondence between operations and events it is usual to identify them by the same name. This makes the correspondence between the event and the operation obvious, and allows the operation to be omitted from the corresponding transitions in the statechart diagrams. When there is a many-to-one mapping between events and operations (for example, when an operation can be accessed from various points in an interface) or when the event and the operation are allocated different names, the correspondence between them must be indicated on the statechart by including the operations as well as the events in transition labels. The only exception to this rule is when a statechart table is used to augment the statechart diagram (see 4.1.3.2).

As a convention, if an event has no effect (i.e. the corresponding operation cannot be invoked) when the system is in a particular state this is modeled by the fact that the associated state has no exit transition for that event.

Figure 4.6 shows an example of the statechart diagram for the *LoanManager* Komponent. The diagram defines the major states of the *LoanManager*, the operations that can be executed in each state, and the events that lead to state transitions. Note that apart from the operations for setting acquired Komponents (i.e. *setItemManager(), setMessageHandler()* and *setReportWriter()* the operations in the diagram are those defined in the functional model described in Section 4.1.2, and listed in the class representing the subject Komponent in the class diagram in Figure 4.2. The operations for setting acquired Komponents are not shown in the behavioral model since they are one-off operations executed as part of the system configuration process immediately after a *LoanManager* is created. Such operations could be shown explicitly by introducing an additional state such as "Unconfigured" to define the initial state of a *LoanManager*, but if there are a lot of them they can sometimes obscure the interesting behavior of the Komponent once it has been initialized. Therefore it is often better to leave such initialization operations out of the behavioral model.

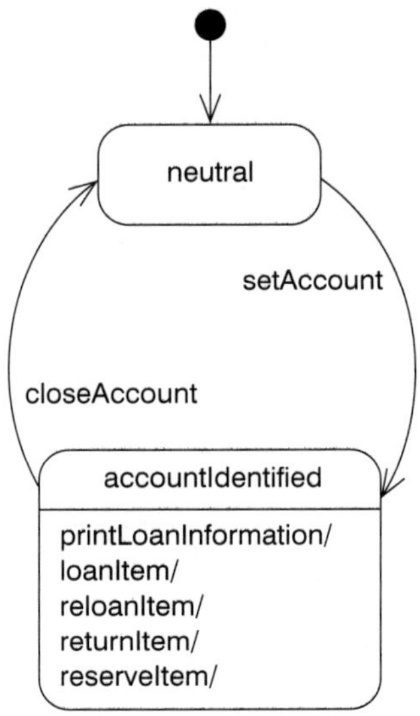

Figure 4.6
LoanManager statechart diagram

Since *LoanManager* Komponents only have two states, the statechart diagram in Figure 4.6 is relatively simple. When the behavior of a Komponent is more complex, states can be hierarchically nested to keep the number of transitions and states low. Also, state attributes and conditions can be used to simplify the statechart by reducing the number of nodes. However, several consistency rules must be observed when using them. For example, conditions guarding transitions from a state can refer only to the attributes of that state and the parameters of the corresponding event.

As always, the statechart diagram must be in agreement with the other models making up the specification. Primarily, this means that the behavioral model must agree with the other two models on the Komponent's operations and externally visible states.

Ideally, operations should be made as "robust" as possible. This means that they should accommodate as many invocation circumstances as possible. In the best case, when all the invocation circumstances are handled, the corresponding assumes clause is always *True*, and the operation will catch any undesirable situations and report error messages. However, such behavior then becomes part of the functionality of the operation. Any conditions expressed in the assumes clause are circumstances in which the operation should not be called. This must be checked in each of the collaboration diagrams in which the operation is invoked (see Chapter 5).

Note that the more robust a Komponent becomes, the fewer pre-conditions its operations have, and the fewer the number of states. However, the definition of states may still be useful to help simplify the specification of operations.

4.1.3.2 Statechart tables

The information in a statechart diagram can also be captured in a tabular form known as a statechart table.[25] There are two ways of using a statechart table. First, as a supplement to statechart diagrams, and, second, as an alternative representation of the behavioral model. When a statechart diagram is too complex, the information overload can be reduced by moving the transition strings into a supplementary statechart table. Only information on the possible state transitions then remains in the statechart diagram.[26]

Tables have both strengths and weaknesses compared to diagrammatic representations of a state machine. The advantages are that tables can be more concise, and can make it easier to see whether transitions are missing. The disadvantages are that tables can become quite large if the number of states is high and the number of transitions is low, because then most table cells are empty. Moreover, hierarchically nested states cannot be represented in a table as well as in a diagram. The choice ultimately depends on the cognitive preference of individual developers or the project manager.

The basic idea behind statechart tables is to use UML transition strings, of the form used to label transitions in a statechart diagram, to define the allowable transitions between the possible states of the system. As shown in Table 4.6, the rows of the table represent the starting states, the columns represent the resulting states, and the cells contains a list of transition strings defining the allowable transitions between each pair of states.

A UML transition string is defined to have the following general form:

event-signature '['guard-condition']' '/' *action expression*

where an action expression has the form

event-name '('*comma-separated-parameter-list*')'

Table 4.6 General form of statechart table

	....	StateN	...
....			
StateM		TransitionString 1 TransitionString 2 .... TransitionString X	
...			

25. The phrase "statechart table" is somewhat oxymoronic in English, but no more so than statechart diagram. We use it for symmetry with the latter.
26. This approach was also taken in the OCTOPUS method by Awad et al. [AKZ96].

Table 4.7 LoanManager statechart table

	neutral	accountIdentified
neutral	....	set Account()
accountIdentified	closeAccount()	loanItem() printLoanInformation() reloanItem() reserveItem() returnItem()

The equivalent statechart table to the statechart diagram in Figure 4.6 is illustrated in Table 4.7. The first row of this table indicates that the only transition possible from the *neutral* state is triggered by an invocation of the *setAccount()* operation and leads to the accountIdentified state. The second row indicates that if a *LoanManager* Komponent is in the *accountIdentified* state, invocation of the *closeAccount()* operation will leave it in the *neutral* state. The invocation of any other operation, however, will leave the *LoanManager* in the same state (*accountIdentified*).

The table form shown in Table 4.6 is useful when there is no restriction on the width of the table or the number of states is low. Otherwise, the table form shown in Table 4.8 can be applied. In this table, each row describes one transition from a source state to a target state.

Table 4.8 Alternative form of statechart table

Source state	UML Transition string	Target state
State1	TransitionString1	Statel
...	...	...
StateN	TransitionStringM	StateK

The drawback of this table form compared to the table form in Table 4.6 is that one cannot detect missing transitions as easily. The benefit is that it does not contain empty cells, i.e., it can be used for any number of states and transitions. The alternative statechart table to that in Table 4.7 is illustrated in Table 4.9. If the states are hierarchically nested then the hierarchy information can be included in a statechart table using a hierarchical numbering scheme for states. Top-level states are numbered 1, 2, 3 etc. Substates of the top-level state 1 would be numbered 1.1, 1.2, 1.3, etc.

Table 4.9 Alternative LoanManager statechart table

Source state	UML Transition string	Target state
neutral	setAccount()	accountIdentified
accountIdentified	closeAccount()	neutral
accountIdentified	loanItem() printLoanInformation() reloanItem() reserveItem() returnItem()	accountIdentified

A statechart table can be used as a supplement to a statechart diagram when mentioning all the operations would clutter the diagram. Table 4.10 shows the variant of the full statechart table in which the resulting states are omitted. This variant is useful since it describes the situations under which different operations are executable, something which is important for user interface descriptions.[27] For example, the statechart table in Table 4.10 specifies that an item can be loaned only if the current state is *accountIdentified*. In OCTOPUS [AKZ96] this type of table is called an action table. The information about the target state reached by a transition remains in the statechart diagram.

Even if the Komponent does not have interesting states at all, a statechart table can be useful to describe the event-operation mapping for the Komponent. In this case, both the source and the target states are omitted. For example, the statechart table in Table 4.11 indicates what logical events

Table 4.10 Statechart table with omitted target states

Source state	UML Transition string
neutral	setAccount()
accountIdentified	closeAccount() loanItem() printLoanInformation() reloanItem() reserveItem() returnItem()

27. Theoretically, it would be possible to omit the source states in a statechart table and to leave the UML transition strings and target states. This form could be useful to see whether all states can be reached, but they are likely to be less useful than the previous forms illustrated.

(e.g. menu item selections) are used to trigger the execution of the Komponent's operations, and defines the mapping of these logical events to the operations. In particular, this shows that two different logical events can lead to the invocation of the same operation. For example, *requestLoan* and *requestBorrow* are both events that lead to the invocation of the *loanItem()* operation.

Table 4.11 Statechart table with omitted source and target states

UML Transition string
requestLoan / loanItem()
requestBorrow / loanItem()
requestPrint / printLoanInformation()
requestReserve / reserveItem()
requestReturn / returnItem()
requestReloan / reloanItem()

Statechart tables offer a powerful complement to statechart diagrams for representing state machines. If there are no significant differences between statechart diagrams and statechart tables in the available tool support (i.e. if the applied CASE tool does not support the execution of statechart diagrams), or no CASE tool is available then statechart tables may be a sensible alternative to statechart diagrams for representing the behavioral model.

4.1.4 Auxiliary artifacts

In addition to the primary artifacts identified above, a Komponent specification can have up to three auxiliary artifacts. Since these are primarily concerned with quality and product line issues, however, these are discussed in the associated parts of the book. The non-functional requirements specification, quality documentation and decision models, which are only used in the case of product line engineering, will be discussed in Part 4.

4.2 Specification process

In this subsection we describe the activities used to develop the specification artifacts described above.

4.2.1 Structural modeling

The goal of the structural modeling activity is to develop the structural model. In general the structural model can have multiple class diagrams and multiple object diagrams, but since it is very rare to have more than one of each, in the following we describe the development activity as if only one class diagram and one object diagram is developed. The steps involved in developing the class diagram are as follows:

1. Create a new sub-Komponent together with a first draft of its specification class diagram based on the containment rules (see Section 6.3.3). Ensure that all the operations of the subject are listed in its operation compartment, and that it is adorned with the stereotype <<subject>>.
2. Ensure that acquired Komponents are adorned with the <<Komponent>> stereotype, and remove any operations from their operation compartment which the subject Komponent does not invoke.
3. Add appropriate logical attributes to all classes.
4. Add essential associations between all classes.
5. Once a draft of the functional model and behavioral models are available, compare these to the structural model to:
   a. check that the functional and behavioral models only refer to model elements in the class diagram (other than locally defined parameters)
   b. add any missing logical attributes required in any of the operation specifications or any of the statechart states
   c. add any missing associations needed in any of the operation specifications to navigate between instances
   d. remove any model elements not needed in the functional model or behavioral model.
6. Repeat 5 on successive drafts of the structural model until the intra-diagram, inter-diagram, clientship, and containment rules are satisfied (see Appendix A).

Object diagrams are optional, but can help to simplify the functional model by defining the names of server instances used by the subject. They are particularly useful for Komponents with public parts, as explained in Chapter 8. They are also useful for large Komponents, or Komponents with multiple server instances of the same Komponent type:

1. Prepare a draft object diagram containing one instance of the subject Komponent, and instances of all the other Komponents of which the subject is a client. The focus should be on stable instances (usually Komponent instances) and links, not transient objects and links.
2. Compare the object diagram with the class diagram to ensure the objects are instances only of the classes in the class diagram, and that multiplicity constraints are satisfied.
3. Compare the object diagram to the functional model to ensure that the instance names match.
4. Repeat 2 and 3 until the intra- and inter-diagram consistency rules are satisfied (see Appendix A).

4.2.2 Functional modeling

The functional model comprises a set of operation specifications, one for each operation of the Komponent. The recommended process for creating them is as follows.

1. Identify the operations of the Komponent by examining all the messages sent to instances of the Komponent by its clients. All clients must be considered (see Chapter 6).
2. For each operation, develop a first draft of the assumes and result clauses
   a. Identify the parameter types and return types of each operation.
   b. Identify the intended effects of the operation in terms of elements in the object diagram, class diagram, statechart diagram/table and parameters.
   c. Snapshot object diagrams[28] can be used to model the desired configurations before and after the operation [DW98]. A snapshot diagram is an object diagram that captures a particular configuration of objects at a particular point in a systems execution. They are therefore useful for depicting the changes caused by an operation.
   d. Identify all appropriate assumptions.
   e. Formulate result and assumes clauses.
   f. Add relevant class diagram invariants.
   g. Ensure that the specification is satisfiable. An operation specification is satisfiable if for all possible initial values of the assumes clause there exist final values that satisfy the result clause [Col94].
3. For each operation, create the receives, returns, reads, changes, sends, and rules clauses from information in the result and assumes clauses
   a. Summarize in the receives clause the items that are passed to the operation by the client.
   b. Summarize, in the returns clauses, the items that are returned to the client after the completion of the operation.
   c. Summarize in the sends clauses the operations of the acquired server Komponents that are invoked by the subject.
   d. Summarize information about the non-parameter items (i.e. the items that are accessed directly) that are changed in the changes clause and items that are only read in the reads clause.
   e. If the results clause is overly complicated, detail computation rules mentioned in the result clause in the rules clause.
4. Repeat 2 and 3 until the intra-diagram, inter-diagram, and clientship rules are satisfied (see Appendix A).

28. Popularized in Catalysis

4.2.3 Behavioral modeling

The behavioral model consists of one or more statechart diagrams and/or one or more statechart tables. The recommended process for creating them is as follows:

1. Decide on the most appropriate representation (diagram or table). Consider available tool support, cognitive preferences, and the type of resulting documentation. If diagrams are chosen, decide whether supplementary statechart tables could be useful for capturing operations that would otherwise clutter the statechart diagram.
2. Identify all externally visible logical states of the Komponent. A visible state characterizes the operations that can be performed on the Komponent under particular circumstances.
3. Identify attributes in the class diagram corresponding to state attributes in the behavioral model.
4. Identify the operations that can be executed in each state, and the state transition that occurs when the corresponding invocation occurs. Operation invocations that leave the system in the same state correspond to internal transitions.
5. Repeat steps 3 and 4 until the intra- and inter-diagram consistency rules are satisfied (see Appendix A).

4.2.4 Quality control

Quality control plays an important role in the Komponent specification activity. Three main quality-control techniques are applicable, inspection, quality modeling and testing, although the latter only partially.

The most important of the specification quality-control activities is inspection. This is performed at the end of the Komponent specification activity to ensure that the following sets of constraints are satisfied:

- specification intra-diagram rules (i.e. rules on a single diagram)
- specification inter-diagram rules (i.e. consistency rules between specification diagrams)
- clientship rules (contract with client Komponent instances)
- containment rules (relationship with the parent).

A detailed listing of these constraints is provided in Appendix A, while a detailed description of the KobrA inspection approach is described in Chapter 18. Inspection is a mandatory part of Komponent specification.

Another effective quality-improvement activity that can be applied during Komponent specification is quality modeling. This applies statistical models to concrete measures derived from the UML specification artifacts to determine concrete quality attributes. A full description of this approach is described in Chapter 19.

Finally, although testing cannot be applied in full until an executable version of the Komponent has been created, one important part of testing – test-case generation – can be performed earlier. Since the operation specifications provide a full description of each operation's parameters and effects, it is possible to develop full functional test cases as part of the specification activity. When combined with a rigorous testing regime, as in KobrA, numerous studies show that the early generation of test cases can improve not only the quality of the test cases themselves, but also the quality of the software artifacts. KobrA, therefore recommends the development of functional test cases, and their subsequent inspection, during the Komponent specification activity.

Realization 5

The specification of a Komponent can be viewed as defining *what* requirements the Komponent is expect to fulfill. The realization, in contrast, can be viewed as the private design of the Komponent, describing *how* it fulfills these requirements. Just as in regular software development, the basic goal is to ensure that the realization faithfully fulfills the requirements laid out in the specification. From the perspective of the Cleanroom approach [MLH87], a Komponent realization is analogous to a clear box, since it defines how the properties of a Komponent are realized in terms of its interactions with other "boxes" (i.e. Komponents).

This chapter describes the nature of the artifacts making up a Komponent realization and the form of the associated development activities. In the next subsection we define the artifacts of a Komponent realization. This is followed by a description of the activities used to create these artifacts.

5.1 Realization artifacts

A Komponent realization may contain up to six distinct artifacts:

1 structural model

2 activity model

3 interaction model

4 data dictionary

5 quality documentation

6 decision model.

Figure 5.1
Primary Komponent realization models

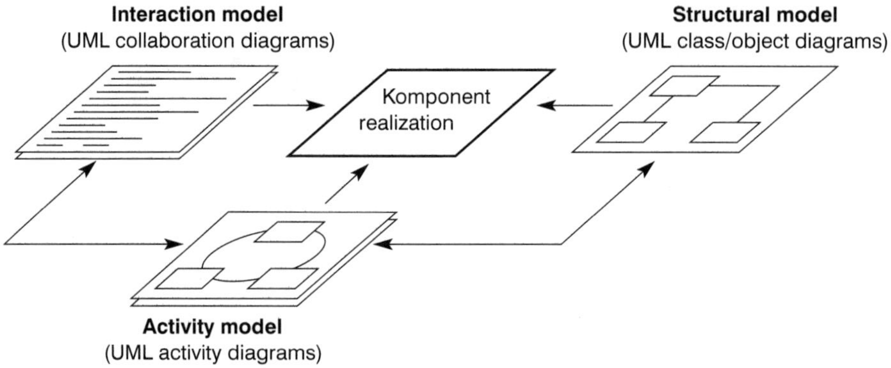

The first three are referred to as the primary artifacts and collectively describe how the Komponent realizes its specification in terms of interactions between other Komponents and objects. Figure 5.1, provides a schematic illustration of how the primary models combine to form the realization of a Komponent. As with the specification models they overlap to a sufficient degree to facilitate their consistency checking, while still providing complementary views of the design. The way in which collaboration diagrams are used in KobrA is based on the design strategy of Fusion [Col94], while the role of the structural model is motivated by the Fusion-variant, FuML [Atk97].

The three remaining artifacts, known as the auxiliary artifacts, describe information related to quality and product line engineering, and will be described in Parts 4 and 5 of the book. The decision model is required only when developing a product line.

5.1.1 Structural model

The purpose of the realization structural model is to describe the nature of the classes and relationships from which the Komponent is realized and the architecture of the Komponent. In general, the structural model is composed of one or more UML class diagrams and one or more UML object diagrams. However, because multiple class and/or object diagrams are rare, we describe the structural model artifacts in terms of one class diagram and one object diagram.

5.1.1.1 Class diagrams

Realization class diagrams describe the classes, attributes and relationships from which the Komponent is realized. As in the specification, the Komponent being realized is the focal point of each realization class diagram, and is known as the subject. It is distinguished from the other classes using the stereotype <<subject>>.

The realization class diagram is a refinement of the specification class diagram. This means that it contains a superset of the information in the specification class diagram. All elements presented in the specification class diagram are still of relevance to the Komponent, and thus have a role to play in the realization class diagram. However, they may be described in more detail. In addition to the element presented at the specification level, a realization class diagram also contains elements, including other Komponents, which are not visible at the specification level, but are important in the realization of the Komponent. From a package ownership perspective, these new model elements are still owned by the package corresponding to the Komponent but just have private visibility.

Every element in the realization class diagram should play some role in the fulfillment of the requirements, or should not be included in the diagram. As with the specification class diagram, therefore, a key goal in creating the realization class diagram is to include the minimum amount of information (i.e. classes, attributes and relationships) required to fully realize the Komponent. This can be determined by the relationships between the class diagram and the other realization diagrams.

Figure 5.2 shows the realization class diagram for the *LoanManager* Komponent. Notice that all the information in the specification class diagram is shown in this diagram as well, but with additional information added to facilitate the description of how the Komponent works. For example, the *LoanStore* class has been introduced as the data structure for storing *Loan* objects, and the *ReservationManager* Komponent has been added to manage reservations. In addition, all of the classes in the diagram list the operations that are needed by the subject Komponent to fulfil its requirements. In the specification class diagram, only the server Komponents can list operations. In general, there are no restrictions on the use of UML class diagram features in the realization class diagram. The only constraint is the principle of parsimony (i.e. the diagram should contain just enough information).

The relationship between the subject and the *ReservationManager* is an example of private composition. An instance of *LoanManager* creates an instance of the *ReservationManager* shortly after its creation, and uses it privately. An alternative form of composition, public composition, will be explained in Chapter 8.

5.1.1.2 Intermediate class diagrams

For small Komponents it can be fairly easy to create the realization class diagram, but the larger the size of the Komponent, the more difficult this becomes. Therefore, it is possible to define additional intermediate class diagrams that provide an incomplete view of the structural aspects of a Komponent. As explained in Section 5.2 below, such intermediate diagrams can play a useful role in the decomposition process. For example, an intermediate class diagram for the *LoanManager* realization may just show the new classes *ReservationManager* and *LoanStore* as simple class boxes. It is important to recognize that such intermediate class diagrams are not intended to replace the final realization class and

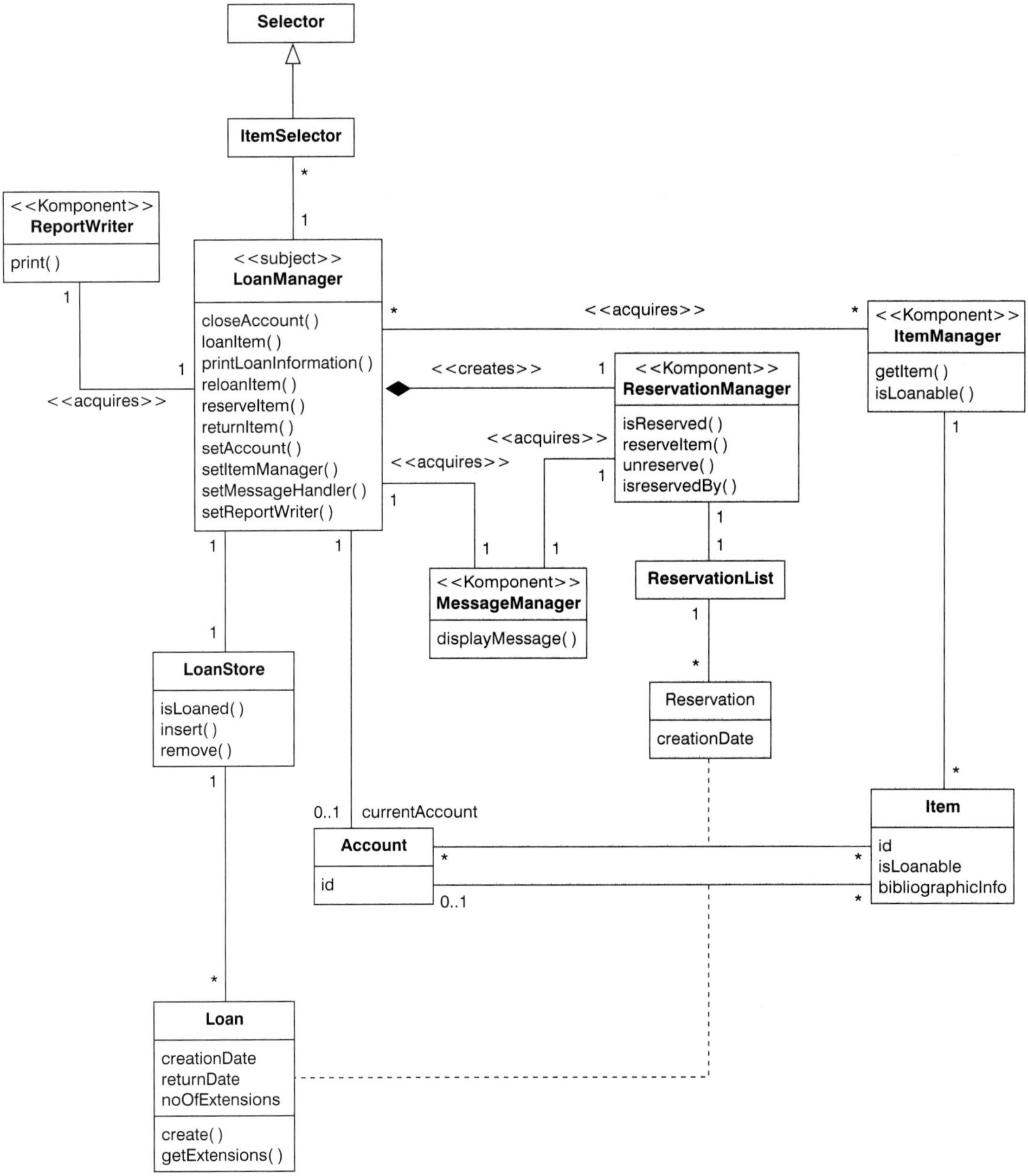

Figure 5.2
LoanManager realization class diagram

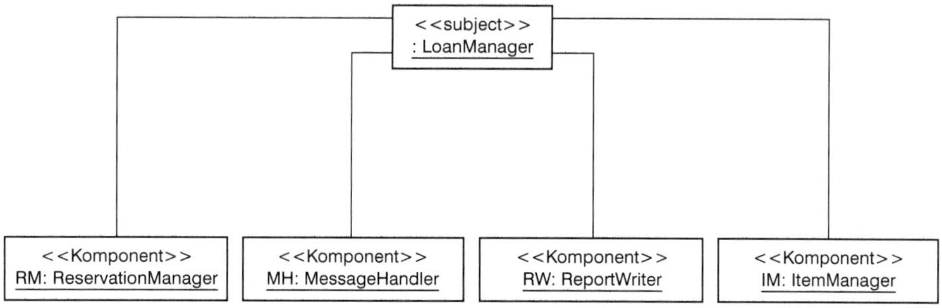

Figure 5.3 LoanManager realization object diagram

object diagrams, and thus are a means to an end rather than an end in themselves. They must therefore be clearly labeled with an appropriate term such as intermediate or preliminary.

5.1.1.3 Object diagrams

Class diagrams describe the nature of the types of elements from which a Komponent is realized. Object diagrams, in contrast, describe the actual instances of these elements that are involved in realizing the Komponent. As such, an object diagram can be interpreted as describing the structural architecture of the Komponent in terms of the major Komponents (i.e. subsystems) that it contains and interacts with.

The actual configuration of a Komponent in terms of object instances, attribute values, and links is rarely totally fixed, but usually changes over time. The goal of a realization object diagram is therefore to capture the relatively fixed, static architecture of the Komponent, and to provide a picture of a typical configuration of the instances with which an instance of the subject Komponent interacts. These are usually Komponent instances, but they can also be simple class instances.

Figure 5.3 gives an example of an object diagram for the *LoanManager* Komponent. This indicates that an instance of the *LoanManager* interacts with one instance of the *ReservationManager* (in fact, it is a part of the *LoanManager*), one instance of *ReportWriter*, one instance of *ItemManager*, and one instance of *MessageHandler*. The last three are all acquired servers.

5.1.2 Activity model

The purpose of the activity model is to describe the algorithms by which the operations of the Komponent are realized. It does this primarily in the form of activity diagrams.

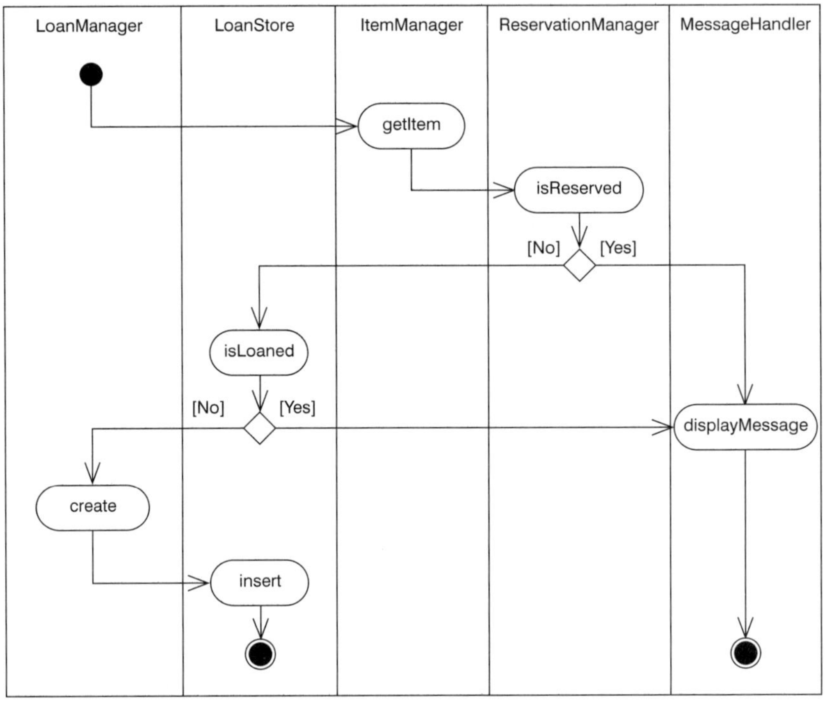

Figure 5.4
loanItem() activity diagram

5.1.2.1 Activity diagrams

A UML activity diagram is viewed as a special kind of statechart diagram in which all the states are action states and all transitions are triggered by the completion of the prior state. In practice, they can be used in the style of flowchart diagrams and/or data flow diagrams to show the flow of control or data in an algorithm.

For simple operations it is possible to capture the entire algorithm used to realize the operation in a single activity chart, as illustrated Figure 5.4. This shows the algorithm used to realize the *loanItem()* operation.

The activity diagram in Figure 5.4 represents a particular form of activity diagram containing swim lanes. These are used to indicate which operations belong to which objects. For example, since the *getItem()* activity appears in the swim lane for the *ItemManager* Komponent, it can be deduced that it is an operation of this Komponent. It is also possible in activity diagrams to show the exchange of object instances in a particular state. Naturally, any of the objects and classes named in the activity diagram, either at the top of the swim lanes, or in exchanged objects, must correspond to those identified in the structural model.

The activity chart for the *reserveItem()* operation, illustrated in Figure 5.5, is simpler because the *ReservationManager* is responsible for managing reservations, and therefore the details about how reservations are handled are embedded within the realization of the *ReservationManager*.

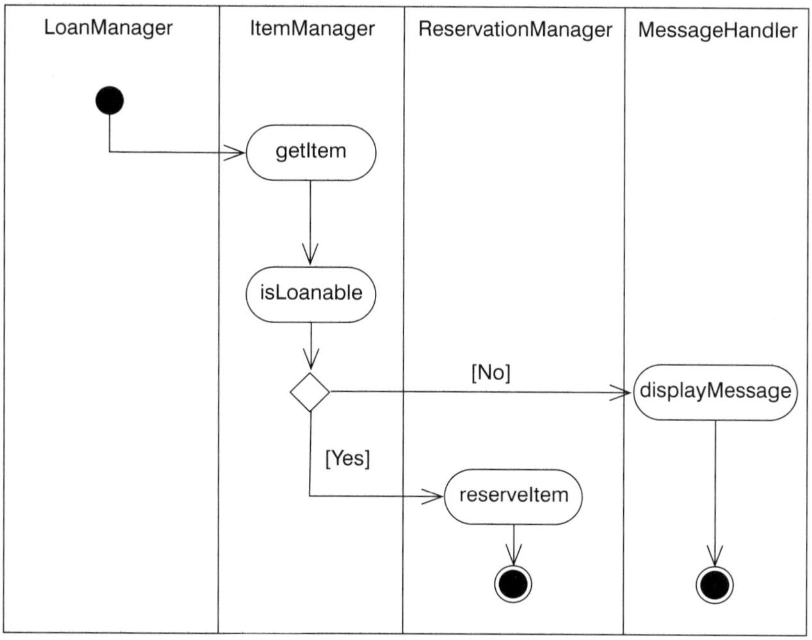

Figure 5.5
reserveItem() activity diagram

Ideally, activity diagrams used to document the algorithms for operations should have swim lanes indicating which of the other Komponents and objects are responsibile for which subactivities. However, this is not a necessity.

5.1.2.2 Intermediate activities

In the same way that the structural model may benefit from intermediate class diagrams for large and/or complex Komponents, the activity model may likewise be simplified by the definition of intermediate activities. KobrA therefore makes it possible to define intermediate activities that capture parts of the overall algorithm used to realize an operation.

To support this, in general, two distinct kinds of activities are identified in KobrA:

- **Free activities** are independent activities that are not allocated to any particular data type. Conversely, free data types have no activities allocated to them. All free activities can use (i.e. access) the free data types without restriction. In programming language terminology, free data types are regarded as "global" to free activities.
- **Allocated activities** are activities that are the responsibility of (i.e. belong to) specific classes or Komponents, but do not necessarily represent operations of the Komponent. Operations are allocated activities, but not all allocated activities are operations. An activity may be allocated to a Komponent (meaning that the Komponent is responsible for realizing it), but may actually be realized in terms of several operations of the Komponent.

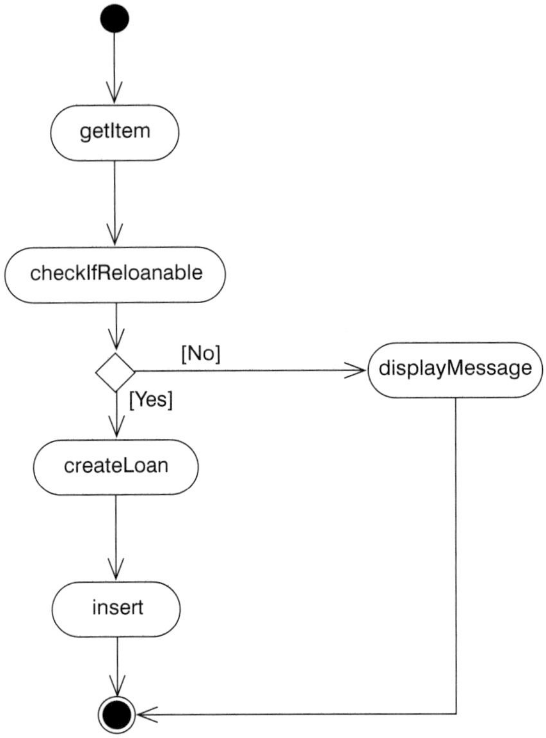

Figure 5.6
reloanItem() activity diagram

Recognizing these different forms of activities allows the realization of a complex operation to be organized in the form of a hierarchy of nested activities. Those at the top of the hierarchy are free activities, describing the general steps involved in achieving the effects of the operation without concern for which data types are responsible. As these activities are decomposed into lower-level activities, decisions eventually need to be made about who is going to be responsible for activities. These decisions allocate activities to data types to define objects. The decomposition process terminates when the leaves of the tree are identified consisting of allocated activities which are operations belonging to other objects (e.g. Komponents).

Naturally, all types of activities can appear in, and be described using, activity diagrams. However, since a free activity is not associated with an object, it cannot be assigned to a swim lane. Thus, an activity diagram with swim lanes can only contain allocated activities (including operations). The converse is not true, however. An activity diagram with no swim lanes can contain a mixture of all kinds of activities.

Figure 5.6 shows an activity diagram for the *reloanItem()* operation of the *LoanManager*. This diagram describes the algorithm used to realize *reloanItem()* in terms of nested subactivities, some of which are "free." These activities can in turn be decomposed into lower-level activities, until eventually a description in terms of operations is attained.

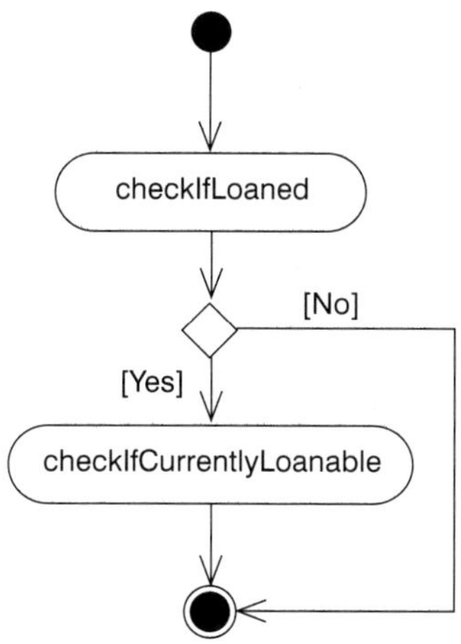

Figure 5.7
checkIfReloanable() activity diagram

Figure 5.7 shows a lower-level activity diagram for the *checkIfReloanable()* subactivity depicted in Figure 5.6 as a subactivity of *reloanItem()*.

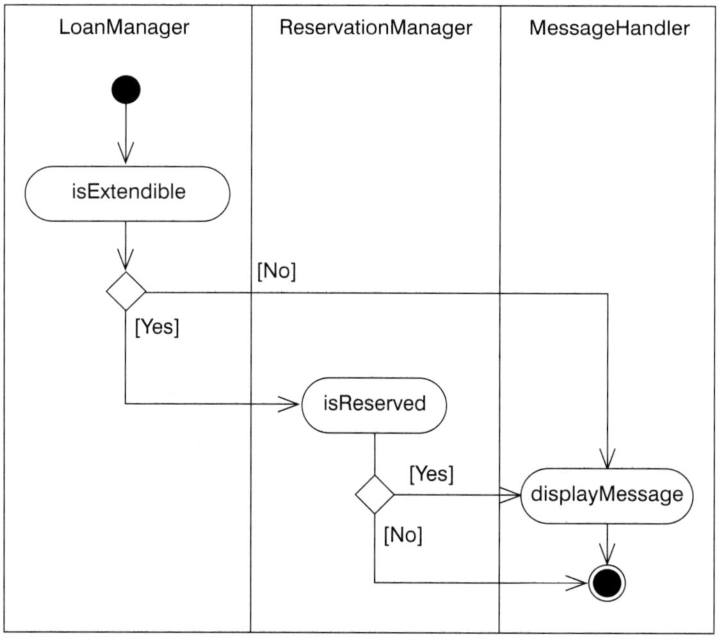

Figure 5.8
checkIfCurrently loanable() activity diagram

Finally, Figure 5.8 shows an activity diagram for the *checkIfCurrentlyLoanable()* subactivity of *checkIfReloanable()* from Figure 5.7. This time, however, all subactivities are operations of server objects or Komponents, and thus no further decomposition is necessary.

5.1.2.3 Activity specifications

Because they can be so helpful, it is also possible for activities which are not operations to have associated specifications. As illustrated in Table 5.1, activity specifications basically take the same form as operation specifications, but with one important exception: a specification of a free activity cannot have a sends or reads clauses. This is because free activities do not interact with other activities through messages, but only through the data flow described in their receives and returns clauses. Thus, all information that a free activity accesses or affects must be passed as parameters in the receives and returns clauses.

Table 5.1 Free activity specification schema

Name	**Name of the activity**
Description	Identification of the purpose of the activity, followed by an informal description of the normal and exceptional effects.
Constraints	Properties that constrain the realization and implementation of the activity.
Receives	Information input to the activity by the preceding activity.
Returns	Information returned to the succeeding activity by the activity.
Rules	Rules governing the computation of the result.
Assumes	Pre-condition on the inputs (in receives clause) that must be true for the activity to guarantee the post-condition (result clause).
Result	Strongest post-condition on the returned entities (returns clause) that becomes true after execution of the activity with true assumes clause.

Table 5.2 illustrates a specification for the *checkIfReloanable()* activity, while Table 5.3 shows a specification for its *checkIfLoanable()* subactivity.

5.1.3 Interaction model

Activity diagrams in the activity model provide a flowchart-oriented picture of the algorithm used to realize an operation (or sub-activity thereof) and thus emphasize the flow of control. The interaction model provides similar information but from the perspective of instance interactions rather than flow control. It consists of a set of interaction diagrams, each describing how a group of instances collaborate to realize an operation (or subactivity thereof). Either form of interaction diagram is acceptable (i.e. collaboration diagram or sequence diagram); by convention we use collaboration diagrams.[29]

29. This form of interaction diagram was popularized by Fusion.

Table 5.2 Activity specification for checkIfReloanable()

Name	checkIfRelonable()
Description	Checks to see if an item can be reloaned.
Receives	item: Item MH : MessageHandler
Result	If item is not loaned printMessage ("NotLoaned") has been sent to MH false has been returned else checkIfCurrentlyLoanable() and return its result

Table 5.3 Activity specification for checkIfCurrentlyLoanable()

Name	checkIfCurrentlyLoanable()
Description	Checks that an item is not prohibited from being loaned because it is either reserved or the MaxExtensions limit has been exceeded
Receives	item: Item MH : MessageHandler loan : Loan
Rule	A Loan is extendable if the number of extensions is less than or equal to maxExtensions (=3) and is not reserved
Result	If loan containing item is not extendable printMessage ("OverExtension") has been sent to MH False has been returned else if item is reserved printMessage ("Reserved") has been sent to MH False has been returned else True has been returned

For small activities it is usually unnecessary to create both an activity diagram and an interaction diagram, because the algorithm is usually clear. However, for larger activities, having both viewpoints is often helpful. Interaction diagrams become notoriously cumbersome for large algorithms, although it is possible to nest them hierarchically like activity diagrams to separate concerns [Col94]. On the other hand, activity diagrams with many swim lanes become equally unwieldy.

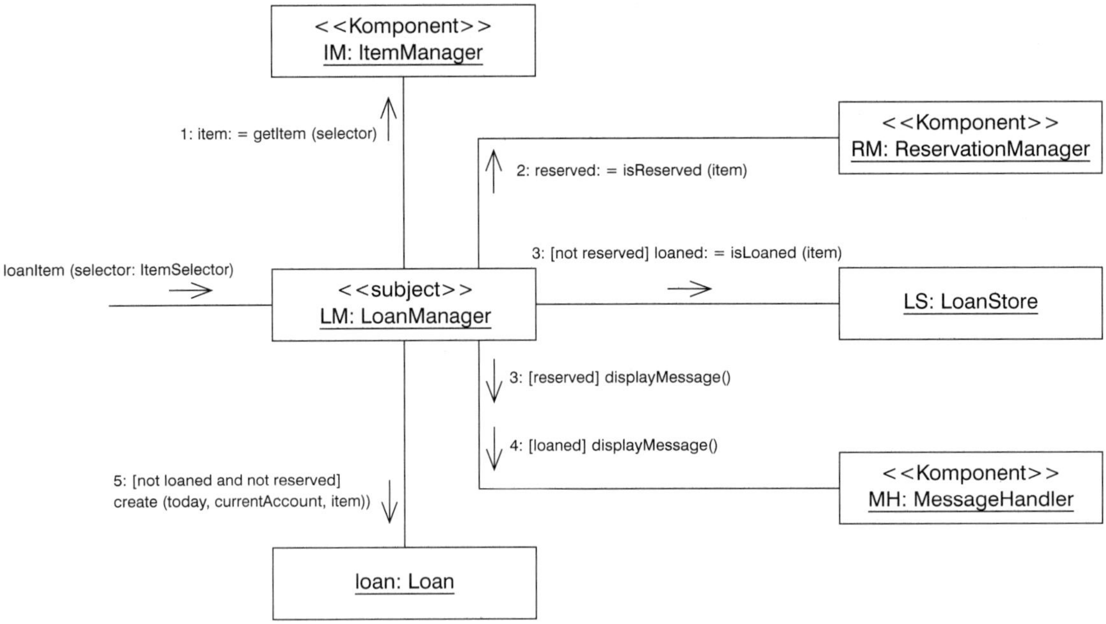

Figure 5.9
loanItem() collaboration diagram

Figure 5.9 illustrates the collaboration diagram for the *loanItem()* operation. Obviously this has to be consistent with the other models in various ways. First, the objects in the interaction diagram have to be instances of classes declared in the structural model. Second, the messages sent to objects in the complete set of collaboration diagrams must be reflected in the appropriate classes in the class diagram. Third, if there is an object model describing the structural architecture of the Komponent, the instances defined there should match with those in the collaboration diagram. Fourth, if there is an activity diagram for the operation, the algorithm described there should match that in the collaboration diagram. Fifth, the names of the operations appearing in the activity diagram should be the same as the messages appearing in the collaboration diagram. Figure 5.10 shows the collaboration diagram for *reserveItem()*.

5.1.4 Auxiliary artifacts

Like Komponent specifications, Komponent realizations have three auxiliary artifacts. The quality documentation artifact is concerned with quality assurance issues and so is described in detail in Chapter 18, while the decision model is concerned with product line engineering and is therefore described in Part 4 of the book. In the following subsection, however, we discuss the data dictionary, the third auxiliary artifact associated with a Komponent realization.

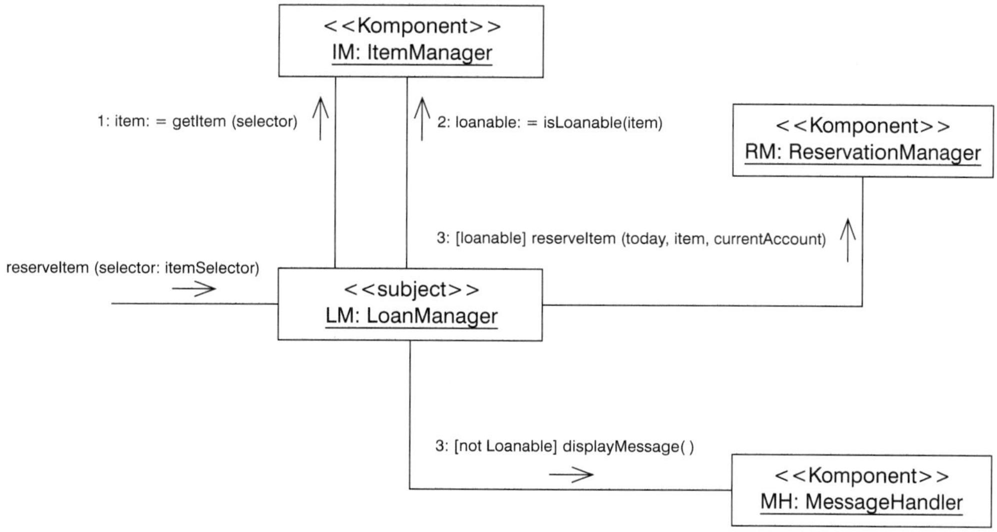

Figure 5.10
reserveItem() collaboration diagram

5.1.4.1 Data dictionary

Given that the realization of a Komponent plays the role of a design in classic development methods, it can be helpful to leverage well-established design techniques in their representation. One such technique is the use of a data dictionary to provide a central and readily accessible description of all the entities appearing in the models [Col94]. Conceptually the data dictionary is a big table with one row for each modeling concept used in the Komponent realization. At

Table 5.4 Data dictionary

Name	Kind	Description
LoanManager	Komponent	a Komponent responsible for creating and managing loans
Account	class	represents the accounts which customers needs in order to borrow items
LoanItem	operation	operation of LoanManager responsible for arranging the loan of items
.....	.....	.....

a minimum a dictionary must have three columns, the first giving the name of the entity, the second identifying its type (what kind of thing it is), and the third giving a brief description of the entity's role in the realization. Table 5.4 shows a simple example of a data dictionary for the LoanManager Komponent.

5.2 Realization process

This section describes the activities used to develop the realization artifacts described in the previous subsection. The main goal of the realization process is to provide an appropriate strategy for realizing the operations supported by the Komponent. This is a highly creative process which, among other things, must identify and characterize the sub-Komponents needed by the Komponent. The basic strategy for achieving this is based on the Fusion [Col94] design approach, which in turn is based on the concepts of responsibility-based design [WWW90]. The main creative step is to invent appropriate inter-object collaborations to realize each of a Komponent's operations, and through this to invent new kinds of objects and sub-Komponents where required.

In Fusion, creating a suitable collaboration for an operation is a single step activity. For small-grained Komponents, with fairly small operations, developing good collaborations for a complete operation in one go is a feasible intellectual leap. However, for large-scale Komponents, whose operations encapsulate significant functionality, this is much more difficult, if not impossible.

The KobrA realization activity addresses this problem by allowing the collaborations that realize a Komponent to be developed in several smaller steps rather than in one intellectual leap. It does this by avoiding a direct jump between object-oriented views of the systems (i.e. from an operation of a Komponent at one level to interactions between Komponents at the level below), and allowing the data and activity ingredients of the software to be analyzed and decomposed separately for a while. This not only splits the creative leap required in realizing an operation into smaller step, but also allows time-proven development techniques to be exploited, namely, the structured techniques of step-wise functional decomposition and data decomposition.

The philosophy embraced by the realization process is thus captured by Figure 5.11. The basic idea is that activities and data can initially be considered independently, during which time they are characterized as being "free." They are then gradually integrated to achieve a fully object-oriented perspective at the level of the Komponent's realization. This strategy also has a natural mapping to the use-case approach, as explained in Chapter 7.

In terms of concrete activities, this approach is realized in KobrA as shown in Figure 5.12. The activity-modeling and structural-modeling activities are responsible for the decomposition and modeling of activities and data, initially independently. The interaction-modeling activity is responsible for integrating the separate activity and data-oriented perspectives and determining which activities are the responsibilities of which data types. This defines the objects (i.e. the entities around which the data and activities are unified).

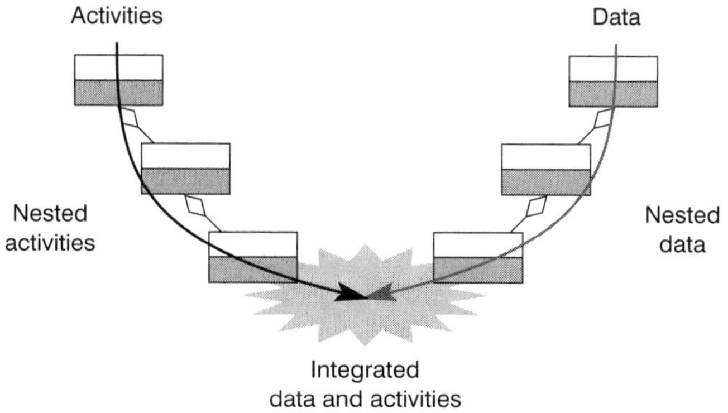

Figure 5.11 Realization process philosophy

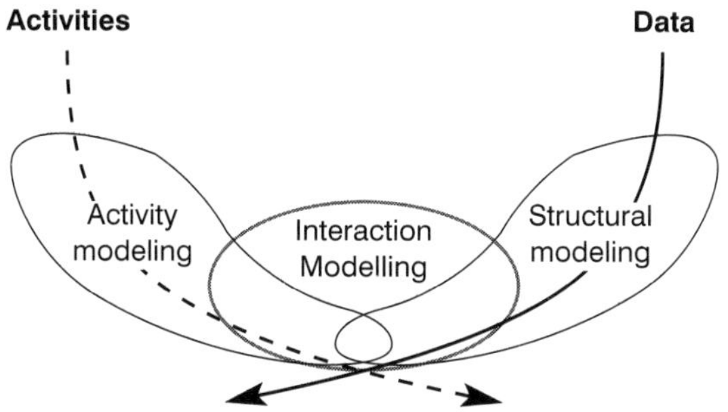

Figure 5.12 Focus of realization activities

5.2.1 Structural modeling

The purpose of the structural modeling activity is to develop the class diagrams and objects diagrams making up the realization structural model. At least one class diagram is required but no object diagrams are necessary. In general there can be several class diagrams and object diagrams, but since there is usually at most one of each in the following we assume that this the case.

5.2.1.1 Class diagrams

The realization class diagram is a refinement of the specification class diagram. This means that all the model elements in the specification class diagram are in the realization class diagram with at least the same set of attributes and operations.

The basic process for developing the realization class diagram can be summarized as follows:

1. **Insert specification class diagram elements.** The specification class diagram is the starting point for the realization class diagram. All diagram elements in the specification class diagram should be copied into the realization class diagram.
2. **Add free data types.**[30] Based on the functionality and behavior identified in the Komponent specification, add data types for any entities needed to store

30. We use the term "data type" rather than class here to emphasize that the concept should be considered independently of operations at this stage.

persistent state or data. Indicators of the need for such data types are states in the behavioral model or post-conditions of the form ".... x has been stored" in the functional model. Also, patterns should be considered at this stage, and if selected their constituent roles modeled. In general, add any "obviously" useful data types at this stage. The result can be captured as an intermediate class diagram (see 5.1.1.2). This step may be unnecessary for small Komponents.

3 **Consolidate classes.** The concepts in the class diagram needs to be reconciled with those in the activity model. The interaction-modeling activity is primarily responsible for performing this. Classes used in the interaction and/or activities models which are not already in the class diagram (or have a different name) should be added (or renamed), while classes already in the class diagram that are not used, or usable, by these models should be removed.
4 **Consolidate associations.** Objects can only communicate if there is an appropriate link from the client to the server. The associations in the class diagram must support the creation of these links as required by the interaction model. Logical links between data types must also be facilitated.
5 **Consolidate attributes and operations.** Every message sent to an object in an interaction diagram must correspond to an operation of the corresponding class in the class diagram. These operations must be added to the class diagram. Get and Set operation pairs can be modeled as logical attributes of the associated classes.
6 **Repeat 3 to 5** until the intra-diagram, inter-diagram, and realization rules are satisfied (see Appendix A).

5.2.1.2 Object diagrams

Object diagrams can be useful in the structural-modeling activity to capture an overview of the structural architecture of the Komponent, and show the instances and client/server relationships needed to support the required interaction patterns.

1 **Model structural architecture.** In the early stages of structural modeling, object diagrams can help provide a structural view of the architecture or design of a component. For large Komponents, therefore, techniques used in architecture design can be applied. In particular, this means that proven architectural styles/patterns can be used (see Chapter 10). Object diagrams are useful for visualizing the concrete application of such patterns.
2 **Model typical instance configurations.** In the later stages of structural modeling, object diagrams can help provide an overview of the instances and interaction patterns in a related group of interaction diagrams. Interaction diagrams using the same group of objects must agree on the number, names and links of the objects involved.

5.2.2 Activity Modeling

The goal of the activity model is to define, for each operation of a Komponent, a hierarchy of activities that collectively realize the operation in terms of the operations of server Komponents. The hierarchy may contain free activities at the top and intermediate levels in the hierarchy, but the leaves of the activity hierarchy need to be operations.

Initially, large operations are typically decomposed from a purely functional perspective and are then integrated with the structural model in conjunction with the interaction-modeling activity.[31]

1. **Identify subactivities.** For each operation, identify the subactivities needed to fulfill the operation specification effects. Classic structured decomposition techniques can be applied at this stage to determine the subactivities.
2. **Analyze flow.** For each sub-activity, identify the data types needed in the course of the activity's execution and the resulting data types produced. For each activity, determine the predecessor and successor activities that produce or consume the respective types. Ensure that they are compatible. Create free activity diagrams to document flow control and/or object exchange.
3. **Allocate activities.** Analyze the coupling of the activities to data types that they use. Allocate activities to the data types to which they have the greatest affinity, or identify them as sub-activities of the subject Komponent. In collaboration with the interaction modeling activity, ensure that the structural model reflects the same objects as the activity model.
4. **Determine operations.** Decide whether the allocated activities are of the appropriate granularity to represent operations of sub-Komponents. If not, further decompose them into smaller-grained activities.
5. **Consolidate operations.** Repeat 3 and 4 until the intra-diagram, inter-diagram, and realization rules are satisfied (see Appendix A).

5.2.3 Interaction modeling

Interaction modeling is the critical activity which reconciles the structural and activity-oriented viewpoints developed in the structural- and activity-modeling activities. Its inputs are the set of subactivities and data types identified in these two activities.

1. **Identify collaborations.** Once tentative allocations have been determined, identify potential collaborations between objects for realizing the unresolved activities (i.e. those currently at the leaves of the activity hierarchy). Consider modifying the data types, activities and allocations.
2. **Reconcile structural and data models.** Iterate between 1 and 2 until all the collaborations have been determined and all activities and data types have been allocated to one another.

31. In a manner analogous to use-case analysis.

3 **Consolidate objects and collaborations.** All the messages within collaborations represent operations of objects. Collect the messages from collaboration diagrams to identify all the operations of classes. Ensure that all objects affected in the operation specifications are appropriately accessed or modified in the operation's collaboration diagram(s).

4 **Consolidate operations.** Repeat 2 and 3 until the intra-diagram, inter-diagram, and realization rules are satisfied (see Appendix A).

5.2.4 Quality control

The same three quality-control techniques are used in the Komponent-realization activity as in the Komponent-specification activity, namely, inspection, quality modeling, and test-case generation. In the case of the inspection activity, performed at the end of the Komponent realization activity the following consistency rule are checked:

- realization intra-diagram rules (i.e. rules on a single diagram)
- realization inter-diagram rules (i.e. consistency rules between realization diagrams)
- realization rules (rules to ensure that the realization fulfills the specification).

Again, a detailed listing of these constraints is provided in Appendix A, while a detailed description of the KobrA inspection approach is described in Chapter 18. As with specification, inspection is a mandatory part of Komponent realization.

Since the full algorithms used to realize operations are documented in the realization models, structural test cases can be developed to complement the functional ones developed during the specification activity. Again they should be inspected along with the other realization artifacts during the realization inspection activity.

Containment 6

In the previous two chapters we described the role and nature of Komponent specifications and Komponent realizations, and explained how they combine to provide a precise but abstract description of what a Komponent does and how it works. We also explained the nature of the realization relationship that ties a Komponent's realization to its specification.

In this chapter we focus on the two other kinds of relationships that relate Komponent specifications and realizations within a Komponent tree. One, the containment relationship, defines the nesting of Komponents within one another based upon their module-like properties. The other, clientship, defines the "uses" dependencies between Komponents that arises when instances of one (the client) invoke the services of instances of another (the server).

Both play a central role in KobrA. Containment, in particular, provides the basis for a hierarchic, fractal-like product and hence the foundation for a recursive development process. As mentioned in Chapter 3 there are actually two main forms of the containment relationship: public and private. In this chapter we concentrate on the private form of containment, since this is the most common and fundamental. Public containment is discussed in detail in Chapter 8.

Before describing the effects of the containment and clientship relationships on the individual artifacts making up Komponent specifications and realizations, we first discuss how Komponent trees should be understood in relation to UML package structures. We then discuss containment and clientship in relation to the artifacts in a containment tree followed by the activities for creating them.

6.1 Package structure

Because of the principle of locality, in all but the simplest containment trees no individual Komponent "knows" about the shape of the whole tree. An individual Komponent only knows who its ancestors and its immediate children are. Because containment relates to the module-like properties of Komponents, the UML depiction of a containment tree uses packages to represent Komponents. For example, the top part of the library system containment tree is depicted in Figure 6.1.

In this representation, the anchor symbol is used to depict the name-space nesting of Komponents within one another. Notice that the stereotype <<Komponent>> is used as a label for all but the top package. The top package does not have a <<Komponent>> stereotype because it is not a real Komponent. It represents the context of the library, and thus only possesses a realization.

Another interesting point to note about this diagram is that it indicates that *LoanManager* acquires an instance of a Komponent called *StockManager* rather than *ItemManager* as defined in the *LoanManager's* specification. In fact, there is no *ItemManager* Komponent in the whole tree. What is happening is that

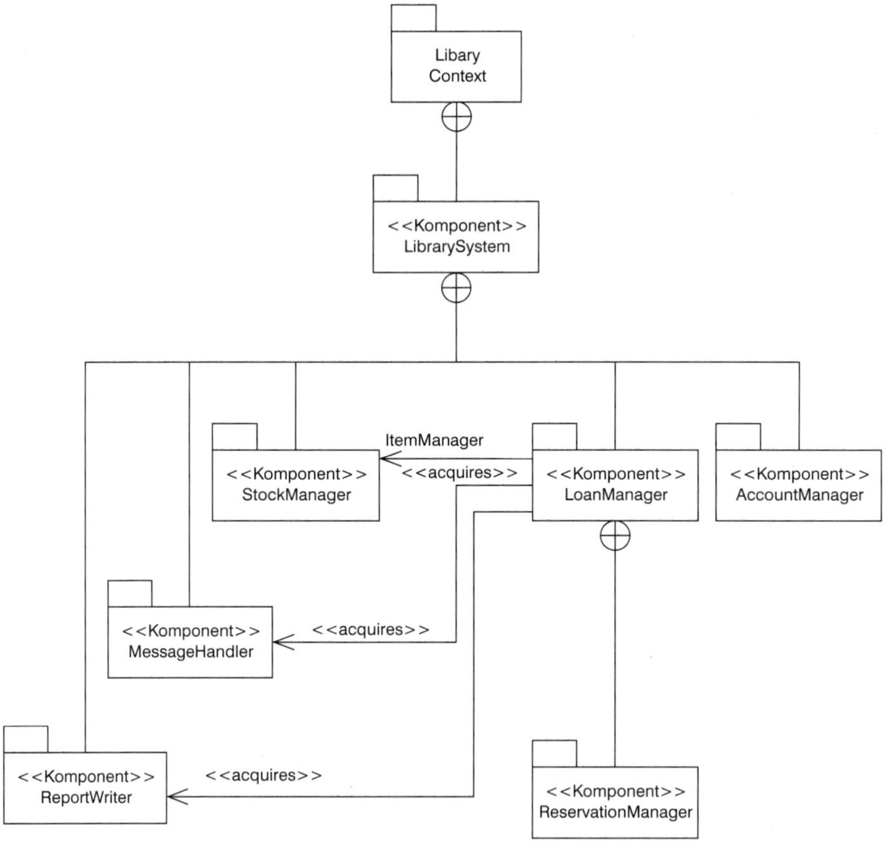

Figure 6.1 LibrarySystem containment tree

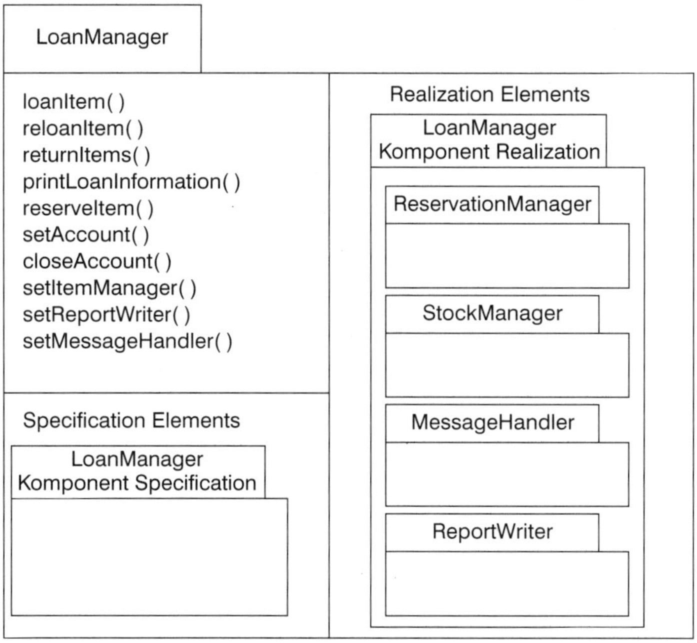

Figure 6.2 Sub-system depiction of LoanManager

StockManager is playing the role of an *ItemManager* from *LoanManager's* perspective. In other words, *LoanManger* thinks it is acquiring an *ItemManager* but in fact it is being given a *StockManager* by *LibrarySystem*. Of course this is only possible if *StockManager* is conformant to *ItemManager*.

As mentioned in Chapter 3, because of the similarity between Komponents and subsystems, it is possible to represent a Komponent using the subsystem notation. An example is illustrated in Figure 6.2. A problem with this type of presentation is that it packs a great deal of information into a single diagram and can become cluttered and difficult to read as a result. In KobrA we prefer to present the modeling information in separate, self-contained models.

It is possible to represent the information in Figure 6.1 in a way that emphasizes the nesting of packages corresponding to Komponents. Figure 6.3 shows such a representation in terms of nested packages.

As well as showing the nesting of packages within one another, this diagram also shows the logical location of various model elements (in addition to packages). Notice how every package representing a Komponent also defines a class of the same name. For practical modeling situations this is the way in which KobrA achieves the class/package duality using the current version of the UML. Every Komponent that has visibility of a package representing a Komponent also has visibility of the class with the same name that can be used to view and model the Komponent as a class. When the extended name of this class appears in a model it has the full name *Komponent:Komponent*.

In Figure 6.3 we have only indicated the location of the class model elements, but all model elements belonging to a Komponent (e.g. operations, states etc.) belong in its corresponding package. The elements that are in the

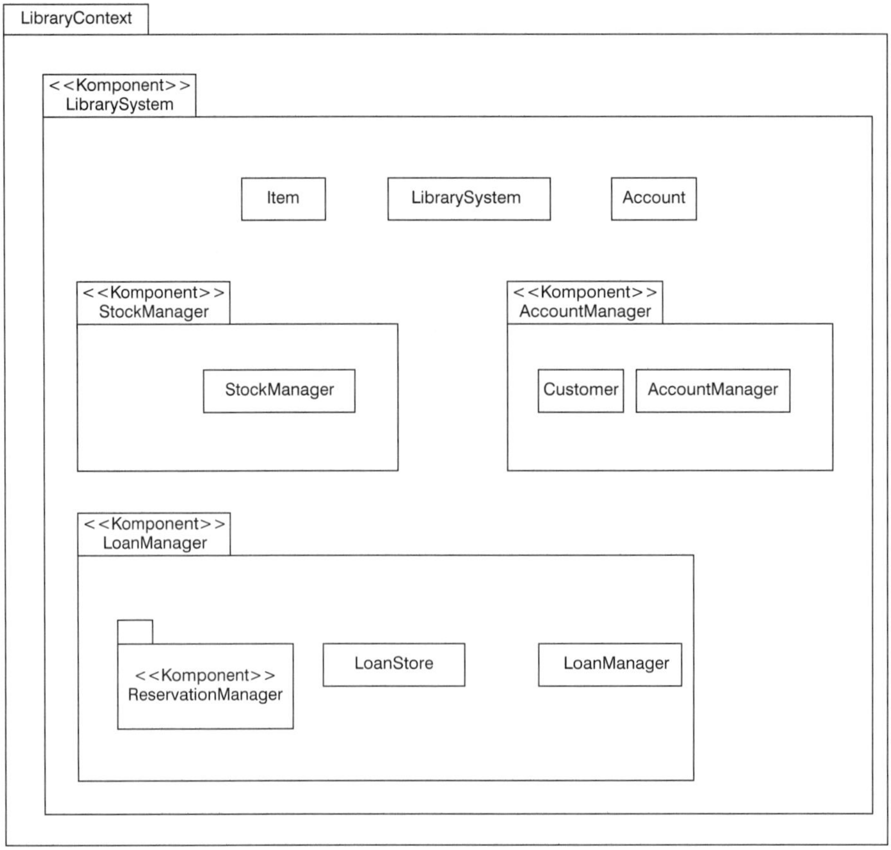

Figure 6.3
Packaging nesting view of library hierarchy

specification have public visibility in the package, while those in the realization have private visibility.

It is important to appreciate that the location of elements shown in Figure 6.3 concerns only the logical model elements. These never actually appear in a UML diagram. UML diagrams only contain "presentation elements" that present subsets of the properties of the logical model elements. There are various ways in which UML diagrams and presentation elements could be organized into packages, with the optimal approach for a particular project depending to a large extent on the idiosyncrasies of the CASE tool to be used. For example, many tools create default packages or "models," and these defaults often cannot be changed.

Figure 6.4 illustrates one typical approach for organizing UML presentation elements in a way consistent with KobrA. Basically subpackages are defined for each of the models and diagrams required in KobrA. Subpackages are defined for the specification and realization of the Komponent, and these in turn have subpackages for each of the respective models. These packages contain subpackages for each diagram, such as each class diagram etc., and these ultimately contain the

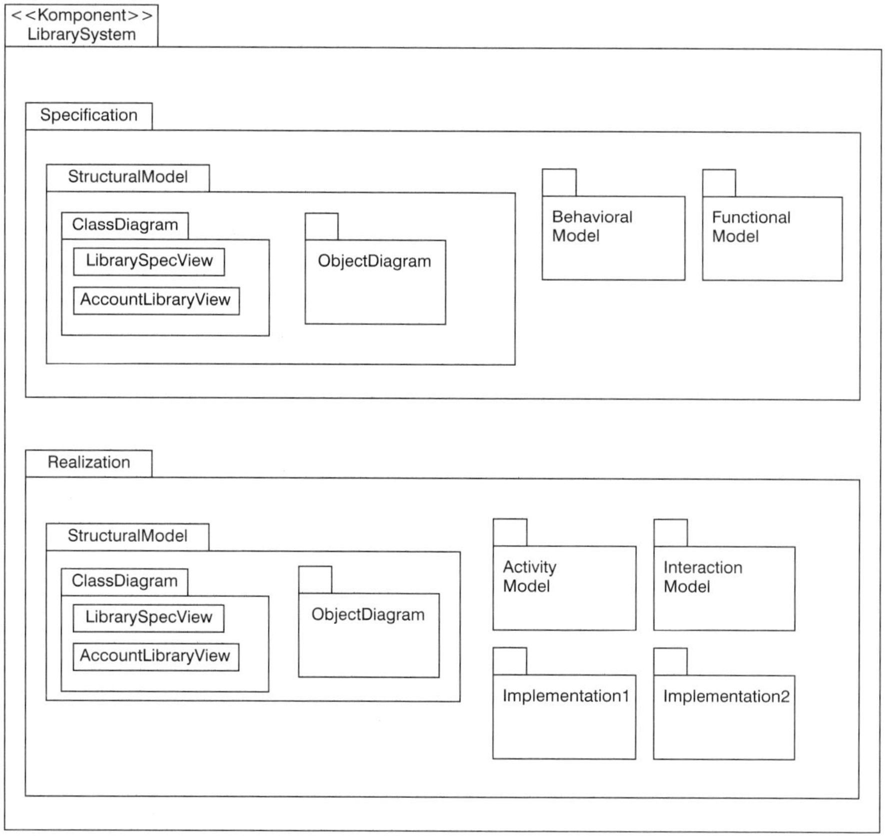

Figure 6.4
Organization of PresentationElements

presentation elements. Since the location of presentation elements has no effect on the path names of logical elements, the use of packages to organize the diagrams and their contents can be tailored to the tools used by a particular project.

6.2 Artifacts

As a basis for discussing the effects of clientship and containment on the individual specification and realization artifacts, Figure 6.5 illustrates a sample run time composition tree and associated development time containment tree. These are actually slightly modified versions of the composition tree and containment tree illustrated in Figures 3.5 and 3.6 respectively. The differences is the insertion of an acquires relationship between instances *b* and *c*, and thus between Komponents *V* and *W*. Recall, that although *d* and *e* are both instances at the leaves of the composition tree, because they are instances of the same Komponent type, that Komponent occupies a higher location in the corresponding containment tree.

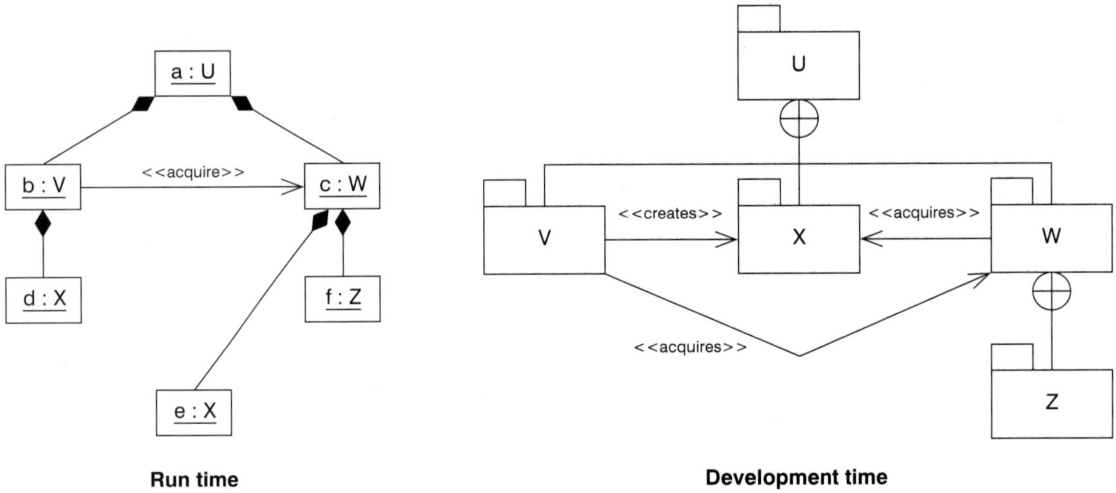

Figure 6.5
Sample composition and containment trees

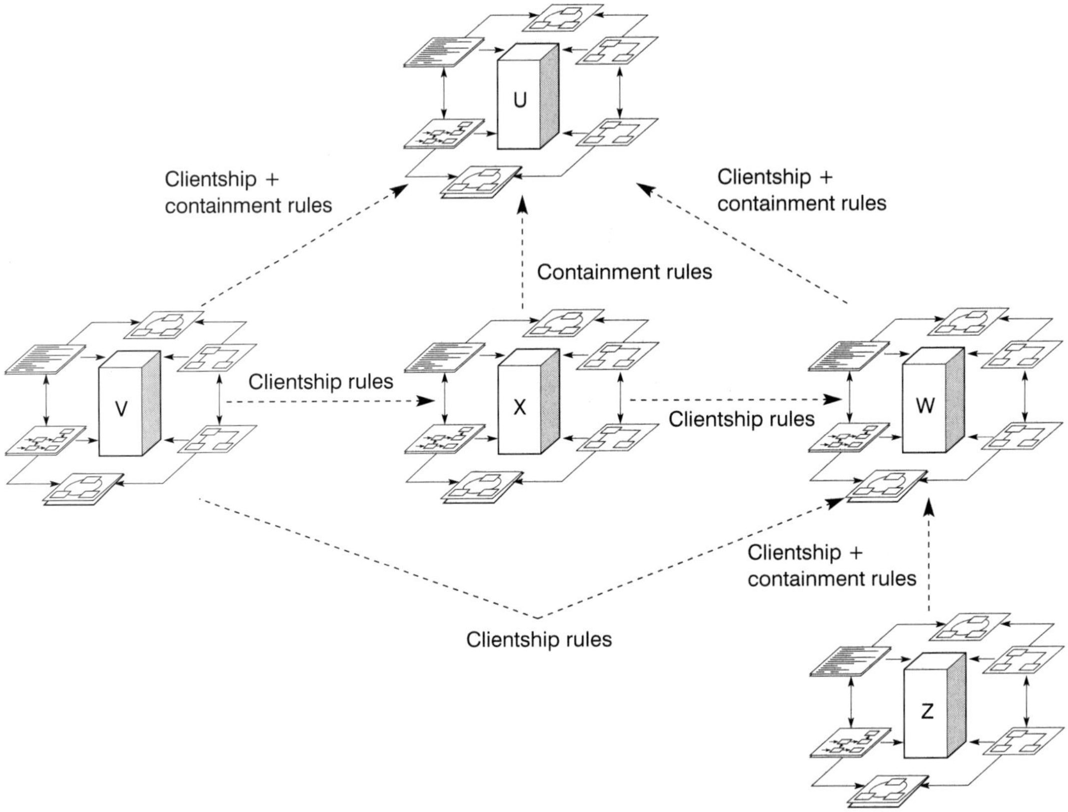

Figure 6.6
Inter-specification and realization relationships

Figure 6.6 provides an alternative illustration of the tree in Figure 6.5 that schematically represents the specification and realization artifacts associated with Komponents and the relevant consistency rules between them.

Typically a Komponent is both a client and a container of its sub-Komponents. This is the case for Komponents *V*, *W* and *Z* in Figure 6.6. Such sub-Komponents are said to be *server sub-Komponents*, because they have both a containment relationship (by virtue of the fact that they are sub-Komponents) and a clientship relationship (by virtue of the fact that they are servers) to their parent.

However, a Komponent does not have to be a server of its parent. This is illustrated by Komponent *X*, which is defined to be a sub-Komponent of *U* so that it is visible to both *V* and *W*, not because an instance of *U* wants to be a client of an instance of *X*. In this case, the only relationship that such a non-server Komponent has with its parent is the containment relationship, controlled by the containment rules. Of course, such non-server sub-Komponents have, by definition, clientship relationships with other Komponents that use them.

In general, a Komponent can have a clientship relationship with any other Komponent that it can see, or that can see it, in the containment hierarchy. The server can either be acquired by the client, as for example *W* is acquired by *V*, or it can be created, as for example, instances of *X* are created by instances of *V* and *W*, instances of *V* and *W* are created by instances of *U*, and an instance of *Z* is created by an instance of *W*. In all cases, the clientship rules ensure that the parties adhere to the clientship contract.

These different kinds of relationship affect the precise way in which the models making up specifications and realizations depend upon one another, as explained below.

6.2.1 Specification relationships

The specification of a Komponent is always subject to the containment rules relating it to its parent Komponent's realization (also referred to as the containing realization). In addition, since it defines the services offered by a Komponent, and a Komponent must always participate in at least one clientship relationship as a server (otherwise it has no reason to exist), the specification is constrained by the clientship relationships in which it participates as a server. In general, a Komponent can be a server to numerous clients, all governed by the clientship consistency rules. Of course, for this to be feasible, all the clients of the Komponent must agree on the set of services it provides.

If the Komponent acquires server Komponents, and thus participates in clientship relationship as a client, the clientship rules also constrain the specification of the Komponent, but this time from the perspective of a client.

6.2.1.1 Containment

Except possibly in the case of public containment (see Chapter 8), the containment relationship between a Komponent and its parent only directly affects the Komponent's specification structural model. The basic idea is that the containing realization (i.e. the parent realization) defines the context for the definition of the sub-Komponent's specification class diagram. In terms of UML packages, a sub-Komponent's packages is a subpackage of its parent's package. Thus, a subset of the model elements visible in (but not necessary owned by) the parent package is imported by the sub-Komponent package. Precisely which subset is imported depends on the clientship relationships as explained in the next subsection.

Again it is important to appreciate the distinction between model elements, which are the logical concepts contained by the model, and the presentation elements which are the images of the model elements presented in different diagrams. The sub-Komponent's specification class diagram contains presentation elements for the publicly visible model elements in the sub-Komponent's package. This includes the model elements imported from the parent package, new model elements that the sub-Komponent defines, and the class representing the Komponent itself (i.e. the subject). According to the principle of locality, the corresponding presentation elements in a sub-Komponent's specification class diagram contains only the information (attributes, operations etc.) required from the perspective of that Komponent.

Figure 6.7 illustrates this principle applied to the specification of the *Reservation Manager* Komponent, which as indicated in Figure 6.1 is a sub-Komponent of the *LoanManager*. This includes the subsets of elements from *LoanManager's* realization class diagram (Figure 5.2) that have a bearing on the *ReservationManager*, such as *Reservation*, *ReservationList* etc. Because the *LoanManager* is the only client of *ReservationManager* (in fact *ReservationManager* is private to the *LoanManager*), all the operations of *ReservationManager* are used by the *LoanManager*. This means that the set of operations presented in the operation compartment of *ReservationManager* is the same as in *ReservationManager's* specification class diagram and *LoanManager's* realization class diagram. If *ReservationManager* had other clients, however, it would be possible for it to have additional operations not used by *LoanManager*. The presentation of *ReservationManager* in its specification class diagram must show all the operations.

The other elements in the class diagram are included because they represent either:

- parameter and return types of the Komponent's operations
- associations needed to navigate between objects when describing the operations, or
- acquired Komponents.

For example, the class *Account* is present in the *ReservationManager's* specification class diagram because it is a parameter of one or more operations of the Komponent. Similarly the relationship between *Account* and *Item* is present since it is necessary for navigating between *Account* and *Item* objects participating in a *Reservation*.

6.2.1.2 Clientship

Clientship relationships affect all specification models. Because they determine what it is that a Komponent is expected to do, and what the Komponent

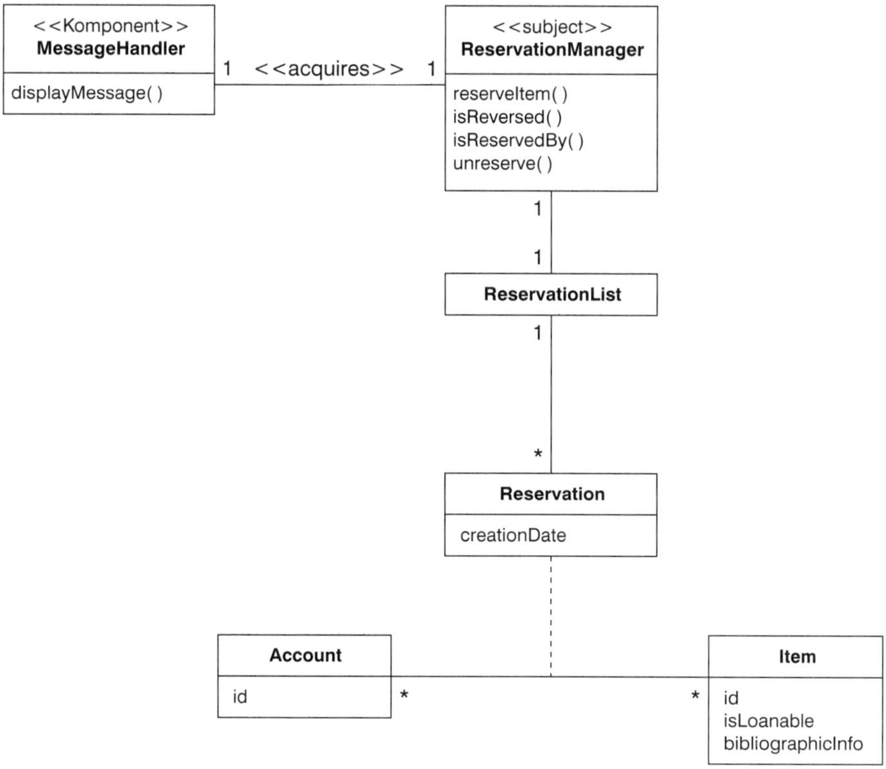

Figure 6.7 ReservationManager specification class diagram

expects, they determine the content of the models. Two different kinds of clientship relationships must be distinguished: clientship relationships in which the Komponent participates as a server; and clientship relationships in which the Komponent participates as a client.

The clientship relationships in which the Komponent participates as a server determine the nature of the services that the Komponent must provide, and the nature of the information passed via its interface. They therefore have a big effect on all the specification diagrams. Generally speaking the Komponent must supply the cumulative set of operations required by all its clients, with the understanding, naturally, that where different clients use the same operation they must agree on its signature and its semantics.

The way in which a client uses a service of the Komponent is documented in the realization of the client Komponent. More specifically, the collaboration diagrams in the client realizations document all the messages sent to instances of the subject Komponent. Collectively, these messages define all the operations of the Komponent (affecting the class diagram and the set of operation specifications), the logical effects expected by the clients (affecting the operation specifications) and the signatures expected by the clients (affecting the class diagram and operation specification). The definitions of the signatures of the operations affects the class diagram because it defines what types (i.e. classes) are exchanged over the interface.

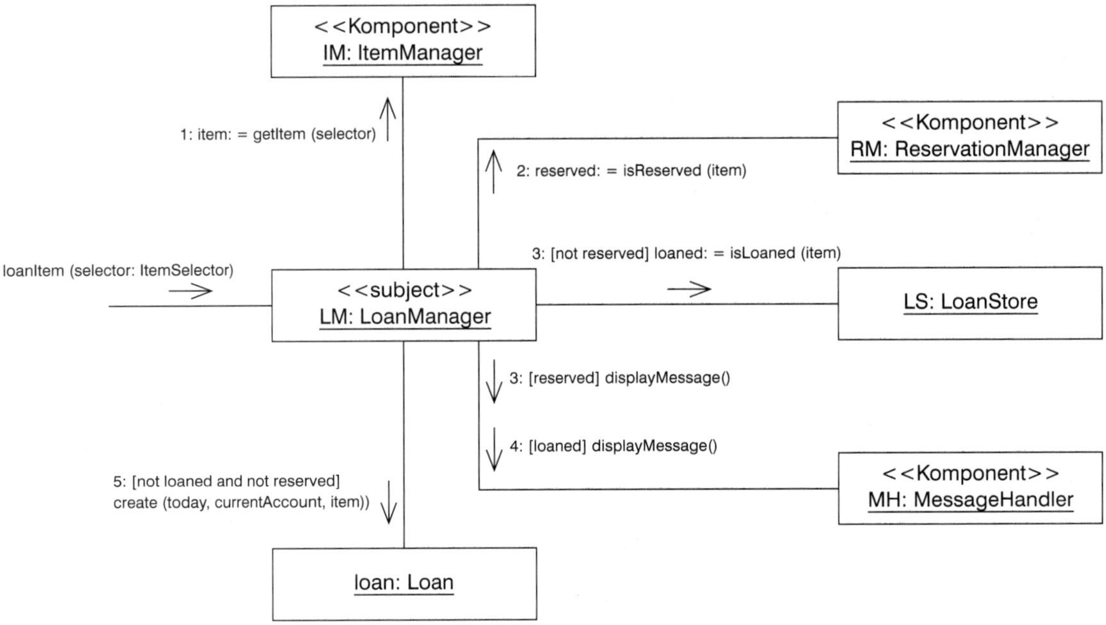

Figure 6.8
Collaboration diagram for LoanManager's loanItem() operation

Table 6.1 isReserved() operation specification

Name	**isReserved()**
Description	Returns true or false depending on whether an item is reserved
Receives	item: Item
Returns	Boolean
Result	If item is contained in one of the reservations stored by the reservation manager, True is returned Otherwise False is returned.

Consider, for example, the various interactions with ReservationManager instances described in the LoanManager collaboration diagrams in Section 5.1.3. For example the collaboration diagram for the loanItem() operation involve a message to an instance of ReservationManager called isReserved(). This collaboration diagram is shown again in Figure 6.8 for convenience. As can be seen from this diagram, the isReserved() message sent to RM takes an Item as a parameter and returns True or False depending on whether the item is currently reserved.

Table 6.1 shows the specification of the *isReserved()* operation in the specification of *ReservationManager*. The operation specification must not only faithfully match the signature of the method call in the *loanItem()* collaboration diagram (in terms of parameter numbers and types, and return value type), but must also describe the intended semantics. Of course, all invocations of a given operation must assume the same semantics. The types received or returned by the operation must also of course appear in the specification class diagram, as illustrated in Figure 6.7.

The clientship relationships in which a Komponent participates as a server also have a big impact on the nature of the behavior (i.e. the externally visible states, events and state transitions) that instances of the Komponent are expected to exhibit. Because the externally visible states and transitions must take into account all the messages sent to an instance of the Komponent (by all its client instances), and their patterns of invocation, determining the behavior is a much trickier task than determining the semantics and signature of an individual operation. The basic goal is to define ordering constraints between specific operations and to identify commonly recurring sequences of invocations. These are both good indicators of externally visible states that can be used to control when operations are invocable.

6.2.2 Realization relationships

Containment relationships have no direct effect on Komponent realizations. The visibility and presentation rules governing a Komponent's realization are defined instead by the realization rules relating it to the Komponent's specification. The realization rules essentially have the inverse effect of the containment rules. Whereas the containment rules reduce the set of presentation elements from a parent's realization class diagram that appear in a child's specification class diagram, the realization relationship expands the set. In other words, the realization class diagram contains all the presentation elements in the specification class diagram, but usually with more information, and adds additional presentation elements for the private model elements developed during the realization activity.

6.2.2.1 Clientship

Clientship rules certainly do have an important impact on the realization model, since they govern the clientship relationship in which the subject Komponent participates as a client. It is important to remember that both acquired and created servers appear in the realization and so must be handled according to the clientship rules.

Every server of the Komponent must be represented as a class in the realization class diagram, and all the operations of the server Komponents invoked by the subject must be listed in the corresponding operation compartments. This information can be gathered from the Komponent's collaboration diagrams, which identify all the messages that its send to its servers. Basically these invocations must adhere to the clientship contract, which means that the pre-condition for an operation must be satisfied and the invocation must have the correct parameters. Naturally, the class types of the parameters must appear in the realization class diagram.

6.3 Process

The previous two chapters have described the two key development subactivities that form the foundation of the KobrA development approach:

1 Komponent specification

2 Komponent realization.

In this section we describe how these activities are combined together in KobrA to provide a top-down, recursive process for software development. This process is not only the basis for incremental development in KobrA, but can also be viewed as an incarnation of the influential spiral model of software development [Boe88] as explained in Chapter 13. Most importantly, however, a recursive process is a basis for a systematic, step-wise development approach akin to the Cleanroom method [MLH87].

Recursive decomposition is a process for the systematic development of a Komponent containment tree, starting at the root and moving systematically downwards towards the leaves of the tree. Each Komponent in the tree is developed using the same underlying artifacts and activities. Not all kinds of containment are created by a recursive development step, however. Containment involving non-server sub-Komponents (i.e. sub-Komponents that are not servers of their parents) is usually created by a reshaping of the tree after an opportunity for the sharing of a Komponent has been identified. The process for conducting this shape change is described in Section 6.3.4. In the remainder of this section we focus on the recursive development activities associated with containment involving server sub-Komponents. This is the primary direction of development in KobrA because the first version of a Komponent is usually almost always developed as a unique server of its parent.

Figure 6.9
The DNA spiral

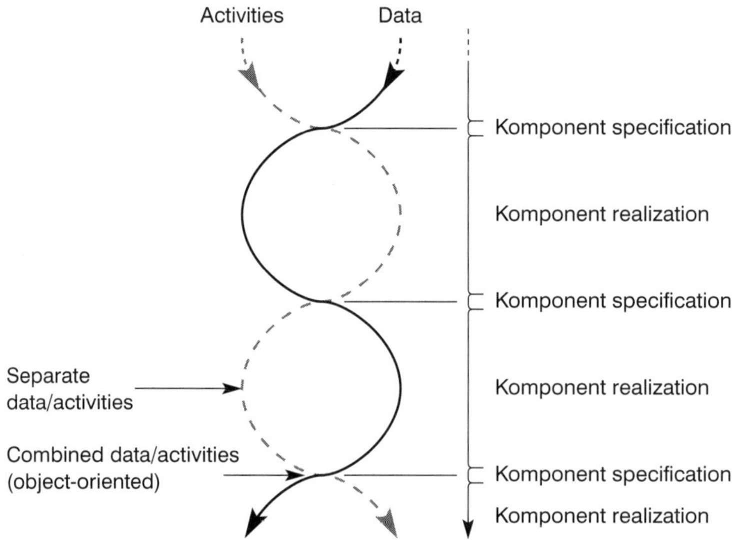

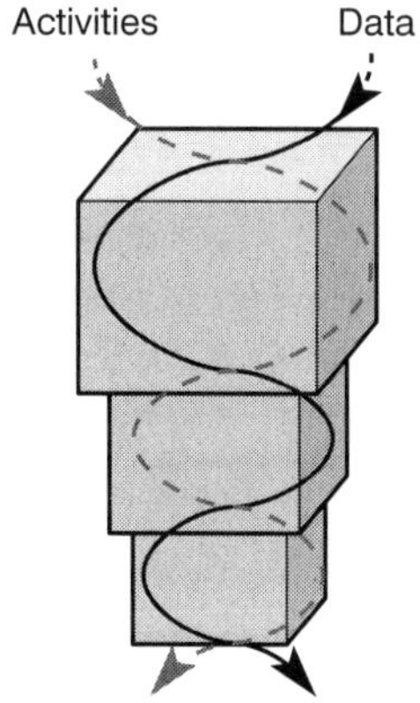

Figure 6.10 Containment tree and DNA spiral alignment

6.3.1 The DNA spiral

Chapter 5 explained how the Komponent realization activity initially focuses on the data and activity ingredients of a software design separately before gradually bringing them to provide a unified, object-oriented view of the sub-Komponents and classes making up the realization of the Komponent. When applied recursively to the development of all Komponents in a containment tree, this leads to a spiral-like view of how the data and activity ingredients of a software system are elaborated during the course of a KobrA development project. We call this model, depicted in 6.9 the DNA (Data aNd Activity) spiral.

To understand this figure it is important to realize that it depicts the relationships between the data and activities as they are elaborated down one particular branch of the containment tree. Conceptually every branch of the tree splits off a new branch of the spiral, descending down that branch. Another important point is that each point of integration of the data and activities is aligned with the realization/specification boundary of a Komponent and its sub-Komponents, and each distinct separation and joining cycle of the DNA spiral moves one Komponent further down a particular branch of the tree. The alignment of the DNA spiral with the Komponents is illustrated in Figure 6.10 for a particular branch of the tree. For the purposes of this figure a Komponent is represented by a cube, with its upper surface representing its specification and its lower surface its realization. A cube attached to the lower surface of another cube is viewed as a sub-Komponent (i.e. contained by) the Komponent represented by the upper cube.

As illustrated in Figure 6.9, the points in the DNA spiral where the data and the activities are unified provide a pure object-oriented view of the system. At these points, all the data items (except primitive data values such as Integers, Characters etc.) have activities (called operations), and all the activities are associated with data. This is the view of the system that is captured in a Komponent realization, and the specifications of the Komponent's children. As explained in the previous sections, the specification of a sub-Komponent is closely related to the realization of its parent, and must agree on the allocation of activities to data at that level of granularity.

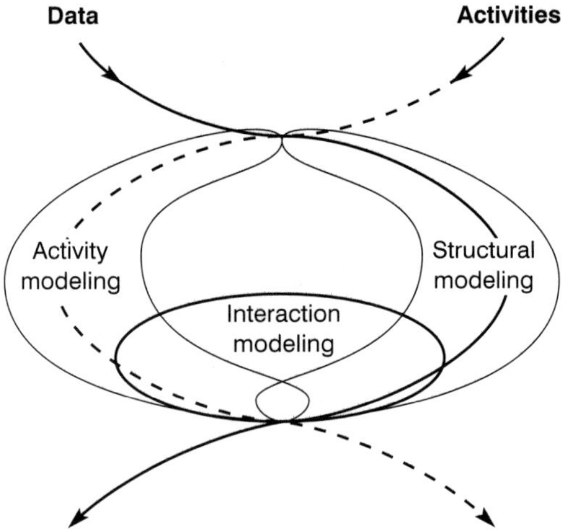

Figure 6.11
Realization activities

In between the unification points, the data and activities are analyzed at varying degrees of independence before being reintegrated. Not only does this separate analysis enable the approach to draw upon some of the most successful and established system-development methods (e.g. recursive step-wise refinement of functions), but it actually mirrors what takes place at certain phases in modern object-oriented development methods. In particular, the use-case analysis techniques which are popular in the early phases of modern object-oriented methods start by focussing on the functional decomposition of the high-level activities before any consideration is given to objects. The DNA spiral supported in KobrA allows these same basic ideas to be exploited at all points and levels of granularity in the development process.

Since Komponent specification is largely a documentation activity, and must conform to the data/activity integration defined in the encapsulating realization, it plays no role in proceeding down the DNA spiral. All the creative work in separately decomposing and integrating the data and activities is performed in the realization activity. Figure 6.11 illustrates the main subactivities of the realization activity, but applied to an arbitrary segment of the DNA spiral. As can be seen from this figure, the activity- and structural-modeling activities play symmetrical roles in decomposing the activities and data aspects of the Komponent. Initially this decomposition may be independent, but towards the end of the cycle the two ingredients are unified with the help of the interaction modeling activity. This comes into play towards the end of the cycle and serves as the main vehicle for merging the two separate views.

This same basic model of decomposition is employed at all phases and levels of granularity in the KobrA approach, starting at the top and working down to the leaves of the containment tree.

6.3.2 Component identification

The detailed substeps of the main realization activities have already been defined in Chapter 5. However, as part of the transition from the realization of one Komponent to the specification of its sub-Komponents, another important activity must be performed – Komponent identification.

The Komponent identification activity is performed at the end of the realization activity in order to identify potential new sub-Komponents. This activity is needed because the primary realization activities do not always discriminate between Komponents and simple classes. When selecting instances of already existing classes or Komponents, the choice between a simple class and Komponent is made automatically. However, *new* classes, introduced into the system for the first time as part of the realization under question, can either be treated as simple classes or Komponents. It is the job of this activity to make the choice.

Komponent identification is a subactivity of the realization activity, because it must be performed before the final version of the structural model can be created with the appropriate application of stereotype <<Komponent>> where necessary. However, it takes place only after the other activities, other than quality assurance, have been completed.

For each new class defined during the realization process, the goal of the Komponent identification activity is to determine whether the class should be regarded as a simple class, or as a sub-Komponent that needs further elaboration. Note we are considering here only new classes that apparently do not yet exist in the part of the containment tree visible to the subject of the realization. Any instances of already existing simple classes or Komponents already have their type defined, and so do not need new class definitions or Komponent specifications.

The choice between a simple class and a Komponent depends on two main factors, multiplicity and granularity. Generally speaking, the number of instances of a Komponent used in the realization of another is very small (usually one). On the other hand, the number of instances of simple classes is often, but not always, quite large. Therefore, if there are a large number of instances of the class under consideration it should typically be treated as a simple class. If there are only a few instances, however, the granularity should be considered. Generally speaking, Komponents are larger than simple classes and encompass more functionality. Simple classes, in contrast, tend to resemble data structures with appropriate access methods, and do not typically encompass a lot of extra functionality. The models involved in specifying a Komponent therefore would typically provide little if any useful information. In short, treating a simple data structure as a Komponent would be overkill.

Because either choice is feasible in the sense that it would lead to a working system, in the end the choice comes down to the properties that are desired or considered most likely for the abstraction. If it may need to be implemented as a physical component, or there is high likelihood that the data structure may in the future be reused for other purposes, the overhead of Komponent specification becomes more worthwhile. If, on the other hand, the class is never likely to be used outside the bounds of the containing Komponent, then it makes much more sense to implement it as a normal class.

Ultimately, the recursive development of the Komponent containment hierarchy will finish when no further sub-Komponents have been identified and all new classes are handled as simple classes.

The basic process is thus as follows:

1. Identify each *new* class in the realization of the Komponent.
2. Count how many instances of the class are created or acquired by the Komponent. If this number is large or subject to large dynamic variation, the class is probably best viewed as a simple class.
3. If there are only a few instances then consider the granularity of the Komponent. If it is large, and encapsulates a great deal of functionality, handle it as a Komponent. Also, if it is known that it needs to be implemented as a physical component, handle it as a Komponent. Otherwise handle it as a class.

6.3.3 Sub-component creation

In Section 4.2.1, the first step in the process for creating specification structural models contained the general statement: "Prepare a first draft of the class diagram based on the containment hierarchy." Now that we have explained the nature of the containment relationship, and the way it affects the specification and realization artifacts, we can define this step in more detail. This step is triggered when a new class introduced as part of the realization of a Komponent is identified as sub-Komponents according to the Komponent identification process in Section 6.3.2:

1. Create a new package with the name of the Komponent as a sub-package of the parent Komponent.
2. Change the ownership of the class corresponding to the sub-Komponent (i.e. with the name of the sub-Komponent) to the new package. In other words, move this model element to the new packages. Note, this means that the parent realization package must now import this element.
3. Identify all the model elements in the parent's realization that are in any way related to the sub-Komponent. This information should be explicit in the parent's realization class diagram, but it is worth checking the collaboration diagrams again to see if any interactions involving instances of the sub-Komponent were missed. Once identified, these model elements should be imported into the sub-Komponents package.
4. Create the specification class diagram to display all the model elements in the sub-Komponent package. Reduce the information in the corresponding presentation elements so that they only show the minimal information of concern to the Komponent
   a. All classes should only show the logical attributes of relevance to the sub-Komponent.
   b. Simple classes should have no operation compartments, while the operation compartments for acquired Komponents should only contain the operations invoked by instances of the sub-Komponent.

The first draft of the class diagram should only contain an initial guess about (a) and (b). When the functional model is elaborated, firm decisions can be made about what the sub-Komponent actually needs.

6.3.4 Tree refactoring

Ideally, when developing the realization of a Komponent, the instances collaborating in the realization of operations should be instances of existing classes or Komponents within the tree that are visible to the subject. This allows the realization to directly import the corresponding class definition or Komponent specification and use it as needed.

If an instance is needed for which there is no appropriate visible class or Komponent specification, there are only three basic options:

- Define the new class or Komponent from scratch using the regular specification and realization approaches discussed in Chapters 4 and 5.
- Find an appropriate external component outside the current tree by applying the reuse activity explained in Chapter 12.
- Find an appropriate class or Komponent in an invisible part of the tree and move it to a visible part. This is called refactoring and is discussed in this subsection.

Like the visibility rules in most programming languages, the visibility rules in KobrA are logical rules designed to help organize the shape of the containment tree in a sensible way. The actual ability of KobrA developers to detect suitable Komponents in logically invisible parts of a tree is largely an organizational issue, and depends on the project management and configuration control strategy being applied. In some organizations, it may make sense to really hide the sub-Komponents of a Komponent from others that should not logically be able to see it, and to provide formal protocols for sharing potentially reusable Komponents. More often, however, all developers have the right to look at all the Komponents in a tree, and can raise a refactoring request if desired.

The basic goal of the refactoring process is to determine the new location for the Komponent to be shared where it will be visible to all the Komponents that need to see it, and to change the models of the Komponents affected by the change until they are once again consistent. The Komponents affected by such a change are the original parent of the moved Komponent, the new parents of the Komponent, the clients of the Komponent and of course the Komponent itself. Basically the realization of the original parent[32] has to be changed so that the moved Komponent is now just a server of the parent and no longer a sub-Komponent. This only affects the class diagrams, and then only the name of the class representing the moved Komponent.

The models of the new parent of the moved Komponent have to be changed to reflect its presence. Only the realization class diagram is affected, and this must have the classes in the moved Komponent's specification added, with appropriate associations.

32. Assuming private containment (see Chapter 8).

The clients of the moved Komponent have to be changed in a similar way. If the client acquires the Komponent the specification class diagram will have to be updated to reflect the new full name of the moved Komponent. However, if it creates instances of the Komponent, the realization class diagram will have to be so changed. All other diagram types should be unaffected.

Finally the models of the moved Komponent may need to be changed to reflect their new location. The basic process is thus as follows:

1. Determine the new location of the Komponent or class to be shared. The new location should be the lowest place in the tree that provides visibility to all the Komponents that need to see it. All the Komponents which the shared Komponent needs to see must also be visible to it in its new location.
2. Change the realization class diagram of the new parent to reflect the presence of the moved Komponent. This means inserting all the classes in its specification class diagram if they are not already present.
3. Change the realization class diagram of the former parent to reflect the absence of the moved Komponent. This means removing all the classes in the moved Komponent's specification class diagram, unless they are needed for some other reason.
4. Change the class diagrams of all clients of the moved Komponent to reflect its new location. For those acquiring an instance of the moved Komponent, the specification class diagram will need to be changed. For those creating an instance, the realization class diagram will have to be changed.

6.3.5 Quality control

Although this chapter has not added any new models to those defined in the previous two chapters, it has introduced two important new types of dependencies between them, those based on Komponent clientship and on Komponent containment. These are of significant importance for quality assurance.

Clientship dependencies, in particular, are the source of many of the consistency rules that are the focus of inspection. Clientship can also be helpful in defining test cases for Komponents in the body of a containment tree because it is the source of their usage profiles. In effect, the usage profile of a Komponent is determined by the interweaving of the collaborations in which the Komponent participates as a server. Analysis of these collaborations, and through this of the usage profile, can help optimize the selection of test cases. Coupling measures used for structural measurement are also derived from the clientship relationships.

Containment is essentially a development-time concept and thus has less value for testing. However, it is also the source of important consistency rules for inspection. The containment and clientship consistency rules are all listed in Appendix A.

7 Context realization

In the previous chapter we saw how, when applied in alternation, the Komponent specification and realization processes form the basis for an iterative development process based on recursive decomposition. Every specification is defined with respect to an encapsulating realization, and every realization is defined with respect to an associated specification. The obvious question that arises, however, is: How does the process start? In other words, where does the first specification or realization come from? This is the role of context realization.

The purpose of the context realization process is to define the top-level realization without the benefit of an associated specification. Although it bears the name "realization," it actually leans heavily on front-end development techniques. The basic goal of the context realization activity is to describe the salient properties of the environment of the Komponent representing the system to be constructed. This is achieved by looking in detail at the business context and the usage context.

The term "realization" is used in the name of this activity for two reasons. First, the set of artifacts produced in context realization is a superset of the normal realization artifacts, and thus is a realization in this very concrete sense. Second, it is well known that introducing a software system to improve an existing business process invariably changes that process. Thus, the description of the environment of a software system invariably reflects the new way in which the business processes will operate (i.e. be realized) once the system has been introduced. The term "context" is used in the name since this is the term best understood as referring to the environment in which a software system operates.

In a sense, the "context" can be viewed as a pseudo-Komponent at the root of the development tree. The system can then be treated as a regular Komponent, just like any other in a containment hierarchy, since the context provides the encapsulating realization against which it can be specified. Of course, the "context" pseudo-Komponent has the unique characteristic that it has no specification. Therefore, since the specification from which the realization process normally begins is missing, it has to be substituted by something else capable of defining what the Komponent is expected to do and how it is expected to behave. In KobrA this process is known as enterprise modeling. The enterprise models are the input for structural, usage and interaction modeling, where usage modeling is a special variant of activity modeling with a focus on the users. However, because of the complexity of the "context," there have to be several feedback loops between enterprise, structural, usage, and interaction modeling. The foundations and details of such a process are described in [Paech00]. In the following, only the major artifacts and steps are sketched.

Thus, the main differences between context realization and normal realization are:

- the additional artifacts and activities related to enterprise modeling
- the increased emphasis on usage modeling and user interfaces
- more feedback loops between enterprise, usage, structural, and interaction modeling as a substitute for the missing specification.

Otherwise, the artifacts and activities involved in context realization are, deliberately, very similar to those of Komponent realization.

7.1 Context realization artifacts

A context realization can have up to seven distinct kinds of artifacts, six of which are the same as those used for a Komponent realization. In addition to the Komponent realization artifacts a context realization also has an enterprise model. The full set of artifacts that can make up a context realization is thus as follows:

1. enterprise model
2. structural model
3. activity model
4. interaction model
5. data dictionary
6. quality documentation
7. decision model.

The first four are referred to as the primary artifacts and collectively describe the environment (i.e. the context) in which the Komponent to be developed operates. The others auxiliary artifacts are described in other parts of the book. The nature of data dictionaries is described in Chapter 5, quality documentation is described in Chapter 18 and decision models are described in Chapter 15.

7.1.1 Enterprise model

KobrA is deliberately much less prescriptive about the enterprise model, and the related process, than it is about the other models. This is because the enterprise model is intended to provide a starting point for the creation of the other realization artifacts rather than to represent an important part of the final realization documentation. As mentioned above, the enterprise model essentially serves as a substitute for a full-blown specification at the top of the development tree.

The goal of the KobrA enterprise modeling process and artifacts is to accommodate mainstream enterprise and business modeling approaches used in industry today. The precise nature of the artifacts making up the enterprise model is therefore relatively unimportant. In fact, most enterprise modeling approaches use artifacts very similar in nature to those making up a KobrA realization – static structure diagrams to capture the informational aspects, and activity diagrams to capture the process aspects.

The difference between the enterprise model and the other realization models is its emphasis rather than its substance. The other realization models focus on describing the system's environment from the perspective of a computer system. The purpose of the enterprise model is to describe the nature of the enterprise for which the system to be built is needed, what general role the system should play, and what business rules should govern its behavior. The enterprise model therefore describes the relevant aspects of the enterprise in a way that completely ignores the fact the system to be built will be realized as a computer system.

This is not to say that the enterprise model is not intended to describe the new way in which the enterprise will work once the system has been created. On the contrary, the enterprise model is intended to capture the new "to be" form of the enterprise once the new system has been introduced. Some enterprise and business process modeling approaches also suggest the definition of an "as is" model, as a strategy for getting to the "to be" model, but this is beyond the scope of KobrA. In particular, KobrA does not require a specific form of business-IT alignment,[33] but stipulates that the results of enterprise modeling reflect this alignment.

Although they may vary in their details, all business and enterprise modeling approaches essentially focus on three main issues – the activities (or processes) that create value, the items involved somehow in creating value, and the rules that govern how the two interact. In order to provide a concrete starting point upon which to base the other models, we assume that the following two core artifacts are created:

1. enterprise concept diagram
2. enterprise process diagram.

The first of these is essentially a class diagram that describes the enterprise concepts of importance. The second is a diagram that shows the nesting of enterprise activities.

33. Information Technology

Figure 7.1
Library enterprise concept diagram

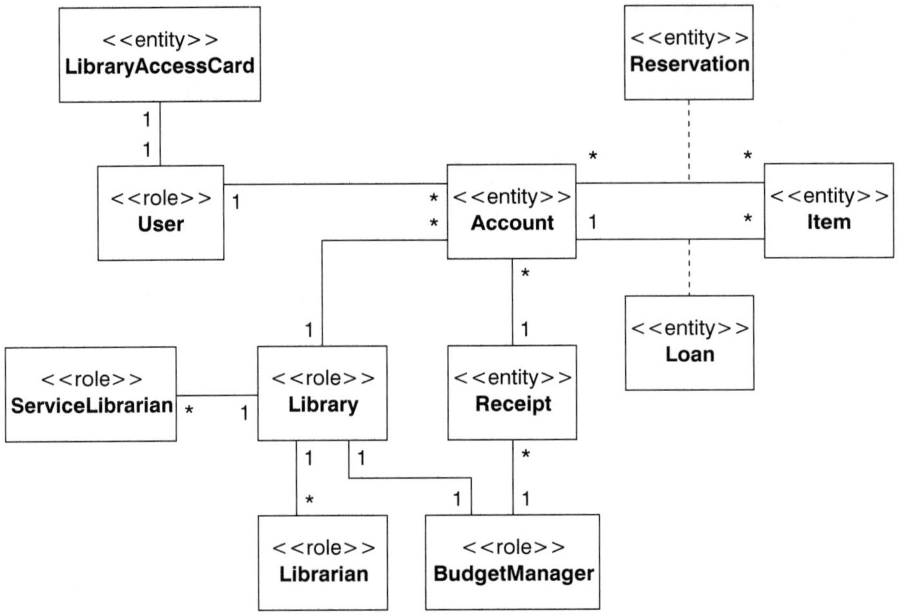

Figure 7.1 illustrates a concept diagram for the library enterprise. This shows the organizational units and roles acting in the enterprise (e.g. *Librarian, User, BudgetManager*) and the major entities for which these roles are responsible (e.g. *Account, Item*). There are two ways to include the computer system to be built in such a diagram. One possibility is to show the computer system as a separate role interacting with the other roles. The other possibility is to draw a boundary around all entities and roles supported by the computer system. Typically, the first kind is a preliminary step to deriving the second kind. The first kind is similar to the context diagrams used in structural analysis. KobrA does not specify whether the first or the second approach should be used since

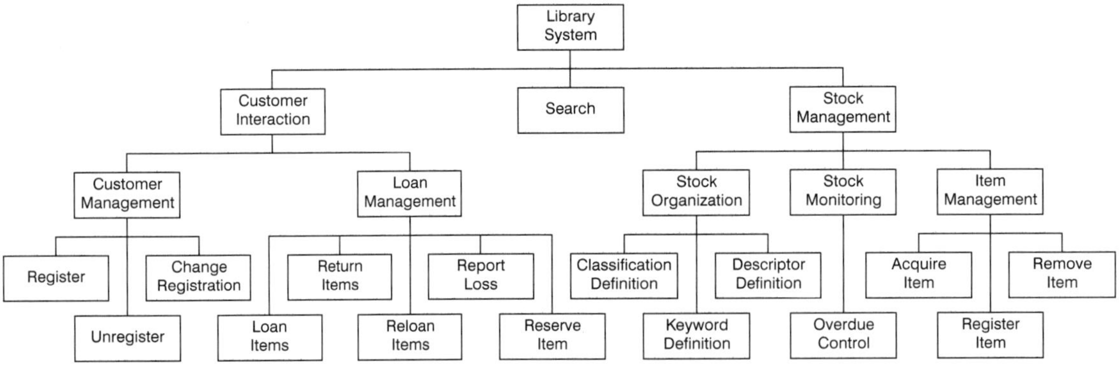

Figure 7.2
LibrarySystem enterprise process diagram

the detailed responsibilities of the computer system will be derived through the usage modeling activity. In Figure 7.1 the first possibility with an explicit *Library* role is chosen.

Figure 7.2 illustrates the hierarchy of activities involved in the library enterprise. The activities at the leaves of this hierarchy represent the starting point for the usage modeling activity. Specifically, the smallest-grained activities (at the leaves of the tree) constitute the starting point for the functional analysis of the system. They reflect the important tasks for the potential system users. Therefore, they are called user tasks in KobrA.

KobrA does not prescribe how the enterprise process hierarchy is developed. Typically, business process diagrams are created in a manner very similar to activity modeling, and the identified activities are collected in a process diagram.

Ideally every activity in the enterprise process hierarchy should be specified by an activity specification, but if not, at least the user tasks for which the system is responsible should be. Table 7.1 shows such a specification for a user task.

Another important goal of the enterprise model is to capture the business or enterprise rules that relate the various activities and concepts together. These are recorded in two places:

1 invariants (constraints) on the enterprise concept diagram

2 assumes clauses, result clauses and rules in the activity specifications.

7.1.2 Structural model

Whereas the goal of the enterprise concept model is to describe the concepts of importance at the enterprise level, the purpose of the structural model is to describe the nature of the classes and relationships, and also possibly objects, which characterize how the system, as a computer system, interacts with its environment. It

Table 7.1 Activity specification for user task loanItems

Name	**loanItems**
Description	After identification of an Account and several Items, the LibrarySystem registers Loans of these Items to the Account.
Constraints	This should be computer supported.
Rules	An Item is loanable if it is not an item that must always stay in the library (e.g. antique books) and not loaned and reserved by another user.
Returns	Loans
Result	For each Item that is loanable to the account, a Loan has been created and stored to register the fact that the Item is loaned to the Account.

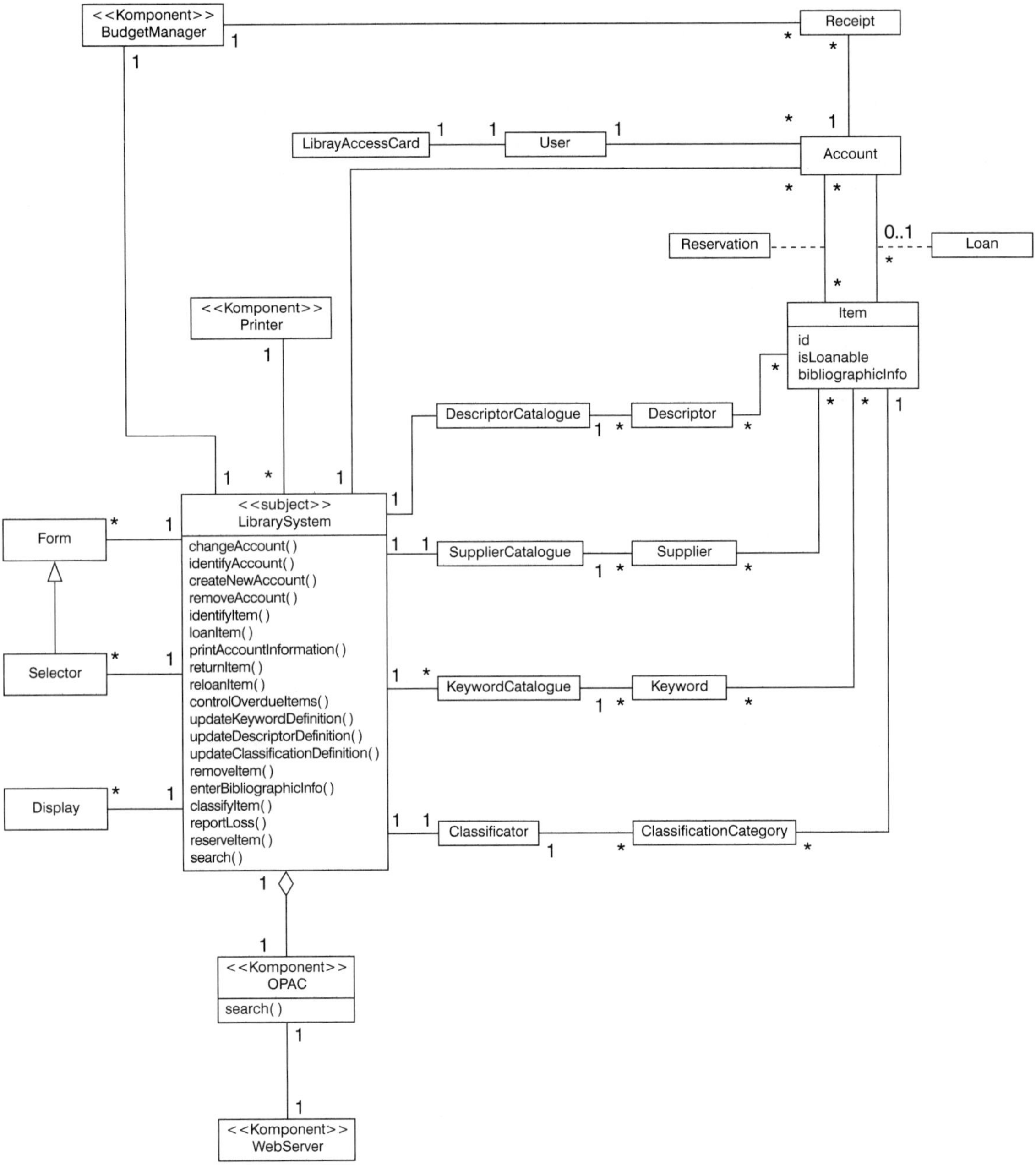

Figure 7.3
LibrarySystem context realization class diagram

uses precisely the same artifacts as the structural model for a normal Komponent realization model described in Chapter 5, namely, class diagrams and object diagrams. Since it is unusual to have more than one class diagram or one object diagram we describe these artifacts as if there is only one of each.

7.1.2.1 Class diagrams

The realization class diagram describes the classifiers that characterize the system's working environment, and the types of information that it needs to interact with this environment. Figure 7.3 illustrates the class diagram for the context realization of the library system.

It shows the computer system to be developed as a Komponent using the <<Komponent>> stereotype. Notice therefore that the name of the class representing the library has changed from *Library* to *LibrarySystem*. It also shows other computer systems, IT devices, and human actors with which the system interacts, using actor and Komponent stereotypes where appropriate. It also shows the major entities involved in any interactions. The entities as well as the roles are taken from the enterprise concept diagram while the IT systems and devices are added as more detail is developed.

One of the main things that distinguishes context realization from normal realization is the consideration that must be given to user-interface artifacts. The nature of the user/system interface is a critical aspect of how the environment expects the system to behave. An important part of the context realization class diagram is therefore a definition of the UI artifacts through which the system and its users interact. The library system uses three very high-level abstractions of UI artifacts:

- **Forms** – windows that allow information to be input and options to be selected, before the information or option is sent to the system
- **Displays** – windows which display information generated by the system
- **Selectors** – a specific kind of form which allows objects from a set of objects stored by the system to be selected.

In Figure 7.3, the corresponding classes appear on the left-hand side of the context realization class diagram. These capture the abstract artifacts that the library system uses to interact with its users, and as discussed in Section 7.1.5, are used in the interaction models to represent the information exchanged over the UI.

7.1.2.2 Object diagrams

As with a regular realization, object diagrams can be used to show a typical configuration of instances in which an instance of the system is required to operate.

Figure 7.4 shows an object diagram for the *LibrarySystem* context realization. Object diagrams are intended to focus on the relatively static or permanent instances, rather than on capturing transient instances such as GUI artifacts or data items. The boundary can be fuzzy, but generally speaking this means that

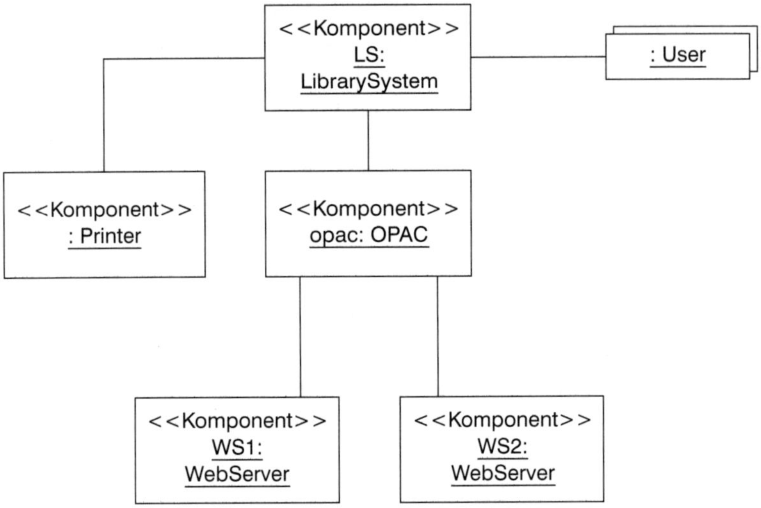

Figure 7.4 Library context realization object diagram

the diagrams should contain instances of Komponents rather than informational classes. However, the multi-object notation can be used to show important actors, such as users in this example.

The figure shows that an instance of the *LibrarySystem* Komponent is expected to interact with one *Printer* and one *OPAC* Komponent. It will also interact with numerous users during its lifetime. This object diagram can be viewed as capturing the structural architecture of the environment in which instances of the *LibrarySystem* Komponent are expected to operate. It is therefore similar to the context diagrams that were recommended in traditional software development methods.

7.1.3 Activity model

The activity model plays the same role in context realization as it does in a regular realization, namely, it describes the nature and relationships of the activities that have a bearing on the operations of the system. In particular, it describes a nested hierarchy of activities that explains how user tasks are realized in terms of one or more operations of the system.

7.1.3.1 Activity diagrams

The highest-level activity diagrams in the activity model usually correspond to the user tasks in the enterprise process diagram that are the responsibility of the system. Figure 7.5 illustrates an activity diagram for the *loanItems()* user task identified in the enterprise process model in Section 7.1.1. It indicates that the *loanItems()* activity is comprised of two subactivities *loanItem()* and *printAccountInformation()*, with the former being repeated multiple times for each item to be loaned.

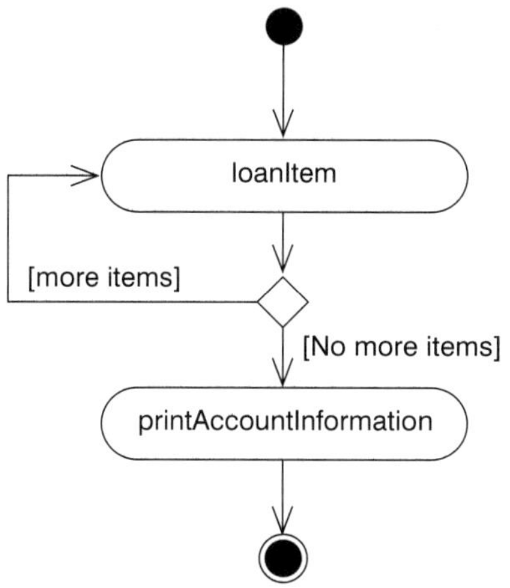

Figure 7.5
loanItems activity diagram

Figure 7.6 continues the nesting by showing the sub-activities making up the *loanItem()* subactivity from Figure 7.5.

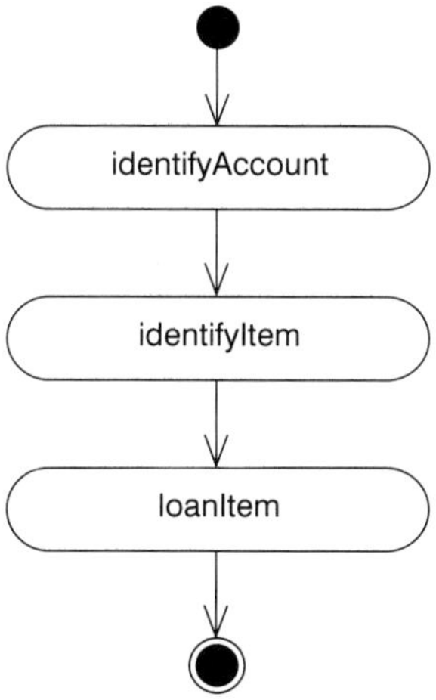

Figure 7.6
loanItem activity diagram

The precise nature of the activity diagrams near the bottom of the activity hierarchy depends on the amount of detail one wishes to include about the user interface. Most modern graphical user interfaces adopt a "two-phase" approach to the servicing of a request by the user. When a user requests a service (e.g. opening a file in Microsoft Word) they are first provided with a form, or dialogue box, to fill in. Only when this form has been filled in and sent back to the system (e.g. "OK" has been clicked), is the actual request satisfied. This means that many user services actually involve two system operations – one dealing with the provision of the appropriate form, and the other with the actual implementation of the service.

It is of course possible to treat all system operations uniformly, and to describe this level of detail within the activity model and/or interaction model. However, since this essentially doubles the number of system operations of a system (one for sending a form and one for processing it) it can often obscure the interesting properties of the system. Since all the form-sending operations are basically the same, it can be tedious to model them all explicitly. In general, therefore, there are three strategies for handling the details of GUI interactions with the user:

- **Explicit form-sending.** Form-sending operations are explicitly modeled. This means they are regarded as normal operations of the systems and should therefore appear in all the appropriate diagrams.
- **Implicit form-sending.** The form-sending mechanism is simply assumed as a service of the GUI, so no explicit form sending operations are required. Users are regarded as automatically having forms available and can start an interaction with the system on the basis of a completed form.
- **Parameterized form-sending.** The set of form-sending operations are abstracted into a single parameterized operation (e.g. sendForm()) which is parameterized with a flag to identify what type of form is required. This can be a useful compromise between the two extremes above.

The form of the activity diagrams near the bottom of the activity hierarchy is determined by which of these approaches is taken. For example, if the implicit form-sending assumption is used, the activities shown in Figure 7.6 would constitute system operations, and would thus represent the leaves of that part of the activity hierarchy. For uniformity with the Komponent realization activity described in Chapter 5, Figure 7.6 could therefore be presented in the form of a swim-laned activity diagram, such as that shown in Figure 7.7 in which the subactivities are assigned to the objects involved, in this case the *ServiceLibrarian* and the *LibrarySystem*.

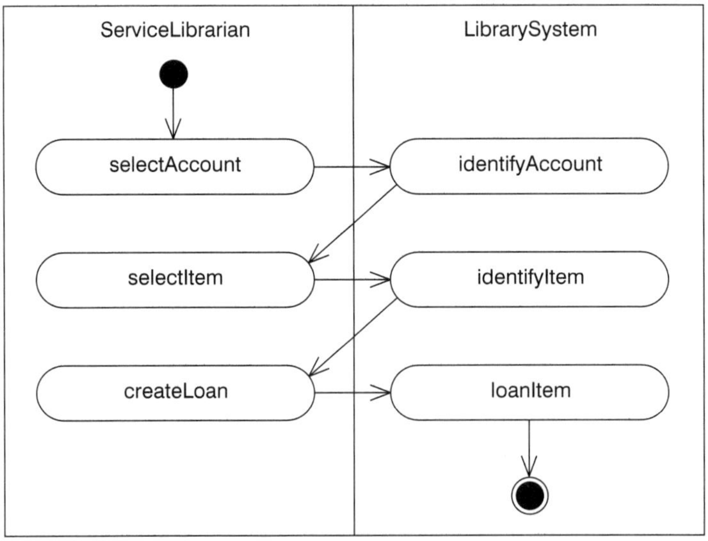

Figure 7.7
loanItem() activity diagram with implicit form-sending

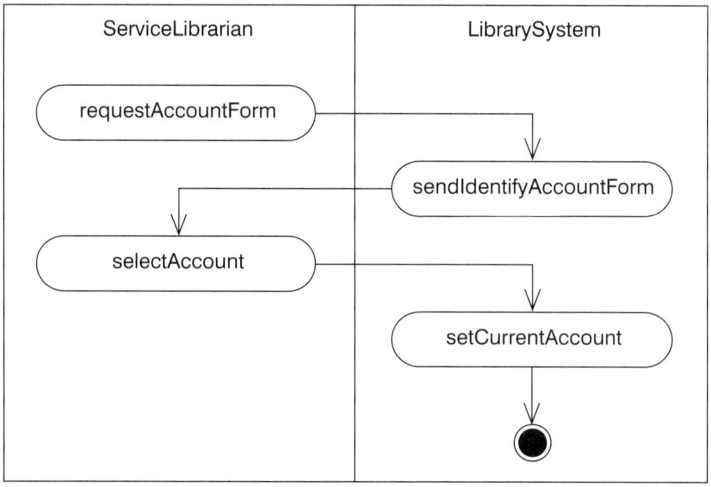

Figure 7.8 identifyAccount() activity diagram

If the explicit-form sending model is used, however, the activities in Figure 7.6 would not be the leaves of the activity hierarchy, and further lower-level diagrams of the form illustrated in Figure 7.8 would be developed. The activities in this diagram now represent leaves of the activity hierarchy, and the activities allocated to *LibrarySystem* represent its system operations.

7.1.3.2 Activity specifications

As in Komponent realization, activity specifications can be used to document the logical effects of activities in the activity hierarchy. System operations need operation specifications as well, of course, but these are contained in the system specification rather than the context realization.

In view of the above discussion, three levels of activities can in general be distinguished [Paech98a, Paech98b]:

- User tasks capture the basic responsibilities of the user. The description clause in the specification typically identifies the major subactivities as they are depicted in free-activity diagrams (see Figure 7.5).
- UI activities capture the interaction between users and the system. The description clause in the specification typically identifies the UI artifacts involved and the transformations on the data as they are depicted in the swim-laned activity diagrams or interaction diagrams.
- System operation activities capture the detailed effects on the data. The description clause in the specification identifies the classes involved. These activities are described during Komponent specification.

Table 7.1 gives an example of a specification for the *loanItems()* activity.

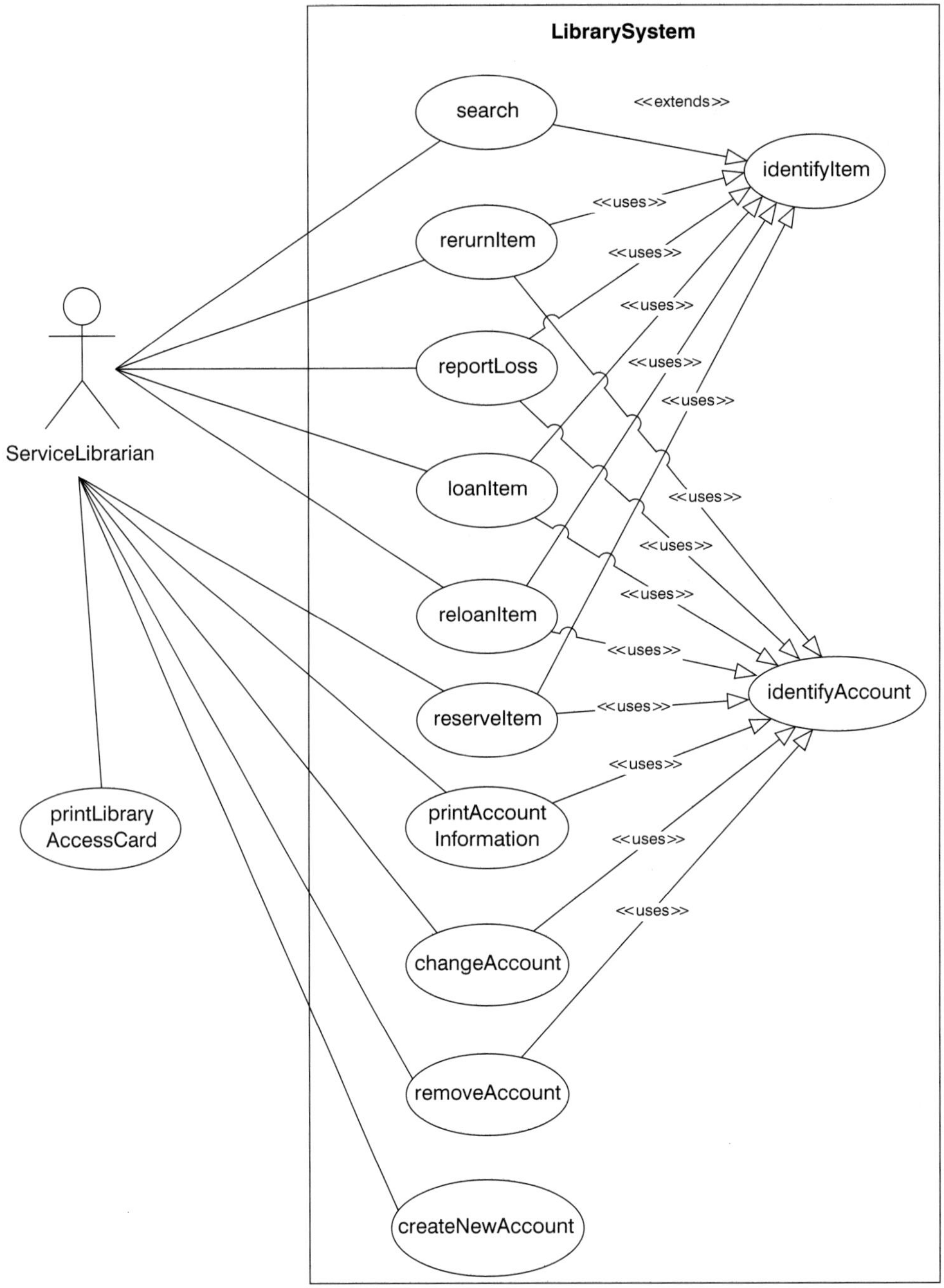

Figure 7.9
LibrarySystem use-case diagram

7.1.3.2 Use-Case model

Although the term "use-case" *per se* is not a core concept in KobrA, use-cases have a direct counterpart in KobrA, namely, user tasks which encapsulate a self-contained service. Developers who are familiar with and prefer the use-case idea can therefore include use-case diagrams as part of the activity model. Use-case diagrams document the relationships between different use-cases, the system and actors. They also show the containment relationships between activities in the activity hierarchy, and therefore, provide a useful overview of the system boundary and which activities are important for which users.

Figure 7.9 illustrates a use-case diagram based on the activity hierarchy developed under the implicit form-sending assumption discussed above. In accordance with the usual concept of use-cases [Jac92], activities are selected as use-cases from this hierarchy if they encapsulate a well-defined service which the user can complete in one visit to the computer. As well as showing the various kinds of use-cases, a use-case diagram shows actors, the system boundary and relationships between actors and use-cases.

Following the spirit of KobrA, the concepts shown in a use-case diagram should be consistent with the other diagrams. Figure 7.9 shows the *loanItem()* activity from Figure 7.5 and its subactivities *identifyItem()* and *identifyAccount()*, from Figure 7.7 all as use-cases because they are all useful, self-contained services in their own right. However, the <<uses>> dependencies between them captures the fact that *loanItem()* is logically one level higher in the activity hierarchy than the other two.

7.1.4 Interaction model

Just as for a normal realization, a context realization can contain interaction diagrams that tie the activities and data items together. Interaction diagrams can provide an alternative view of the way in which user tasks are realized in terms of system operations. For historical reasons, sequence diagrams are

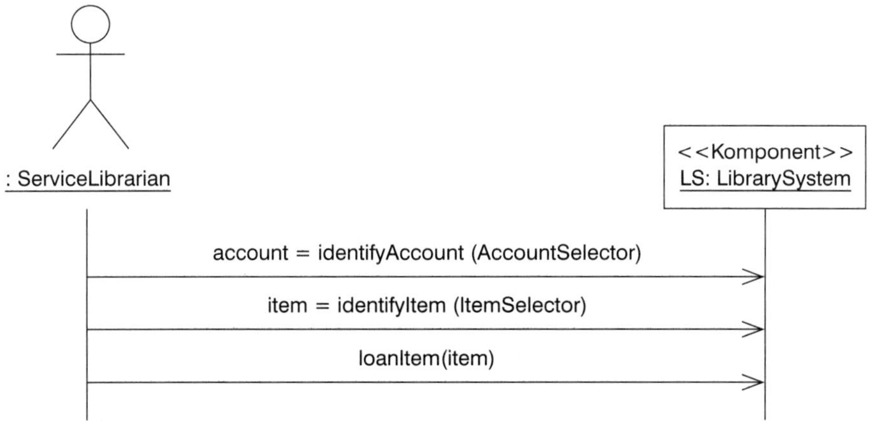

Figure 7.10
loanItem() sequence diagram with implicit form-sending

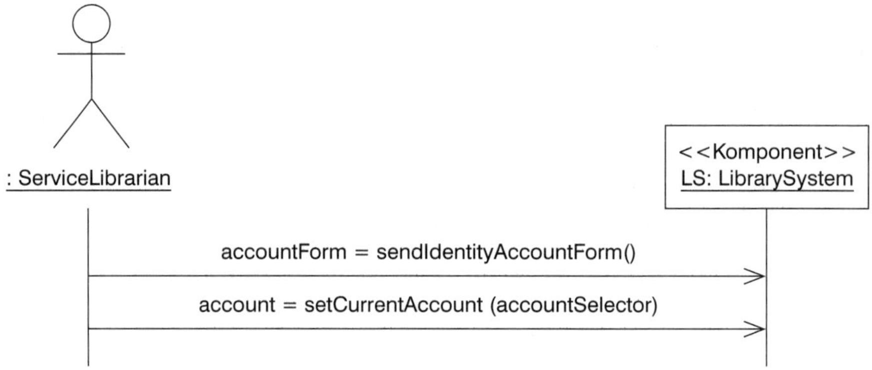

Figure 7.11 IdentifyAccount sequence diagram

typically used for context realization interaction models, and collaboration diagrams for normal realization interaction diagrams, but this is only a convention. Any kind of interaction diagram can be used.

Figure 7.10 illustrates the sequence diagram for the *loanItem()* activity which corresponds to the activity diagram in Figure 7.7. Figure 7.11 illustrates the sequence diagram corresponding to Figure 7.8 based on the explicit form-sending assumption.

7.2 Context realization process

As illustrated in Figure 7.12, the activities involved in context realization have much in common with those for a normal realization. The exception is the enterprise-modeling activity, which is responsible for the front-end requirements elicitation.

As can be seen from Figure 7.12, the context realization activity employs the same model of activity and data independence/integration as the other processes. In fact, it represents the top of the DNA spiral discussed in the previous chapter. At first, activities and data are considered separately and they are gradually decomposed and integrated to create a unified object-oriented view.

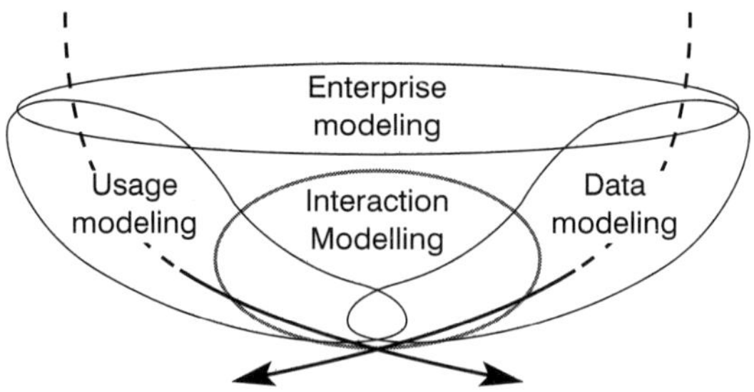

Figure 7.12 Context realization activities

7.2.1 Enterprise modeling

As explained above, KobrA is not very prescriptive about the nature of the enterprise-modeling process. This is deliberate since it is the intention that KobrA be usable with the user's preferred enterprise modeling approach. In fact, often this activity is not carried out at all. However, this places the onus on the usage and structural-modeling activities to identify the activities and entities the system should support. Furthermore, without such an explicit activity business-IT alignment is not properly conducted resulting in suboptimal, not to say *ad-hoc*, system support for the enterprise. Therefore, KobrA recommends an explicit enterprise modeling step such as that of Jacobson [JEJ94], Marshall [Marshall00] or Allen [Allen00] for e-business.

7.2.2 Structural modeling

Like structural modeling within Komponent realization, the purpose of structural modeling within context realization is to develop the structural model. Its main input is the enterprise concept diagram.

7.2.2.1 Class diagrams

The process for developing the context realization class diagrams can be summarized as follows:

1. **Select enterprise concept diagram elements.** The enterprise concept diagram represents a good starting point for the realization class diagram. All roles interacting with the system to be built as well as all entities involved in this interaction are taken from the enterprise-concept diagram. This includes, of course, the entities the system will be responsible for.
2. **Add new classes.** The enterprise concept diagram only shows the major roles and entities. During usage modeling further roles and entities will be identified. The entities typically correspond to information or material exchanged between the activities. For the structural model and the activity model to be consistent, every entity used in the activity diagrams must be an instance of a class type appearing in the realization class diagram. Therefore, the names of these classes, their relationships, and their attributes must be added, if they are not already defined. Through interaction modeling even more entities are uncovered, in particular UI artifacts. Again, these must be added to the structural model.
3. **Add operations.** During usage or interaction modeling the system operations are identified. These are added as operations to the system Komponent. It is neither necessary nor desirable to add operations to other Komponents or entities of the structural model, since the major goal of context realization is to understand the context and not to prescribe its behavior on an operational level. The business processes identified during enterprise modeling and the activity diagrams identified during usage modeling suffice to describe the behavior of all other roles involved in the enterprise. Similarly, operations for the entities are not used, but attributes suffice to describe entities.

7.2.2.2 Object diagrams

Object diagrams can be useful in the structural modeling activity to capture an overview of the instances of Komponents and actors with which an instance of the system interacts. The process for creating them is the same as for regular realizations (see 5.2.1.2).

7.2.3 Usage modeling

The purpose of usage modeling is to create the activity models, similarly to activity modeling within Komponent realization. This activity is termed "usage modeling" rather than "activity modeling" to emphasize the fact that it focuses on the way in which the system is used by its environment.

As discussed above, there is no sharp boundary between the enterprise model and the activity model. They both contain descriptions of activities in the environment of the system that affect the way the system is expected to operate. The difference between them is one of emphasis rather than substance. When performing context realization in practice, therefore, the usage and structural modeling have to pick up where enterprise modeling left off. For example, if the specification of the user tasks which the system has to support (such as *LoanItems*) is not available, these must be created. Alternatively, the enterprise model may already contain these specifications, and so they do not have to be created in the usage modeling activity. For the purposes of this section we will assume the completion of an enterprise modeling activity resulting in the artifacts identified in Section 7.2.1.

As within activity modeling in general, two kinds of activity diagrams can be created during usage modeling: swim-laned activity diagrams with allocated activities, and free-activity diagrams with free activities. The first four steps of the process below describe the creation of free-activity diagrams; step 5 describes the derivation of swim-laned activity diagrams from free activity diagrams; and the last two steps describe the creation of the activity hierachy and the activity specifications:

1. **Select enterprise activities.** Identify which of the user tasks in the enterprise process model are supported by the system. These represent the starting point for usage modeling.
2. **Identify subactivities.** For each user task, identify the subactivities to fulfill the user task specification. They can be identified by looking at the description clause or at the information or material entities named in the receives, changes or returns clause of the specification. Each read access or change to an entity is a candidate for an activity.
3. **Identify flow.** For each subactivity, identify the information or material needed in the course of the activity's execution and the resulting information or material. These entities need to be consistent with the specification structural model. For each activity, determine predecessor and successor activities providing or consuming the respective entities. To emphasize sequentiality constraints introduce control flow.
4. **Refine free-activity diagrams.** Repeat steps 1 to 3 until each subactivity is either completely within the responsibility of the system to be built, or within the responsibility of another role.

5 **Create swim-laned activity diagrams.** Determine the Komponents involved in the execution of the user task. They need to be consistent with the structural model of the context realization developed in parallel. The subactivities are allocated to the Komponents. Typically, the subactivities dealing with certain entities are allocated Komponents or roles responsible for these entities. Sometimes, Komponents or roles without special responsibilities for entities have to be introduced. Their main purpose is to coordinate other Komponents or roles in the activity execution.

6 **Create activity hierarchy.** In parallel with the development of the activity diagrams all subactivities are collected into an activity hierarchy starting with the user tasks supported by the system.

7 **Create activity specification.** In parallel to the development of the activity diagrams activity specifications are created. A specification is only required for allocated activities.

7.2.3.1 *Use-case modeling*

As is evident from the preceding subsections, use-cases *per se* do not feature directly in the context realization activity. It is perfectly possible to perform a KobrA project without ever considering use-cases. However, this is more due to terminology rather than substance. The key ideas embodied by modern use-case modeling techniques are available in KobrA.

In principle, any user tasks, UI activities or system operations could be modeled as a use-case in a use-case diagram. However, the usual approach is to show only use-cases which encapsulate a well-defined service that a user can complete in one visit to the computer. Any activity from the activity hierarchy with this property can therefore be viewed and modeled as a use-case.

The modeling artifacts used in traditional use-case modeling techniques are also available in a fairly direct form. Three main artifacts are usually described in the use-case literature: use-case diagrams; textual use-case specifications and sequence diagrams (variously called use-case instances or scenarios). The information conveyed by use-case diagrams is already available in the structural and activity models of a KobrA realization, but use-case diagrams are still acceptable for developers who prefer them. The textual use-case specifications correspond to activity specifications, and the sequence diagrams, of course, are found in the interaction model.

The essential development activities found in use-case modeling approaches closely resemble those in the left-hand branch of Figure 7.12. The initial focus of use-case modeling is the purely functional decomposition and specification of a high-level use-case into lower-level use-cases, as in usage modeling, while the later focus is on the documentation of these lower-level activities in terms of sequence diagrams, as in interaction modeling.

7.2.4 Interaction modeling

The goal of this activity is to integrate the activity and structural models of the system's environment, and to describe precisely how the entities in the system's environment interact with it. There is no clear boundary between usage modeling and interaction modeling. For simple user tasks, an interaction model can be directly created from the user-task specification. For complex user-tasks, activity diagrams are developed and interaction diagrams can be derived from them. Without explicit usage modeling, the onus is on interaction modeling to identify subactivities and the information and material flow between the system users and the system. In the following we only describe the steps necessary to define system operations, the control flow and UI message exchange. Thus, we stipulate the activities corresponding to UI use-cases as well as activity diagrams for these activities as input to interaction modeling. The output is a set of interaction diagrams. By convention, sequence diagrams are typically used in context realization.

The basic steps involved are as follows:

1. **Identify actors.** For each UI activity, take the actors from the corresponding activity diagram. These are the actors of the interaction diagram.
2. **Identify UI interactions.** For each UI activity, design the detailed UI interactions in terms of forms and displays. The messages exchanged correspond to the input and output for the subactivities of the activity diagram of this UI activity. The control flow between the subactivities is added so that every interaction between the system and the user is triggered by the user. For each message, the forms and displays used are described. These forms and displays must be part of the structural model.
3. **Identify the system operations.** Each message corresponds to one system operation.

7.2.5 Quality control

Since a context realization has all the artifacts of a normal Komponent the inspection, quality modeling, and quality-control activities can be performed in the same way. The only difference to the inspection of a normal Komponent is that since context realization has no specification, the realization-consistency constraints make no sense. The two types of consistency rules that should be checked in an inspection are therefore:

- realization intra-diagram rules (i.e. rules on a single model)
- realization inter-diagram rules (i.e. consistency rules between realization models).

In contrast with a normal Komponent realization, there is no value in generating test cases for the context realization since, by definition, it is not part of the system to be developed. The context realization can be useful as a means of developing the system test harness, but this is beyond the scope of this book.

Instead of test-case development, however, other forms of quality assurance are applicable that do not make sense with normal Komponent realizations. The main quality attribute of importance for the context realization is the satisfaction of the procurers and the users of the system. There are several possible ways of ensuring this satisfaction through, for example, workshops, early prototyping, or task analysis. However, their description is also beyond the scope of this book. A first introduction can be found, for example, in [Macaulay95].

Public containment 8

Chapter 6 described a form of containment known as private, or black-box containment, in which sub-Komponents are completely encapsulated by their parent. This corresponds to the semantics of private ownership for UML packages. Private containment is the most fundamental and commonly used form of containment in KobrA.

However, another form of containment is also useful in component-based development – public (or white-box) composition. With this form of containment, a sub-Komponent is visible at the interface of its parent, allowing the sub-Komponent to be accessed by other Komponents outside the parent. From the perspective of UML packages this corresponds to public ownership.

Public containment, which concerns the development time relationship between Komponents based on their module-like properties, has a counterpart at run time. Public composition is a relationship between Komponent instances in which all the properties of UML composition hold except privacy. In other words, one instance is still viewed as being a part of another (the composite), and still has its identity and lifetime bound to the composite, but is visible to other external Komponents.[34]

Public composition is useful because it allows Komponents to be organized in ways that more closely match the typical usage styles of physical component technologies (e.g. CORBA, .NET/COM+, J2EE/EJB). Public containment, on the other hand, is useful because it is a common way to support public composition, and also because it allows the shape of Komponent containment trees to more naturally match the run time organization of Komponent instances.

34. Public composition achieves the same effect as public instance variables, or data members, in programming languages like Java and C++. They allow external classes to directly access a class's features.

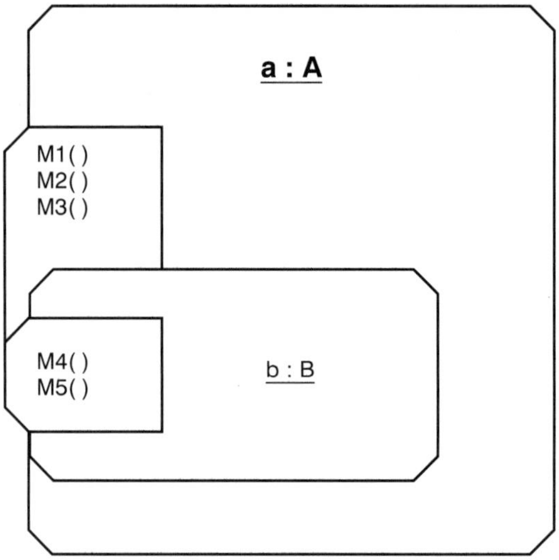

Figure 8.1
Public composition example

In the following subsection, we explain the motivation for, and the properties of, public composition and public containment in more detail, and clarify the differences between them. We then explain their effects on the specification and realization artifacts described in the previous chapters.

8.1 Public containment versus public composition

Although related, public composition and public containment are distinct concepts that can be used either together or separately. The former changes the nature of the interaction of Komponent instances at run time, while the latter changes the nature of Komponent dependencies at development time.

8.1.1 Public composition

Public composition is a variant of the regular composition relationship (as defined in the UML) in which the part is visible outside the composite. As such, it is a relationship between run time Komponent instances. A schematic representation of the semantics of public composition is provided in Figure 8.1. This shows that *a*, an instance of Komponent *A*, is composed of a part *b*, an instance of Komponent *B*, and that *b* is visible to clients of a. In other words, the operations made available to clients of *a* include the operations of *b*.

This is one way in which it is possible for services supplied by a Komponent to contain the services of one or more smaller Komponents.[35] In UML terminology, this is one way in which a Komponent can support multiple interfaces. Figure 8.2 shows how the encapsulated services provided by public parts in the

35. The other way is by Komponent specialization explained in Chapter 9.

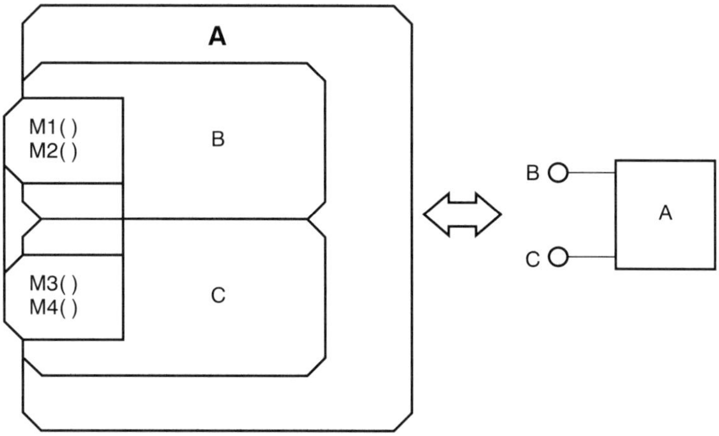

Figure 8.2
Komponent supporting multiple interfaces

composite Komponent can be understood in terms of the UML interface concept. An instance of the Komponent *A* is composed of two public instances, one of the Komponent *B* and one of the Komponent *C*. In UML terms, this can be understood as Komponent *A* supporting the interface *B* and the interface *C*. Due to the principle of parsimony, KobrA does not include the "interface" as a formal concept, since the capabilities that it provides are already supported by the specification concept (when used with public composition). However, users who prefer interface diagrams of the kind shown in Figure 8.2 are of course free to use them, with the semantics illustrated in the figure.

An important corollary of the fact that the specification of a Komponent with public parts incorporates the specifications of those parts is that the composite Komponent is conformant with (i.e. can be substituted for) any of its public parts. In terms of Figure 8.2, wherever an instance of Komponent *B* or *C* is expected, an instance of *A* can be supplied in its place. This is because an instance of *A* contains a publicly visible instance of *B* and of *C*.

It is important to realize that the only difference between public composition and private composition is the visibility of the parts. All other characteristics of composition are the same including the fact that the lifetime of the part is tied to that of the composite, and the fact that the composite creates the part. These properties distinguish Komponent instances with public parts from, for example, factory Komponents which also create instances and export them (or make them visible to) external components. In the case of factory Komponents, however, the lifetime of the created instances are not tied to those of the factory. On the contrary, a factory Komponent only has a relationship to its products for a short time.

8.1.2 Public containment

Public containment is the analog of public composition at development time. In terms of the module facet of Komponents, it means that a sub-Komponent has public visibility rather than private visibility. As illustrated in Figure 8.3, public composition is represented by the stereotype <<public>> adjacent to the containment relationships. Figure 8.3 therefore indicates that of the three sub-Komponents of Komponent *P*, *R* and *S* are public sub-Komponents while *Q* is a regular private sub-Komponent.

Figure 8.3
Public containment

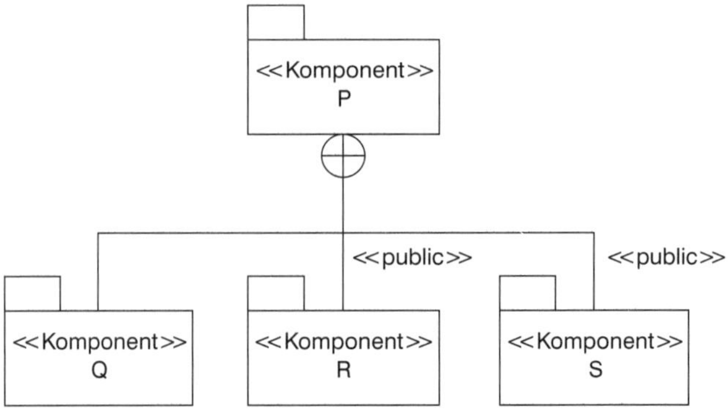

As explained in Section 3.4.2, public containment extends the normal tree-visibility rules with an additional rule that allows a Komponent that has visibility of a Komponent *K* to also have visibility of public sub-Komponents of *K*, in a recursive manner. Thus, in terms of Figure 8.3, assuming that the tree in the figure is a sub-tree of a larger tree, any Komponent that has visibility of *P* also has visibility of *R* and *S*.

This is useful because it provides an alternative way of making Komponents visible for sharing. As explained in Chapter 6, with normal private containment the only way to make a Komponent available for sharing is to elevate it to a point in the tree where it is visible to all Komponents that wish to use it. However, this has the disadvantage that sometimes the new owner of the Komponent may have little logical relationship to the Komponent. Public containment avoids this problem by allowing a Komponent to be made shareable without changing its position. Thus, shared Komponents can be located with (i.e. owned by) the Komponent with which they have the closest logical affinity.

To illustrate this principle, consider again the sample composition and containment trees from Figure 6.5. This figure is duplicated in Figure 8.4 for convenience. Recall that in order to allow instances of *V* and *W* to create instances of *X*, *X* has to be defined as their peer at the same level of the containment tree.

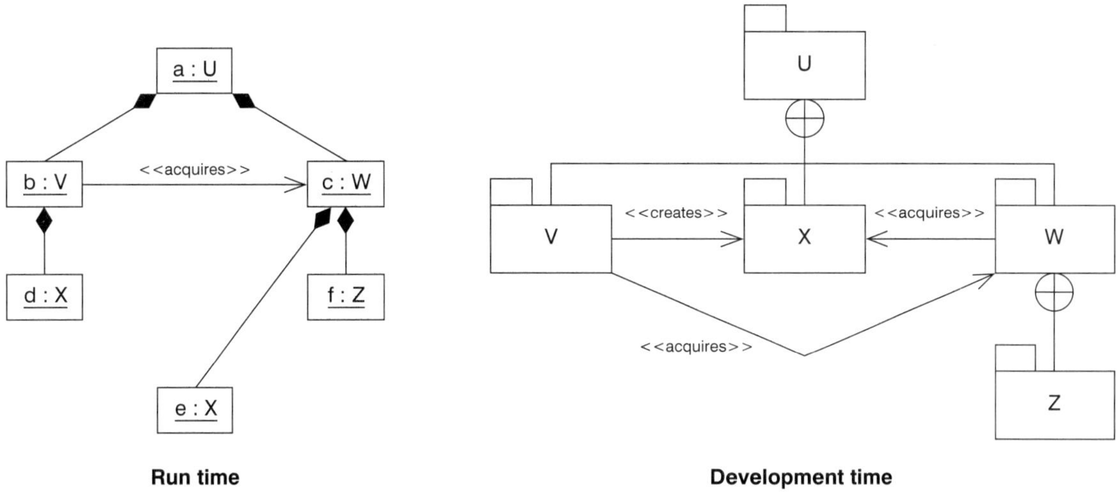

Figure 8.4
Composition and containment trees from Figure 6.5

Figure 8.5 shows a modified version of the containment tree in which the required sharing is achieved by public containment rather than by Komponent elevation. The run time configuration of instances is identical in both cases – only the containment tree is different. In Figure 8.4, Komponent *X* has been elevated to the same level as Komponents *V* and *W* that wish to instantiate it. In Figure 8.5, on the other hand, Komponent *X* is made a public sub-Komponent of *W*, thereby making it visible to *V* also. This has the advantage that the shape of the containment tree more closely matches that of the composition tree, and *X* does not need to be moved from the Komponent it is originally owned by, or has more natural affinity with. Chapter 10 presents some architectural patterns for exploiting these properties of public containment.

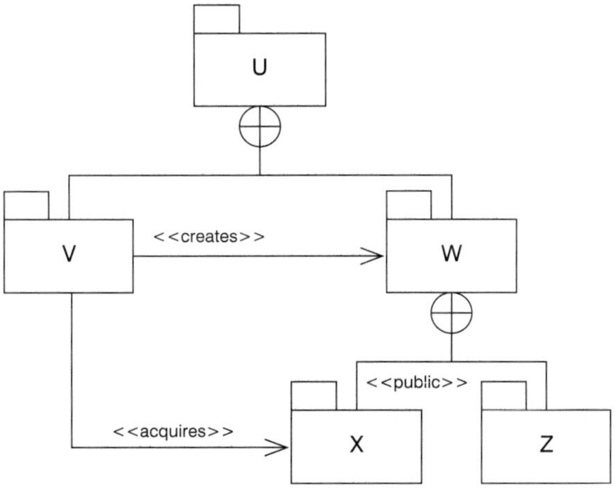

Figure 8.5
Sharing through public containment

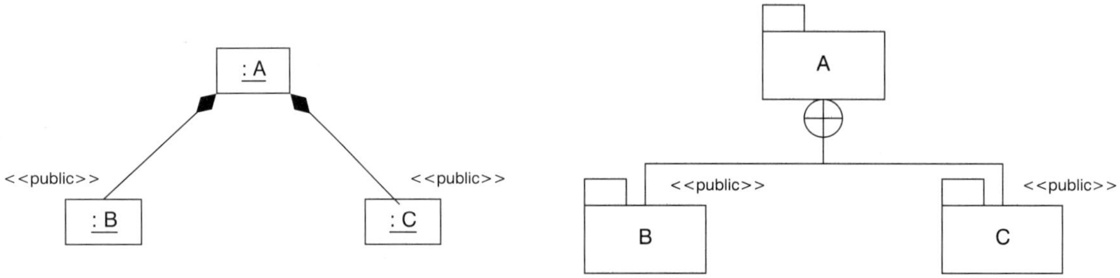

Figure 8.6
Public containment for public composition

As explained above, public composition and public containment are both useful independently of each other. However, they are also naturally complementary. The most usual way of providing the visibility needed to support public composition is to use public containment, and the most common use of public containment is to support public composition. Figure 8.6 shows a containment tree using public containment to support the public composition in Figure 8.2. The right-hand side shows the containment tree, with Komponents *B* and *C* as public sub-Komponents of *A*, and the left-hand side provides a tree-based representation of the public containment in Figure 8.2. The correspondence between the two is obvious.

8.2 Artifacts

Given that public composition and public containment are independent of one another, they have different sets of effects on the Komponent modeling artifacts described in the previous chapters. These effects are explained in the following two subsections, first the effects of public composition, then the effects of public containment.

8.2.1 Public composition

Composition arises when an instance of one Komponent, the composite, is "made up of" an instance of another Komponent, the part. In the UML, composition is viewed as a form of aggregation with strong binding between the composite and the part. The UML requires that a part be strictly encapsulated by (i.e. private to) the composite, but this particular requirement is relaxed in KobrA for public composition.

To establish composition between Komponent instances at run time, a corresponding composition relationship must exist between their classes at development time, one playing the role of the composite and one playing the role of the part. Composition is a form of clientship relationship, with the composite acting as the client and the part as the server. The Komponent defining the role of the part may or may not be a sub-Komponent of the composite. If it is, it must be a public sub-Komponent of the composite (i.e. related to the composite by public containment),[36] otherwise clients of the composite could not see it.

The use of public composition does not alter any of the fundamental consistency rules described in the previous chapters of the book. The main effect of public composition is to change the nature of the information displayed in the various models and diagrams. Moreover, the scope of these effects is limited. When one Komponent, *P*, is a public part of another composite Komponent, *C*, only the specification of *C* and the realizations of the clients of *C* are affected (that is, have a different form than they would have had). The part, *P*, is not affected at all, because as a server it has no knowledge of whether it is being used as a public or private part.

8.2.1.1 Composite specifications

Unless they are acquired from outside, the servers of a Komponent are normally encapsulated within its realization. The main difference in the case of a Komponent, *C*, with a public part, *P*, is that *P*'s existence and properties must be made visible within the specification of *C*. This is to allow clients of *C* to gain access to *P*. All of the models and diagrams in a composite specification are potentially affected.

8.2.1.1.1 Structural model

The structural model of a composite specification must include a description of the structural properties of the public part by including the relevant aspects of its specification structural model. It is also possible for the specification object diagram to include a description of the public part's identity (i.e. the name by which it is known to the composite), by including a representative instance of the public part. In a sense this makes a segment of the architecture of the composite visible at its interface.

To illustrate how this works, suppose the *LoanManager* were a public part of the *LibrarySystem*. The specification of *LibrarySystem* therefore becomes a composite specification and includes information from *LoanManager's* specification, as illustrated in Figure 8.7. The class representing *LoanManager* is included in the class diagram, along with numerous supporting classes from *LoanManager's* specification class diagram. The presence of public composition is represented graphically by the fact that *LoanManager* is connected to *LibrarySystem* (the subject of the diagram) via a composition relationship.

Not all modeling elements in the part's specification class diagram are necessarily included in the composite's specification class diagram.[37] Acquired server Komponents need not be shown. Some of the methods of the part may also be omitted. The things that must be shown are the essential associations and the simple classes representing parameter types for the operations of the part.

36. One must be very careful with terminology when discussing composition. Although a part is a child of its composite in the run time composition tree, and thus could reasonably be called a sub-Komponent, we reserve the word sub-Komponent for a child in the development time containment tree.
37. Of course the model elements themselves remain in the package corresponding to the part Komponent. Only new presentation elements are included in the composite's diagram.

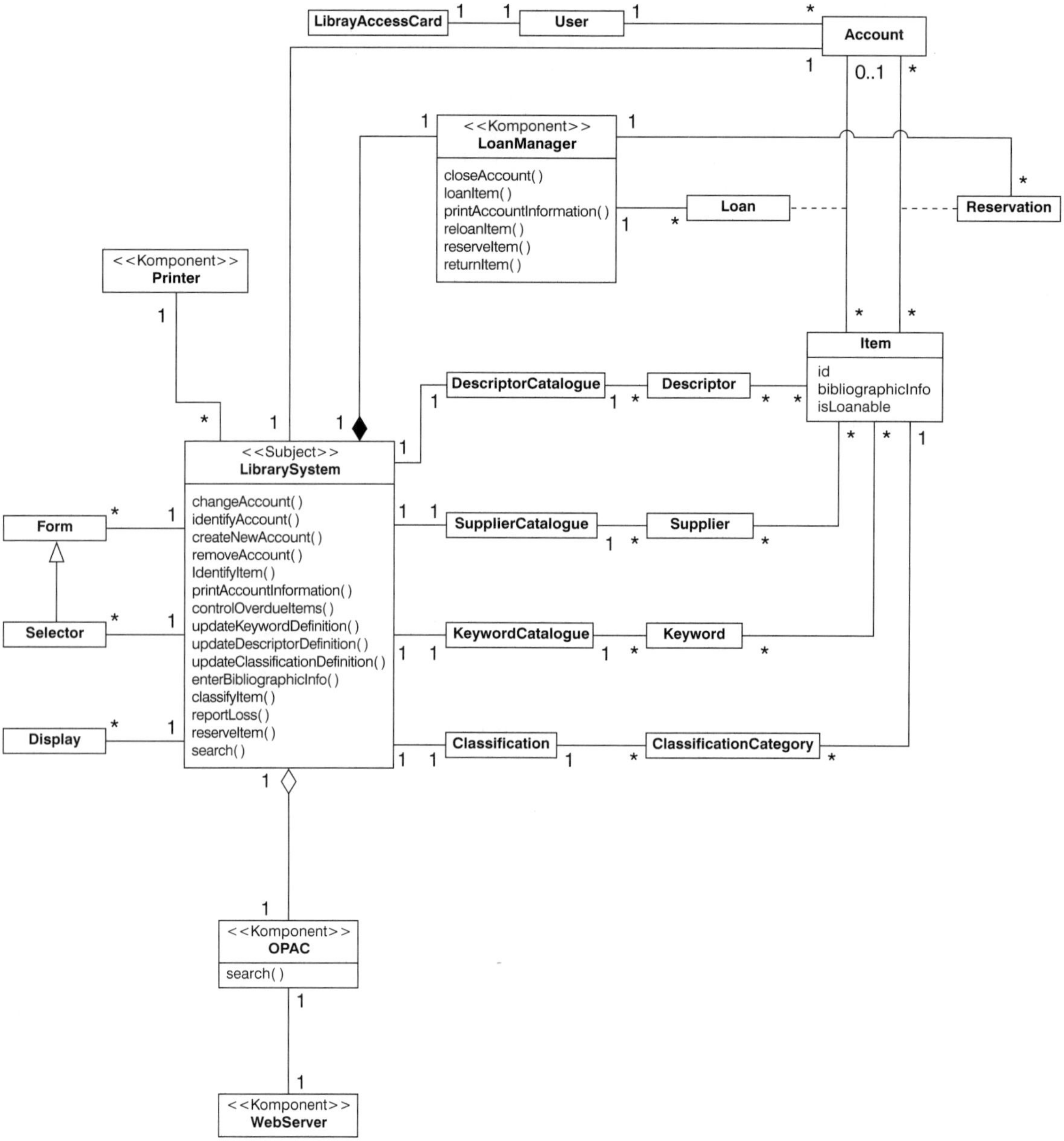

Figure 8.7
LibrarySystem with public LoanManager

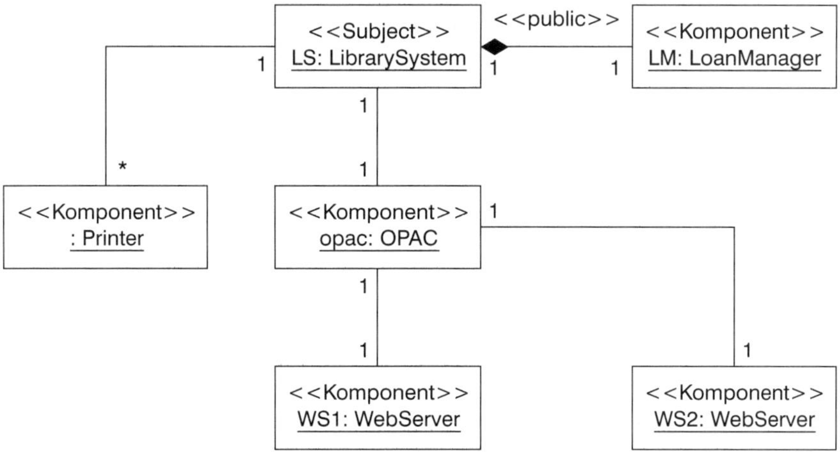

Figure 8.8 LibrarySystem specification object diagram with public LoanManager

The inclusion of servers acquired by the part Komponent in the specification of the composite depends on whether or not they are also acquired by the composite. If so, then obviously they are also included as acquired servers of the composite. However, it is possible for the composite to generate its own instance of a server needed by the public part (i.e. as a private part). When this is the case, the configuration of the parts of the composite is a matter for the composite's realization.

Only the operations of the private part that are supplied to clients of the composite need be included in the composite specification class diagram. Any other operations can be omitted. However, in contrast with simple client-server relationships, where only the operations invoked by the client are included in the client's class diagram (either specification or realization), in the case of a public part further operations not invoked by the client (i.e. the composite) can be included. This is because they may be intended for invocation by the clients of the composite, and therefore must be included in its specification just like its own operations.

Figure 8.8 illustrates how the object diagram of a composite specification can expose a public part's identify and role. Here the specification object diagram for the *LibrarySystem* indicates that an instance of the *LibrarySystem* has one instance of the LoanManager as a public part, and that it refers to it by the name *LM*.

8.2.1.1.2 Functional model

The functional model of a composite Komponent is affected in two ways. Again the basic well-formedness and consistency rules do not change. Only the information content changes with respect to the case when there is no public composition.

The first effect is that the number of operation specifications is likely to be fewer. Because the operations of the public part are visible to clients of the composite Komponent, the composite usually has to supply fewer operations itself. When a part is private, the only way in which the composite can allow its clients to benefit from the part is by providing "forwarding" operations that forward request to the part's operations. When the part is public, however, such forwarding operations become redundant. This is one of the main benefits of public composition. In the case of the *LibrarySystem*, for example, all of the operations supplied by *LoanManager* have to be supported by the *LibrarySystem* Komponent as well as the *LoanManager* Komponent if the latter is not a public part of the former.

The other effect is that a composite's operation specifications have visibility of the public part (since it is defined in the specification) and can thus refer to its operations and attributes. In this respect, a public part is treated in exactly the same way as an acquired server. An operation specification can refer to the instance via its name and specify that its logical attribute values are changed, or that its operations are invoked.

8.2.1.1.3 Behavioral model

The behavioral model of a composite Komponent is also potentially affected in two ways by the presence of a public sub-Komponent. The most common way is by the definition of a statechart table to define a mapping between the operations logically supported by the parent Komponent, and those supplied by the sub-Komponent. If there is a one-to-one correspondence between the two, then no statechart table is needed because it is obvious how invocations of an operation on *LibrarySystem* should be handled. However, if there are possible invocation events whose names do not precisely match the operation names of the part, a statechart table can be used to define the correspondence.

Statechart tables used in this way only establish a connection between the *events* at the level of the composite Komponent and *events* at the level of the part. However, it is also possible to establish a connection between *states* at the two levels. For example, publicly visible states of the composite could be composed of one or more states of its public parts. For example, the library system as a whole is in a state where it can loan books (library-level state) when the number of current loans is not equal to the number of loanable books (LoanManager-level state).

8.2.1.2 Client realizations

As mentioned above, the realizations of clients of a composite Komponent are also affected by the fact that it has public parts. The fact that the composite object has a public part has to be captured in the client's realization. In fact, the public part of the composite is treated no differently to any other server Komponent of the client. This means that it has to be represented as a class in the client's realization class diagram along with the other supporting classes and associations, and the invocation of its operations have to be modeled in the appropriate interaction and activity diagrams, just like those of any server of the client.

Figure 8.9 (a) illustrates the situation that is common when a composite Komponent *C* has a private part, *P*, with an operation that *C* would like to make

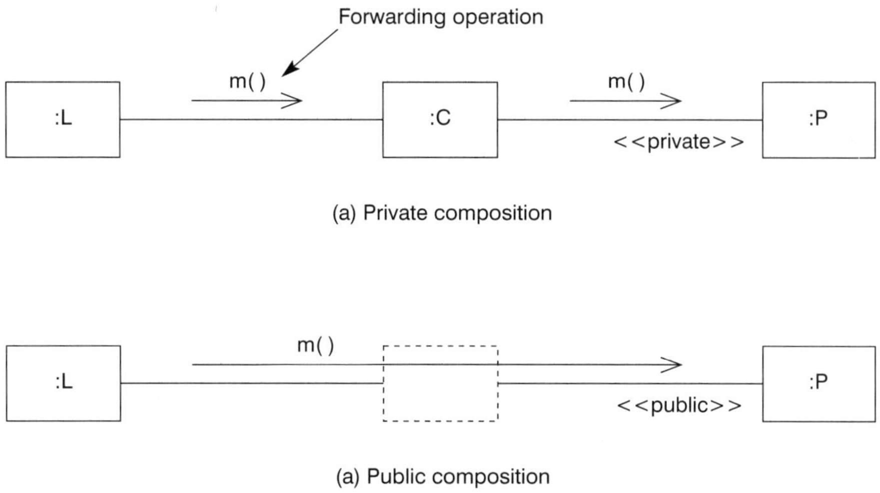

Figure 8.9
Client operation invocation with public composition

available to a client *L*. When *P* is a private part, *C* must define its own operation *m()* whose job is simply to forward requests to the corresponding operation *m()* of the private part. The client, *L*, thus invokes the operation *m()* of *C* which then invokes the corresponding operation *m()* of *P*.

Figure 8.9 illustrates how public composition can reduce the number of methods and invocations involved by allowing *L* to invoke the operation of *P* directly. Since *P* is now visible to *L*, *L* can invoke the operation *m()* of *P* directly. This not only avoids the redundant definition of methods at both the composite level and the part level, but also improves the run time performance.

Figure 8.10 illustrates how the sequence diagram in Figure 7.10 might be modified if *LoanManager* were a public part of *LibrarySystem*. The *identifyAccount()* operation is still handled by *LibrarySystem*, because its realization involves several

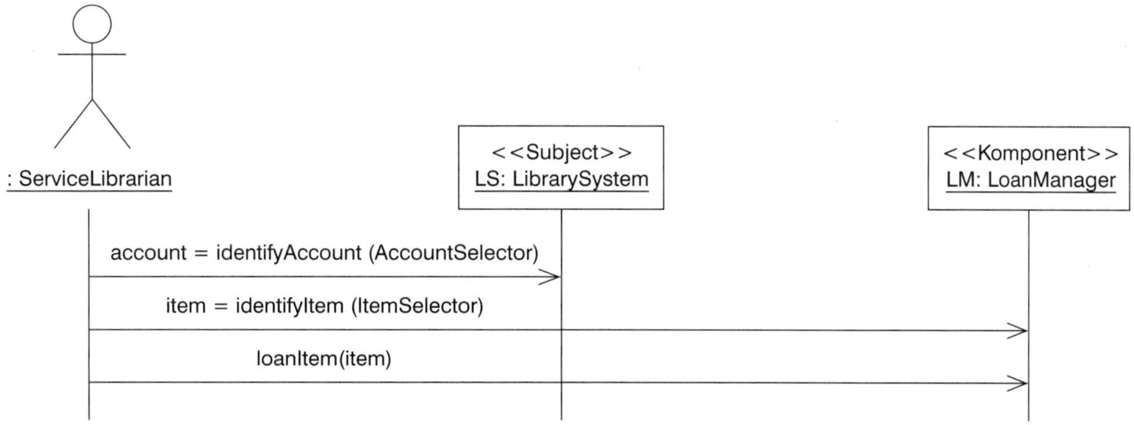

Figure 8.10
LoanItem() sequence diagram with public LoanManager

server Komponents in addition to the *LoanManager* instance *LM*. However, the *IdentifyItem()* and *LoanItem()* message are sent directly to *LM*, thereby obviating the need for *LibrarySystem* to have trivial forwarding operations.

8.2.2 Public containment

In contrast with public composition, the effects of public containment are not directly visible at the level of Komponent specifications and realizations and their diagrams. Instead, the effects of public containment manifest themselves in the overall shape of the Komponent containment tree rather than in the nature of the individual diagrams. Therefore, it has a major indirect effect on the nature of the diagrams in a containment tree. As illustrated in Figure 8.3, in a containment (i.e. package) diagram, public containment is indicated by annotating a containment relationship with the stereotype <<public>>.

8.3 Process

As explained in the previous section, the main effects of public composition and public containment, whether applied separately or in tandem, are on the general information content of the Komponent models and the overall shape of the containment tree. The well-formedness and consistency rules that govern the nature of individual modeling artifacts are basically the same in both public and private situations. In other words, once the need for public composition and containment have been identified, the detail-modeling activities are essentially the same as described in the earlier chapters of the book. The main difficulty in using public composition and containment is deciding when and where to apply them rather than on developing the models. Therefore, in this section we discuss the trade-offs involved in choosing between the public and private versions of composition and containment in particular scenarios, and provide some guidelines about when and where to use them.

8.3.1 Public composition

Although the specification activity must take public composition into account once it has been selected, it is during realization that the opportunities for public composition are recognized and evaluated. During the realization activity, opportunities for public composition can arise in both a top-down and bottom-up manner. In other words, the realization of a Komponent *C* can uncover opportunities for sub-Komponents of *C* to have public parts, or for *C* itself to have public parts.

In general, there are many different scenarios in which public containment may be preferable over private containment. Therefore, it would be impossible to enumerate them all. Several situations are quite common, however, and are discussed in further detail below. Once public composition has been selected the usual specification and realization methods are (re-)applied, taking the additional constraints of public composition, identified above, into account.

8.3.1.1 User-interface parts

One of the most frequent applications of public composition is in connection with user interfaces. When designing a user interface during the context realization process it may become apparent that it is advantageous to make one or more parts visible to the user. This arises, for example, when the user wishes to invoke operations on a specific business object rather than just generic operations on the system as a whole. The benefits of public composition can be revealed by any of the realization activities, but is particularly noticeable during interaction modeling.

This scenario is an example of the top-down determination of public composition. Once the decision has been made to make parts of the system public, this should be reflected in the context realization using the approach explained in 8.2.1.2. The specification of the system can then be subsequently developed to capture the precise nature of the public composition.

8.3.1.2 Pre-determined parts

The interface scenario described in the previous subsection is a special case of a more general scenario in which a part of a Komponent is identified before the properties of the Komponent itself are elaborated by the encapsulating realization. As well as the user-interface scenario just discussed, this can also occur when a non-functional requirement dictates that a particular Komponent must be used. For example, it is common for the use of a database, or some other standard Komponent, to be dictated by project management as part of the non-functional requirements.

When this is the case, the developer of the realization encapsulating the composite Komponent with the pre-determined part must decide whether to make direct use of this part, or to wrap it up completely within the composite Komponent. The choice depends on a trade-off between the benefits of reduced complexity and run-time overhead offered by public composition and the benefits of enhanced maintainability (i.e. modifiability) offered by private composition. Once a part of a composite has been made public it is very difficult to change the way in which the Komponent works. In contrast, keeping a part private allows the realization of a Komponent to be changed without affecting its clients.

Again this is an example of the top-down determination of the opportunity for public composition. Once the decision has been to make the pre-determined part public, this must be modeled accordingly in the encapsulation realization, and then subsequently in the specification of the composite Komponent.

8.3.1.3 Server affinity

An important characteristic of the previous two scenarios was that the presence of a part was pre-determined, and the issue was whether to make it private or public. Sometime, however, an opportunity for public composition may become apparent due to the tight coupling of a client Komponent to one of its servers.

One of the key issues to be decided during the development of a Komponent is whether a server of the Komponent should be acquired or created. When a server is completely private to a client it is clear that private composition is the most natural relationship. However, the sharing of a given server by more than one client does not automatically mean that it should be acquired by them. Sometimes a shared server has much more affinity to one of its clients than its other. When this is the case, it may be beneficial to make the server a public part of the Komponent to which it has the closest affinity.

Close affinity does not necessarily mean tight coupling because a composite does not necessarily interact significantly with its public parts. Affinity here means that the server has high cohesion with its potential composite, and has a lifetime bounded by it. The benefit of making the server a public part of the composite is that the latter is much more self-contained and thus reusable. The less a Komponent has to rely on Komponents supplied by the environment in order to function correctly, the more robust and reusable the Komponent will be.

Like the previous two scenarios this is an example of the top-down determination of the opportunity for public composition. Assuming that the coupling and cohesion between components is analyzed as part of the realization of the encapsulating Komponent, the composite can be developed accordingly from the start.

8.3.1.4 Excessive forwarding

An example of the bottom up determination of public composition is provided by the phenomenon of excessive forwarding. If, after completing the realization of a Komponent it become apparent that a large proportion of the Komponent's operations essentially forward invocations to the private part, there is a good case for making this part public. An operation can be characterized as "forwarding" if the bulk of the work in its implementation is done by another Komponent.

Elevating a private part to the interface and making it public obviously has a major impact on the Komponent's clients. More specifically, it will require the realizations of all the clients to be modified to interact with the public part. This should therefore be performed only if the benefits to the Komponent's realization of making the part public outweigh the disadvantages to the realizations of its clients.

8.3.2 Public containment

Just as with public composition there are many different scenarios in which public containment can be valuable, and it is impossible to describe them all here. However, in the following subsections we discuss three of the most common reasons for using public containment.

8.3.2.1 Increased cohesion

In order for a Komponent to be usable by other Komponents it has to be visible to them. As explained in Chapter 6, one way of doing this is to elevate a Komponent to a point in the tree where the normal visibility rules enable it to be seen by every Komponent that wishes to use it. This approach has advantages and disadvantage, however. The advantages are greater conceptual simplicity, and the fact that only the original and new parents of the moved Komponents and no other Komponents in the tree, are affected. The disadvantage, however, is that Komponents can be removed from a place where they naturally belong, to a place where they do not belong. In other words, it can lower the cohesion of Komponents.

Public containment can solve this problem by enabling a Komponent to be made visible to all the Komponents that wish to use it without moving it. Komponents can thus be made available for shared access without moving to an unnatural location in the tree. In short, sharing can be achieved with optimum cohesion. The disadvantage of public composition is that attaining general visibility of a Komponent low down in a deep branch of a tree requires all the intermediate Komponents in the path to co-operate in the public containment. This can therefore force an unnatural form of containment on other Komponents in the tree therefore.

8.3.2.2 Refactoring avoidance

Another important motivation for using public containment to support shared visibility of Komponents is to avoid the refactoring effort that is required in the Komponent elevation approach (see Section 6.3.4). Under the right circumstances public composition can allow a Komponent to be made generally visible without any effort, whereas making the same Komponent generally visible by elevation would require significant refactoring effort.

Public composition, therefore, offers a very pragmatic way of making Komponents generally visible in an effortless way until, and if, the tree is refactored. In other words, even for the cases when elevation makes the most sense, visibility changes can be achieved easily in a temporary way and "saved up" until the tree is refactored and all outstanding visibility issues are handled in an optimal way.

8.3.2.3 Public composition

The other commonly occurring motivation for public containment is to support public composition. As explained above, a public part of a composite Komponent does not have to be an instance of one of its sub-Komponents, but can be an instance of any of its servers. Often, however, it is desirable for a public part to be an instance of a sub-Komponent. This is the case when a need is recognized for a public part of a Komponent type that has yet to be defined.

In any event, since the specification of a composite's public part must be visible to its clients, a public part must be an instance of a public sub-Komponent.

8.3.3 Quality control

Since public containment and composition are essentially concerned with visibility issues rather than functionality issues they do not have a major impact on quality-control activities. The same basic inspection, testing, and measurement techniques are applied as before, while, of course, taking into account the changes to the distribution of information and visibility implied by public composition and public containment.

Generalization 9

The Komponent generalization (or specialization) mechanism plays the same basic role in KobrA as it does in the UML and object-oriented programming languages.[38] It provides a way of exploiting commonalities within Komponent specifications in order to avoid unnecessary replication, and also to define explicit conformance (i.e. substitutability) relationship between Komponents.

The KobrA generalization mechanism only affects Komponent specifications. In the terminology often used for programming languages, it supports type inheritance rather than implementation inheritance. This means that only the externally visible properties of a Komponent are inherited, not its realization. A Komponent specialized from another Komponent must provide a brand new realization, just as if its specification had been defined in the normal way.

Because Komponent specifications can be related by generalization, as well as by containment and clientship, they can play a useful role in a containment tree even if they do not have an associated realization. Therefore, it is possible to define "free" Komponent specifications without associated realizations. Such specifications serve only to define a common set of Komponent properties that can be specialized (and subsequently realized) by other Komponents. They therefore play a similar role to abstract classes in object-oriented programming languages (and interfaces in Java).

38. In the context of programming languages, terms such as "inheritance", "subclassing", and "extends" are also used to refer to this concept.

9.1 Artifacts

In the package-oriented representation of a containment tree, Komponent generalization is represented by the usual UML generalization symbol. This is illustrated in Figure 9.1 that shows a new version of the *LibrarySystem* tree in which the *LoanManager* and *ReservationManager* Komponents are both specializations of a new *Manager* specification. The aim of introducing this specification is to exploit the commonality that exists between the various kinds of *Manager* Komponents in the system. All managers maintain information about some form of association between some form of item and some form of account. Moreover, they all need the services of some form of database. Figure 9.1 also illustrates that package symbols representing "free" specifications intended only for specialization rather than direct realization are labeled with the stereotype <<specification>>.

As far as the module facet of Komponents is concerned, Komponent generalization has the semantics of generalization between packages in the UML. The items in the general Komponent are added to the name space of the specialized Komponent and become visible to its clients. The relationship also has the very important effect of declaring that a Komponent conforms with any Komponent from which it is specialized. This means that instances of a specialized component can be supplied wherever instances of its generalizations are required (in short, they can play the role of instances of the generalization).

As far as the class facet of Komponents is concerned, the semantics of generalization is defined by the consistency rules that relate the specification diagrams of a specialized Komponent to the specification diagrams of the Komponent from which it is specialized. These are described in detail in the following subsections.

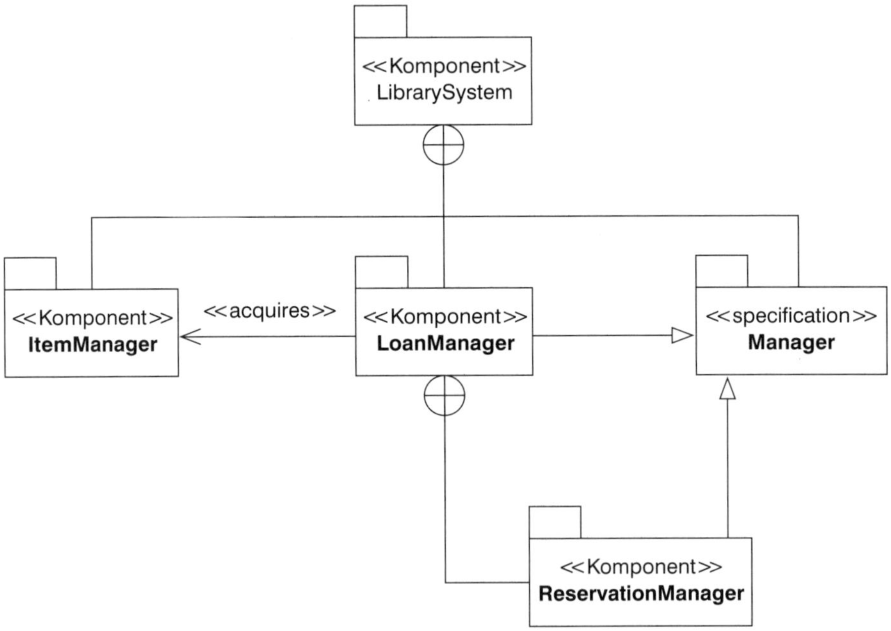

Figure 9.1 Komponent tree featuring specialization

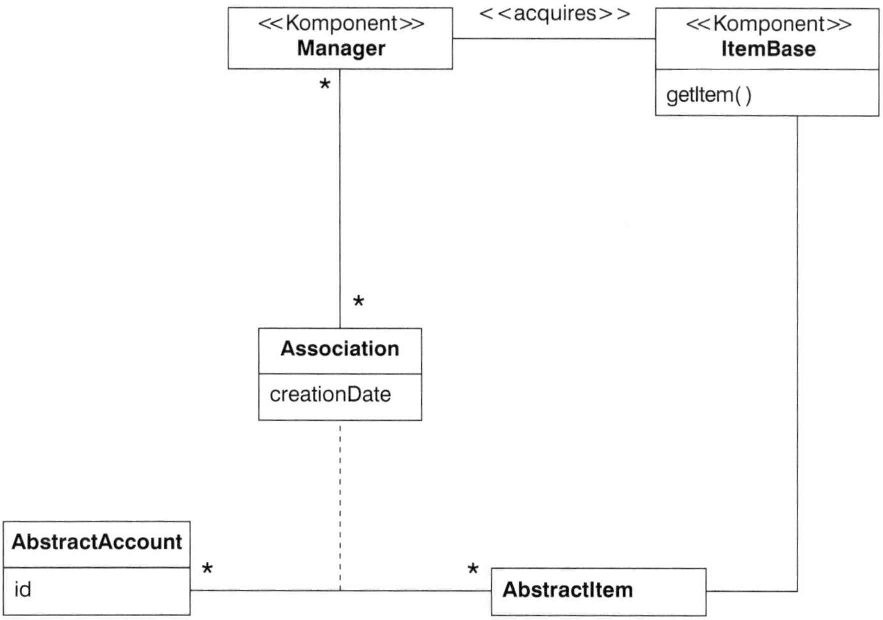

Figure 9.2 Manager specification class diagram

9.1.1 Structural model

The specification structural model is the most significantly affected by Komponent generalization. The basic idea is that everything described in the specification structural model of the general specification must also be described in the structural model of the specialized specification, subject to the possibility that new classes can play the roles of inherited classes. This applies to the object diagram as well as the class diagram.

9.1.1.1 Class diagram

Specialization between Komponents has the effect of setting up multiple, pair-wise specialization relationships between individual classes within the respective class diagram. Each specialization class is viewed as playing the role of its generalization within the specialized Komponent.

Consider, as an example, the class diagram of the Komponent *Manager* from Figure 9.1. As shown in Figure 9.2, this contains the specification information that is common to the various manager Komponents used in the system.

When a class, such as *ReservationManager*, is specialized from *Manager*, the model elements in *Manager's* class diagram are added to *ReservationManager's* class diagram, and are specialized by at least one class. This effect is illustrated in Figure 9.3.

In Figure 9.3, every class inherited from *Manager* (whether a simple class or a Komponent) has a specialization class that plays its role in the specialized Komponent. All other model elements including associations are also inherited

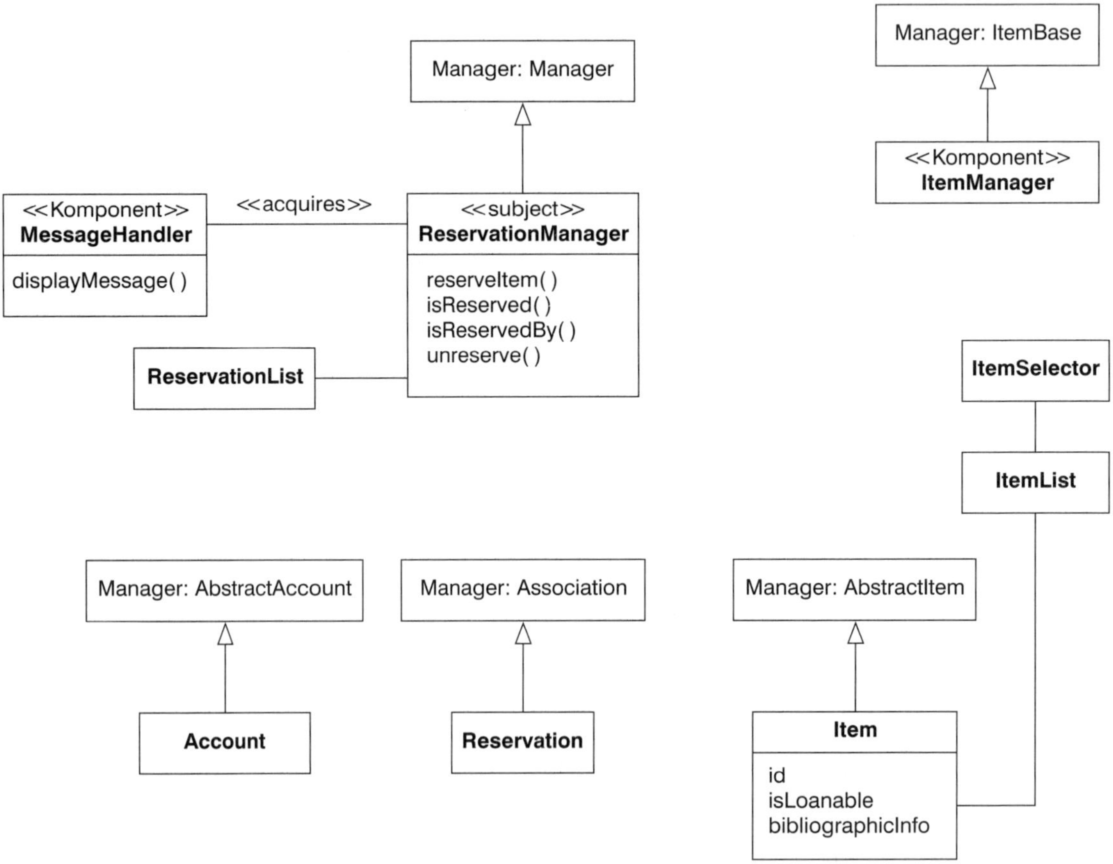

Figure 9.3
Pair-wise specialization semantics of specialization

and so do not need to be respecified in the specialization class diagram. Of course, additional model elements can be added as required in the new class diagram.

The semantics of the pair-wise generalizations in Figure 9.3 are exactly those of generalization in the UML. However, at the level of the Komponent as a whole there is the additional idea that the specialization classes in the new class diagram plays the same role as their generalization class in the old class diagram. To better convey this idea, and avoid the tedious specialization of multiple classes, Komponent specialization is not normally represented in the way shown in Figure 9.3, but instead is depicted using the UML pattern notation, as indicated in Figure 9.4.

This diagram is intended to have the same semantics as Figure 9.3 at the level of individual pairs of classes, but also makes it clear which classes in the specialized class diagram play the role of classes inherited from the general class diagram. The class diagram in Figure 9.4 is intended to have precisely the same meaning as the directly created specification class diagram illustrated in Figure 6.7.

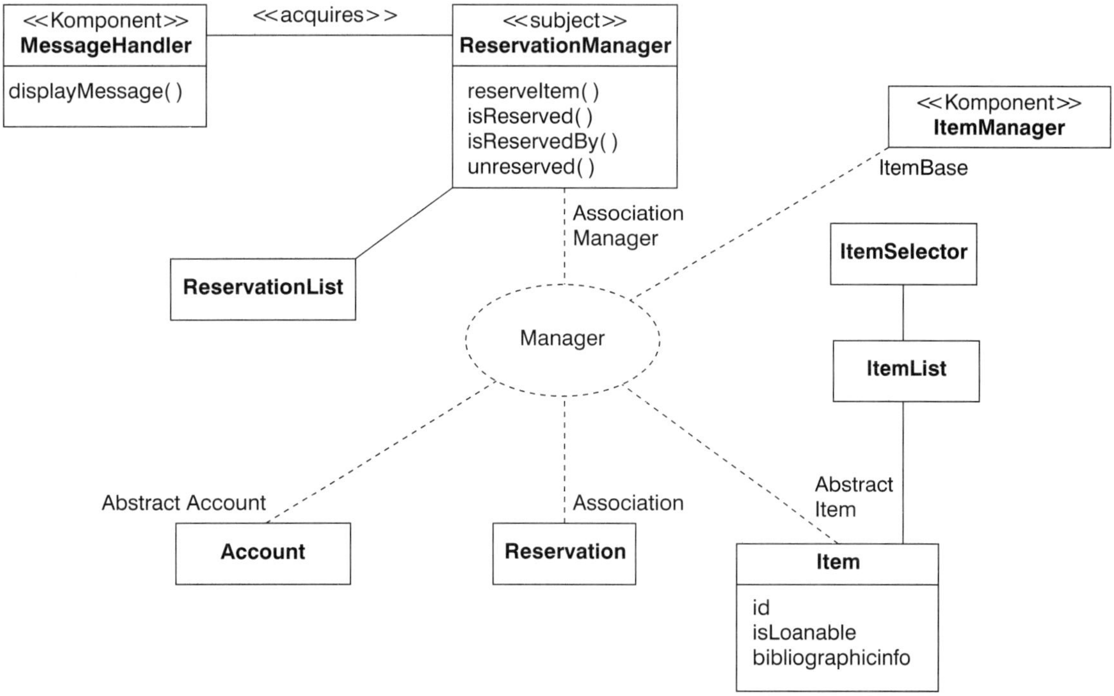

Figure 9.4
Specialization of ReservationManager from Manager

One of the advantages of Komponent genereralization is that the information defined in a specification can be easily reused to define new specifications. Figure 9.5 shows how the specification of the *LoanManager* Komponent can be specialized from *Manager*, in a similar way to *ReservationManager*.

Although the general rule is that more information can be added to a specialized specification class, one important exception is that additional constraints (i.e. invariants) cannot be added. The reason for this restriction will become apparent when the effects of specialization on the functional model are explained.

9.1.1.2 Object diagrams

The general principle of diagram extension does apply to object diagrams, however. Basically any information described in the general object diagram, if one exists, must also be present in the specialized object diagram. Moreover, there is no reason why further information cannot be added at the specialization level.

9.1.2 Functional model

In order to conform to its generalization, and satisfy the requirement of substitutability, a specialization must support all the operations of its generalization, and these

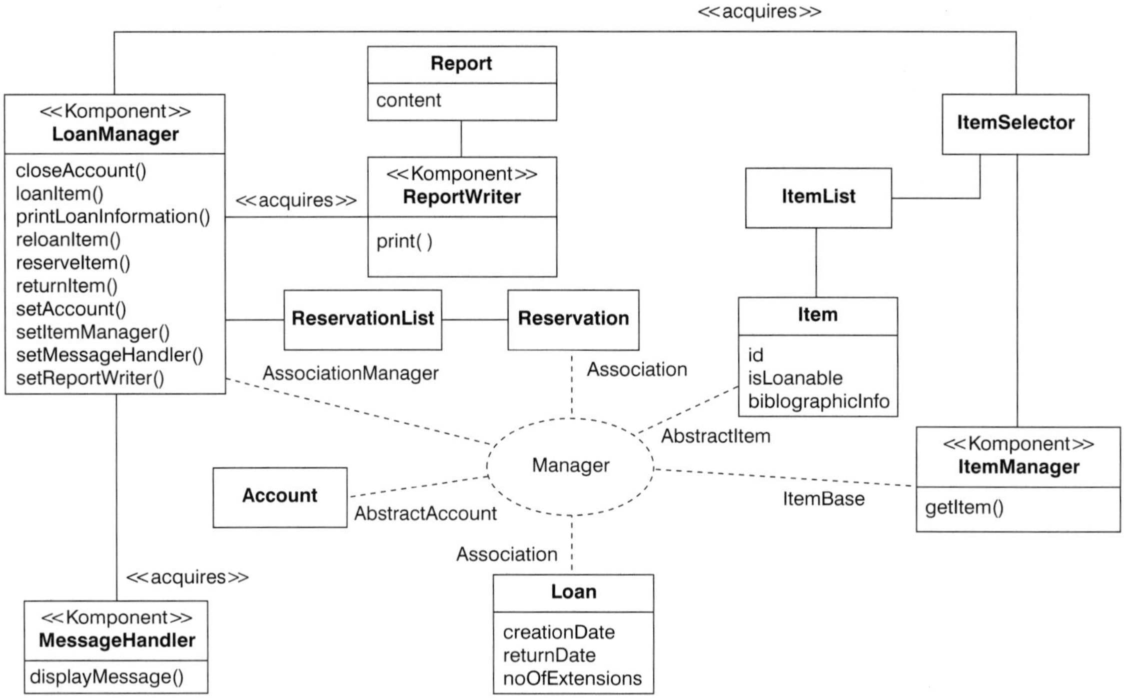

Figure 9.5
Specialization of LoanManager from Manager

must all have at least the same effect. As the default, therefore, a specialized Komponent inherits all the operation specifications of its generalization subject to the understanding that elements of the generalization's class diagram mentioned in the operation specifications are substituted for the corresponding elements in the specializations specification. Thus, for example, any reference to the class *AbstractItem* in an operation specification of the *Manager* would be changed (logically) to a reference to Item in the corresponding operation specifications for *LoanManager*.

These changes are implied, and provided that there are no other modifications to an inherited operation, there is no need to physically copy and modify the operation's specification. If no new specification is defined for an operation at the level of the specialized Komponent, the inherited semantics are assumed. However, it is also possible to change the semantics of the operation by defining a new specification for it at the level of the specialized Komponent.

Unlike most programming languages, KobrA provides no mechanism for reusing a part of the definition of an inherited operation and either extending or modifying its functionality (e.g. super, etc.). In KobrA, if the semantics of an operation are extended in a subclass the whole operation specification must be rewritten. Moreover, to ensure the logical conformance of the specialization to the generalization, the assumes clause of the new operation must be the same as, or weaker than, the assumes clause of the old operation, and the result clause must be the same as, or stronger than, the result clause of the old operation. The first constraint is intended to ensure that if a client satisfies the pre-condition for the operation at the level of the generalization it does so for the specialization as well. The second constraint is intended to ensure that if an instance of the specialized Komponent satisfies the post-condition for the operation, it fulfills the expectations placed on instances of the general Komponent.

Since the other clauses in an operation specification are based on the assumes and result clauses, these rules are sufficient to define the precise relationship between the old and new specifications of an operation. The rules also indicate why it is impossible to add any constraints to a specialized class diagram. Constraints in a class diagram correspond to invariants that are logically part of the assumes and result clauses of every operation of the Komponent. Adding further constraints would therefore violate the requirement that the assumes clause of an operation at the level of the specialized Komponent be weaker than, or the same as, that at the level of the general Komponent.

9.1.3 Behavioral model

In order to satisfy the general goal of substitutability, the behavioral model of a specialized Komponent *B* must be *protocol* and *structural conformant* to its generalization Komponent *A*. Protocol conformance means that a Komponent *B* can be substituted for a Komponent *A*. This substitutability idea implies that *B* accepts all events that *A* accepts and that *B* produces all the actions that *A* produces. Of course, *B* can accept additional events that *A* does not care about. *B* can also produce additional actions that *A* does not produce. However, there is one restriction in the latter case. *B* can invoke only those operations that are provided by the server Komponents related to *A*.

Structural conformity requires that the specialized Komponent *B* contain the same states and the same transitions as the general Komponent *A*. However, *B* can contain additional transitions and states. States and transitions can be added in several ways. First, concurrent states and corresponding transitions can be added. Second, sequential states and corresponding transitions can be added. Third, a simple state can be refined into a composite state by decomposing it into concurrent or sequential substates. Also, the guards and the destination states of existing transitions can be changed. However, it is not possible to remove states and transitions.

9.2 Process

As with public containment and composition, explained in the previous chapter, a generalization relationship between Komponents can be generated in both a top-down and a bottom-up manner. When more specialized Komponents are generated from one or more existing Komponents in a top-down way, the process is generally referred to as specialization. In contrast, when more general Komponents are generated from one or more existing Komponents in a bottom-up way, the process is typically referred to as generalization. As explained below, generalization is more complex than specialization.

9.2.1 Specialization

Specialization is a simpler way of creating a generalization relationship between Komponents than generalization because it leaves the existing Komponents unchanged. Assuming that the specification or specifications to be specialized from have been identified, the process of defining the new specification is fairly straightforward:

1 **Specialize class diagram.** For each class in the general class diagram identify the class in the specialized class diagram that will play its role. This class must be a specialization of the corresponding general class according to the normal rules of generalization in the UML. Essentially this means that it can add additional attributes and operations (for classes representing Komponents). The UML pattern notation is used to indicate which class in the specialized class diagram play the role of which class in the general class diagram.
2 **Add additional classes.** The class diagram for the specialized specification can have additional classes and associations. Once the specialized classes have been defined, any necessary additional classes to describe the extended features need to be added. No additional constraints can be added, however.
3 **Create Extended Object Diagram.** If an object diagram is deemed necessary, a complete new object diagram must be defined. This should indicate the new classifiers of the objects in the diagram, according to the role correspondence defined in the class diagram, and should add any new objects utilized by the Komponent.
4 **Identify changed operations.** Identify which of the inherited operations are unchanged in the specialized specification and which need to be respecified. Any change to the functionality of an operation, including extensions to the functionality, necessitates respecification.
5 **Create new operation specifications.** Create operation specifications for the inherited operations that need to be respecified and for any new operations. All operations are specified in the normal way using the model elements defined in the specialized class diagram.
6 **Identify additional states.** Determine whether the specialized specification has additional states or transitions over and above those of the general Komponents. All inherited states and transition must still be valid, but additional states and transitions can be added in a consistent way.
7 **Create extended behavioural model.** Extend the statechart diagram and/or statechart table to define the new states and/or new transitions.

Since a specification can be specialized from several Komponents (i.e. multiple inheritance) the usual name-clash problem can also arise. This can be handled fairly straightforwardly in the various models by qualifying the name of an inherited element with the name of the specification from which it is inherited. Given that Komponents have an associated package, the usual UML hierarchical naming scheme can be used.

When there is a name clash between inherited operations it is necessary to define which one is actually invoked when a client issues a call using the overloaded name. This can be resolved using a statechart table or a statechart diagram to indicate which operation is invoked in response to the event. Suppose, for example, that a Komponent *S* inherited two operations called m() from specifications *A* and *B*. The uncertainty about which of the inherited methods is invoked when the message m() is sent to an instance of *S* can be resolved by the following entry in a statechart table:

```
m/A::m()
.....
```

This indicates that it is A's version of m() that is invoked when an invocation event, *m*, is received.

9.2.1.1 *Subtyping*

One of the main motivations for specialization is to explicitly make one Komponent conformant with another. In programming language terminology this is known as subtyping. If a Komponent *S* is defined to a specialization of another Komponent *T*, *S* is conformant with, or is a subtype of *T*. This means that instances of S can be used in the place of instances of *T*.

Since specifications serve the role of interfaces in KobrA, the fact that a Komponent can specialize several others provides a mechanism by which a Komponent can realize several separately defined interfaces. The classic object-oriented strategy of using abstract classes to define interfaces and inheritance to assign interfaces to concrete classes is therefore supported in KobrA. In fact, since a Komponent specification can contain no realization information, this strategy for combining interfaces is more like the Java approach based on interfaces.

9.2.2 Generalization

Generalization provides a bottom-up way of creating a generalization relationship in which a generalized specification is generated from one or more existing specializations. However, because the generalization notation is asymmetric (as in programming languages) it is only possible to document generalization relationships within specialized Komponents. Thus, as well as resulting in the creation of a new generalized specification, the generalization mechanism also requires the original specifications to be modified to indicate that they are now specializations. The generalization process therefore essentially subsumes the specialization process described in the previous subsection.

9.2.2.1 *Factorizing out commonality*

The main motivation for generalizing from a set of existing Komponents is to factor out their common features and define them in a new self-contained specification. This can be valuable if instances of the pre-existing classes need to be polymorphically exchanged for one another, or if it is expected that new Komponents to be added in the future will need to specialize the factored-out properties and behavior.

Since generalization is essentially orthogonal to containment, subject to the constraint that specialized Komponents must be able to see their generalizations, Komponents in a generalization graph can occupy any locations in the containment tree. However, introducing generalized Komponents or specifications always involves a refactoring of the generalization tree, just as it does in programming languages.

9.2.2.2 Interface definition

Another common reason for generalization is to allow a public part to expose only a part of its full interface (i.e. specification) to the clients of its encapsulating Komponent instance. In other words, generalization is useful for allowing parts to be only partially public. This is achieved by defining the part to be a specialization of a new specification that presents on the publicly visible subset of its properties. By hiding the full definition of the part in the realization of the composite, and making its inherited specification visible in the specification of the composite, clients of the composite can only see the desired subset of features.

Since the full specification is usually known before the publicly visible generalization, this scenario is usually reached in a bottom-up manner through generalization. As always, however, the exact process by which the various Komponents involved are generated is irrelevant as long as the appropriate consistency rules are adhered to at the end.

9.2.3 Quality control

Generalization (or inheritance) traditionally raises some challenging issues for quality control in the context of object-oriented systems, and this is also the case in KobrA. Its impact on inspection is limited to the introduction of additional consistency rules that must be checked whenever a specialized Komponent is inspected. In the case of testing, however, the question arises as to whether it is possible to reuse the test cases developed for a generalization class when testing a specialization class. In other words: Is it possible for a specialized class to inherit test cases as well as the functional code? The problem is less severe in KobrA than in regular object-oriented languages because KobrA does not allow the realization of a method to be inherited and/or modified. Provided that the operation specification for an inherited operation is unchanged, therefore, the test cases for the operation defined at the level of the generalization should be reusable. However, additional test cases have to be defined for new or modified operation specifications.

Generalization is also the source of additional complexity from the point of view of quality measurement because it must be considered in the definition of several of the common internal quality attributes such as size and complexity. However, it also extends the range of coupling measures that can be applied. In conjunction with the quality approaches outlined in Chapters 18 and 19, an organization can investigate if, in its environment, the use of Komponent generalization (or other Komponent assembly techniques presented in Chapters 6 and 8) is particularly conducive or detrimental to the quality of the resulting system. From such findings, locally tailored recommendations for the use of these modeling techniques can be derived.

Components and patterns 10

Components are one of the leading software reuse technologies. They define, implement, and package reusable software building blocks that can be efficiently assembled into new software systems. As such they embody accumulated knowledge about what good reusable components are and how they should be implemented. However, this is only one part of the picture. Regardless of how good individual building blocks may be, their effective reuse requires additional knowledge and experience about good ways of assembling them. Such knowledge transcends individual components, but must also take their individual characteristics into account.

In the object-oriented community, frameworks and patterns have become popular vehicles for capturing successful assemblies of components. Frameworks are semi-complete systems that can be instantiated (to the extent anticipated by the framework developers) to meet the needs of a particular customer, while patterns are abstract reusable solutions that can be applied to new problems. Frameworks are therefore typically of a larger scale than patterns, and tend to be more focussed and specialized in nature. Patterns, in contrast, are more general purpose, and typically describe solutions and approaches that can be used in multiple software domains. As a reuse technology, they therefore bridge the gap between individual components and large-scale frameworks.

Since frameworks are discussed extensively in Chapter 15, in this chapter we focus on patterns. We first explain the nature of pattern technology and how it is supported in the UML, and discuss how well-known published patterns can be exploited in a KobrA project. We then describe some KobrA-specific patterns that capture good strategies for leveraging the idiosyncrasies of KobrA's approach to component modeling.

10.1 What are patterns?

The idea of formulating generic solutions to recurring problems was first proposed by Christopher Alexander [AIS77] [Alex79], a building architect. He proposed a pattern language that provided workable solutions to recurring problems in the design of buildings and introduced a three-part schema (context, problem, and solution) for describing them. His vision was for the inhabitants of a building to be able to design and construct buildings using this pattern language. Even though this ambitious goal was never fully realized, his ideas were highly influential and even crossed from the building architecture domain to software engineering where they had a much bigger impact.

Kent Beck and Ward Cunningham [BC87] first introduced Alexander's idea of a pattern language to the computer science domain. They proposed a system of patterns, focussed on Smalltalk windows, to help software developers design the interfaces their programs. However, it was not until the publication of the "Gang of Four"[39] pattern collection in 1994 that patterns really took off in the software domain [GHJ95]. The "Gang of Four" introduced the notion of design patterns as descriptions of communicating objects and classes that are customized to solve a general design problem in a particular context. Since then, numerous books on patterns have been published, including the Pattern Languages of Program Design series [CS95], [VCK96], [MRB97], Pattern-Oriented Software Architecture [BMR96] and Analysis Patterns [Fowl97] to name a few. Many web-based pattern repositories have been established (e.g. the Portland Pattern Repository [Port] or the Hillside Patterns Home Page [Hill]) and approaches for supporting patterns with tools are also emerging [BFV96].

The basic idea of a pattern is applicable at almost all levels of abstraction and phases of development in the software life-cycle. Prime examples include:

- **analysis patterns** – general reusable object structures
- **architectural patterns** – the overall structure or architecture of a system
- **design patterns** – the "classical" patterns that help in the design of object-oriented applications
- **code idioms** – recurring coding strategies that help facilitate implementation.

At whatever level of abstraction or scale patterns are used, however, they consist of the same basic elements:

- **name** – the name of the pattern
- **problem** – the contexts in which the pattern is helpful
- **solutions** – the strategy by which the pattern solves the problem
- **consequences** – the trade-offs embodied in the pattern.

39. Erich Gamma, Richard Helm, Ralph Johnson, and John Vlissides.

In Gamma et al [GHJV95], patterns are categorized as creational, structural or behavioral. However, other categorization approaches are also used, such as, for example classifying patterns according to the context in which they can be applied [BMR96].[40]

10.1.1 Patterns in UML

In view of the importance of patterns, the UML provides an explicit notation for depicting the application of patterns within a model. As an example, consider the mediator pattern presented in [GHJ95]. The mediator pattern describes the definition of a role (the mediator) which encapsulates how a set of other roles (the colleagues) interacts. It promotes loose coupling among the colleagues and enables the variation of their interaction. A panel consisting of a number of user-interface elements can be designed using this pattern. Figure 10.1 shows the mediator pattern applied to three classes. A *PanelManager* class plays the role of the mediator, and the *ListBox* and *TextField* classes are the colleagues. The mediator's task is to notify all of the colleagues whenever one colleague has been changed (e.g. the text in the text field is updated when an entry in the list box has been selected). This decouples the different user-interface elements.

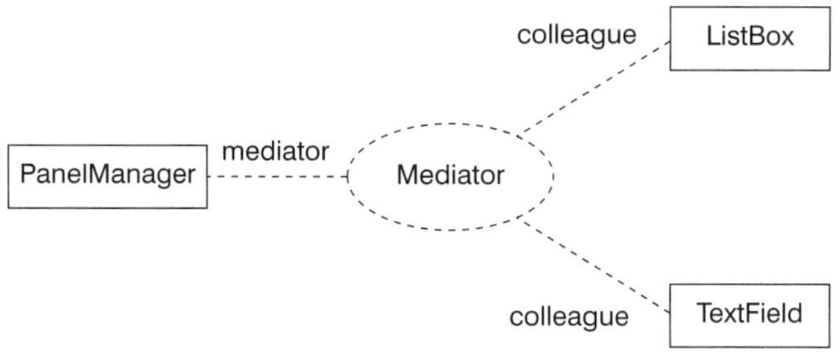

Figure 10.1
Mediator pattern in the UML

Figure 10.1 illustrates the UML notation for applying this pattern within the context of a class diagram. As can be seen from this diagram, the name of the pattern appears in a dashed oval connected to each of the concrete classes participating in the pattern via a dashed line. The name of the role played by each participant is shown in the form of a role name adjacent to each participant. Other than describing the name of the pattern, and the mapping of roles, the UML provides no support for capturing the other elements of a pattern, such as the context or the problem solved by the pattern.

As explained in Chapter 9, this pattern notation is also used in KobrA as a shorthand way of representing Komponent generalization. The name of the generalization Komponent is placed where the pattern name usually goes, in

40. Classic patterns of the kind discussed in this chapter define "good" object-oriented structures and behaviors. In Chapter 11 on Komponent implementation we introduce transformation patterns which represent "good" mappings between such structures.

the centre of the oval, and the role names are used to indicate which classes in the specialized Komponent specification substitute for (i.e. play the role of) which classes in the general specification. Using this pattern notation therefore avoids the need to define many pair-wise generalization relationships. Although it uses the UML pattern notation, however, Komponent generalization does not involve the application of patterns in the usual sense.

10.2 Leveraging existing patterns in KobrA

There are two inter-related ways in which conventional, published patterns can be exploited in KobrA. Since the different representations of a Komponent (i.e. specification, realization, and implementation) resemble the typical artifacts of a traditional development method (i.e., specification, design and implementation), all the main categories of patterns are potentially useful during the various stages of Komponent development. Analysis patterns may be useful during the specification of a Komponent (especially those close to the root) and in the development of the context realization. Architectural and design patterns are likely to be useful in the development of Komponent realizations, the former particularly near the top of a Komponent tree, and the latter more so near the leaves. Idioms are likely to be of value in the development of Komponent implementations.

As well as applying patterns at the level of an individual specification or realization, however, it is also possible to apply them at the level of the overall containment tree. This offers a way of organizing containment trees in a way that matches well-known, successful architectural patterns. In fact, because of the hierarchical way in which the overall architecture of a containment tree is determined by the structure of the individual realizations, the component-level and tree-level applications of patterns are actually closely related. For example, patterns applied in the realization of an individual component will have an impact on the nature of its children, and thus ultimately on the shape of the tree. Similarly, patterns applied at the top level of the tree will place constraints on the children of Komponents and thus on their individual realizations.

In the following subsections we consider how well-known architectural patterns can be used to help determine the structure of a KobrA containment tree. We first discuss the general purpose architectural pattern of layering, and then consider various leading architectural patterns used in the domain of business software.

10.2.1 Layered architecture pattern

One of the most well-known and ubiquitous architectural patterns is the layered pattern. This defines a way of organizing a system in terms of several distinct layers, each encapsulating a certain subset of the overall system's functionality. Two distinct versions of this pattern are generally distinguished – strict layering and loose layering.

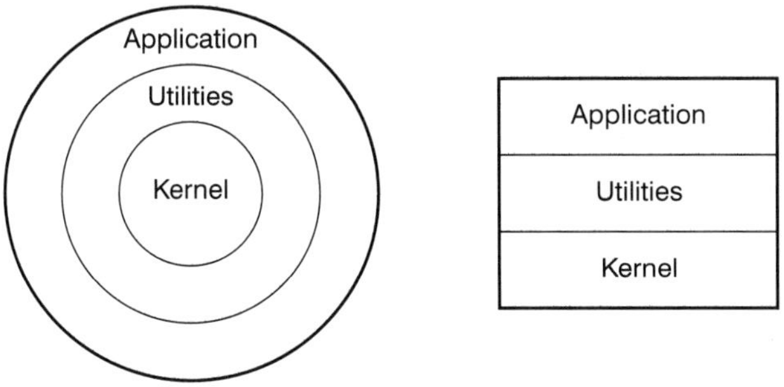

Figure 10.2
Strict layering

10.2.1.1 Strict layering

With strict layering, a strict hierarchy is enforced so that each layer can only interact with its immediately adjacent layers. Each layer therefore builds upon the layer below to provide enhanced facilities to the layer above. Figure 10.2 shows two ways of representing such a layered architecture. In this example, a system consists of three layers called kernel, utilities, and application. Each layer can only access the immediate layers below, making it impossible for the application layer to directly invoke anything in the kernel layer. It can only utilize the services of the kernel layer indirectly, via the utilities layer.

KobrA's composition and containment concepts provide a natural way of representing such a strictly layered architecture. Because a given layer instance has exclusive access to the layer instance below, at run-time layer instances correspond to Komponent instances related by composition, and at development time layers correspond to components related by containment.

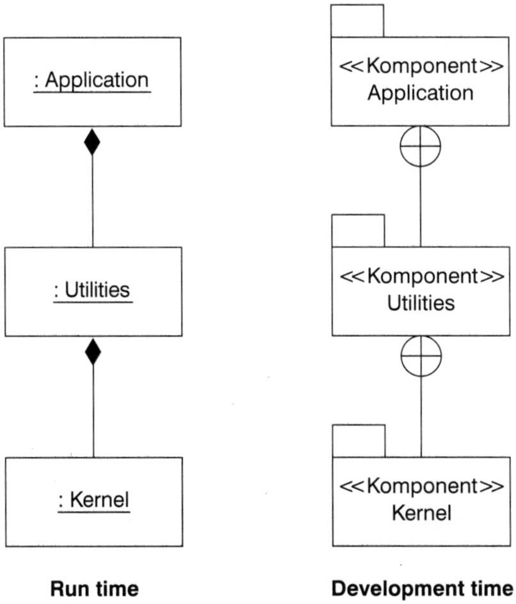

Figure 10.3
Komponent representation of strict layering

Figure 10.4.
Strict layering using pattern notation

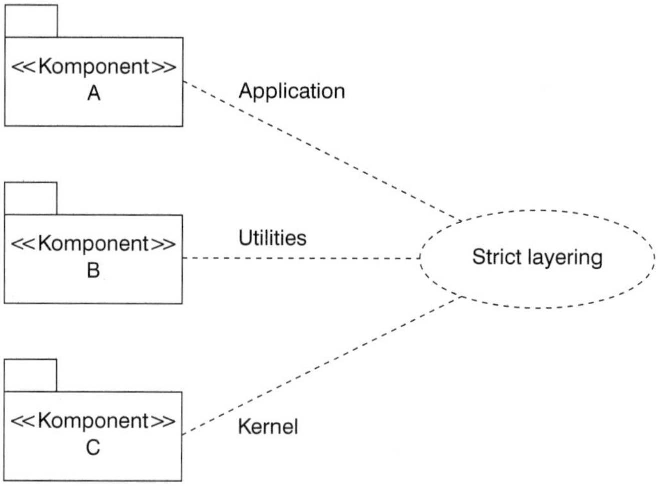

Figure 10.3 illustrates how a KobrA containment tree can most directly represent the layered architecture depicted in Figure 10.2. The left-hand side of Figure 10.3 shows the situation at run time and the right-hand side the situation at development time. At run time, each layer corresponds to a component instance that is a private part of the component corresponding to the layer above. As explained in Chapter 3, this run-time arrangement is most naturally supported by a containment tree with the same shape.

The names of the Komponents organized in a strict layering scheme do not have to carry the names of the layers, as in Figure 10.3. The fact that a strict layering pattern is being applied can be documented using the UML pattern notation discussed in Section 10.1.1. The use of this pattern to assign Komponents in a containment tree to logical layers is depicted in Figure 10.4.

10.2.1.2 Loose layering

Loose layering differs from strict layering in that individual layers are not constrained to interact with their immediately adjacent neighbors. In general, arbitrarily complex interaction patterns can be created. When layers are modeled as Komponents, loose layering is obviously more difficult to handle in KobrA because it corresponds to a more complex arrangement and interaction pattern. However, although the KobrA representation may appear to be more complex, this is only because KobrA illustrates more of the underlying complexity inherent in the architecture. Furthermore, the more complex the interaction pattern becomes (i.e. the "looser" the layers become), the less well the layered pattern fits as a characterization of the system's architecture.

As an example, consider the loosely layered architecture depicted in Figure 10.5. This is intended to show that *Layer7*, the top layer, can interact directly with the bottom layer (*Layer1*) and second-lowest layer (*Layer2*) without going through the intermediate layers (*Layer3* to *Layer6*).

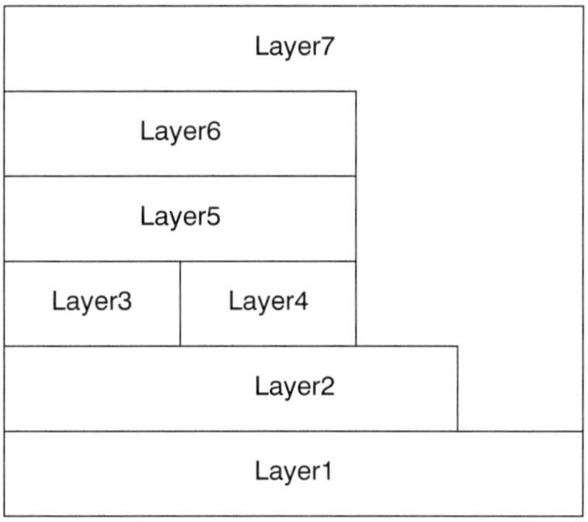

Figure 10.5 Loosely layered architecture

The simplest way of representing this situation in KobrA at run time is to view interacting layers as peer Komponent instances in the creation tree, and to raise each Komponent instance to a point in the creation tree where it can be seen by all its clients. This is illustrated in Figure 10.6. In this tree, *Layer7* is responsible for creating an instance of *Layer6*, *Layer2* and *Layer1*, and for passing references of these around as needed to create the desired communication pattern. Thus, an instance of *Layer1* has to be passed to *Layer2* and to *Layer6*. *Layer6* in turn passes the reference to *Layer1* to its subcomponent, *Layer5*, which in turn passes it to its two sub-Komponents, *Layer3* and *Layer4*.

A containment tree supporting the run-time structure depicted in Figure 10.6 using only private containment is illustrated in Figure 10.7. Notice that since *Layer2* has to be visible to *Layer3* and *Layer4* as well as *Layer7* it has to be placed at the second level of the containment tree. This is the only way in which the desired visibility can be achieved using private containment. Once the position of *Layer2* has been established, it follows that *Layer1* must be at the same level, because it needs to be visible to both *Layer7* and *Layer2*.

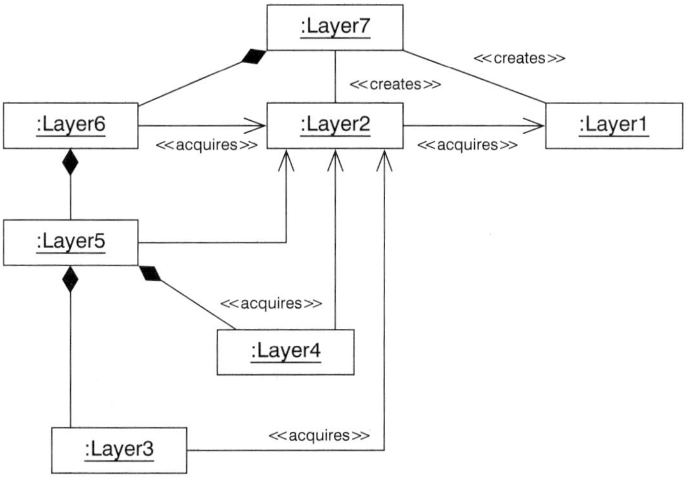

Figure 10.6 Komponent instances in a loosely layered architecture

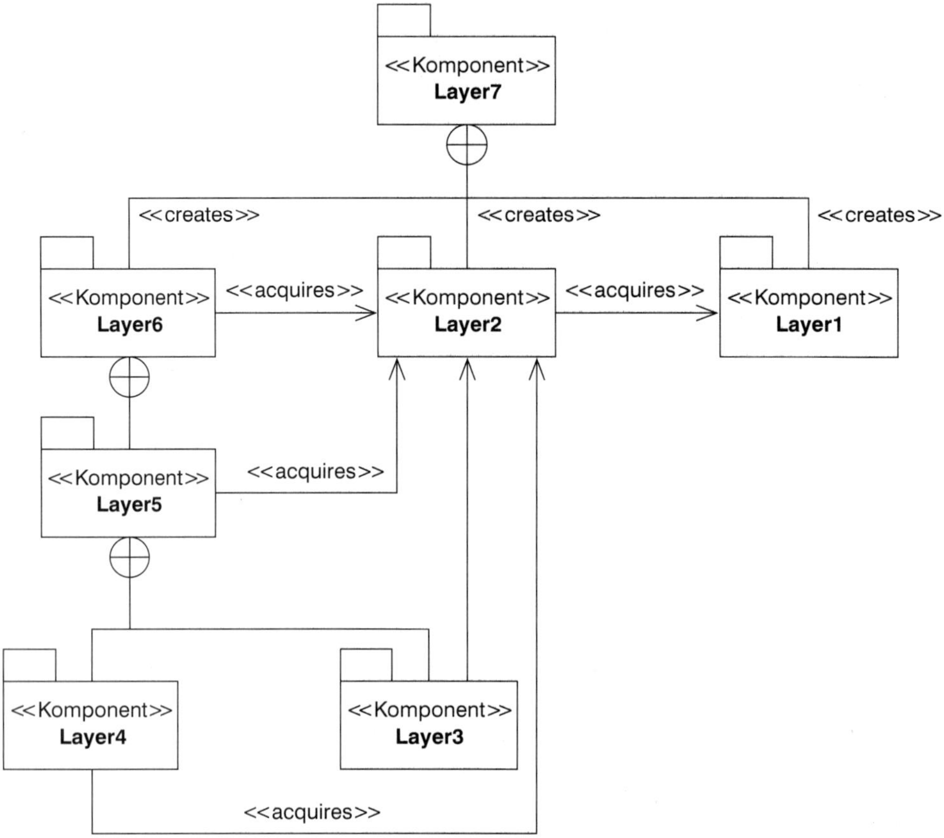

Figure 10.7
Loose layering containment tree

The containment visibility rules ensure that the Komponents in the containment tree in Figure 10.7 have precisely the right level of visibility needed to support the run-time configuration depicted in Figure 10.6. However, some developers may prefer to define a containment tree in which the location of the Komponents more closely matches that of the layers in the block diagram. This can be achieved by using public containment rather than private containment for lower layers that are accessed by non-adjacent higher layers, such as *Layer1* and *Layer2*.

As illustrated in Figure 10.8, by making *Layer1* a public sub-Komponent of *Layer2*, it can be placed at a lower level in the containment tree than *Layer1*, and the whole shape of the tree can be made to match the arrangement of the layers in the block diagram in Figure 10.5. The correspondence between the two is more than just cosmetic. It shows that the KobrA containment model cleanly subsumes the abstraction and encapsulation techniques offered by layering.

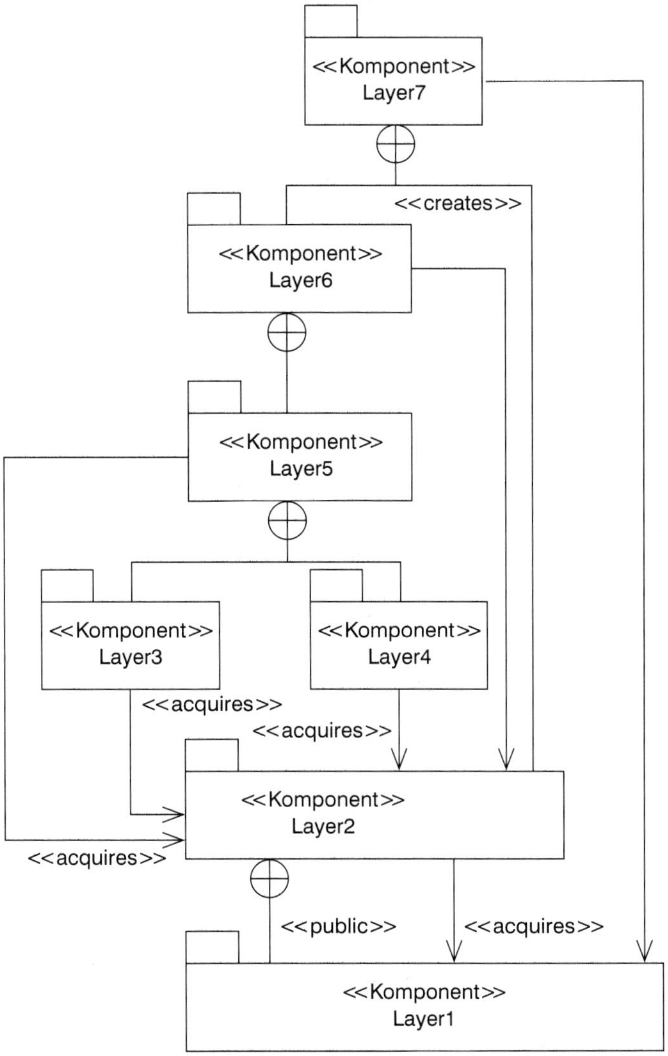

Figure 10.8 Layered architecture with public containment

10.2.1.3 White-box layering

Another important difference between layering models is whether the layers themselves represents "black-box" abstractions, whose internals are invisible to other layers, or whether they represent "white-box" abstractions whose internals are partially, or even completely visible. The scenarios discussed in the previous subsections are examples of black-box layering, because each layer was viewed as a Komponent whose internal structure is unknown. With white-box layering, however, the internal structure of some of the Komponents would be visible to other layers. The extreme, but common, form of white-box layering is when the layers are purely logical, and serve only as a convenient grouping of components.

Figure 10.9
White-box layers

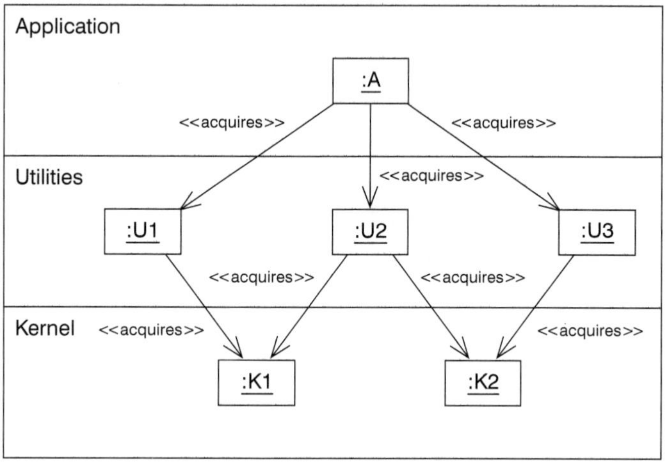

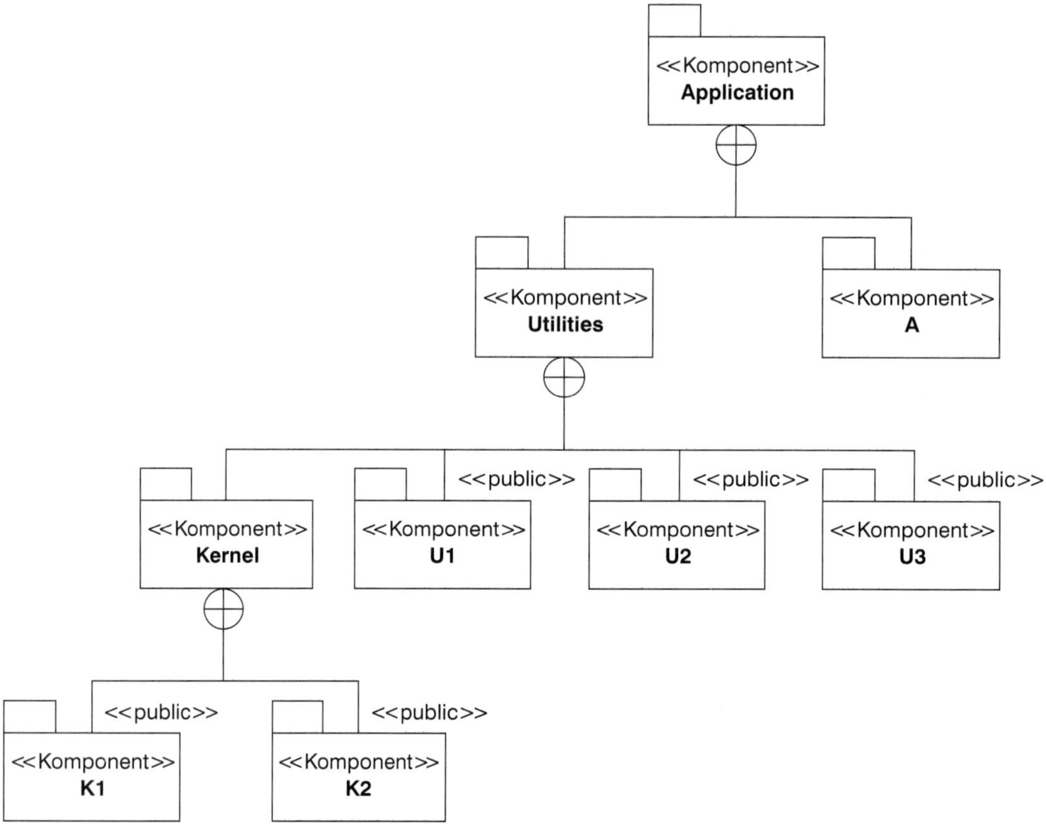

Figure 10.10
Komponent tree for White-box layers

Public containment is even more important for supporting white-box layering in KobrA than it is for black-box layering. Figure 10.9 illustrates a variant of the strict-layered architecture in Figure 10.2 in which the layers have internal component instances that have visibility of one another and interact directly.

The KobrA containment tree that provides exactly the right level of visibility for the white-box architecture in Figure 10.9 is illustrated in Figure 10.10. The intention of the architecture depicted in Figure 10.9 is that Komponent instances within one layer can only interact with Komponent instances within the immediately adjacent layers. This can be captured quite neatly in a Komponent containment tree by the careful use of public and private containment, as illustrated in Figure 10.10.

Recall that a Komponent in a containment tree always has visibility of its peers, and anything higher up in its branch of the containment tree. Thus, the Komponent A contained by the Komponent Application has visibility of its peer *Utilities*. Furthermore, its also has visibility of *Utilities'* sub-Komponents *U1, U2* and *U3* by virtue of the fact that they are public sub-Komponents. Importantly, however, it does not have visibility of *Kernel,* as required by the strict-layering semantics, because this is a private sub-Komponent of Utilities. *U1, U2,* and *U3* do have visibility of *Kernel,* and its sub-Komponents *K1* and *K2,* since they are also public sub-Komponents. The visibility semantics of the strict white-box layering are therefore precisely satisfied.

10.2.2 Distributed system architectures

The various versions of the layering architecture pattern discussed in the previous section are all general purpose patterns applicable in virtually all domains. In this section, we examine the application of these ideas to distributed software architectures of the kind typically used in the domain of business software, the main focus of KobrA. We describe some of the basic architectural concepts used in this domain, and illustrate how they can be supported in KobrA using the concepts of Komponents and containment trees. All the discussed architectures rely on the layering concepts just elaborated, but usually employ the term "tier" rather than "layer."

10.2.2.1 Client-Server architectures

The simplest form of a distributed business architecture is the client–server architecture. In its most primitive form this consists of a server, which is often a file store or database, and one or more clients on remote machines that access this data. Typically most of the "intelligence" is located in the clients.

Figure 10.11(a) provides a schematic representation of this architectural pattern, in the way it might be presented in the literature, while parts (b) and (c) show the development-time and run-time equivalents in KobrA, respectively. Figure 10.11(a) indicates that multiple clients typically access a single server that manages the data of importance to the enterprise. Although the interaction between the clients and the server is two-way, as indicated by the arrows, the server is usually unaware of the identity of the clients.

Figure 10.11(b) shows how this basic scenario can be represented in terms of a Komponent containment tree. The *Client* and *Server* types are naturally represented as Komponents, but in addition *a system* Komponent has been added as the root of

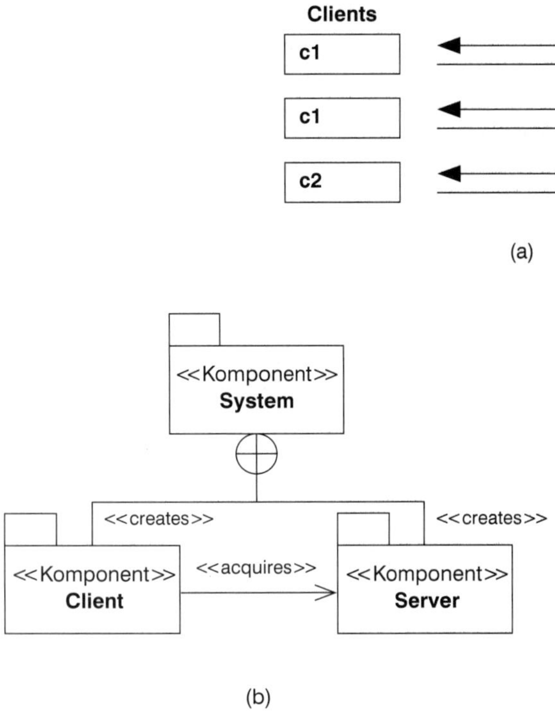

Figure 10.11
Client-Server architectures in KobrA

the tree. This is needed because there is no creation relationship of any kind between the client and the server. Both come into existence independently, and somehow the client has to gain visibility of the server in order to invoke its services. This is indicated by the fact that the client <<acquires>> an instance of the server. The *System* Komponent at the root of the tree may not actually have any concrete instances in the run-time system. If this is the case its role would have to be fulfilled by some other agent, such as a human operator, responsible for creating the clients and the server. This is in fact the usual situation. The strategy of adding "phantom" Komponents to the root of a containment tree is useful whenever there is no actual software Komponent instance that serves as the root of the creation tree. This is common in distributed systems.

Figure 10.11(c) is an object diagram illustrating a typical run-time scenario in which three *Client* Komponent instances acquire access to a single *Server* Komponent instance.

10.2.2.2 Three-tier architectures

Three-tier architectures represent a refinement of the basic client–server architecture in which the business logic is separated from the graphical user interface (GUI). The business logic is thus contained in a third tier between the client and the server containing the data.

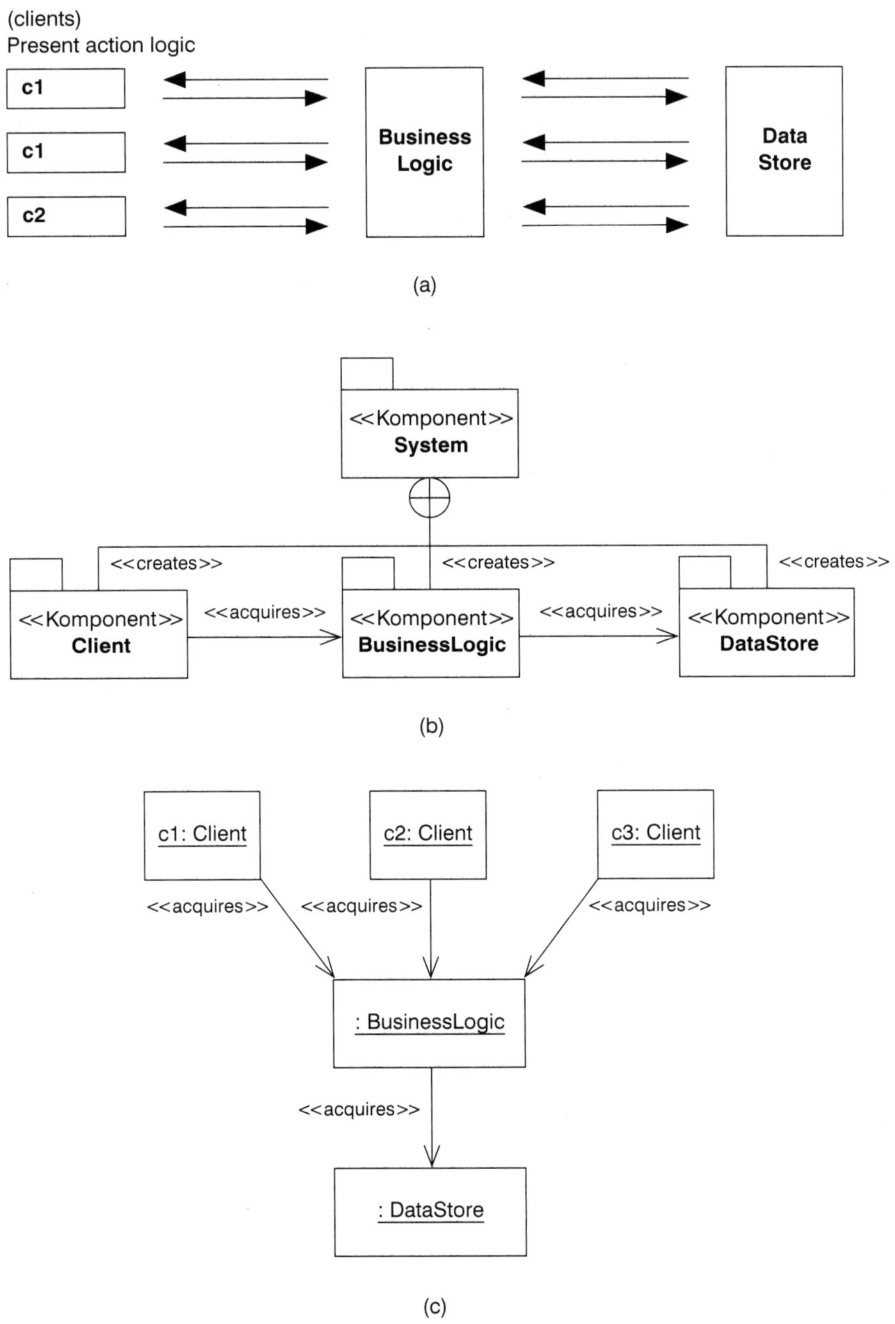

Figure 10.12 Three-tier architecture in KobrA

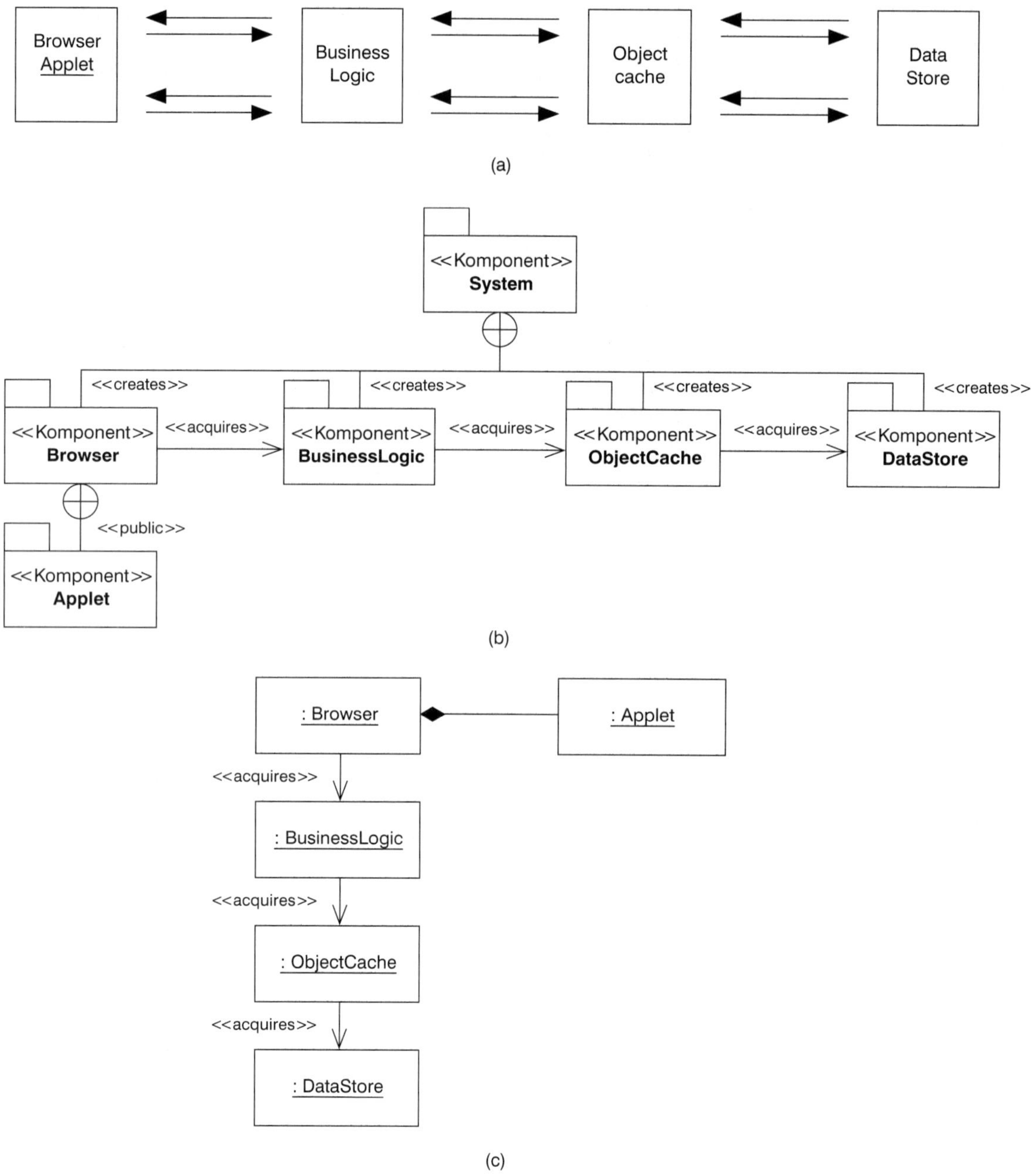

Figure 10.13
N-tier architecture in KobrA

Figure 10.12(a) provides a schematic representation of this architectural pattern, while parts (b) and (c) show the development-time and run-time equivalents in KobrA. The basic features are the same as for the client–server architecture, but with an extra Komponent representing the business logic. As before, a phantom *System* Komponent is needed to serve as the root of the containment tree, since the instances of the *Client*, *BusinessLogic*, and *DataStore* are usually created by extra software means.

10.2.2.3 N-tier distributed architectures

The most general variant of the client–server architecture which takes the principle of encapsulating concerns to its logical conclusion are N-tier architectures [GOW01]. This is a common strategy for supporting so-called thin clients that contain a small part of the business logic as well as the GUI. Applet-based web applications are a prime example.

As illustrated in Figure 10.13, the main enhancement to the three-tier structure shown in the previous subsection is the introduction of an object cache manager between business logic and the data store. This optimizes access to the raw data in the data store by maintaining interpreted representations of the data in a high-level exchangeable representation such as HTML or XML. The principles of modeling in KobrA are the same as for the previous versions of the client–server approach. The only difference is the explicit modeling of *Applets*. In the development-time containment tree (Figure 10.13(b)) *Applets* are represented by a public sub-Komponent of *Client*, although the only real requirement is that it be globally visible. In the run-time development tree this is shown as a private sub-Komponent of the *Browser* Komponent instance. This is because logically the browser creates the *Applet* (Komponent instance) and is the only Komponent instance that accesses it.

10.2.2.4 Distributed-object architectures

Although they exhibit growing levels of sophistication, the variants of the client–server architecture discussed above all have the property that the ultimate suppliers and users of data play only a single role – either solely as clients, or solely as servers. In the most general model of distributed computing, distributed objects can act as clients and/or servers in a completely symmetrical way, and the overall system is viewed as a community of communicating peer objects. A term sometimes used for this is "a multi-client/multi-server architecture."

The best known model for such a general distributed object architecture is the Object Management Architecture (OMA) standardized by the OMG. As illustrated in Figure 10.14, this assumes the existence of a standard communication backbone, supported by Object Request Brokers (ORBs), and populations of objects classified as *Application*, *CORBAFacilities* or *CORBAServices*. The first of these are intended to represent end-user distributed objects, while the latter two represent infrastructure objects that provide standard capabilities to application objects. These range from basic system-software services, known as *CORBAFacilities*, to more sophisticated often domain-specific capabilities, known as *CORBAServices*.

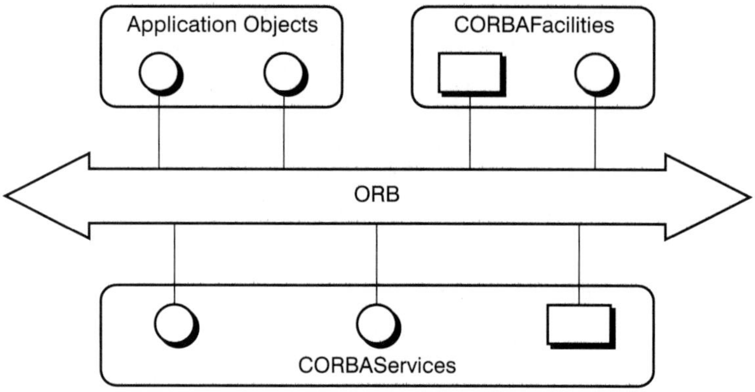

Figure 10.14 OMG Object Management Architecture

Figure 10.15 shows a simple way of representing the OMA in the form of a Komponent tree. Basically, since Komponents have visibility of each other (reflecting the visibility rules in the OMA) the simplest strategy is to make them all children of a single phantom-root object. Figure 10.15(a) illustrates this situation, and shows a representative sample of *Applications* (*A* and *B*), *CORBAFacilities* (*MOF* and *AnalysisDesign*) and *CORBAServices* (Persistence). Notice also that an explicit Komponent is included to represent the ORBs. A phantom root is used because, as with the entities in the client–server architecture (see 10.2.2.1), distributed-objects instances are usually created manually, and thus have no actual creator in a software form.

Figure 10.15(b) illustrates a representative configuration of Komponent instances corresponding to a system of distributed objects. Three distinct ORBs are shown, supporting various application Komponent instances and *CORBAFacilities* and *CORBAServices* instances. Notice also the prevalence of <<acquires>> relationships between the Komponent instances. This reflects the fact that distributed objects typically have to take a certain amount of time during their initialization to register themselves and discover who their neighbors are.

10.3 Component containment patterns

The previous section described how well-known, published architecture patterns could be leveraged in the development of KobrA systems. In this section, we take the opposite point of view, and focus on distinct patterns for leveraging the unique features of KobrA – especially its emphasis on Komponent containment. These patterns only have relevance for methods which, like KobrA, adopt a recursive approach to component-based development, and containment as the basis for organizing the development-time description of Komponents. We identify three distinct patterns for organizing the Komponents in a tree:

- universal visibility pattern
- need-to-know pattern
- component library pattern.

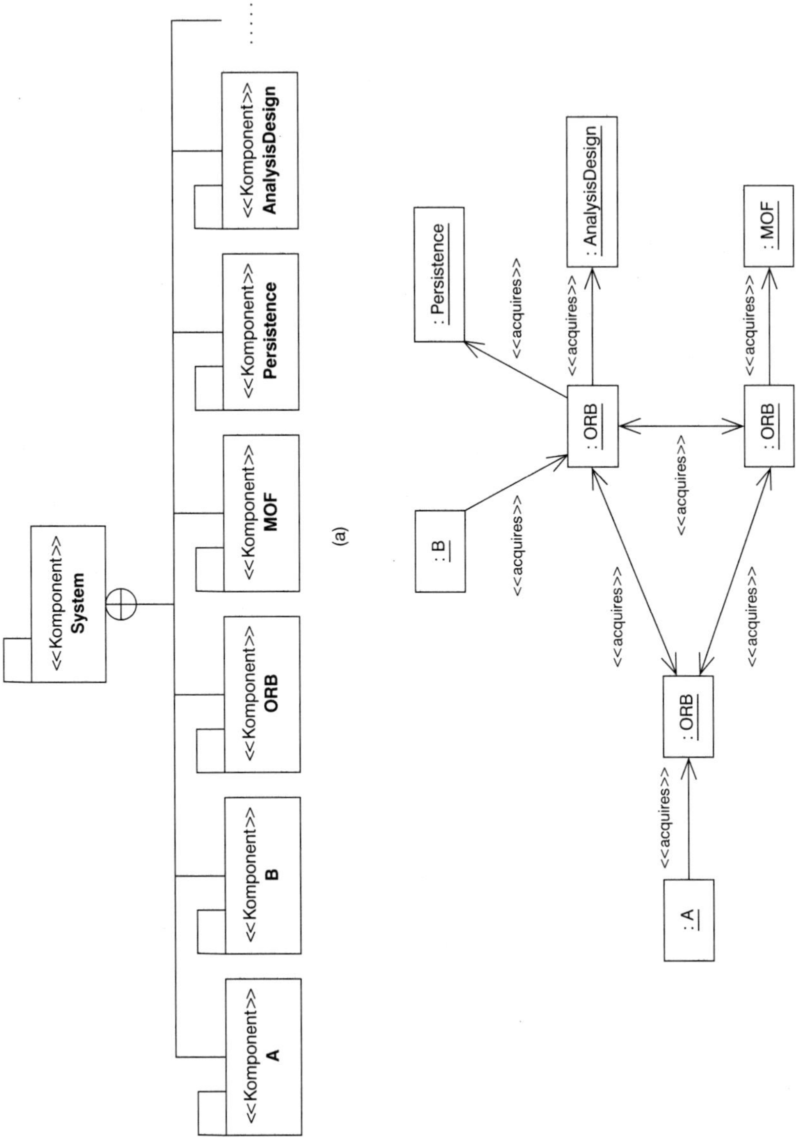

Figure 10.15
Simple OMA architecture in KobrA

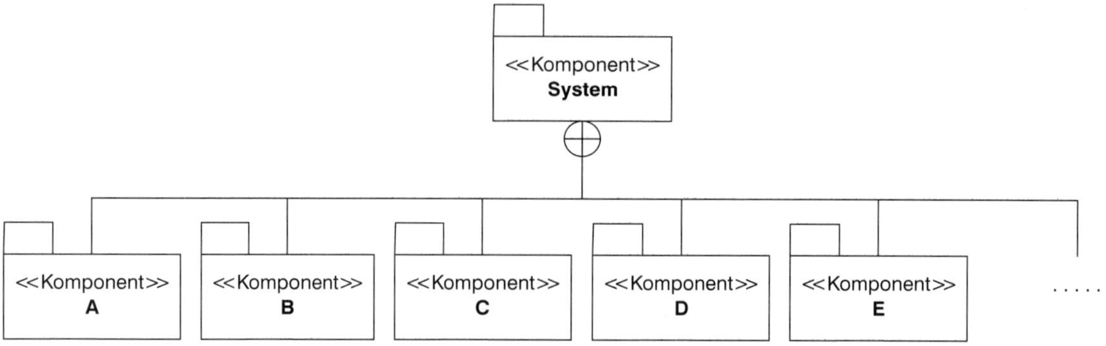

Figure 10.16
Two-level tree providing universal visibility

10.3.1 Universal visibility pattern

Universal visibility is a containment tree pattern in which all Komponents are given visibility of all other Komponents by restricting the tree to two levels – the root and all the rest. This pattern was actually used in Figure 10.15 as a way of capturing the OMG's OMA distributed-object architecture. Figure 10.16 provides an abstract representation of this pattern.

According to the KobrA visibility rules, every Komponent in this tree has visibility of every other Komponent, allowing the creation of arbitrary run-time clientship and composition structures. This architecture is therefore compatible with all run-time configurations.

Since it is the simplest to understand, and supports the widest rage of run-time configurations, this architecture pattern is likely to be useful for enterprises that are introducing KobrA for the time, and need to become comfortable with the modeling concepts before mastering the organization of containment trees.

10.3.2 Need-to-know pattern

Although universal visibility has the advantage that it can support all possible run-time configurations of Komponent instances that can be created from a given set of Komponents, it goes against the long-established principles of software engineering that entities should only be given the minimum visibility of other entities that they need to do their job. In the military, this is known as the "need-to-know" principle. The motivation for applying this principle in software engineering is to maximize the cohesion and minimize the coupling of Komponents, and through this to improve their independence and reusability.

The strict layering architecture discussed in 10.2.1.1 is a good example of this principle in practice. In a strictly layered architecture, each layer (i.e. Komponent) is only allowed to see, and hence access, immediately neighboring Komponent instances.

Figure 10.17 provides a general representation of this principle in its extreme form when used to organize a containment tree. Instead of giving rise to a flat

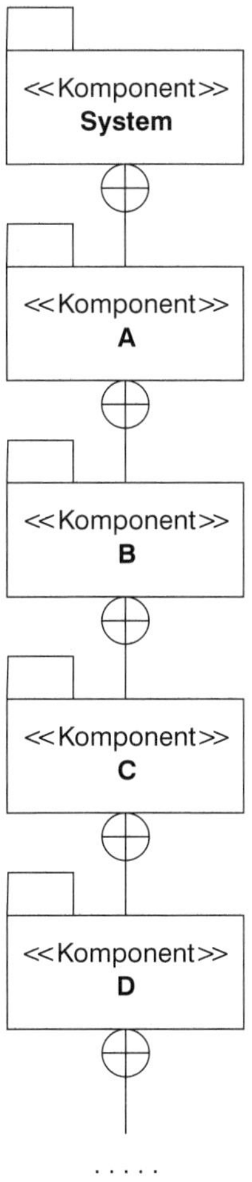

Figure 10.17
Deep, multi-level containment tree

tree with a maximum number of branches, like the universal visibility pattern, this pattern gives rise to a deep tree with the minimum number of branches. Trees of this nature only support a small number of possible run-time component instance configurations, so much more care is needed to make sure that they support the desired situation. Generally, deep containment trees of this kind usually reflect deep composition trees at runtime.

Creating deep containment trees based on the need-to-know principle is much more difficult, but has a high return on investment in terms of increased quality and reduced maintenance problems.

10.3.3 Component library pattern

The two previous patterns differ in the way that they organize the Komponents in a tree to satisfy the required run-time visibility requirements. A powerful pattern that lies somewhere between the extremes just described is the so-called

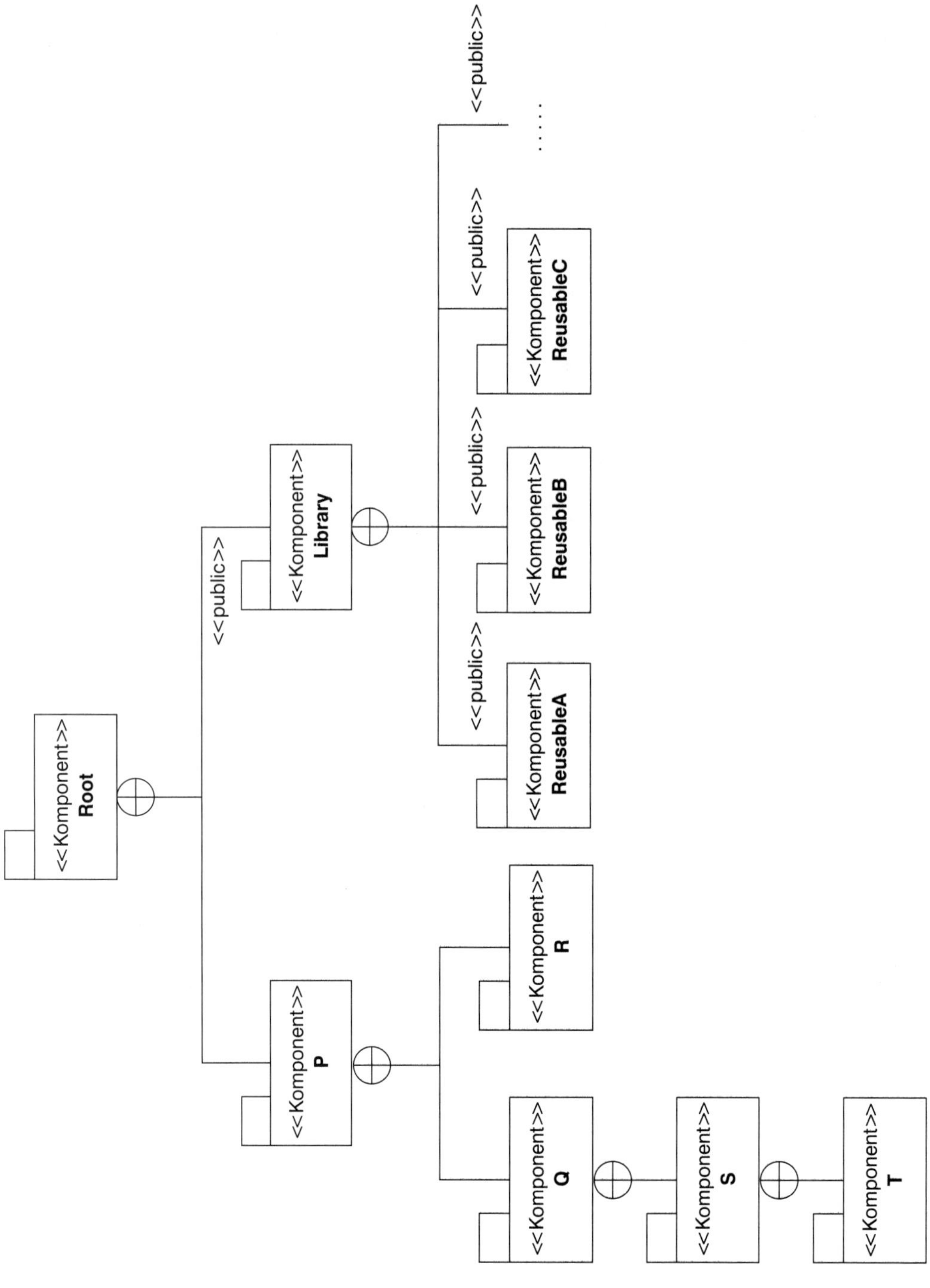

Figure 10.18
Komponent Library pattern

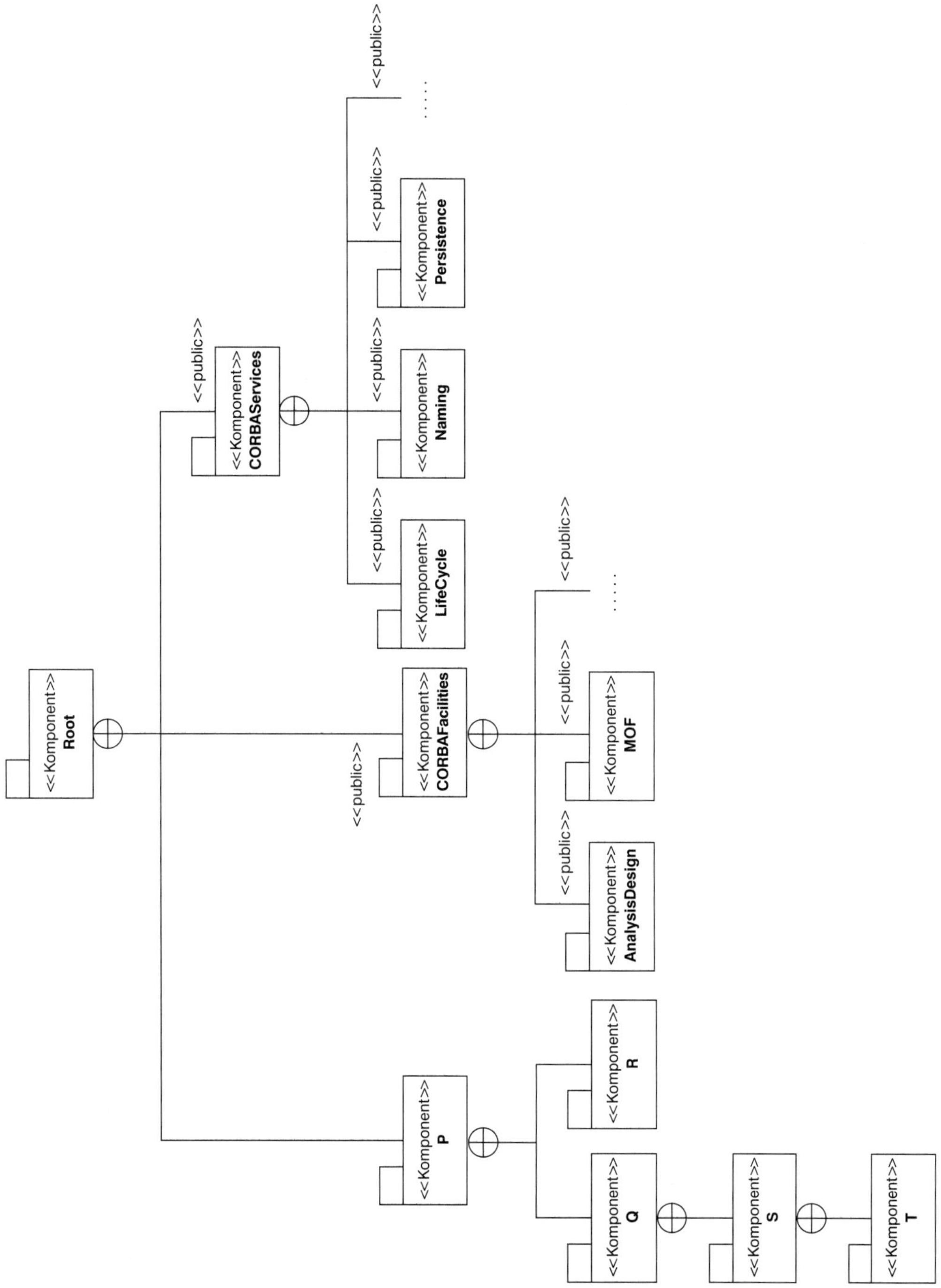

Figure 10.19
Komponent Library pattern applied to OMG's OMA

Komponent library pattern. The idea behind this pattern is to divide the containment tree into two distinct parts. One part contains the reusable Komponents that need to be universally visible since they are used by a relatively large number of Komponents, and the other part contains the "user" Komponents that are not themselves so widely used by others.

As illustrated in Figure 10.18, which provides a general representation of this pattern, the left-hand branch of the tree is organized according to the "need-to-know" pattern, while the right-hand side is organized according to the universal visibility pattern. The Komponents on the left-hand side, identified by single letters, are intended to represent arbitrary user Komponents (or applications in CORBA terminology).

The viability of this pattern depends on the use of public composition. The Komponents in the left-hand part of the tree can only see the sub-Komponents of Library because they are public sub-Komponents.

This is a useful pattern because it allows reusable Komponents to be collected together in a well-known, globally visible part of the containment tree, and thus supports a more bottom-up, assembly-oriented style of development characteristic of the component paradigm. As illustrated in Figure 10.19, this pattern can be used to provide a more sophisticated representation of the OMG's Object Management Architecture than that represented by Figure 10.15. Using this pattern, Komponents representing *CORBAServices* and *CORBAFacilities* can be organized in appropriate repositories (on the right-hand side and in the middle of the tree in Figure 10.19), leaving the application objects to be organized in a more deeply nested way (on the left-hand side).

PART 3

Embodiment

The artifacts generated by the component modeling activities are at a high level of abstraction akin to analysis and design. In fact, most of the artefacts are UML models. Naturally, these are only of value if they can be transformed into an executable form in an efficient and verifiable way. In KobrA, the set of activities involved in developing executable and deliverable incarnations of logical Komponents are encompassed by the concept of Komponent embodiment. Since the entire system is represented as a Komponent in KobrA, Komponent embodiment also addresses the implementation and packaging of deliverable artifacts. Part 3 of the book explains how Komponent embodiment is supported in the KobrA method, and how the activities involved can be combined with the component modeling activities to support an incremental approach to development.

Chapter 11 begins by describing a systematic strategy for mapping Komponent realizations into a form that can be processed by automated compilation tools. The approach is based on the principle of separating refinement steps (changes in abstraction level) from translation steps (changes in representation form), and involves a special implementation-oriented profile of the UML.

Chapter 12 complements Chapter 11 by describing how existing components, such as COTS components, can be used to realize and implement a KobrA component rather than implementing everything from scratch.

Chapter 13 rounds up by discussing various strategies for sequencing the component engineering and implementation or reuse activities, with a view to supporting the incremental development of the final system.

Implementation 11

The models and artifacts generated by the component modeling activities describe the properties of components at a level akin to analysis and design. Therefore, before any software can be deployed, these must be transformed into semantically equivalent executable forms. In practice this involves a manual step, in which the artifacts are translated into an intermediate representation amenable to processing by automatic tools (e.g. source code), and an automated step in which these tools are applied to generate the truly executable (e.g. binary) forms. In common with prevailing terminology, this intermediate representation is termed an "implementation," and the manual activity responsible for transforming a realization into an implementation is known as the implementation activity.

The only constraint on a Komponent implementation is that it faithfully embodies the semantics of the Komponent realization in a way that conforms to the non-functional requirements. Any implementation technology capable of satisfying this constraint can be used to implement Komponents, including high-level programming languages (e.g. Java, C++, C, Ada, Python), physical component technologies (e.g. JavaBeans, EJBs, COM, CORBA), and databases. Because of the properties of Komponents, physical component technologies are likely to provide the most flexible and natural implementation of Komponents, followed by programs written in object-oriented programming languages.

Although a Komponent's implementation can be thought of as being closely coupled to its realization, it is not regarded as being part of the Komponent containment hierarchy. Komponent implementations exist in a separate dimension, orthogonal to the Komponent containment hierarchy. This dimension is concerned with the abstraction level and representation form of components rather than their containment relationships. The orthogonality of these dimensions is important because it means that, in general, any process capable of producing a

Figure 11.1 Komponent implementation artifacts

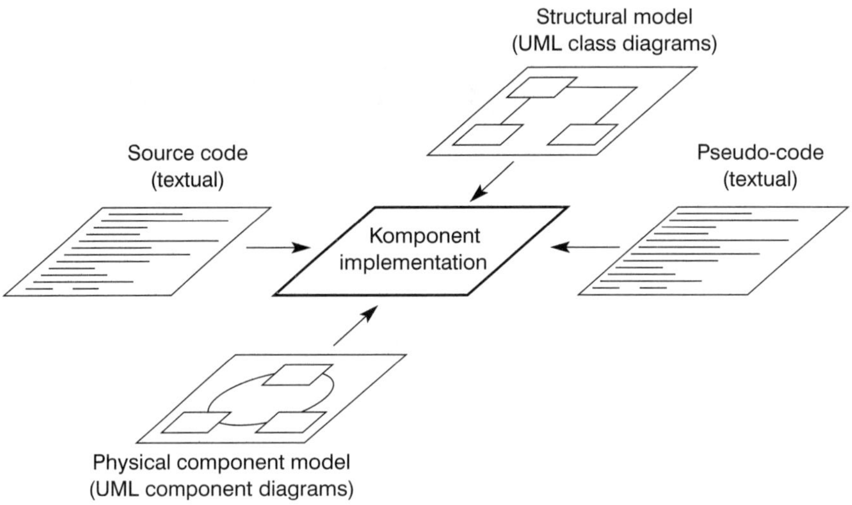

faithful implementation of a Komponent can be used in conjunction with the KobrA component modeling activities. However, KobrA defines its own recommended implementation activity that fits in with the spirit of simplicity and systematic development that underpins KobrA. This is the topic of this chapter.

The KobrA implementation approach is primarily based on the SORT approach described in [Bun01]. In general, it allows an implementation to have up to four different artifacts, as illustrated in Figure 11.1:

1 **source code** – a textual representation of a Komponent using a programming language, or some other processable implementation
2 **structural model** – a class diagram which represents the structural entities of a Komponent at an abstraction level close to source code
3 **physical component model** – maps logical Komponents to physical components
4 **pseudo-code** – describes the algorithmic implementation of a Komponent.

In addition, Komponent implementation within framework engineering makes use of the Komponent's decision model which describe the variabilities embodied within the Komponent.

11.1 Separating refinement and translation

The general goal of the implementation activity of any object-oriented development method is to translate an abstract, informal representation of an artifact (commonly referred to as a design) into a more concrete, formal[41] representation that can be processed by automatic tools. As illustrated in Figure 11.2, in

41. Formal in the sense that it can be processed by automated tools, rather than in the mathematical sense.

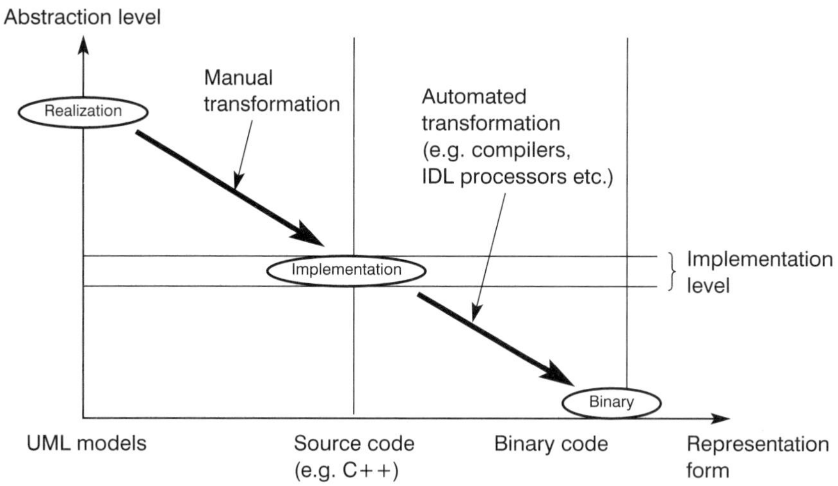

Figure 11.2 Goal of the implementation activity

KobrA the abstract, informal representation is a Komponent realization, and the concrete, formal representation (i.e. the source code) is an implementation in some programming language or other processable representation. Figure 11.2 also emphasizes that the implementation is only really an intermediate representation of the Komponent. The truly executable representation has to be created by a further, largely automated step. In KobrA, this step is part of the construction activity which is explained further in Chapter 13.

Most object-oriented methods provide little if any guidance on how to perform the implementation activity. Moreover, any guidance that is given almost always tries to span the gap between the abstract, informal representation and the more concrete, formal representation in one big step [BA99a]. High-level design models, and even analysis models, are typically mapped directly into

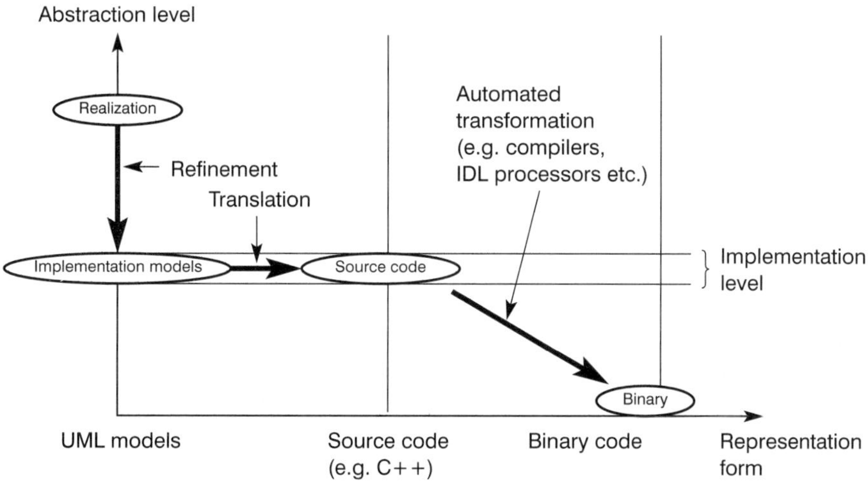

Figure 11.3 Separating refinement and translation

code with or without the help of CASE tool code-generation facilities [Web95]. Developers trying to perform such transformations therefore have to simultaneously concern themselves with different representation forms and different abstraction levels. As a consequence, the implementation activity is typically one of the weakest aspects of many contemporary object-oriented methods, and is the source of many defects.

As shown in Figure 11.3, in practice the step from an abstract, informal representation to a more concrete, formal one (i.e. from design to code) involves two distinct transformations: *refinement* and *translation*. The first, refinement, is the description of a given phenomenon at a lower level of detail (e.g. Komponent realization versus implementation models). In the context of the UML, a refinement step takes one arrangement of pieces and rearranges or expands them to create another group connected via refinement relationships [DW98]. The second transformation, translation, is the description of a given artifact in a different form (e.g. from implementation models to Java or C++ source code). In contrast to refinement, transformation describes a relation between two descriptions at the same level of abstraction. Both descriptions have the same subject, and contain the same amount of detail, but they use different languages or representational forms.

If refinement and translation are mixed, the relationship between a Komponent realization and its executable description is messy and potentially misleading. For example, an association between two classes can be represented as a simple line in a UML diagram. The implementation of such an association requires not only a change of notation, but also many implicit decisions (i.e., an association may be implemented by reference via an array of pointers, by a dictionary object, etc.). Therefore a developer implementing such an association has to make several, possibly undocumented, decisions which make the link not only hard to understand but also difficult to maintain.

The KobrA implementation approach is based on the principle of cleanly decoupling refinement from translation in the transformation of a Komponent realization to a Komponent implementation [BA99a]. Instead of performing both activities in one big step they are performed sequentially. The KobrA models in the realization process are first refined into a more concrete form at the implementation level of abstraction (i.e. the implementation level in Figure 11.2 and Figure 11.3) and are then translated directly into a tool comprehensible form (e.g. source code). This approach represents a further separation of concerns within the overall KobrA development method.

Separating refinement and translation steps reduces the likelihood of errors during the transformation process, since individual steps are smaller and only address one dimension of concern (i.e. either abstraction level or representation form). It also greatly improves traceability in the implementation process. The latter is achieved by cleanly defined relationships between the different levels of abstraction. This in turn increases the maintainability of the resulting implementation.

11.2 UML implementation profiles

Successfully separating concerns for refinement and translation in the implementation process is contingent upon:

1. the precise definition of the level of abstraction at which refinement should stop and translation should begin (the implementation level)
2. the provision of concrete assistance for the enactment of refinement and translation transformations.

The first of these requires the definition of a set of UML constructs, together with associated constraints on their use, that match the concepts in the chosen formal implementation representation (e.g. programming language). For this the UML profile mechanism can be used. This is intended specifically for the purpose of defining a tailored set of modeling concepts together with rules for their use. In KobrA, a UML profile is created to define the implementation level at which refinement ends and translation is performed. This is known as an *implementation profile.*

UML implementation profiles can be defined not only for each different kind of formal implementation language, but also for different company-specific ways of using these implementation languages. For example, a Java implementation profile could be created to capture the standard concepts of Java within the UML, but in addition, organization-specific variants of this profile could be defined to capture a particular organization's policy for using Java (e.g. a DoD-Java implementation profile). In general, therefore, implementation profiles can be tailored to best meet the needs of the project or organization applying KobrA.

The best implementation profile for a given project is not necessarily the one that provides the closest fit to the chosen implementation language and usage policies. The ubiquitous trade-off between generality and reuse applies to the implementation activity as well as other activities in KobrA. An implementation profile that closely matches the implementation language (and associated usage policies if they exist) typically leads to the most efficient usage of language features, and thus ultimately to performance, but it also lowers the reusability of the implementation models. The more general the implementation models the larger the set of implementation languages to which they can be translated. As usual, therefore, the trade-off is between performance and reusability.

Given the ubiquity of object-oriented languages, and the fact that component technology has its root in the object paradigm, it is likely that object-oriented languages such as Java, C++ and Ada will be used to implement KobrA components. Moreover, because of their common heritage, most mainstream object-oriented languages are based on the same core set of principles. In other words, the implementation profiles for specific object-oriented languages have a large intersection. Rather than defining a set of distinct implementation profiles for each distinct language therefore, KobrA offers a pre-defined implementation profile, called the Normal Object Form (NOF), which captures the common features of object-oriented languages. This prevents the redundancy of defining these common features multiple times for each language-specific implementation profile.

The NOF has the advantage that its features can be translated into any of the mainstream implementation languages. Thus, NOF implementation models are more reusable than language-specific implementation models. More importantly, however, the NOF can serve as the basis for defining language-specific and/or organization-specific profiles. The UML has the notion of specializing profiles, so language-specific profiles and/or organization-specific profiles can be defined as specializations of the NOF. This means that the NOF may be used directly to control the content of implementation models or can be tailored to the needs of specific projects and organizations using the UML profile mechanism.

11.2.1 The Normal Object Form

The Normal Object Form is a pre-defined UML profile that captures the core concepts of object-oriented programming [Bun01, BA99b]. Its features do not correspond to a strict intersection of the features of the main object-oriented programming languages, however, because these often present the same underlying concepts in different ways. On the contrary, the NOF's features have been defined by a careful analysis of the core concepts of object-oriented languages and other implementation-oriented notations such as ION [AI95].

As a UML profile, the NOF specification consists of:

1. a UML subset which contains only those modeling elements which are closely related to the implementation level
2. the definition of new model elements using the built-in UML extensions mechanism (i.e. stereotypes)
3. the definition of constraints on the use of the new and old modeling elements.

NOF models are consequently valid UML models. The difference between NOF models and regular UML models are that the former only contain model elements that are semantically "close" to the concepts of object-oriented programming languages. NOF implementation models therefore provide a graphical representation of the "as is" structure of an implementation, such as a program, in a tool-comprehensible form. The NOF can thus be thought of as defining a "normal form"[42] to which the models of a Komponent must be "reduced" before they are transformed into language-specific source code. This is obviously the origin of its name.

A description of the exact set of model types and model element types that make up the NOF, as well as a discussion of why they are part of the NOF, is lengthy and highly technical. In this chapter we therefore only provide a general overview of the make up of the NOF. Full details can be found in [Bun01].

At first sight it might appear that a *subset* of the existing UML modeling concepts would satisfy the needs of implementation models. Certainly the UML already contains many low-level features that have a very close correspondence to object-oriented language concepts, such as classes, methods, packages, etc. However, it turns out that there are also numerous other fundamental object-

42. Akin to the normal forms of the relational data model.

oriented programming language features that are not represented directly within the UML. For example, the various choices for the implementation of an association are not well supported by current UML modeling constructs. Therefore, as well as ruling out the use of many existing UML modeling elements, the NOF profile also adds several new modeling elements.

Because object-oriented programs focus on a structural presentation of how a software system is to run, it is the structural model which is primarily affected by the NOF. Naturally, the fundamental elements of object-orientation are present, such as classes, objects, attributes, links, and inheritance. However, the NOF requires that these concepts appear only at the level of detail in which they appear in an object-oriented program. Thus, for example, when a class appears in a NOF diagram, its precise implementation stereotype must be defined – whether it is an abstract class, a template class, or a utility class. If a class is not marked as belonging to any of these categories in a NOF diagram, this indicates that a decision has been made to implement it as a normal class. In contrast, from a normal UML diagram is not possible to determine whether the absence of specific markings is because the information has simply been omitted or because the corresponding decisions have not yet been made. A general goal of the NOF is to rule out such ambiguities. In this spirit, all attributes of a class must have visibility markings. In other words, it must be clear whether they are to be implemented as public, private or protected members (in the case of C++).

The NOF includes model elements from other kinds of UML diagram types, but, apart from class diagrams, the only one utilized in KobrA is the component diagram. This is because the information that would be provided by other UML diagram types is already contained in the realization models, or is captured in other artifacts such as the pseudo-code. The role of component diagrams within a Komponent implementation is to describe the mapping of Komponents (i.e. logical components) to physical components.

11.2.2 Transformation patterns

Implementation profiles define the level of abstraction (i.e. the implementation level) at which refinement should end and translation should begin. In other words, they define the output of the refinement activity and the input to the translation activity. However, they provide no guidance on how refinement and translation steps should be performed. In KobrA, the required support is provided in the form of transformation patterns. There are naturally two forms of transformation patterns – refinement patterns supporting refinement steps, and translation patterns supporting translation steps.

Since mathematically formal transformations cannot be defined between informal languages (e.g. between UML diagrams), transformation patterns present mappings in the most concrete way that remains – as patterns. Like the well-known patterns of Gamma et al. [GHJV95], they define general solutions to a recurring problem, as well as the context in which the solution is relevant, the trade-offs involved in applying the solution, and appropriate practical guidelines for using the solution. However, whereas the regular design patterns of Gamma et al. [GHJV95] and Buschman et al. [BMR96] define "good" object-oriented structures and behaviors, transformation patterns define "good" mappings between them. In other words, typical patterns describe the solution to an abstract problem, and thus may be reused as part of a transformation pattern (i.e. the transformation target structure) where the remaining parts (i.e. start structure and needed context information for such a transformation) have to be added.

Since the transformation patterns used in a project are oriented towards the applied implementation profile, they, like implementation profiles, come in many flavors, and can be tailored to the needs of specific organizations and/or implementation technologies. The transformation patterns defined in KobrA are naturally focussed on supporting the NOF implementation profile. A more detailed description of the transformation patterns available for use within KobrA is provided in [Bun01].

11.2.2.1 Refinement patterns

Refinement patterns define sound refinement mappings from regular (i.e. non-NOF) UML modeling elements to NOF-UML elements. To keep track (traceability) of refinements the refinement relationships between such elements are made explicit by using the UML stereotype <<refinement>>. This enables developers to traverse the refinement structure in order to identify those places where further work is needed. However, as mentioned above, simply defining the mapping is not sufficient. If there is more than one pattern for a specific construct it must be possible to choose the pattern which is most suitable for a given context. An essential part of a refinement pattern is therefore a list of the primary quality factors which the refinement influences (e.g. performance) and the level of this influence. An example refinement pattern, describing the refinement of a composition relationship, is shown in Figure 11.4.

Basically a refinement pattern consists of three parts:

1. The top area is used for storing administrative information (i.e. unique pattern number, pattern name, and pattern family). These are used to help developers find a specific pattern quickly, and to support the use of tools for maintaining the complete set of patterns.
2. The area in the middle of each pattern contains a graphical description of the actual refinement. This description depicts how a specific UML construct (e.g. a unary association) can be refined to a UML-NOF construct (e.g. a clientship relationship) by presenting both constructs and their refinement relationship.
3. Finally, the bottom area characterizes the context in which the pattern is applicable, augmented by textual descriptions of the refinement itself. The context characterization provides a subjective estimate of the pattern's influence on various quality factors. In addition, the refined construct (i.e. the UML-NOF construct) and the refinement itself is explained textually to provide a better understanding.

Figure 11.4
Refinement pattern

# 7	Pattern "Composition (1)"	Category: Structure

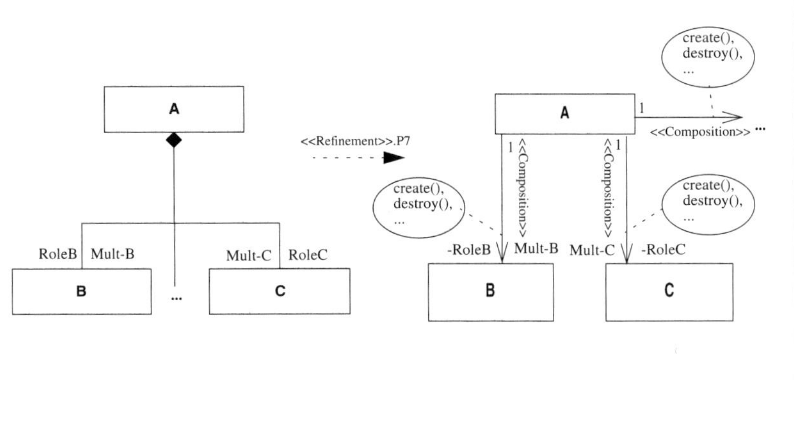

Context:

	Influence (- -, -, o, +, ++)
Maintainability	o+
Performance	++
Reliability	o
Reusability	o
Security	-
Space	-
...	

Description:

- A composition relationship is refined by clientship-relationships.

How-To:

1. Each composition relationship becomes a <<composition>> clientship directed from the whole to the part.
2. The multiplicity at the client side is one and the multiplicity at the server side is taken from the original construct.
3. Roles of the original construct become private roles of the clientships. If no roles are given they are created for the clientship.
4. The create operation of class A creates objects of class B and C.
5. The destroy operation of class A creates all objects of class B and C.
6. Other adornments of the composition relationship are unchanged.
7. The clientship relationships have to be refined further (i.e., additional Call-Specifiers etc.).

Trace: - Refinement Patterns: 1, 9-18, 55, 57, 59, 60
- Translation Patterns: 18-20

11.2.2.2 Translation patterns

Translation patterns perform a similar role to refinement patterns, but for translations instead of refinements. Although the two are strictly separate, the relationship between the lowest-level refinement patterns and the associated programming constructs must be clear. In other words, translation patterns are not solely oriented towards implementing a NOF model element, but also towards continuing the process of seamlessly developing code from models. Then, and only then, it is possible for a developer seamlessly to traverse a software system from early analysis to code.

Figure 11.5
Translation pattern

# 1	Pattern "Clientship <<Composition>> (1)"	Target: C++	Category: Structure

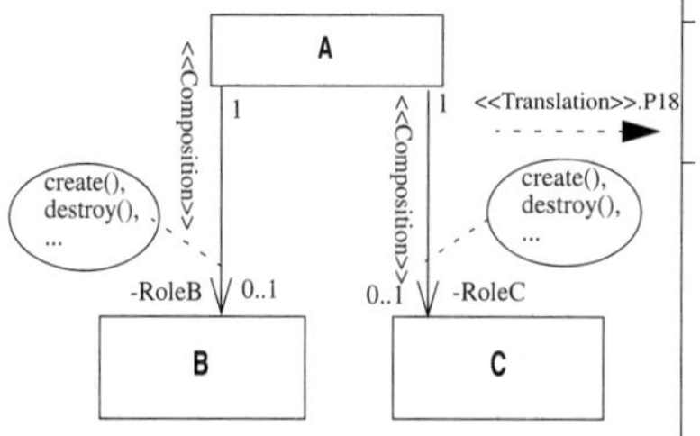


```
class B {
friend class A;
...
};
```

```
class C {
friend class A;
...
};
```

```
class A {
public:
   A();
   ...
private:
   B b;
   C c;
   ...
...
};
```

```
A::A():b(),c()
{
}
```

<u>Context</u>:

	Influence (- -, -, o, +, ++)
Maintainability	o
Performance	+
Reliability	o
Reusability	o
Security	--
Space	-
...	

<u>Description:</u>

- Another way of implementing an <<Aggregation>> clientship is to use embedded objects. The composite class A defines a variable which holds zero or one objects of a composite part (i.e., variables b and c). The variables b and c are initialized during the creation method of class A. The lifetime of the composite parts of object A, namely objects of class B and C, is restricted to the lifetime of the composite object due to the nature of composition. Therefore A's creation method instantiates class B and C in an inheritance like manner. If an object of class A ceased to exist the embedded instances of classes B and C are automatically destroyed. In the field of programming this strategy is known as embedded objects.
  Please note that the implementation of constraints (e.g., ordering), or -specifiers is discussed by pattern: ***25-27***

<u>Trace:</u> - Refinement Patterns: 2, 3, 7, 20, 21,-23
- Translation Patterns: 1-7, 8, 25-27

Like refinement patterns, translation patterns are usually not applicable in all circumstances. Usually there are several applicable translation patterns for a given NOF element, because the context (e.g. non-functional requirements) plays a major role in determining which translation is the most suitable (e.g. a system needing high-speed will be implemented in a different way from a system where security and safety are the major concerns). Consequently, a translation pattern has to provide this context information to allow developers to choose the pattern most suitable for their needs. As for refinement patterns we use subjective estimates ("—, -, o, +, ++") of the degree to which the translation pattern influences certain quality factors, and these can differ from those of related refinement patterns. Translation patterns are also presented using the same template as refinement patterns (see Figure 11.5).

11.3 Implementation artifacts

The previous sections described the basic supporting mechanisms of KobrA concerning the implementation of models (i.e. mapping graphical models to source code). This section introduces the models and their encompassing UML diagrams as utilized by the KobrA implementation activity.

11.3.1 Implementation structural model

The implementation structural model is derived from a Komponent's realization structural model which contains class diagrams that describe the structural properties of a Komponent. The implementation structural model is a refinement of the realization structural model composed only of NOF elements. It therefore describes the structural properties of the Komponent at a low level of abstraction, close to a programming language, and can be viewed as a graphical representation of the structure of a program.

11.3.1.1 Class diagrams

A realization class diagram describes the structural properties of a Komponent at a level of abstraction akin to design. This provides insight into how the structural entities of a Komponent should be realized, but does not provide sufficient information about how these entities are mapped onto code. Thus, it can contain fairly abstract concepts, such as association classes, ternary associations, etc. which have no direct counterpart in programming languages. In other words, decisions which have to be made when implementing such diagrams are not part of these diagrams (e.g. how associations are implemented by reference, by value, etc.). The implementation class diagram provides a more concrete description of the structural properties of the Komponent in terms of the concepts within the NOF implementation profile of the UML. Therefore an implementation diagram has to model the structural entities of a Komponent in such a way that it can be mapped onto code without making implicit decisions. It is produced by the application of the refinement patterns mentioned in the previous section to the realization class diagram, and thus embodies the results of the refinement decisions that have been taken to refine the higher level concepts into NOF-level concepts.

The creation of the implementation class diagram can be automated to a reasonable degree (application of refinement patterns), but the decision about the "best" refinement, taking into account the non-functional requirements, remains a creative activity.

An example of a realization class diagram for the *LoanManager* Komponent is illustrated in Figure 5.2. It includes all the information from the corresponding specification class diagram, but also contains additional information. The level of information is akin to design in the traditional software development process, and is not yet detailed or complete enough for implementation. Open questions are:

1. How can the association class *Loan* and the corresponding association between the *Account* and *Item* classes, be efficiently implemented? If the expected number of links is small, it may be sufficient to implement the association class as a normal class and connect all three classes by individual associations. In contrast, if the number of links is expected to be high the association class may become a dictionary class, which can use hash tables to provide quick access.
2. Which parameters and data types are going to be used for attributes and operation parameters of the single classes? Without specifying the concrete interface of a class it is nearly impossible to correctly implement access to that class. Even languages which do not need the specification of types (e.g., Smalltalk) need a carefully specified interface to guarantee correct responses at run time.
3. Are there any regulations concerning visibility and is there a need for get/set operations? A typical example is attributes which should not be manipulated directly from the outside to prohibit unwanted side-effects (i.e. a private attribute), but which can be accessed and updated in a clean way (i.e. access is granted by get/set operations).

A simple example of an implementation class diagram is illustrated in Figure 11.6. This diagram shows that the association class construct is refined into several uni-directional associations between the *Loan* and *Account* classes, as well as between the *Loan* and *Item* classes. Each association now has clearly defined roles, as well as a clientship stereotype which determines its implementation (i.e. the <<Standard>> stereotype, as defined by the NOF, depicts a relationship where the client holds a direct reference to the public parts of its server in its main data structure). All attributes are declared to be private (information hiding), which required the addition of several get/set operations, and all operations are declared to be public. For each operation the parameters and types are defined to be close to their implementation. In this example the class *LoanStore* holds a list of *Loans* in its data structure. This list can either be implemented as a linked list or via hash tables.

11.3.2 Component diagrams

There are many different implementation technologies that can be used to implement components, ranging from programming languages through to databases and script files. Not surprisingly, one of the most natural and flexible ways of implementing a KobrA Komponent is to use one of the well-known physical component technologies such as JavaBeans, Enterprise JavaBeans, CORBA or COM.

The end product of a software development project is a set of executable modules that can be loaded onto a computer and executed. Each separately

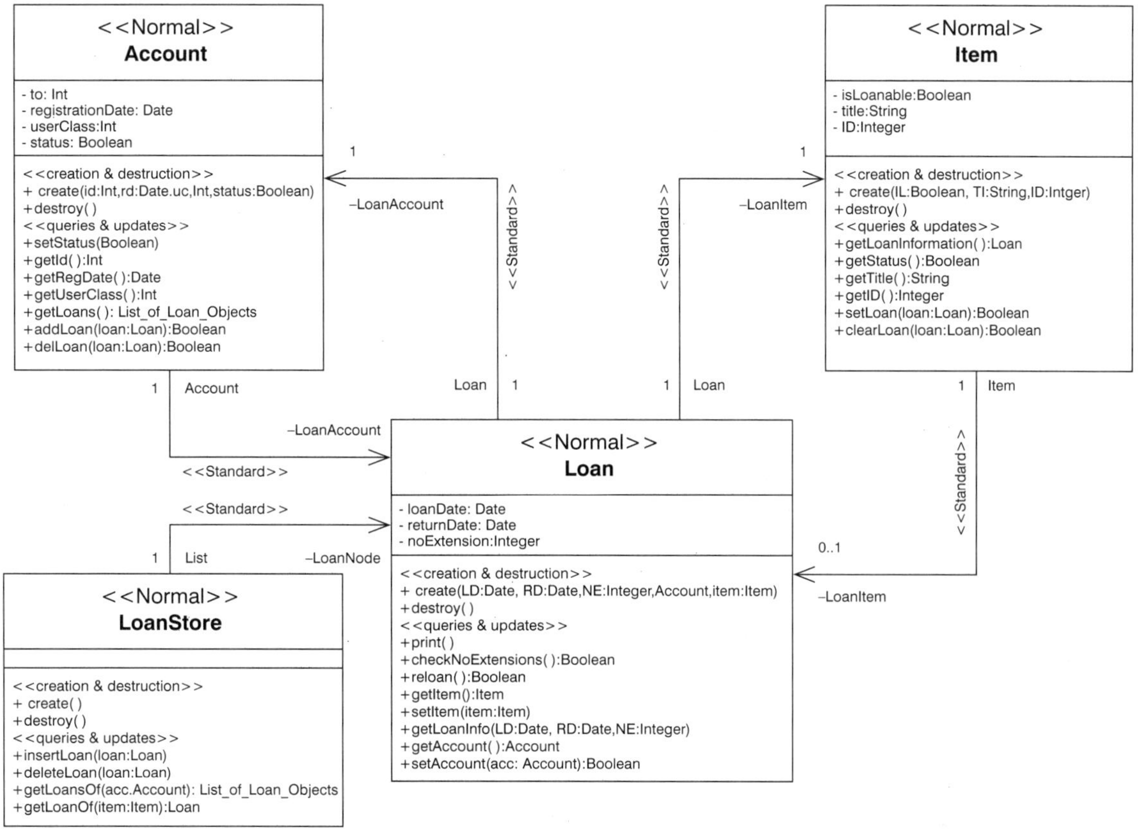

Figure 11.6
Implementation class diagram

deployable unit is called a physical component in KobrA, regardless of the implementation technology used to generate it. This is the component concept in the UML. With most implementation technologies, the physical components have to already be defined (at least partially) in the source code or other machine comprehensible representations. For example, regular programs require a main program or similar construct, JavaBeans require the appropriate interface elements and CORBA components require the appropriate interface definition. The purpose of the component diagram is to define the mapping of KobrA Komponents to physical (i.e. separately executable) components as embodied in the source code.[43]

43. This usage of component diagrams differs from that presented in [Bun01] where component diagrams are applied to depict the physical structure (i.e. file structure) of the final program. However, this is not sufficient for KobrA since a Komponent can be implemented in various ways and not only by an object-oriented programming language (i.e. component technologies such as CORBA, JavaBeans, etc.). Therefore a mapping of "logical" to physical components is required.

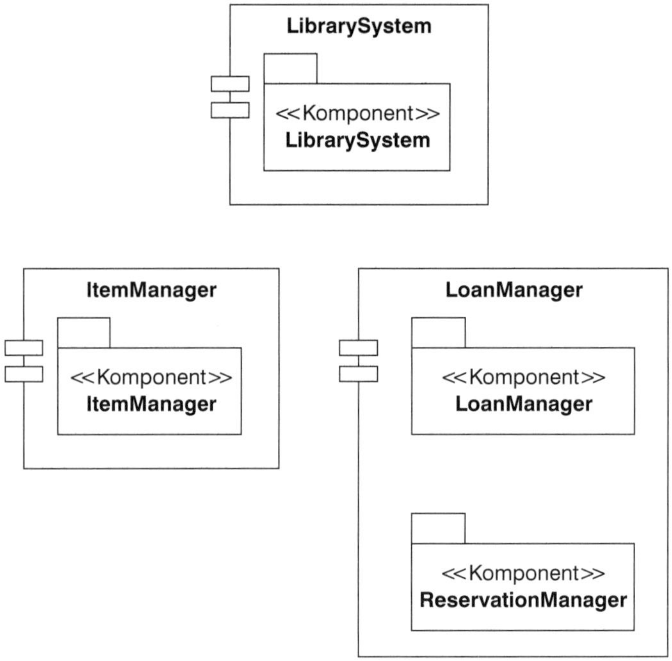

Figure 11.7
Library component diagram

Not every logical component has to be implemented as a separate physical component. For reasons of performance, or technology limitations, it may be desirable to map a group of KobrA components into a single physical component (see Figure 11.7). This figure shows one way of mapping the Komponents in the *LibrarySystem* tree into physical components. *LibrarySystem* and *ItemManager* have been mapped into their own individual physical components, while the *ReservationManager* has been included, together with the *LoanManager* Komponent, into a single physical component. The process of folding several components into a single physical component is known as "flattening" and is explained further below.

Obviously, as an extreme case it is possible to flatten a complete Komponent tree into a single physical component as illustrated in Figure 11.8. This may be the best policy when performance is a major concern, or when the available implementation vehicle is a simple programming language. The folding of sub-Komponents into their parent's physical component in this way represents a trade-off between deployment flexibility and performance which must be made in view of the execution platform, the implementation vehicle and other relevant factors such as the predicted changes to the system.

By and large, component diagrams conform to the principle of locality, like all other KobrA artifacts. However, it does not make sense to override decisions, made at one point in a KobrA tree, lower in the tree. If a given Komponent has been assigned to a physical component by one of its ancestors, then it has no component diagram. It will still have a class diagram and source code, however, because the assignment of Komponents to physical components usually only affects a small part of the implementation.

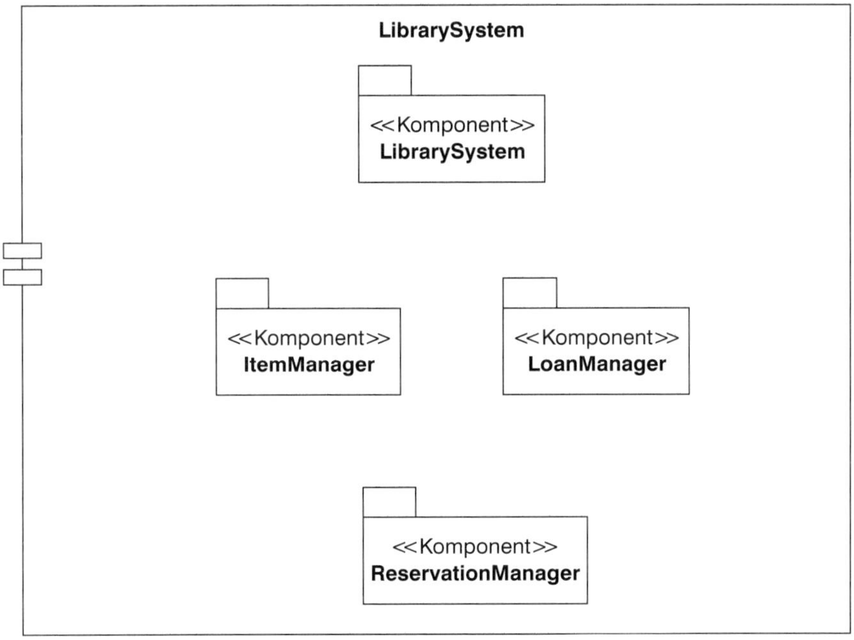

Figure 11.8 Flattening a complete tree

It is possible for component diagrams to make partial statements about the mapping of logical components to physical components, and to defer parts of the mapping to the diagrams of sub-Komponents. Consider Figure 11.9 for example. This is the component diagram for the implementation of the *LibrarySystem* Komponent that assigns the *LibrarySystem*, *ItemManager* and *LoanManager* Komponents to separate physical components. This means that the implementation of *LibrarySystem* expects to interact with instances of *ItemManager* and *LoanManager* as if they were separate physical components (e.g. via an ORB). However, since the component diagram does not include the *ReservationManager* sub-Komponent of *LoanManager*, the decision about how this is implemented is deferred to *LoanManager*. The implementation of this Komponent can either flatten the tree and include it with its children into a single physical component, or it can map them all to separate physical components.

11.3.3 Pseudo-code

The pseudo-code descriptions, created during the implementation activity, are developed manually to highlight the algorithms used to implement a Komponent's operations. Often the information captured by the pseudo-code description is already present in the form of a Komponent's interaction or activity model but the pseudo-code provides a compact and low-level description of these models for the purpose of implementation. The creation of pseudo-code is therefore optional, since a developer may use the realization models directly.

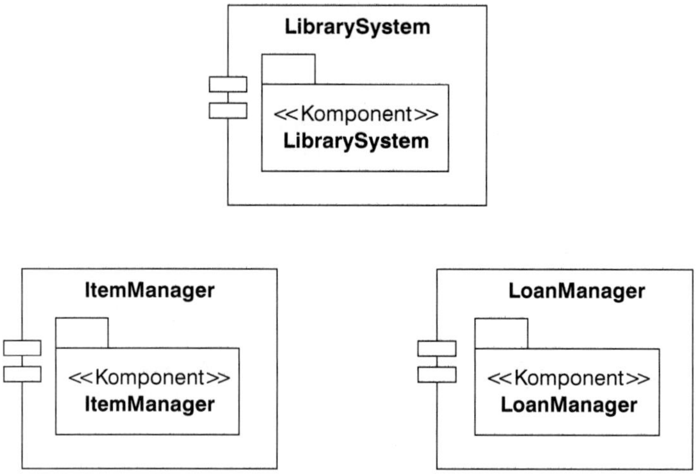

Figure 11.9 Partial logical to physical node mapping

11.3.4 Source code

The source code is the final result of the KobrA implementation activity. It contains a description of the component in a high-level language such as C++, Java, or CORBA IDL, but is not limited to programming languages (e.g., using COM or DCOM components).

Figure 11.10 C++ header file for LoanStore class

```
class LoanStore {
    public:
            // creation & destruction
            LoanStore( );
            ~LoanStore( ) {};
            // queries & updates
            Loan getLoanOf(Item &item);
            Boolean insertLoan(Loan loan);
            Boolean deleteLoan(Loan loan);
            LoanNode* getLoansOf(Account acc);
    private:
            struct LoanNode {Loan Element; LoanNode *Next, *previous;};
            LoanNode *Header;
}
```

Figure 11.10 visualizes how the *LoanStore* class is implemented in C++. The example displays only an excerpt of the header file to indicate the implementation. Additionally the translation patterns or the pseudo-code artifacts have defined the prototypes of the methods which are then stored in a body file.

```
Loan::LoanStore getLoanOf (Item &item)
{
LoanNode *Ptr = Header->Next;
while( Ptr != Null)
{
  Ptr=Ptr->Next;
  if (Ptr->Element.getItem( ) == item)
   {
        return Ptr->Element;
   }
   return Null;
}
```

Figure 11.11
C++ code for getLoanOf() method

Figure 11.11 shows how the methods of the *LoanList* class may be implemented. The example displays only the complete implementation of one method. All subsequent methods are implemented in the same style.

11.3.5 Sample implementations

To complete the picture of how Komponents can be implemented, in this section we show how the Komponents of the *LibrarySystem* could be mapped into the three leading component technologies. We also illustrate how data classes in the case study could also be represented using XML. The illustrated mappings are not intended to be complete, but rather are intended to illustrate the principles by which the information in the UML models is mapped into the different component standards. All the implementations assume a one-to-one mapping of Komponents to physical components.

11.3.5.1 COM+ Mapping

COM+ is a binary standard, which means there is usually no source code associated with a COM+ implementation. Nevertheless, there is a very close match between the principles of KobrA and the principles of COM+ Like KobrA, COM+ views the client–server relationship as the fundamental interaction mechanism. A client is an object that accesses the functionality and contents of other objects. Like KobrA, a client is also called a container when it also maintains the persistent states of object instances. A server in COM+ is a provider of a component and its constituent objects and represents a binary standard for a demand-loaded code module such as a DLL or an EXE that makes a component and its objects available to the outside world.

Komponents can be mapped individually or group-wise into a COM+ component. Simple classes used in the realization of a Komponent are handled as local objects in the COM+ model. Figure 11.12 illustrates a situation in which the *LibrarySystem* Komponents *AccountManager*, *LoanManager*, and *StockManager* are all implemented as individual COM+ components, and the *Account* class in the *AccountManager* is implemented as a local object.

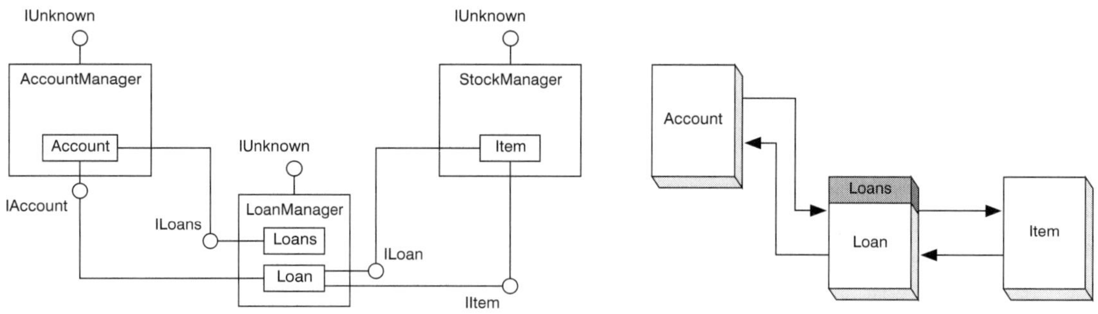

Figure 11.12
LibrarySystem Komponent and relations

A Komponent's specification functional model is most naturally transformed into a COM+ CoClass interface and the ATL-Implementation class captures the class with outside visibility. Furthermore a component factory is added to the COM+ component so that the component can be instantiated by a call to a COM+ system (see Figure 11.13)

The mapping of a Komponent relationship to COM+ relationship is more sophisticated. Simple one-to-one relationships can be directly transformed. Thus the relationship in a Komponent structural model will be mapped to a reference

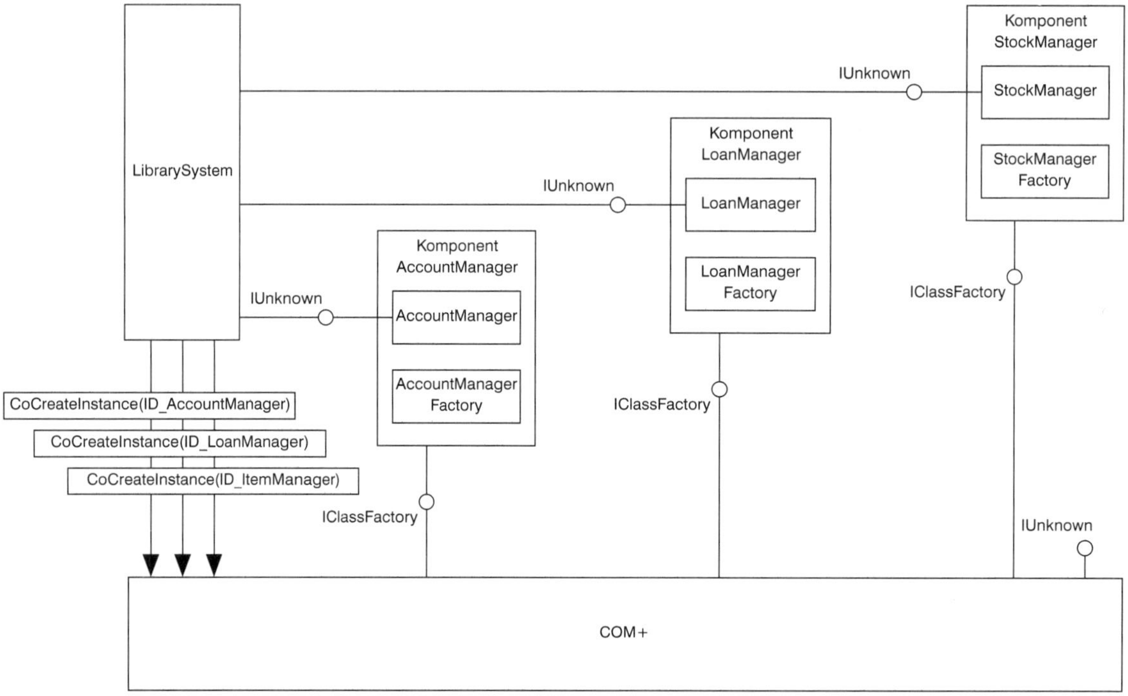

Figure 11.13
COM+ LibrarySystem creation

between the interface of the corresponding local COM+ server object and COM+ client object. The mapping of a one-to-many relationship will change the inner structure of a component. The object at the target side (multiplicity many) of the relation has to be split into a normal COM+ object and a COM+ collection. Existing one-to-one relationships will then be assigned to the new COM+ object, whereas the one-to-many relation will be assigned to the new COM+ collection. For example, the *Loan* inside the *LoanManager* is split into a new *Loan* object and a *Loans* collection (see Figure 11.12). The new *Loan* object references the *Account*, whereas the *Account* references the new *Loans* collection.

At run time, the COM+ components will be created according to the properties of the Komponent creation tree. A client component calls the COMt CoCreatelnstance() function with the ID of the component. The IUnknown interface will then be returned and the parent component will scan the interface of the child. For example, the *LibrarySystem* will call the CoCreatelnstance() function for the *Loan*, *Account* and *StockManager* and obtains the pointers to the corresponding IUnknown interfaces (see Figure 11.13). After that the *LibraryManager* will determine the interfaces of the components via the IUnknown interfaces.

As well as the structure of Komponents, their run-time behavior also influences the nature of COM+ components. Each <<acquires>> relationship requires the integration of a delivery method inside the acquiring component. This makes it possible to obtain the interface pointer to an acquired component from the component that created it. For example, the *LibrarySystem* turns over the StockManager interface pointer to *LoanManager* via the *setIStockManager()* method that extends the *LoanManager* interface (see Figure 11.14).

11.3.5.2 EJB mapping

Enterprise Java Beans is a component technology based upon the Java programming language. In contrast with COM+, components are described textually as a variety of Java program modules. Two forms of EJB are distinguished, Session Beans and Entity Beans. In this example, the *LoanManager* Komponent is implemented as a Session Bean.

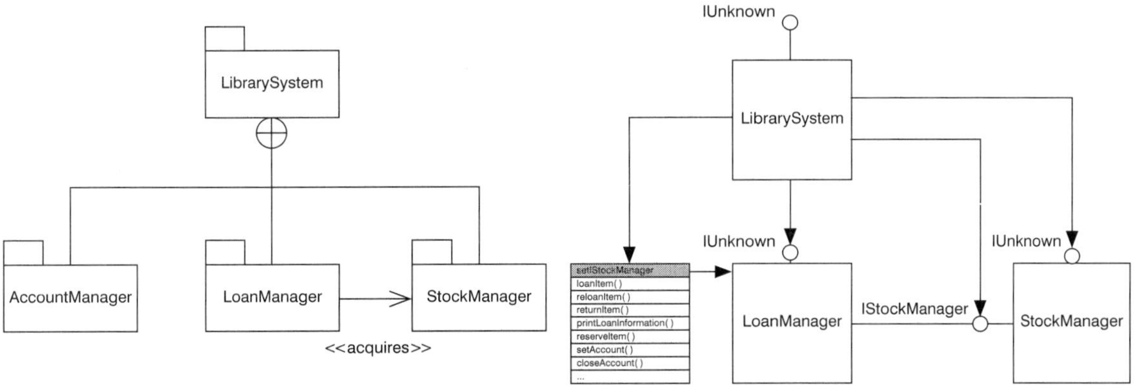

Figure 11.14
COM+ StockManager usage

```
package ejb.LoanManager;

import javax.ejb.*;
import java.rmi.Remote;
import java.rmi.RemoteException;
import java.util.*;

public interface LoanManagerHome extends EJBHome {
   public LoanManager create( ) throws CreateException, RemoteException;
}
```

Figure 11.15
LoanManager – Home Interface

The functional core of an Enterprise Java Bean (EJB) are regular Java methods, which can thus be generated from the implementation models in the same way as for a regular Java program. What distinguishes an EJB component from a simple application is that it implements the Home and Remote interfaces. Figure 11.15 illustrates the EJB Home Interface for the *LoanManager* EJB, and Figure 11.16 illustrates its Remote Interface.

The core code that actually encapsulates the functionality of the Komponent is contained in a class which implements the SessionBean interface as illustrated in Figure 11.17.

```
package ejb.LoanManager;

import ejb.ItemManager;
import ejb.MessageHandler;
import ejb.ReportWriter;

import java.rmi.RemoteException;
import java.rmi.Remote;
import javax.ejb.*;

public interface LoanManager extends EJBObject, Remote {
    public void closeAccount() throws RemoteException;
    public void loanItem( String filter ) throws RemoteException;
    public void printLoanInformation( int currentAccount ) throws RemoteException;
    public void reloanItems( String [] items ) throws RemoteException;
    public void reserveItem( String filter ) throws RemoteException;
    public String [] returnItems() throws RemoteException;
    public void setAccount( String id ) throws RemoteException;
    public void setItemManager() throws RemoteException;
    public void setMessageHandler() throws RemoteException;
    public void setReportWriter() throws RemoteException;
}
```

Figure 11.16
LoanManager – Remote Interface

```
package ejb.LoanManager;

import ejb.ItemManager;
import ejb.MessageHandler;
import ejb.ReportWriter;

import javax.ejb.*;
import java.io.Serializable;
import java.util.*;
import java.rmi.*;

public class LoanManagerBean implements SessionBean {

  private transient SessionContext ctx;
  private transient Properties     props;

  //SessionBean interface methods implemented here
    …..
  //remote interface methods implemented here
     ….
}
```

Figure 11.17 LoanManager Session Bean

```
#include "ItemManager.idl"
#include "MessageHandler.idl"
#include "ReportWriter.idl"

interface Account{
  attribute string id;
};

interface Item{
  attribute string id;
  attribute boolean isLoanable;
  attribute string title;
};

……

interface LoanManager{
    void closeAccount();
    void loanItem( in ItemFilter filter );
    void printLoanInformation( in Account currentAccount );
    void reloanItem( in ItemSeq items);
    void reserveItem( in ItemFilter filter );
    void returnItems( out ItemSeq items );
    void setAccount( in Account account );
    void setItemManager( in ItemManager itemManager );
    void setMessageHandler( in MessageHandler messageHandler );
    void setReportWriter( in ReportWriter reportWriter );
};
```

Figure 11.18 LoanManager IDL interface

11.3.5.3 CORBA mapping

CORBA provides a looser mapping of a Komponent's interface to specific implementation languages. The only requirements is that the interface of a CORBA component, known as an application, be defined in the OMG's standard interface definition language (IDL). This definition can be processed by dedicated IDL Compilers to create the code fragments and stubs needed to make the application independently executable and accessible in a distributed environment. Implementing a Komponent in CORBA therefore amounts to the creation of the appropriate IDL, together with the creation of the functional code in the way described in the previous sections. Figure 11.18 illustrates a CORBA IDL representation of the *LoanManager's* interface.

11.3.5.4 XML Mapping

The previous three subsections have all addressed the implementation of Komponents. However, as explained in Part 2 of this book, simple classes also play an important role in KobrA as a means of storing and transporting data objects. An important standard for representing such data objects is the Extensible Markup Language (XML) mark up standard. Figure 11.19 shows how the simple class, *Item*, which stores the data for loanable items in a library, could be represented in XML.

Figure 11.19 Item XML representation

```
<?xml version="1.0"?>
<!DOCTYPE Item SYSTEM "item.dtd">
<Item>
  <id>42</id>
  <isLoanable>true</isLoanable>
  <bibliographicInfo>Colin Atkinson et al:
Component-based Product Line Engineering with UML.
The Component Software Series.
Addison-Wesley, 2001.</bibliographicInfo>
</Item>
```

11.4 Implementation process

In this section we explain the associated activities by which the concepts of the NOF and its accompanying refinement and translations patterns are the tools needed to systematically develop high-quality object-oriented systems. The NOF defines the theoretical foundation, and the collective set of refinement and translation patterns can be seen as a knowledge archive for choosing the pattern most suitable for project's specific needs.

The KobrA implementation process is aimed at providing support for systematically applying refinement and translation patterns based on the NOF. The process itself is defined in such a way that it is adaptable to the overall development process and notations/languages used. Overall the process consists of four major activities:

- **Flattening** serves to develop the physical component model by mapping logical Komponents onto physical components. The physical components structure can be developed by applying specific refinement patterns, which map static structures onto UML components, or by human experience. In summary, the flattening activity requires careful decisions about the subsequent implementation and has to consider the prevailing non-functional requirements as well as other context factors.
- **Refinement** is aimed at refining the structural model of the Komponent's realization to a level of detail close to source code, and to develop the implementation model which reflects the physical structure of the Komponent. The activity takes the existing structural model of the Komponent as an input and refines this model, by the application of refinement patterns.
- **Translation** focuses on mapping the refined structural model, the implementation model, as well as the execution and interaction model of the Komponent realization to source code. This is the latest point in time to decide on which programming language, component technology, or middleware to use. The result of the translation activity will then be a representation of the Komponent in the chosen implementation technology.
- **Finalization** aims at developing a Komponent implementation which will not only compile but also satisfy relevant non-functional requirements.

11.4.1 Flattening

The flattening activity is used to develop the physical component model by mapping the logical Komponents onto physical components. The main goal is to flatten the KobrA tree and to decide on the composition of physical components.

The physical component model (i.e. the component diagram) is developed by carefully examining the KobrA tree to decide how to set up physical components, and by applying specific patterns that map static structures onto UML components.

The process of developing the physical component model can be summarized as follows:

1. **Analyze tree.** This is performed by traversing the tree and by examining if the actual Komponent and its descendants are already physically located within an ancestor component. In this case an additional physical component model is not required and the flattening activity can be stopped. Furthermore the containment of all descendants within a physical compo-

nent is checked. The result is a list of descendants or physical components, which may be contained within the actual component.

2 Perform flattening.

a Select appropriate patterns.

b For each of the Komponents identified in the previous step, decide if it is to become a distinct physical component or a part of a composite physical component. This is performed recursively by determining the containment of descendants at each level. The exit criteria are either if a lowest-level physical component is reached, if a descendant should be mapped onto a different component, or if the decision on physical composition is deferred.

c Document the performed actions to keep track of decisions.

3 Develop physical component model.

a Describe each identified component by its UML symbol.

b Show subpackages if more than one Komponent is contained.

c Adjust the component structure by adding additional components (e.g. needed for a specific implementation technique).

11.4.2 Refinement

The refinement activity is aimed at refining the structural model of the Komponent's realization to a level of detail close to source code and has to deal with UML class diagrams. Class diagrams are refined from the Komponents' realization models. Therefore, primarily, the structural model is examined in order to identify those elements, which are not at the appropriate level of detail and which have to be refined. However, the refinement activity is not restricted to the structural model. Often, information captured by other diagrams (e.g. collaboration diagrams), is folded into the NOF structural model as part of the refinement activity. Typical examples are the messages exchanged between two classes. Obviously each connection between two objects in a collaboration diagram depends on an association/link in the corresponding class and object diagram. The refinement of such an association takes this information into account and represents it within the NOF structural model (e.g. a method call in a collaboration diagram becomes part of a NOF clientship relationship (Call-Annotation). An additional task of the refinement activity is to develop the pseudo-code descriptions to enrich the dynamic behavior documentation as needed for translation and finalization.

The process for refining realization diagrams can be summarized as follows:

1 Identify the elements to be refined. This is performed by going through the class diagram elements and by collecting all entities that cannot be directly mapped onto source code (i.e. which are not at the NOF level).

2 Select appropriate refinement patterns.

   a For each entity select a refinement pattern from the catalog. If there is more than one possible pattern select that which matches the non-functional requirements.

   b Sort all selected patterns according to a prioritization strategy.

3 Perform refinements.

   a Start with patterns of highest priority.

   b Perform refinement as shown by the pattern.

   c Document the performed refinement to keep track of changes.

4 Identify dynamic entities. This is performed by going through the refined class diagram as well as the realization behavioural diagrams to collect all those operations, classes, etc. which may contain non-trivial behavior and by specifying this behavior in terms of pseudo-code.

5 Update physical component model. This is performed to reflect the actual refinements in the component composition (i.e. a specific refinement may invalidate the actual component composition).

In summary every realization class diagram has to be refined to the NOF level of detail during the refinement activity.

11.4.3 Translation

The translation activity is aimed at mapping the physical component model, implementation structural model, behavioral Komponent realization models and the pseudo-code to source code. Thus translation has to deal with four kinds of UML diagrams: class, component, interaction, and activity diagrams.

11.4.3.1 File structure

The first activity to actually map a Komponent onto source code is to set up its physical structure. In other words the single files, as required by a specific implementation vehicle (e.g. body and header files, or "include" relationships in the case of C++) are created.

In KobrA the implementation models are either translated by applying appropriate translation patterns or by implementation guidelines defined in the context of technologies such as CORBA and JavaBeans. Translation patterns support and precisely guide naming (e.g. of files) activities and the implementation of dependencies between components. One example is the import dependency between a specification and a body file. In C++ such a dependency is implemented via an *#include* statement which itself is resolved by a precompiler.

In addition to creating the physical structure, the infrastructure for actually compiling and executing a physical component has to be defined. In the context of object-oriented programming languages this could be the definition of a Makefile including library paths or the reservation of network resource. In general, the physical structure is generated as soon as the implementation model exists in a reasonably stable state by the construction activity described in the next chapter.

11.4.3.2 Static structure

As soon as the infrastructure is ready the structural model of the Komponent is implemented. The model is analyzed to identify its translation sets (i.e. the model is partitioned into chunks), which are then prioritized. The goal is to define a sequence of translation sets, whereby each set can be handled by a single programmer. For translation the prioritization is not used to ensure that high-risk features are completed early, and that low-risk features are delivered last. Instead prioritization is used to agree on the implementation strategy or better to define the sequence of translation in such a way that each translation can build upon previous translations.

As soon as these preparation activities to identify and group the translation sets are finished, the main task of translating the model to source code can start. Each translation set is mapped onto code by selecting and applying an appropriate translation pattern, taking into account the non-functional requirements.

The process for translating the structural model can be summarized as follows:

1. **Identify translation sets.** Translation sets are developed from the refined structural model, by identifying those parts which can be treated by a single person within a single step (i.e. the refined design is divided into several smaller parts. In the simplest case the structural model is declared to be one translation set. Unfortunately, in practice the model is often too complex for translation in one big step. Therefore, logical sets have to be found, which can then be put into a prioritized sequence for implementation. A good strategy for identifying such logical sets is to go through the functional model of the Komponent specification in order to partition the structural model according to the existing operation specifications.
2. **Prioritize translation sets.** There are many possible strategies for deciding which element to translate first. Choosing the most suitable strategy depends on the context of the project, human experience, and existing programming guidelines within a software organization. One group of strategies are top-down approaches as used by many object-oriented CASE tools, or proposed by hierarchical object-oriented methods as SOUND [ABJ+98] or Hood [Gro94]. Typically such strategies use the divide-and-conquer principle, going down from packages to classes until the lowest level of detail is reached. Another strategy is to follow the guidelines for implementing software systems as described by object-oriented text books [LT97, CN93]. Following

these guidelines, translation sets can be ordered according to this sequence: classes, generalization/specification relationships, class interfaces (methods, parameters, etc.), class attributes, and other class relationships.

3 Select appropriate translation patterns.

a For each entity select a translation pattern from the catalog. If there is more than one possible pattern, select that which best matches the non-functional requirements.

b Sort all selected patterns according to the identified translation sets and the defined prioritization strategy.

4 Perform translation.

a Start with translation set and patterns of highest priority.

b Perform translation as shown by the pattern.

c Document the performed translation to keep track of changes.

11.4.3.3 Dynamic behavior

The previous translation activities have created a source code representation of the structural model. In other words the physical architecture as well as the structural entities (e.g. classes) have been created. To work properly this code has to be augmented by the implementation of the Komponent's dynamic behavior (i.e. the statements which, if executed, result in the desired results).

The implementation of dynamic behavior uses several descriptions of the logical Komponent as an input. On the one hand it uses the implementation structural model to identify where operations and structures are located together with the existing source code. On the other hand, it makes use of the dynamic behavior described by the interaction model (co-operation with other objects), activity model (algorithmic descriptions), and pseudo-code descriptions. The latter serve as an abstract description of the behavior to be implemented and can therefore be directly translated to source code.

In general the implementation of dynamic behavior requires a strategy for implementing the control flow information described in the activity model and the collaboration between classes described in the interaction model. In general, there are three basic strategies for doing so:

- using the location within the program to hold the state (procedure-driven)
- direct implementation of a state machine mechanism (event-driven)
- using concurrent tasks.

A popular technique for implementing the information in the dynamic models is based on the ideas presented in [RBP91] by identifying the main control path, identifying alternate paths, identifying backwards paths, and implementing the states and transitions that remain as exception conditions. Putting these basic steps all together for each method of a class, the source code for the class is readily attained.

11.4.4 Finalization

The finalization activity can be best characterized as a period of carefully analyzing the existing code (translated from the Komponent realization and implementation models), and of adding missing details. Source code developed by mapping graphical models onto code is usually not in a shippable state. It may be necessary to optimize its performance, space usage, etc. During finalization a programmer, responsible for developing the code, checks the models and the existing source code to identify places to be modified/enhanced, performs these changes, and verifies them.

Component reuse 12

The previous chapters have introduced the KobrA development processes in a way that suggests a top-down approach to development. Starting from the context realization, successively smaller Komponents can be developed by applying the specification and realization activities in a recursive manner, until Komponents are reached that have no need for further child Komponents. This support for top-down development is one of the main strengths of the KobrA method, providing the underlying foundation for step-wise decomposition and systematic quality control.

Despite its strengths, however, a pure top-down approach is not only impractical, but the idea of developing everything from scratch based only on system-level requirements is at odds with the component-oriented approach of assembling existing parts. The component paradigm implies a bottom-up strategy, in which sets of existing building blocks are successively assembled into larger entities until the system as a whole has been attained. For this reason, KobrA also supports a bottom-up (i.e., more component-oriented) style to development, yielding an overall development process which is a balance between the top-down and bottom-up strategies.

The bottom-up approach is supported in KobrA in two ways. The development activities introduced in the previous chapters already accommodate a bottom-up style of development (as well as a top-down style) because they allow a Komponent to be realized by the assembly of other Komponents that have already been defined.[44] Therefore, provided that the right Komponents are

44. Note that because of the way in which the containment tree is derived from the creation (or composition tree) if it exists, reused components are usually nearer the top of the containment tree than the bottom. However, the component instances that are actually assembled are at the bottom of the creation tree (and composition tree if it exists).

available and are organized in the appropriate way, the work involved in realizing a Komponent can be heavily balanced towards the assembly of instances of existing Komponents, rather than the development of new functionality. In the extreme case, if the realization of the Komponent representing the whole system is achieved in this way, very little, if any, top-down development is involved in the whole process. Patterns for organizing component trees to support this style of development are discussed in Chapter 10.

In its simplest form, (re)using[45] existing components in the realization of another Komponent is just a matter of instantiating them, and using their services in a way that conforms to the clientship rules. However, this is not always sufficient. KobrA therefore complements this simple form of bottom-up assembly with a more sophisticated mechanism for reusing components in ways for which they may not have been originally intended. This includes components that may not even have been developed using the KobrA approach. This mechanism is the topic of this chapter.

12.1 The role of reuse

Instantiating pre-defined components to contribute towards the realization of new software systems (e.g. components) is one of the cornerstones of the component paradigm. However, the range of abstractions for which this simple instantiation approach works (without modifying them) is limited. It generally works well for smaller-grained components and for components capturing widely known software abstractions, such as user-interface components, or abstract data types such as stacks or queues, for example, but does not scale up well to larger or less general components. For such components, some kind of adaptation is usually needed to (re)use them.

Although it is more work, the return on investment from reusing larger-scale, more specialized components in ways that were perhaps not envisaged by their developers is potentially much greater. This form of "heavyweight" reuse is therefore the ultimate goal of the component movement, with terms such as "business object" and "Commercial Off The Shelf (COTS)" components being coined to capture the concept of large-scale components specialized for specific enterprise roles [McIl68]. Several of the mainstream component technologies are also more oriented towards the packaging of large-scale "server side" components (e.g. EnterpriseJavaBeans, CORBA) than small-scale abstract data types or user-interface components.

Making the vision of a commercial market for large-scale "business" components a reality presents a number of significant challenges, however. Four, in particular, are [GAO95]:

- low-level inter-operability problems
- a lack of components that can be reused

45. Some would argue that simply instantiating an existing component in another part of a program (or system) for the purpose for which it was developed is better called "use" rather than "reuse."

- the difficulty of finding suitable components
- architectural mismatches.

In recent years, physical component technologies such as JavaBeans, CORBA and COM that view components as binary, ready-to-run packages of software interacting via standard communication mechanisms, have gone a long way towards addressing the low-level inter-operability problems. Their growing ubiquity has also led to an increased selection of candidate components and to the emergence of component clearinghouses that make COTS components available for sharing and trading via the Internet.

However, even after the situation improved with respect to the first three challenges, the problem of architectural mismatches has proved more stubborn. The source of this problem are the assumptions a reused component makes about the structure of the system in which it is reused. These assumptions often conflict with assumptions made by other parts of the system, and are usually implicit. This, in turn, makes it very difficult to analyze architectural mismatches before a system is built from reused components. The consequences of architectural mismatches include excessive code, poor performance, the need to modify a reused component, the need to rewrite existing functionality, overly complicated resulting systems, and an error-prone construction process. The four main categories of assumptions that can cause mismatches are the:

- nature of components
- nature of their interactions
- nature of the global architectural structure
- construction process.

To improve the chances of successful reuse, it is essential to minimize the impact of the assumptions made by both parties involved – the reused component and the reusing environment – by making them explicit. KobrA component specifications help significantly in this regard by providing comprehensive but abstract descriptions not only of what a component provides but also what it expects. The elements that provide information about the assumptions are the non-functional requirements specification, the operation specifications, and the dependencies that are given in class diagrams. Komponent specifications therefore provide an excellent foundation for identifying and reconciling the expectations of both parties involved in reuse, whatever the nature of the underlying implementation, and thus for facilitating the more heavyweight forms of reuse envisioned above.

The purpose of the KobrA reuse activity is to bring pre-existing components originally "outside" a particular containment tree inside that tree in a way that is fully integrated and compliant with the consistency rules. KobrA does not address the problem of finding components,[46] but only focuses on introducing components into a containment tree once they have been found.

46. Although KobrA specifications could also be helpful in this regard.

The potential sources of components are many and varied, but three main sources are worthy of mention.

- Komponent trees that have been developed during earlier projects and from which Komponents can be extracted and reused
- third-party COTS components or other components that are available in an organization; there are a number of methods published that support the selection of components for reuse – examples are PORE (Procurement-Oriented Requirements Engineering) [MN98], OTSO (Off-The-Shelf Option) [Kont95], and CAP (COTS Acquisition Process) [OP00].
- existing "legacy" systems, which can either be wrapped up to look like Komponents, or from which potential components for reuse can be extracted using reengineering techniques [BBC99].

The following sections introduce the KobrA reuse mechanism following the usual style of separating products from processes.

12.2 Reuse artifacts

According to the development processes defined in the previous chapters, a new Komponent specification is only added to a Komponent tree if no existing Komponent already provides the required services. If a suitable Komponent is already in the tree this is simply instantiated in the realization that needs it according to the simple bottom-up assembly approach mentioned above. If a suitable Komponent is available, but not in the part of the tree visible to the realization that wants to use it, the simplest solution is usually to refactor the tree, as explained in Chapter 6.

Assuming that a new Komponent is needed that is not available in the tree, and the specification of this Komponent has been developed, there are two ways in which a concrete incarnation of the component can be obtained. One is to realize and then implement the Komponent from scratch according to the processes defined in Chapters 5 and 11. The other way is to try to find an existing external component with suitable properties and bring it into the tree. A component (of any kind) is external to a Komponent tree if it is not part of the tree.

There are basically two different types of external components that can be reused:

- KobrA components – Komponents that have been developed using the KobrA method, but are part of another component tree. In general, therefore, they will have KobrA specification, realization, and implementation.
- foreign components – components that have been developed using a technology other than KobrA. Such components will therefore be documented in different styles, and my not even have any models or other abstract descriptions.

The strategy for reusing both kinds of components is essentially the same, although as might be expected, the reuse of foreign components is sometimes a little more involved. Figure 12.1 shows the main artifacts involved in the reuse of an external Komponent, and how these artifacts change as a result of a successful conclusion of the reuse process.

12.2.1 Desired and offered specifications

The left-hand side of Figure 12.1 illustrates the situation that exists when a Komponent within a containment tree has identified the need for a particular child Komponent, and a Komponent has been found with properties that are reasonably close. The properties that the reusing Komponent desires of its child are described in the form of a normal Komponent specification. During the reuse process this is known as the *desired specification*. The relationship between the reusing Komponent and the desired specification is the normal parent–child relationship governed by the relevant clientship and containment rules. Since the candidate Komponent for reuse is a KobrA component, its properties are also specified by a Komponent specification. During the reuse process this is known as the *offered specification*.

The right-hand side of Figure 12.1 illustrates the situation that exists after the successful conclusion of the reuse approach. Essentially any differences that existed between the desired specification and the candidate specification have been resolved, and a single agreed specification has been established. Unless the desired and candidate specification were identical at the start of the process, which is rarely the case, reaching an agreed specification will require changes in the reusing Komponent or in the reused Komponent or both.

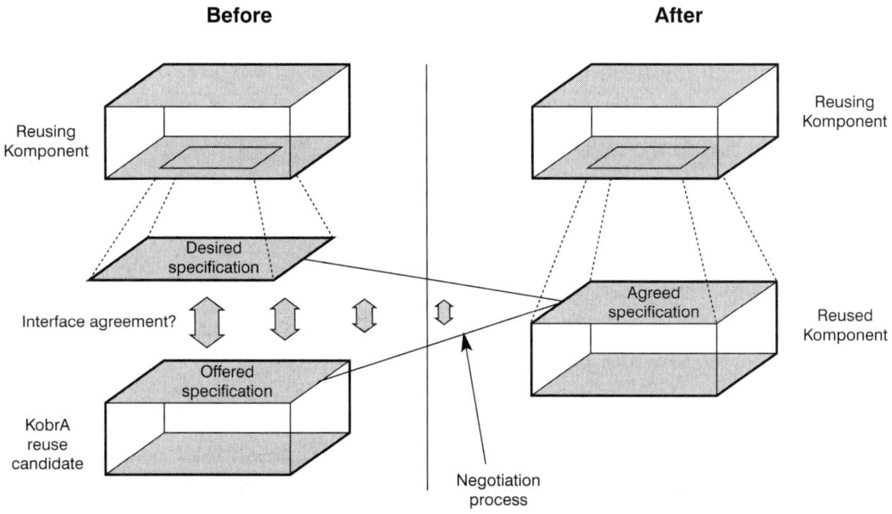

Figure 12.1
Reuse of external KobrA components

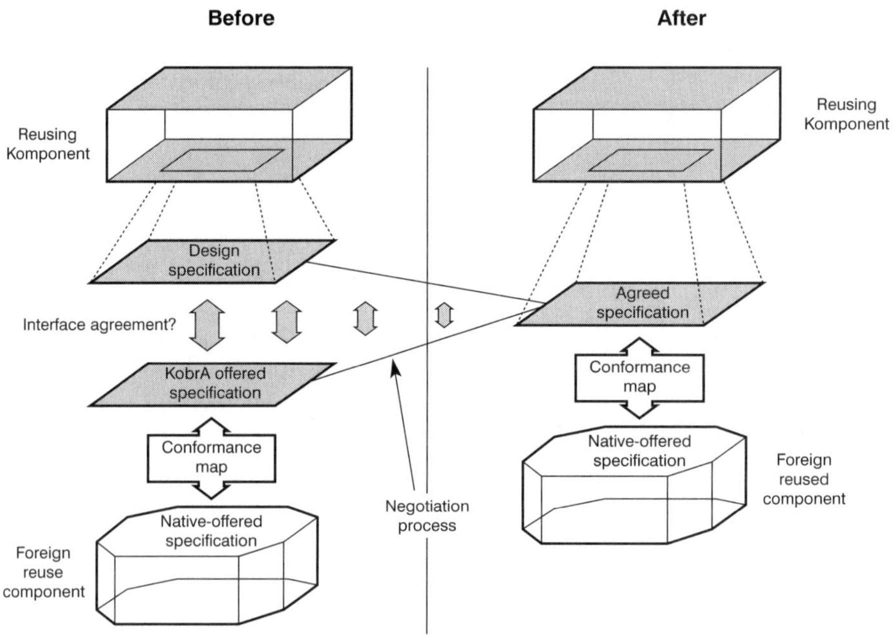

Figure 12.2
Reusing a foreign component

12.2.2 Conformance map

The reuse scenario is a little more complicated with a foreign component, but the principle is the same. The difference stems from the fact that a foreign (i.e., non-KobrA) component will by definition not have a KobrA compliant specification. At best the description of the component will contain a similar set of models providing similar kinds of information, but usually the component is described in a totally different way (if at all). Therefore, to facilitate a satisfactory comparison between the desired specification and the properties offered by foreign reuse candidates it is necessary to present these properties in the form of a KobrA specification. This is the role of the conformance map. Once a KobrA representation of the foreign component's properties has been attained (i.e., the offered specification) the difference between this and the desired specification can be resolved in exactly the same way as for the reuse of a KobrA component.

As indicated in Figure 12.2, the role of a conformance map is to define a mapping between the KobrA representation of a foreign component's externally visible properties and the component's native representation. The former is called the *KobrA-offered specification* and the latter is called the *native-offered specification*. The conformance map transforms – from the viewpoint of the Komponent tree – the foreign component into a Komponent. Therefore, it can be reused in different trees once a conformance map has been created.

Since the KobrA representation of the foreign components' properties supported by the conformance map facilitates the comparison of the desired and offered properties, it is important that the conformance map be complete (i.e., every part of the consumer specification must be mapped to some element of the supplier specification) and consistent (i.e. the mappings must be correct).

The conformance map can thereby only help if the information that is needed to perform the comparison – that is, structural, behavioral, and functional information – is actually present in the foreign component's native specification. In many cases the required information might well be present, but distributed among different models or different views of the component. For example, different object-oriented methods define different ways of specifying the structural, functional, and behavioral aspects of a class or component. In such circumstances, the conformance map maps the views in the native specification to views used in KobrA.

If the KobrA and native specifications used the same names for the fundamental elements of foreign component (i.e. operations, parameters, attributes, etc.), conformance maps actually become quite generic in nature, since they define mappings between different specification approaches rather than between specific instances of the application of these approaches. If multiple foreign components of the same kinds are reused, therefore, the same conformance map can often be used.

Unfortunately, however, it is often the case that the native specification of the foreign components does not provide all the necessary information. Then it is not possible to create a conformance map that provides a complete mapping, and the KobrA representation of the offered properties will be incomplete. An example is the documentation extracted from the source code of a JavaBean component using structured comments (e.g. Javadoc documentations). This kind of documentation contains only structural information about the component but is clearly insufficient to create complete operation specifications or statechart diagrams.

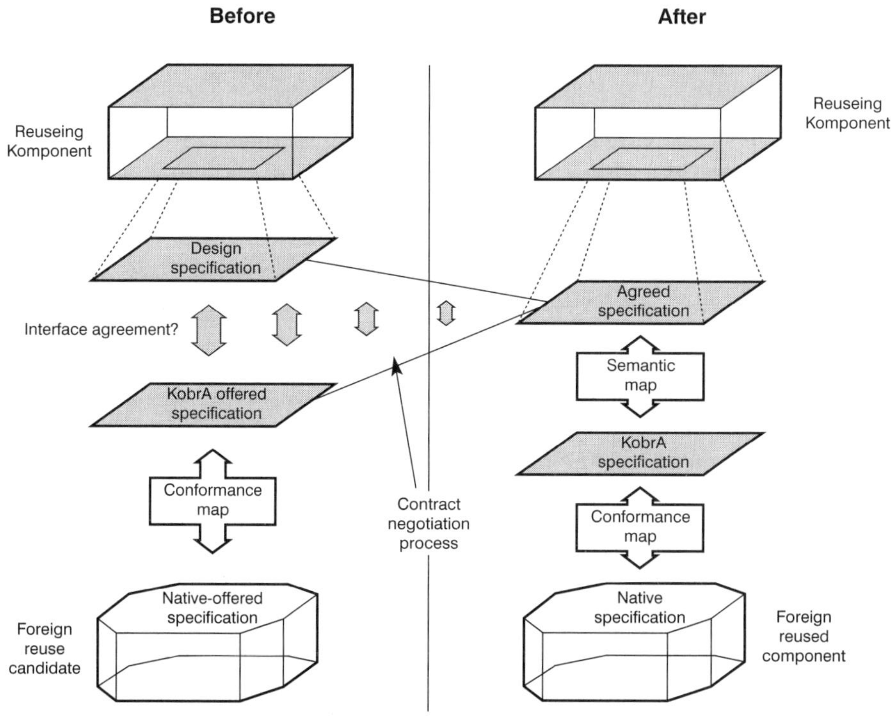

Figure 12.3 Foreign component reuse with semantic map

12.2.3 Semantic map

Because foreign components often have no leeway for changing the properties that they can offer, the only way in which an agreed specification can be attained is if the reusing Komponent tree is adapted to fit the properties offered by the component as presented by the KobrA-offered specification. This is the situation depicted in Figure 12.3. A semantic map is used to create a mapping between the agreed specification and the KobrA-offered specification. However, this is only possible when the difference between the agreed and offered specifications is essentially one of name mappings rather than inherent functionality. In such cases, wrapper Komponents must be used. The use of wrapper Komponents is described in Section 12.3.2.

12.3 Reuse process

The final goal of the component reuse activities in KobrA is to fully integrate into a Komponent tree a component developed earlier outside the tree (i.e., an external component). To achieve this, the specification desired by the reusing Komponent and the specification offered by the pre-existing external component have to be brought into agreement. When such a situation exists, the reused component realizes, and usually also implements, a specification required by the reusing Komponent, and the reused component is then fully integrated into the Komponent tree. Conformance maps and semantic maps support this integration.

Reaching an agreed specification might require changes not only to the reused component and the reusing Komponent, but also to the wider Komponent tree. Component reuse therefore is cost-effective only when the effort involved in arriving at an agreed specification with a pre-existing external component is less than the effort involved in realizing and implementing the child Komponent from scratch. If this is not the case, the required child Komponent should be developed from scratch using the regular realization and implementation activities.

The remainder of this section discusses the activities involved in component reuse in more detail.

12.3.1 Component selection

The first activity that has to be performed to reuse components in KobrA is the identification of candidate reusable components. The desired specification is the starting point for this activity. Existing Komponent trees and component repositories from third-party COTS vendors can be used to search for components. Another source of components are existing applications from which components can be extracted for reuse applying re-engineering techniques. For each potential candidate component, the degree to which the desired specification is satisfied has to be elicited and documented. This degree depends primarily on the size of the semantic difference between the desired and offered specifications and the amount of flexibility that the candidate component and reusing Komponent tree have to adapt to initial differences. In the case of foreign components the nature of the documentation is also a factor, since effort is involved in developing a KobrA representation of its specification.

In the component selection activity, candidate components are sought, for which the "distance" between the desired interface and offered interface is not too large. This means that there has to be a reasonable chance of reconciling differences on both sides and reaching an agreed specification. The chances of reaching an agreement are dependent not only on the actual semantic distance between the desired and offered interfaces, but also on the flexibility of the specifications (i.e. the degree to which they can be adapted).

Various scenarios can occur during component selection depending on the precise situation and the prevailing business constraints. At one extreme is the scenario in which the usage of the pre-existing component is dictated by non-technical considerations. For example, there are often political or business-related reasons why a particular pre-existing component has to be used. This represents the situation where there is zero flexibility on the side of the offered interface. This situation of no flexibility on the side of the offered interface also occurs in the case when binary components are reused. The only way, in which an agreed specification can be reached under such circumstances, is if the client can use the supplied interface "as is." Unless the desired specification is exactly the same as the offered specification, which is extremely rare, this requires the reusing Komponent, and perhaps the tree containing it, to adapt to the offered specification.

The other extreme corresponds to the situation in which the reusing Komponent has zero flexibility as to the specification that it can use. In other words, the specification is fully dictated by the reusing Komponent. In this case agreement can occur only if the offered and desired specifications are the same (as before) or if the candidate component is flexible enough to satisfy the desired interface.

Another issue involved in selecting a component for reuse is the notation in which its specification is written. When the pre-existing component is a Komponent (i.e. a component with a KobrA compliant specification model), the selection process can be based on a direct comparison of the offered and desired Komponent specifications. Each of the specification models is analyzed in turn for closeness-of-fit.

If the pre-existing component is foreign (i.e. it has not been modeled according to KobrA), the compatibility analysis is more complex. A mapping must be created between the native specification and an equivalent KobrA specification. Otherwise, it is not possible to judge how well a component satisfies the needs of the reusing Komponent. The mapping between a foreign component's native specification and an equivalent KobrA specification is the conformance map. An initial version of the conformance map is created when a component is first considered for reuse.

12.3.1.1 Negotiation

Once a set of candidate components for reuse has been found, one of them must be selected for actual reuse. This activity is called negotiation. It involves the investigation of adaptations that would have to be made to the desired and offered specification in order for the reusing and reused component to agree on a common specification. Ideally, this investigation should be performed for all candidates. The result of the negotiation activity is the selection of the component to be reused and a list of agreed changes to either the reused component, the reusing Komponent tree or both. If it is determined that specification agreement is not achievable with any of the candidates without undue (i.e., non-cost-effective) effort, the result of the negotiation activity is the decision that the required component should be developed from scratch.

A pre-existing component can only be reused if an agreed specification can be attained. This often means that both the reused component and the reusing Komponent tree will have to be adapted. It has to be determined whether it is better to change the desired specification (i.e. the requirements of the reusing tree) or the offered specification (i.e. the candidate component is adapted), or both. As a consequence, the pre-existing component whose specification most closely fits the desired specification is not always the best candidate. This is because it may have very low flexibility with respect to this specification (i.e. it is very difficult or impossible to adapt), whereas another component with an initially lower closeness-of-fit can be much more easily adapted. Of course, there are often other non-functional requirements that can dictate which component is actually reused.

Often, both the reusing tree and the reuse component have to be flexible and accept some level of adaptation in order to arrive at an agreed specification. Precisely where this balance should lie, however, depends on a delicate trade-off of many factors, the most significant being the adaptation effort.

The result of the negotiation activity is an agreed specification, as well as a list of adaptations to the reusing Komponent tree and/or to the candidate component. Depending on which kinds of adaptations are required, one or both of the following adaptation activities will subsequently be needed.

12.3.2 Containment tree adaptation

The containment tree adaptation activity is performed, when the desired specification must be adapted in order to reuse a component. There are two ways in which this can be achieved. The first is to change artifacts within the Komponent tree so that it matches the agreed specification. In the best cases this involves changes only to the realization of the reusing component, but often changes to other parts of the tree may be needed as well due to ripple effects.

The second way to adapt a component tree is to introduce a new "wrapper" Komponent that bridges the differences between the original desired specification and the agreed specification. As illustrated in Figure 12.4, a wrapper Komponent is a normal Komponent that is inserted between the original Komponent that needed a specific child Komponent, and the candidate component to be reused.

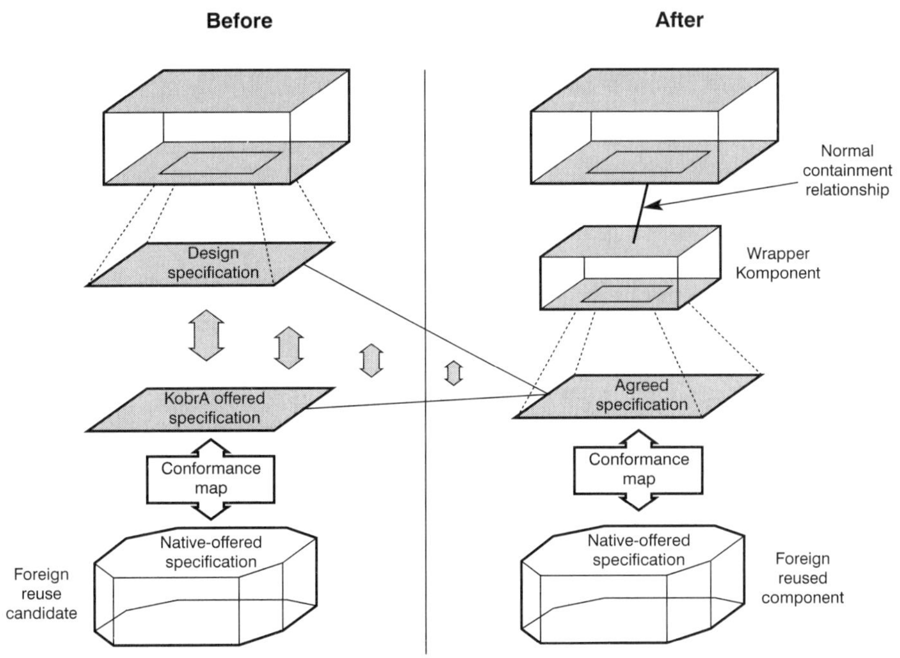

Figure 12.4 Use of wrapper Komponents

The effect of inserting a wrapper Komponent is that the original component that contained the desired specification is now no longer the reusing component – the new wrapper Komponent is. The wrapper Komponent was developed for the exact purpose of satisfying the original desired specification using an agreed reuse specification. Inserting a wrapper Komponent is therefore a good way of adapting the reusing tree without actually changing the existing components of the tree. So, the change is one of addition (of the wrapper component) rather than modification.

Wrapper components are developed just like any other component by applying the realization and implementation activities. In a sense, therefore, this could be viewed as an instance of the "build from scratch" scenario mentioned above. The important difference is that a wrapper class is built for the dedicated purpose of incorporating a specification that can be directly supported by an external component. In fact, arriving at the properties of the wrapper Komponent is part of the negotiation process.

12.3.2.1 Component adaptation

Component adaptation is performed, when the component to be is reused must be changed in order to satisfy the agreed specification. The component must be changed so that it actually implements the agreed specification. Depending on the origin and the type of the component, different techniques exist for adaptation.

The easiest case (from a technical point of view) is the case in which the reused component is a KobrA Komponent. Then, the Komponent can be changed using the KobrA development activities to accommodate the new specification.

For foreign components, the possibilities for adaptation are more limited. If the source code of the component is available and understood, it may be cost-effective to actually change the implementation of the reused component. This requires a detailed understanding of the implementation of the component, and the application of regular software-development and quality-assurance techniques to make sure that the changes are made correctly.

If the reused component is not changeable (e.g. when it is only available in binary format), a semantic map can be defined to provide a mapping between the agreed specification and the original offered interface. Since a KobrA representation of the offered interface should already have been created to facilitate the negotiation activity, it is most convenient to define the semantic map between the agreed specification and the KobrA offered specification, as illustrated in Figure 12.3. A semantic map achieves the same effect as the insertion of a wrapper Komponent but without the need for run-time overhead. However, it is only possible when the difference between the agreed and offered specifications is essentially one of name mappings rather than inherent functionality.

Figure 12.5 Reusable ReservationManager

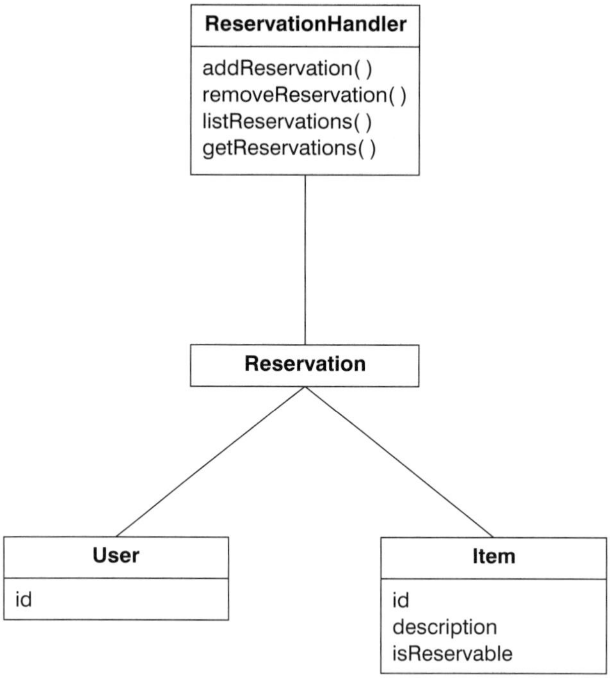

As an example, consider a scenario in which the *ReservationManager* introduced in Chapter 9 is to be realized and implemented by means of a reused component. Figure 12.5 shows the specification class diagram of an existing reservation managing component called *ReservationHandler*, that is candidate for reuse in the library system. The class diagram in the figure indicates that, instead of the concept of accounts, the concept of a user is implemented in the Komponent. This means that it cannot be put directly in the Komponent hierarchy, but that some adaptation has to take place. As mentioned above there are three ways of resolving specification conflicts. The component tree can be adapted, resulting – in the extreme case – in the exchange of the concept of account by users in the whole library tree. The second possibility is the adaptation of the reused Komponent by replacing users by accounts. The last possibility is to introduce a semantic map that resolves the conflict by mapping accounts to users.

Incremental development 13

The ultimate purpose of software development is to generate running computer programs that satisfy the needs of users and sponsors. The activities discussed thus far, in Parts 2 and 3, have led only to the creation of tool-comprehensible representations of physical components. Moreover, the dependencies between KobrA development artifacts have only defined a partial ordering of their associated development activities in the context of a multi-Komponent tree. This chapter introduces the final activities involved in creating executable systems, and discusses optimal sequencing of KobrA development activities.

13.1 System construction and deployment

As explained in Chapter 11, the Komponent implementation activity leads to a tool-comprehensible representation of a logical Komponent, e.g. source code. Furthermore, the mapping of the Komponent containment tree with its logical Komponents onto a set of physical components was discussed. To create a running system, where instances of these physical components are executing, at least two further activities have to be performed: system construction and deployment. Usually a third activity, testing, is also performed in between these two, but this is not strictly essential. Testing is described in detail in Chapter 18. In this subsection we focus on the system construction and deployment activities. If it has been possible to reuse a physical component, system construction and testing can usually be avoided.

Deployment is still necessary, however. Both the system construction and deployment activities are highly dependent on the specific technologies chosen to create, organize and run the binary incarnations of physical components.

Since KobrA focuses on a more abstract description of the logical Komponents making up a system, it is not very prescriptive about how the system construction and deployment activities should be performed and what artifacts they should involve. The goal of this chapter is to provide general guidelines and identify things to consider.

13.1.1 System construction

As noted earlier, the Komponent implementation activity – more specifically, the flattening subactivity – defines the mapping of logical Komponents onto physical components. This mapping is documented with UML component diagrams. However, although physical components are identified during Komponent implementation, they are not physically created. Source code created during the Komponent implementation activity is still associated with the logical Komponent and not yet stored with the physical component.

The system construction activity is responsible for creating the physical components structure based on information in the Komponent implementation artifacts, building all physical components by applying the appropriate tools to create a truly ready-to-run incarnation of the physical component, and releasing the deployable system. The precise number and type of tools involved depends on the implementation technology used. Typically, a programming language compiler is needed, but other kinds of tools are also commonly involved, such as IDL preprocessors, linkers etc.

The configuration management technology also has an important bearing on the system construction activity. As well as the executable incarnation of a physical component, the other artifact that is usually generated by the system-construction activity is some form of build or configuration description. This can range from a UNIX make file to more sophisticated repository-oriented configuration descriptions.

In general, therefore, three subactivites are involved in system construction:

- **construction** – the organization of the physical components and their contents in a way acceptable to the implementation tools (for example, setting up a directory or repository structure, and filling this with source code and build descriptions)
- **component building** – the actual running of tools such as compilers, IDL preprocessor etc. to generate a binary representation of the physical component
- **releasing** – combining the various deliverable artifacts (binaries, deployment information) for a specific release, and arranging them on the distribution medium.

KobrA deliberately makes a clear separation between Komponent implementation and system construction. The idea is that up to Komponent implementation, tools need to provide specific support for KobrA, but for system construction, existing tools can be easily used. Imagine some kind of KobrA workbench that manages Komponents and their associated models, providing dedicated support for KobrA's component engineering activities and automatic checking of KobrA's well-formedness rules. For system construction, existing CASE tools can be easily used. The construction subactivity is then responsible for bridging the gap between the content of the KobrA workbench and the traditional CASE environment.

13.1.2 Deployment

Deployment is the final act of loading executable incarnations of physical components onto computing devices and setting them running as component instances. Again the details of how this is achieved will depend to a large extent on the nature of the target execution environment and the operating system(s) controlling it.

Although the details are technology-specific, some of the decisions that have to be documented are common to all systems. In particular, there is always a need to identify what component instances exist in the final deployed system, what communication paths exist between them, and in a distributed system, which processing nodes they reside upon. UML deployment diagrams are intended for just this purpose. Once these instantiation and allocation issues have been resolved, therefore, KobrA recommends the use of UML deployment diagrams to display the run-time configuration of component instances and nodes.

Figure 13.1 shows a deployment diagram illustrating a possible distribution of physical components to nodes for the library system. In this particular configuration, an instance of the *ItemManager* component and of the *LoanManager* component have been allocated together to a *DatabaseServer* node, an instance of the *OPAC* component has been allocated to an *ApplicationServer* node, together with an instance of the *WebServer* component and the *OPAC* component. Finally, another instance of the *OPAC* component has been added to a *LibraryPC* node.

The only constraint on the deployment of component instances is that the number of components and their clientship relationships at system start-up

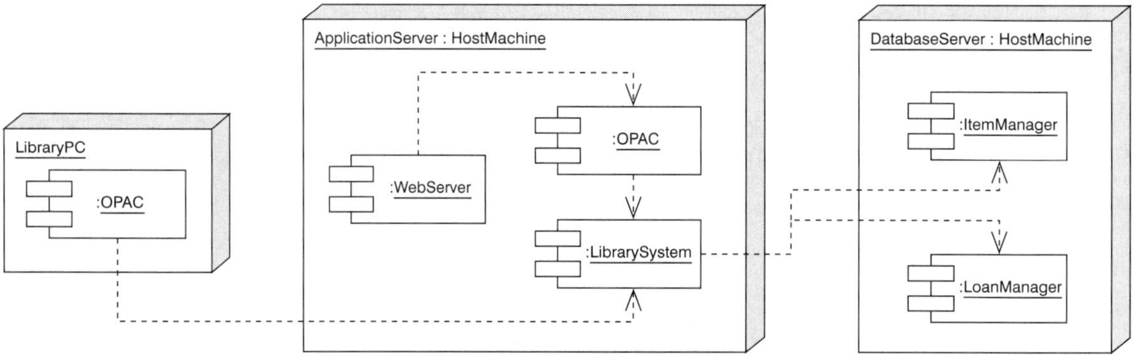

Figure 13.1
Library system deployment diagram

time are consistent with the information embedded within the Komponent tree, including of course, the overall shape of the tree. The only relaxation of this requirement is that creation relationships that span physical models are allowed to be realized manually. In other words, if an instance of Komponent X creates an instance of Komponent Y, but they are mapped to different physical components, it will not always be possible for this to actually take place automatically. However, this can take place manually.

13.2 Activity sequencing

In the context of an individual, "small" component the dependencies between the development artifacts could be interpreted as imposing a strict order on the associated development activities:

specification → realization → implementation → construction → testing → deployment

or when reuse is performed

specification → reuse → deployment

This means that the component is first specified in its entirety, then realized in its entirety, then implemented in its entirety, and so on until one or more physical incarnations are deployed. However, for components that are fairly large, or for systems containing more than one component, this may not be optimal. It may then be better to handle distinct parts of a large component (e.g. operations), or distinct components, separately. This subsection provides some guidelines for organizing the sequencing of activities in such circumstances.

In the early years of software engineering, the waterfall model was the most ubiquitous software process model. This was based on the concept of monolithic development steps, similar to those used in KobrA, each performed to completion on the whole system before the next step started. However, it did not take long for software engineers to realize that such a monolithic and strictly sequential approach bore too high a risk of project failure. The main problem with a strict waterfall model is that little information on the validity of the work performed in the earlier activities is obtained until close to the end of a project[47] once the first executable has been created.

Almost all modern software-development methods consequently adopt an iterative (cyclic) strategy for applying the main development activities, and an incremental approach to the assembly of the final executable incarnation of the software. Such a strategy not only breaks the system into smaller more manageable parts, but also provides much faster feedback on the quality of the work performed in the distinct increments.

The advantages of iterative and incremental development were recognized well before the advent of the object paradigm. Incremental development is not only one of the cornerstones of the Cleanroom approach, but it is also the underlying premise of the famous spiral process model [Boe88].

47. The only exception is feedback from inspections.

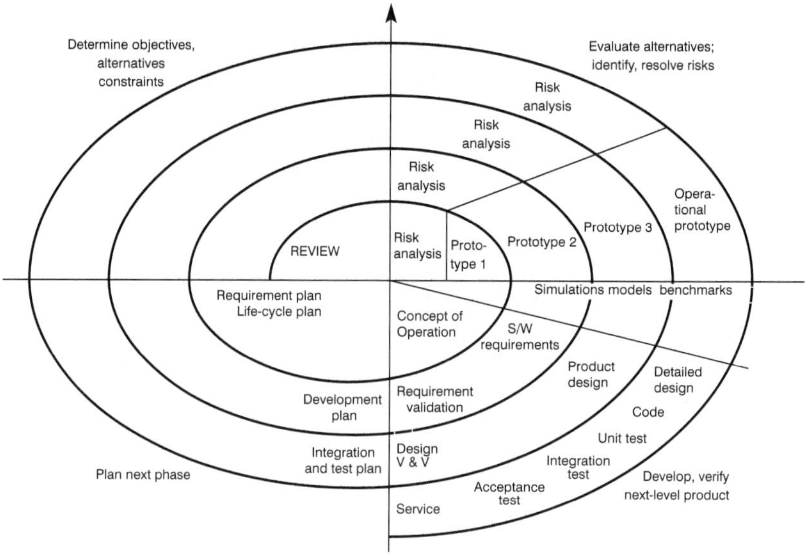

Figure 13.2 Spiral model of software development

As illustrated in Figure 13.2 the spiral model views software development in terms of a an ever tightening spiral of activity, in which a circuit round the spiral corresponds to a mini-project, involving all development activities and resulting in an executable increment of the system.

13.2.1 Function-oriented versus component-based increments

Although the principle of incremental development is simple, the practice is much less so. The big question is how to define which parts of a software system should be addressed in an increment, and how the boundaries of such parts should be determined in a practical way.

The prevailing fashion in object-oriented development methods is to take use-cases as the basis for defining development increments [Kru00] [Cot95]. This has the advantage that it leads to increments that represent well-defined slices of functionality. Users and customers can thus be shown recognizable parts of the system as early as the first few increments in the development process. The disadvantage of using use-cases as the basis of increments is that, at least when applied in a strict sense, it can make the attainment of an optimal architecture more difficult. This is because the function-oriented slices defined by use-cases can cut across (i.e. involve) objects within the architecture in arbitrary ways. In other words, use-cases are in general orthogonal to the objects within a system.

If a single use-case is therefore carried through to implementation before any other use-cases are considered, the development of the architecture of the system (and its subsystems) will have to be initiated on the partial view of the system provided by the use-case. However, architectures do not lend themselves to incremental development. On the contrary the architecture of a system needs to be optimized to take into account all the use-cases and non-functional requirements that the system has to support. This "tension" between the use-case and architecture-centric forces in a development project is well described in the Rational Unified Process [Kru00].

For this reason, a pure use-case-driven approach is rarely used in practice. Instead, initial insight is gained into the nature of the architecture from a consideration of all the use-cases, before any of them are carried through to an executable increment.

Thanks to the simple, hierarchical strategy for organizing components, and the naturally iterative development process, KobrA is able to provide a simple and natural approach to incremental development. The increments are not only more closely tied to the components (and thus more architecture centric) than in most contemporary approaches, but because of KobrA's strict modeling rules, their boundaries are also better defined. KobrA's approach to incremental development is elaborated in the following subsections, using the usual separation of products from process issues.

13.3 Artifacts

Since incremental development represents a strategy for applying the KobrA development activities, it works with normal artifacts of Komponent modeling, namely, Komponent specifications, realizations, and implementations. However, there are two additional concepts of importance in the incremental application of KobrA development activities that are explained in this subsection.

13.3.1 Stub specifications

A Komponent stub is a cut-down version of a Komponent that mimics the behavior of the Komponent in a limited number of situations. The motivation for defining a stub is to create an executable "stand-in" for a Komponent, before the full version is developed, to facilitate the testing of its parent. To do this effectively, therefore, it is essential that a stub performs its limited role correctly. Testing Komponents with incorrect stubs is worse than performing no tests at all, because it provides false confidence of high quality in the tested Komponent. Correctly specifying and checking the behavior of a stub is therefore as important as specifying and checking the behavior of a normal Komponent.

In fact, the difference between a Komponent stub and a regular Komponent really just originates from its role than from its characteristics. Depending on the nature of the system under development, and the demands of the nonfunctional requirements specification, Komponent stubs could actually become quite complex. If this is the case, a Komponent stub can be treated like a normal Komponent, and developed using all the normal development activities.

Usually, however, a Komponent stub has very limited functionality. Essentially, simple versions of the Komponent's operations are required that mimic the functionality of the real operations for a very limited number of possible input values and states. Precisely which set of input values and states

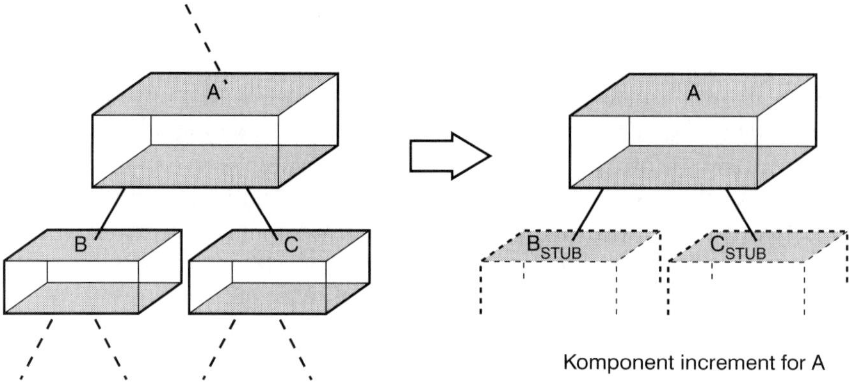

Figure 13.3 Komponent increment

the stub has to mimic the real operations for depends on the test cases chosen to test the parent Komponent.

Even for very simple stubs, therefore, it is valuable to create a Komponent specification that describes the effects of the operations for the limited set of input values and states for which it is expected to mimic the real Komponent. Such a specification is naturally called a stub specification, and consists of a set of operation specifications that augment the full specification with the description of the operation stubs.

13.3.2 Component-based increments

The goal of incremental development is to grow an executable system over time by gradually adding complete and verified increments until the whole system is complete. The word "verified" is important here, because it means that increments must be logically self-consistent. For KobrA, which is an inherently component-oriented method, the most natural increment of development is component-oriented. However, a single Komponent alone does not constitute a readily verifiable (i.e. testable) collection of artifacts. Thus, in KobrA, the fundamental increment of development is defined to be the artifacts for a given Komponent plus the stubs for the server sub-Komponents of the Komponent.

Figure 13.3 contains a schematic representation of a Komponent increment. The left-hand side of this figure illustrates a Komponent A, at an arbitrary place in a Komponent tree with two children B and C. The right-hand side shows the Komponent increment based on A. As illustrated, the increment contains not only A (i.e. all the artifacts for A) but also stubs for A's children. At the Komponent tree level this requires at least specification stubs for B and C (i.e. a set of operation stubs for the operations of B and C, respectively). And at the implementation level this requires corresponding code. In the rare case when the stub is complex enough to need a design, it may also have an associated realization in the Komponent tree.

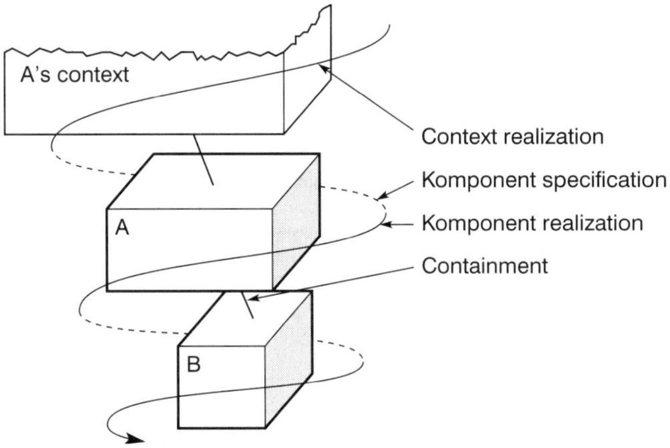

Figure 13.4 Recursive Komponent modeling process

Defining Komponent increments in this way means that they represent self-contained capsules of functionality that can not only be meaningfully tested in isolation, and with the executable code generated during previous increments, but they can also represent meaningful units of inspection. Because the functionality encapsulated within stubs is modeled at more or less the same time as the parent Komponent that uses them, they can be inspected together. This increases the likelihood of correct tests, and thus the confidence in the quality of the Komponent delivered by an increment.

13.4 Process

As demonstrated by the Cleanroom approach, a recursive development process is inherently iterative. It also lends itself naturally to interpretation as a spiral-based process, albeit with multiple branches corresponding to the branches of the development product. Figure 13.4 shows how the primary component engineering activities of specification and realization, when visualized in connection with the hierarchic product that they generate, can be regarded as leading to a spiral-based development process.

Figure 13.5 reinforces the spiral nature of the process by showing a bird's eye view of the development path depicted in Figure 13.4. The analogy with the spiral model of Boehm et al. is self-evident. Of course, Figures 13.4 and 13.5 illustrate the activities applied to a single branch of a potentially multi-branch tree. However, a regular spiral can be retrieved simply by tackling sub-Komponents in a sequential order.

Figures 13.4 and 13.5 consider only the process involved in component engineering, that is, only the activities involved in developing the containment dimension of the tree. As illustrated in Figure 13.6, because implementation represents a different dimension of concern, the order in which implementation activities are performed is in principle independent of the order in which the modeling activities are performed, subject to the constraint that an individual Komponent cannot be implemented before it has been specified and realized.

Figure 13.5
Recursive development: spiral view

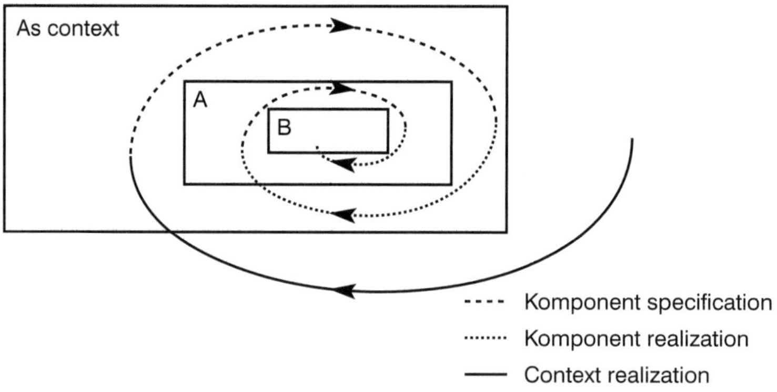

Figure 13.6
Development spiral with implementation

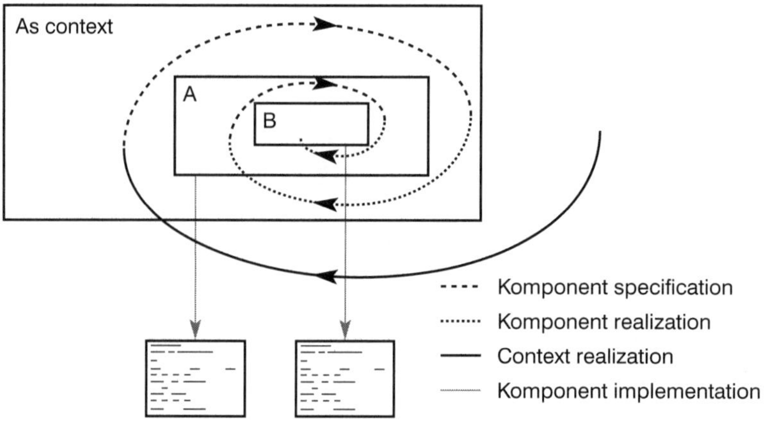

Figure 13.6 is an extension of Figure 13.5 that illustrates how implementations, and the activity which creates them, conceptually lie outside the component engineering spiral in a separate dimension of concern. In general therefore, there are many different ways in which implementation steps can be interleaved with component engineering steps.

At one extreme, it is possible to develop the entire containment tree before performing any implementation steps. This approach corresponds most closely to the classic waterfall approach, in which implementation is only started once analysis and design have been completed. For the same reasons that the waterfall approach has fallen out of use, this sequencing of activities in KobrA is rarely expected to be optimal. For projects with long-term, guaranteed funding and stable requirements (e.g. large government projects), this approach may make sense, because it avoids a lot of implementation and testing work that is strictly unnecessary. However, these conditions are rare.

At the other extreme, it is possible to perform implementation steps at the earliest possible opportunity. This is the approach which, broadly speaking, corresponds to incremental development. The basic idea is to implement, construct, and then test individual components as soon as they have been realized

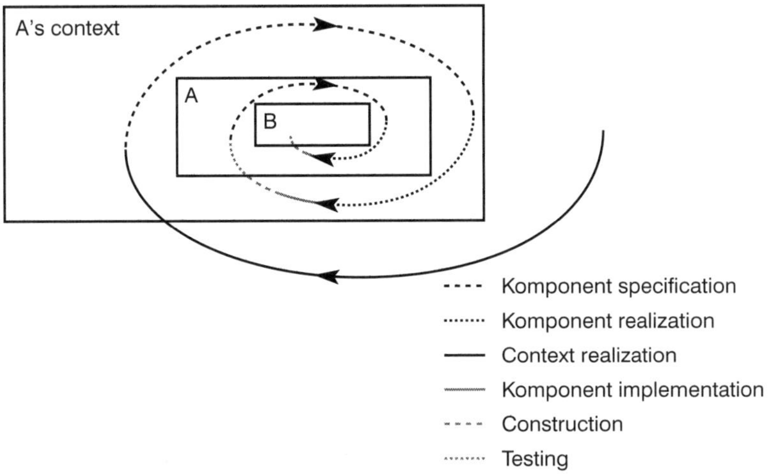

Figure 13.7
Spiral view of incremental process

(and thus also specified). In terms of Figure 13.5, incremental development corresponds to performing each implementation step before progressing deeper into the containment hierarchy. Figure 13.7 illustrates this idea by folding the implementation, construction and testing activities into the development spiral along with the component-modeling activities. This model corresponds to the general spiral model illustrated in Figure 13.2.

13.4.1 Component increments

Figure 13.6 is a little misleading as it implies that the implementation, construction and testing activities are performed on a single Komponent, and nothing but a Komponent. However, as discussed in the preceding subsection, if meaningful testing is to be performed on increments before they are added to the final product, appropriate high-quality stubs also have to be defined and implemented. Thus, unless quality is to be sacrificed, the artifacts developed during each cycle of the spiral must be component-oriented increments, not just as individual Komponents. As explained in Section 13.3.2, a Komponent increment includes the functionality for an individual Komponent plus stubs for the Komponent's children.

For Komponents of reasonable size, therefore, the incremental development approach recommended by KobrA is the cyclic development and completion of successive Komponent increments, based on the shape of the Komponent tree.

13.4.2 Operation increments

For large-scale Komponents near the root of a containment tree, it may be necessary to define smaller-scale increments. The obvious units to choose are activities or operations. In principle, as soon as the Komponent is identified it is possible to choose an individual operation of the Komponent and apply each of the development activities (specialization, realization, implementation, construction, and testing) to this operation in isolation. When applied at the level of the root Komponent, this essentially corresponds to a use-case-driven strategy for choosing increments.

However, because of the tension between the function-oriented and component-oriented approaches discussed above, this strategy is not recommended in KobrA. Adopting individual activities as increments, thereby making them the units of specification, realization, implementation etc., means that the architecture of Komponents has to be elaborated incrementally. Arriving at the optimal architecture for a Komponent or system (which is a Komponent), however, involves a consideration and balancing of all the activities that the Komponent is responsible for.

KobrA therefore recommends that only Komponent increments be considered with respect to the component-modeling activities. In other words, specification and realization should only be applied at the level of Komponents, not individual activities or operations, and should be immediately followed by the creation of the sub-Komponent stubs. This ensures that all the forces that affect the realization of a Komponent are considered when its architecture is established. However, once a Komponent increment has been developed, it can be transformed into executable parts in either component-oriented or operation-oriented increments. In other words, the whole Komponent increment can be implemented, constructed, and tested in one go, or individual operations can be implemented, constructed, and tested separately.

Stated another way, component engineering should always be performed according to Komponent increments, but transformations in the implementation dimension (i.e. the implementation, construction, and testing activities) can be performed using either component-oriented or activity-oriented increments. This provides the best balance between the need for sufficient information during the Komponent realization activity, and the need for the rapid creation of executing parts of a system.

PART

Product line engineering

4

The activities and concepts described in the previous two parts of the book focus on component modeling and implementation or reuse from the context of a single system. This is a perfectly acceptable way of using the KobrA method, but it fails to leverage the advantages of product line engineering, and the enhanced reusability of components this can bring. This part of the book explains how support for product line engineering is added to the concepts in the previous parts.

Part 4 encompasses three chapters. Chapter 14 introduces product line engineering by discussing product line concepts with respect to general reuse problems and their solutions provided by product line engineering. Chapter 15 then extends component engineering with product line concepts. Its result is a framework containing a component hierarchy for a system family. Consequently component engineering using product line concepts is called framework engineering.

The output of framework engineering is not a particular product. It is, rather, a generic framework that needs to be instantiated to meet the needs of specific applications. Hence, an additional activity is needed which builds applications using frameworks. This additional activity is called application engineering and is described in Chapter 16.

Product line concepts 14

As autonomous, self-contained software building blocks, components provide a solid foundation for software reuse. However, simply developing software in terms of components is not, on its own, likely to lead to significant increases in reuse levels. Experience indicates that reusing a component usually involves more than simply plugging it into a new context, and unless components are carefully written to maximize their reusability they tend to be difficult to reuse beyond the application for which they were written.

Despite this fact, many methods that aim to support components assume that simply decomposing software in terms of components is sufficient to promote reuse. They fail to make the distinction between "developing for use" (i.e. developing a component for a specific application) and "developing for *r*euse" (developing a component for numerous applications) [FHA00]. However, recognizing and separating these two concepts is essential if the reuse potential of components is to be fully exploited [Lim98]. Promoting the latter is the underlying motivation for product lines as explained in the following subsections.

14.1 Promoting reuse

The distinction between "developing for use" and "developing for reuse" applies to all software assets, not just components. Figure 14.1 illustrates the traditional "reuse" life-cycle based on software artifacts (e.g. components) initially developed for use with one application in mind. Basically, the artifacts created in one application-engineering project are stored in an asset base so that they can be reused in other applications. Future application engineering projects then attempt to develop parts of the application by reusing assets existing in the asset base. However, as many organizations have experienced, this approach usually fails to achieve the expected improvements in reuse because it does not solve the following problems:

- **Documentation.** Each asset placed in the asset base must be documented to facilitate its reuse. However, determining what information, to what level of detail, is needed by potential reusers is a non-trivial problem.
- **Classification.** All assets and associated documentation must be structured according to a common classification scheme. This classification scheme is the key for accessing the asset base, and must, therefore, be defined in a way that optimally supports the reuse process.
- **Identification.** As Tracz [Trac95] pointed out, reusable software must exist before you can reuse it. Consequently, during the application engineering process an engineer must be aware of assets that are potentially reusable in a given context. In general, this requires some form of systematic technique for identifying reuse candidates, via the classification scheme, for example. Unfortunately, however, existing classification schemes have proven to work only in a limited set of reuse contexts. Moreover, only a small, manageable set of good candidates should be identified, because an engineer will only actually reuse a component if there is a direct reduction in time and effort.
- **Evaluation.** When a set of reuse candidates has been identified, the candidates must be evaluated with respect to adaptation and/or development effort. In order to make confident decisions concerning the selection of a reuse candidate or the development of the needed component without reuse, a systematic and repeatable evaluation approach is required.
- **Coordination.** When an asset has been selected, it will be adapted to match the context in which it will be reused. To benefit from reuse at the enterprise level, an organization must coordinate concurrent application engineering projects to avoid identical adaptations within different projects. Therefore, in addition to the organizational issues, the reuse contexts of projects must be comparable.
- **Integration.** Reused assets must be integrated with the application under development so as to remove conflicting assumptions about the environment and architectural mismatches. Unfortunately, an asset's assumptions about its environment are often implicit and therefore neither documented nor visible.
- **Evolution.** The maintenance of assets reused in numerous applications is more complex than for assets used only in a single application. Basically, the maintenance of the various reusing applications must be co-ordinated to share as much of the maintenance effort as possible, and to improve the reusable asset base.

These issues are as valid for components as they are for regular software development assets, but are only partially addressed in contemporary component-oriented development methods. KobrA's reuse mechanism, described in Chapter 12, addresses the integration of existing components and the evaluation of competing

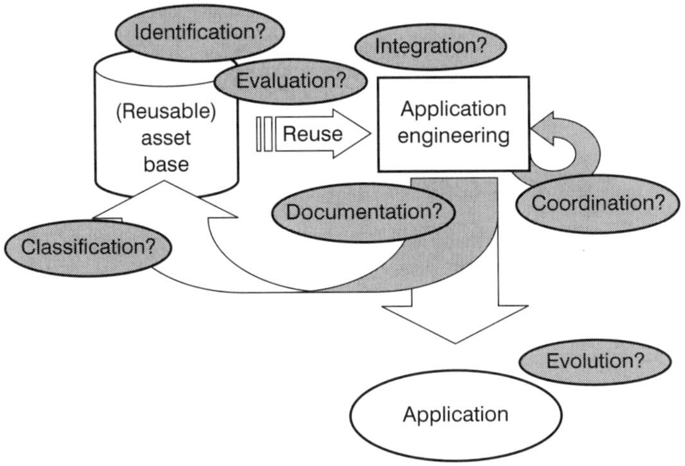

Figure 14.1 Traditional software reuse

reuse candidates, but only in a limited way. When reusing a component in KobrA, a conformance map documents the mapping between the services provided by an existing component and those that the component must fulfill. It therefore documents the reuse contract and enables a rough estimation of the reuse effort needed.

Beyond this limited support, however, the component engineering aspects of KobrA described up to this point do not by themselves address the reuse issues described above. This is the role of the product line engineering part of KobrA. By cleanly integrating component engineering with product line engineering, KobrA provides a powerful and systematic strategy for handling the difficult issues involved in actually making reuse work in practice.

14.1.1 Domain engineering

Product line engineering is basically an approach for planning and developing several similar systems concurrently in order to systematically exploit their commonalities. Historically, it is a descendant of domain engineering, which pioneered the creation of a generic reference model for an application domain from which multiple, similar concrete applications could be derived. The life-cycle of the domain engineering approach is shown in Figure 14.2. As can be seen, domain engineering splits the overall software life-cycle into two main phases: domain engineering (i.e. developing assets for reuse), and application engineering (i.e. using these assets in the creation of new applications). The basic idea is to use the properties of the real-world domain in which the applications are built to operate in order to tackle the reuse problems listed above:

- **Documentation.** The entities and relationships in the domain are used to document the reusable assets. For example, a component would support a specific set of activities that are meaningful in the domain.
- **Classification.** Assets are classified according to the structure of the domain. This is identified by domain experts during domain analysis activities.

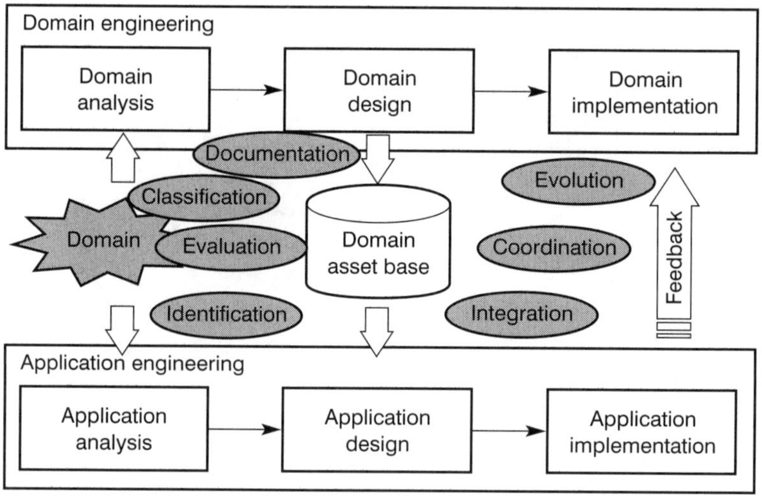

Figure 14.2 Domain engineering life-cycle

- **Identification.** Places that could benefit from reuse often correspond to problems in the domain of interest. Because the contexts for potential reuse are known to engineers, and reusable assets are classified according to domain abstractions, reuse candidates can be identified by searching the asset base for solutions to the domain problem under consideration.
- **Evaluation.** Reuse candidates can also be evaluated using the domain abstractions. Differences are expressed in terms of domain variabilities, so that the "distance" between reuse candidates and the problem to be solved in the domain context can be used as a measure for estimating the reuse effort.
- **Coordination.** The split of the development life-cycle enforces feedback of the application engineering projects to the domain engineering activities. This is also a channel by which similar reuse or change requests can be co-ordinated. Again, domain abstractions are the key for comparing and planning the realization of requested changes.
- **Integration.** Domain engineering also defines a reference architecture for systems in the domain. This simplifies integration because the assets are built to be reusable for this architecture and any associated implicit assumptions concerning the domain.
- **Evolution.** Domain engineering handles the maintenance of reusable assets in a similar way to the maintenance of single applications, but does so at the level of the domain. It therefore coordinates the processing of the many change request originating from users of specific applications in the domain. Since the maintenance of all reusable assets is primarily performed at the domain level, maintenance effort is integrated and minimized.

Although domain engineering represents a significant step forward in the support of reuse, practical experience has shown that the definition of the domain under consideration is problematic. When the domain is chosen to be too small, the domain model and the reference architecture fail to address important issues. As a result, significant changes will typically be necessary in future projects, undermining the value of the whole domain-engineering effort. On the other hand, when the chosen domain is too big, the effort invested in domain engineering is higher than is really necessary, and the already significant investment involved in domain engineering goes beyond that which is cost-effective. In short, although domain engineering provides a powerful set of solutions to the fundamental reuse problem identified previously, the success of a domain engineering effort is highly sensitive to the correct scoping of the domain. If a chosen domain is too big or too small, domain engineering can do more harm than good.

14.1.2 Product line engineering

Product line engineering solves this problem by using the characteristics of a set of concrete (existing, planned, or future) products to scope the domain. Everything required by a concrete product is part of the domain – everything else is outside. This concept of defining the domain through a set of products is depicted in Figure 14.3.

The essential concept of product line engineering is to take common and varying aspects of applications into account at all life-cycle stages and for all kinds of products. Commonalities and variabilities are equally important: commonalities define the skeleton of systems in the product line; variabilities bound the space of required and anticipated variations of the common skeleton.

14.1.2.1 Commonality

product line engineering is only useful when an organization develops several systems in one application domain. By definition, this implies that these systems have at least some features in common, otherwise it would be difficult to view

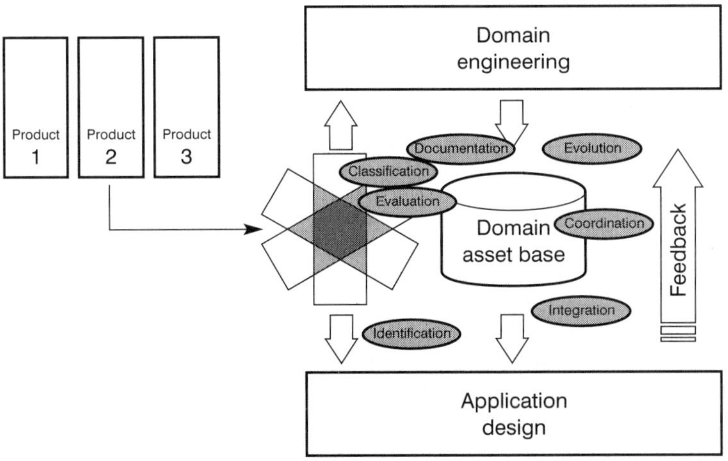

Figure 14.3
Reuse within product line engineering

them as occupying the same domain. In a sense, therefore, the common features of a family of products serve to characterize the domain. Typically, organizations limit themselves to the domain or domains that they have expertise in.

Commonalities are important for establishing a common understanding within an enterprise of the kinds of applications that it provides. The determination of whether a characteristic is a commonality or variability is often a strategic decision rather than an inherent property of the product family. For example, the execution platform can be a commonality when an organization decides to provide only an NT solution.

14.1.2.2 Variability

Variabilities are characteristics that may vary from application to application. In general, all variabilities can be described in terms of alternatives. At a coarse-grained level, one artifact can be seen as an alternative to another artifact. Then during application engineering, the artifact that best matches the context of the system under development is selected. Although simple in theory, providing an effective representation of the variabilities in a product family is one of the most critical, but also problematic, factors in the success of a product line engineering project.

The first issue that must be resolved when analyzing a variability is whether it is to be realized as a run time variability or a development time variability. By their very nature, software systems are inherently full of variabilities because their execution paths are driven largely by run time input. Thus, for example, the fact that a given book in a library is or is not loaned by a user could be considered a variability. Such run time variabilities are an inherent part of software, and are handled effectively by traditional design and programming techniques.

Product line engineering, in contrast, is concerned with development time variabilities that are resolved before a software system is loaded onto its final execution environment. The problem is that there is no strict boundary between development time and run time variabilities. Deciding whether a variability should be resolved at development time, or should be built into a system for resolution at run time, is a strategic decision that can have a large bearing on the success of a product. In an extreme case it is possible to realize all variabilities at run time, that is, build all features of the entire domain into a single system with appropriate parameters to select the desired features at run time. However, this would not be regarded as product line engineering. On the contrary, one of the goals of product line engineering is to avoid the code bloating which results from an attempt to resolve all variabilities at run time.

The second challenge in representing (development time) variabilities is to determine the right level of detail and presentation style. Because they have always had to deal with run time variabilities, most graphical modeling languages already have rudimentary facilities for representing them. In the UML, variabilities are handled by a combination of generality and constraints. A UML model is typically presented at a level of generality that covers all possible run-time variations, and is accompanied by textual constraints (usually in OCL) that specifies which of the many possible combinations are acceptable.

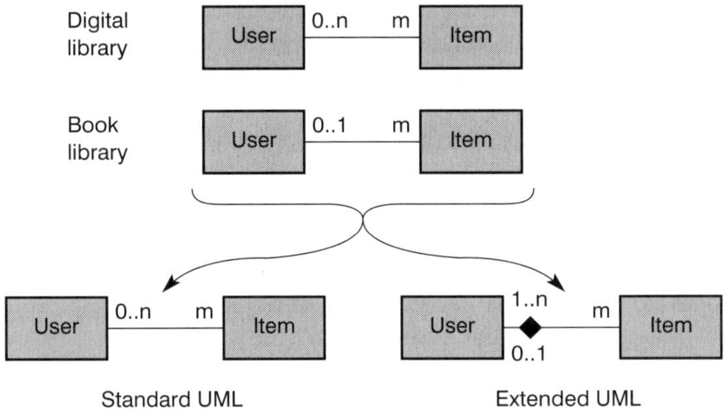

Figure 14.4
Different levels of variability

In theory, development time variabilities could be handled in the same fashion as run time variabilities (as mentioned above, development time variabilities can be realized as run time variabilities instead). This basic approach could still be used to handle product lines. In other words, the UML models could be generalized to the point where they make statements that accommodate all members of the family, and could be supported by OCL constraints describing how the general features change among family members. However, this approach has two major drawbacks. First, it would mean that the information about variabilities is actually captured in the textual constraints rather than in the UML. This defeats the whole object of graphical modeling which is to convey critical information in a concise, compact form. Second, and more importantly, it means that variabilities that distinguish family members in a product line are mixed up with the run time variabilities that are "common" to all members of the product. In short, it suppresses the very information that is essential for effectively understanding and using a product line.

For this reason, for optimal support of product lines it is important to have graphical support for explicitly presenting feature alternatives that characterize distinct members of a product line. For example, suppose that there are two variants of the basic library system, a standard book library and a digital library that permits temporary access to some data. The variability between these two systems is that a digital item can simultaneously be loaned by more than one user, while a physical book can be loaned only by one user. The relationships between the classes *User* and *Item* for the two systems are shown in Figure 14.4. The top two class diagram fragments illustrate the relationship between *User* and *Item* for the digital library and book library separately, while the bottom two diagram fragments show how the difference would be resolved with and without explicit graphical support for product lines. The left-hand diagram, which uses standard UML, simply takes the generalization of the two class diagrams.[48] In contrast, the right-hand side, which uses an enhanced form of the UML, merges the information from the two diagrams in a way that makes the properties of the two variants explicit.

48. In this case, this is simply the class diagram for the digital library since it subsumes that for the book library.

Analogous extensions are needed for all other UML diagrams used in KobrA, such as activity diagrams, collaboration diagrams, etc. In principle, each element of a diagram, such as classes, aggregation, inheritance, may be optional or an alternative within a set of elements. Consequently for each element an extension to capture its variability must be defined.

14.1.2.3 Decision models

Simply identifying and modeling the variabilities among the products in a product line does not define what features are associated with what products. This is the role of the decision model. Essentially a decision model consists of a hierarchy of decisions that relate user visible options to specific system features. At the bottom level, there are simple decisions that are directly related to a variation point within assets. A *simple decision* captures the following information:

- A textual, domain–related question that represents the decision to be made
- the set of possible answers to that question
- references to the affected assets and variation points
- description of the effect on the assets for each possible answer.

Simple decisions themselves neither support application engineers nor reduce the application engineering effort. For this, domain knowledge is needed to resolve simple decisions, as well as dependencies among them. This is achieved by defining decisions at a higher abstraction level, thereby influencing the possible answer set of other decisions. Hence, *high-level decisions* capture the following information:

- a textual, domain–related question that represents the decision to be made
- the set of possible answers to that question
- references to the affected decisions
- description of the effect on the answer set of the affected decisions.

The goal is to minimize the number of decisions that must be resolved during application engineering. Therefore, high-level decisions should be defined in such a way that their resolution automatically leads to the resolution of other decisions.

14.2 Product line engineering in KobrA

Product line engineering is an inherent part of the KobrA method. When pursuing a product line approach in KobrA, the overall software life-cycle consists of two basic product line engineering activities:

1. **Framework engineering**. This applies the Komponent-modeling and implementation activities described in Parts 2 and 3, accompanied by additional subactivities for handling variabilities and decision models, to support a

family of similar applications (i.e. development for reuse). A framework therefore contains a generic Komponent tree that captures the common and variable characteristics of a product line. The use of the Komponent modeling and implementation activities for framework development, as well as the extra subactivities needed to support feature variabilities, are described in the following chapter.

2 **Application engineering.** This uses the framework developed during framework engineering to build particular applications. Since one of the goals of application engineering is to remove the variabilities in the framework, and resolve the decisions in the decision model, Komponent composition trees for applications are very similar to those described in Part 2 (i.e. for a single system). The only difference is that Komponents are accompanied by a decision model instance, which captures the decisions made in resolving the decision model for a particular Komponent. The uses of the Komponent modeling and implementation activities for application development, as well as the extra subactivities needed to support framework instantiation, are described in Chapter 16.

The life-cycle for product line engineering in KobrA is depicted schematically in Figure 14.5. Framework engineering involves all the basic activities discussed in Parts 2 and 3, but applied in a generic way to create generic components. The result is a framework, a generic application capturing the characteristics of all the members of a product line. When applied in a generic way within framework engineering, context realization corresponds to the traditional domain engineering task of domain analysis, and the recursive process of specifying and realizing generic components corresponds to the construction of a domain-specific reference architecture.

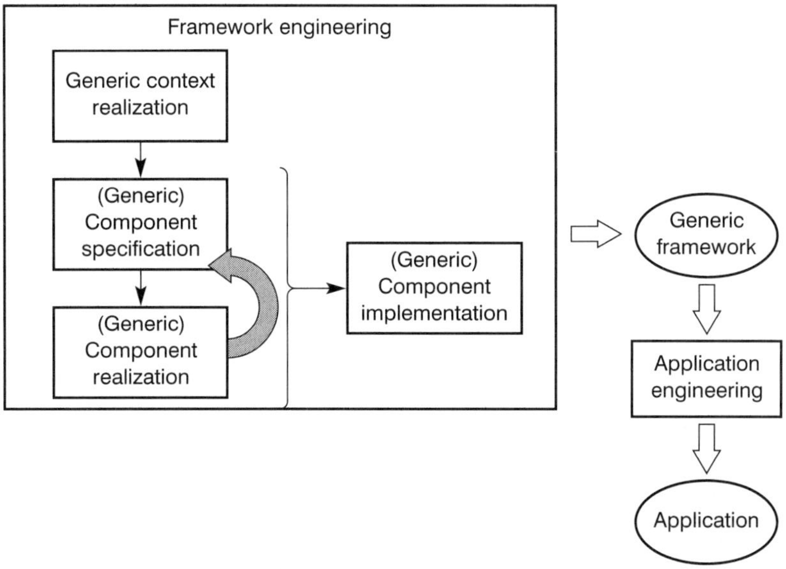

Figure 14.5 Product line life-cycle

During application engineering the framework created during framework engineering is instantiated to build particular applications. Application engineering also involves the basic activities described in Parts 2 and 3, but applied in a specific way to create specific products (i.e. specific components).

When applied in the context of product line engineering, therefore, two distinct variants of the basic component modeling and implementation activities and artifacts described in Parts 2 and 3 can be identified, as described in Appendix B. Basically the products and processes involved in framework engineering are *generic* and belong to a *framework*, while those invloved in application engineering are *specific* and are related to an *application*. These terms can be used wherever necessary to qualify artifacts and activities according to whether they are generic or specific, or associated with a framework or an application. However, in order to avoid clumsy terminology, the framework and generic cases are assumed to be the default. Thus, unqualified terms such as "component" and "context realization" are assumed to refer to the framework/generic case.

14.2.1 Single-system engineering

In KobrA, single-system development is viewed as merely a special case of product line engineering, albeit a special one. The component modeling and implementation activities were described in a product line independent way in Parts 2 and 3 not because product lines are of secondary importance, but in order to introduce concepts in an optimal order from a didactic perspective.

There are two ways in which the development of a single system can be viewed as a special case of product line development. A single system can be viewed either as a special case of a framework which has zero variabilities, and thus requires a trivial (i.e. empty) application engineering activity to create an application, or single-system development can simply be viewed as an instance of application engineering without the benefit of a framework as input (see Figure 14.6). In either case, the result is the same, and the activities and artifacts involved are precisely those described in the previous two parts of the book. In the remainder of Part 4 we describe the different variants of the artifacts and activities involved in framework and application engineering.

Figure 14.6 Single-system development

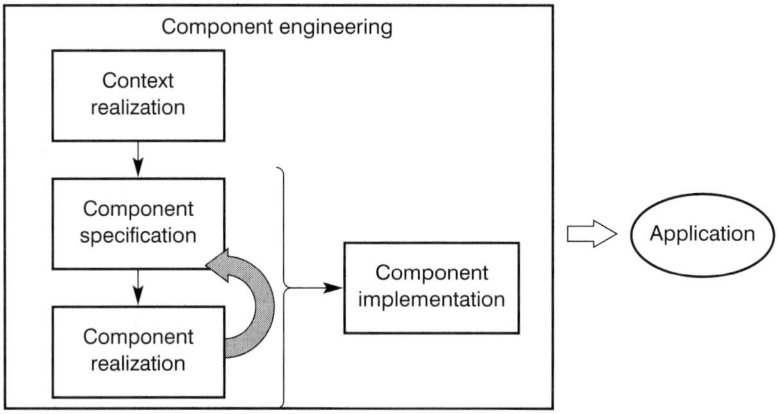

15 Framework engineering

In KobrA a framework is created using the same basic concepts, artifacts and activities as those described in the previous sections of the book, but generalized to cover a family of systems rather than an individual system. The main enhancement needed to achieve this generalization is the explicit consideration and representation of the commonalities and variabilities that relate the members of a product line. KobrA adds support for these issues in a seamless way to the core concepts, thus making product line engineering concerns totally orthogonal to the basic component modeling and implementation concerns.

As explained in the previous chapter, although Parts 2 and 3 introduce the component modeling and implementation concepts independently of product line engineering concerns, it should not be inferred that product line engineering is in some way subservient to single-system development, or represents the minority case. On the contrary, single-system development is viewed as a special case of product line engineering. This chapter therefore describes how the component modeling and implementation activities described in the previous parts of the book are generalized to support frameworks. The following chapter does the same for applications. As in the preceding chapters, we first discuss the artifacts and then the activities.

15.1 The generic component model

As discussed in Chapter 3, every Komponent in a containment tree can be regarded as representing an individual system in its own right, regardless of its position in the containment hierarchy. In a project that is not intending to exploit product lines, each Komponent defines the properties of exactly one system. When generalized to support a product, however, each Komponent in the containment tree can be thought of as defining a family of systems – each member of which represents a different variant. A Komponent that represent a family of systems (or components) as opposed to just one is referred to as a generic Komponent, and the artifacts of which it is composed are referred to as generic artifacts.

Not every Komponent in a framework need be generic. In fact, in the extreme case, none of the Komponents in a framework need be generic. However, such a framework represents a situation in which a product line only has one member. As soon as more members are identified, one or more of the Komponents in the framework must be generic. To distinguish Komponents within a framework from those within an application, the term "framework" can be used. Thus, a "framework Komponent" resides in a framework regardless of whether it is generic. Since a Komponent must either be in a framework or an application, the unqualified term "Komponent" is taken to refer to a framework Komponent.

A generic Komponent differs from a specific Komponent in two main ways

1. The regular artifacts describing the Komponent capture *all* the features for all members of the family represented by the Komponent. In addition, they use the stereotype <<variant>> to highlight features that are not common to all members of the family (i.e. features that represent variabilities between the family members).
2. An additional artifact, known as a decision model, is defined that relates the variabilities in the family to externally visible choices that collectively identify a particular member of the family.

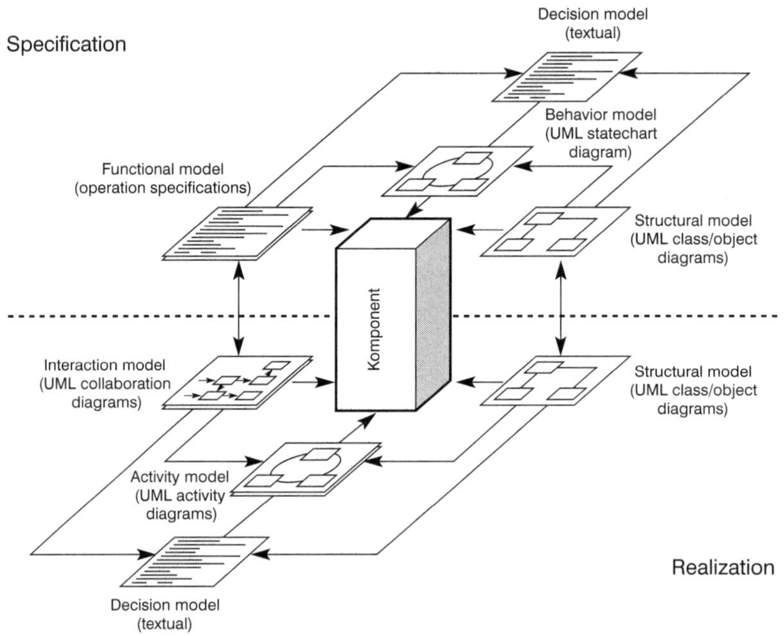

Figure 15.1 Schematic representation of a generic Komponent

This applies to both the specification and realization of a Komponent. In other words, the specification and realization of a Komponent both have their own decision models relating the variable features captured in their respective models. This is important since it allows variabilities in Komponent realizations to be handled separately from variabilities in Komponent specifications. In fact, the variabilities in a Komponent's realization should, by definition, be related to variabilities in the Komponent's non-functional requirements specification (within its specification) since different realizations should only change "how well" a realization supports non-functional requirements.

A schematic representation of a generic Komponent, including the additional decisions models, is shown in Figure 15.1. Figure 15.2 is a Venn diagram illustrating how generic Komponents are used to capture the commonalities and variabilities within a family of system (i.e. component). Circles A, B and C in this figure each represent the set of features possessed by an individual systems (i.e. components) in the family, while the thicker black line represents the set of features possessed by the generic Komponent. The shaded intersection of the three circles represents the set of features that are common to all three components in the family. These are modeled in the normal way using the normal UML features. Everything else outside the shaded intersection corresponds to a variable feature, since it does not belong to every member of the family, and is thus highlighted as such within the UML models using the <<variant>> stereotype.

From the generic UML models it is only possible to discern which features are common (i.e. inside the shaded intersection) or variable (i.e. outside the shaded intersection). It is the job of the associated decision model to define which variable features belong to which components.

There is no requirement for the features of the family members encapsulated by a generic Komponent to overlap. It is possible for the circles in Figure 15.2 to have no intersection at all. If this happens when all features of the components are being represented, including specification level features (i.e. those that are externally visible), something has probably gone wrong because it does not make sense to try to model totally distinct systems as part of a single product line. However, when the circles are restricted to realization-level features, such a

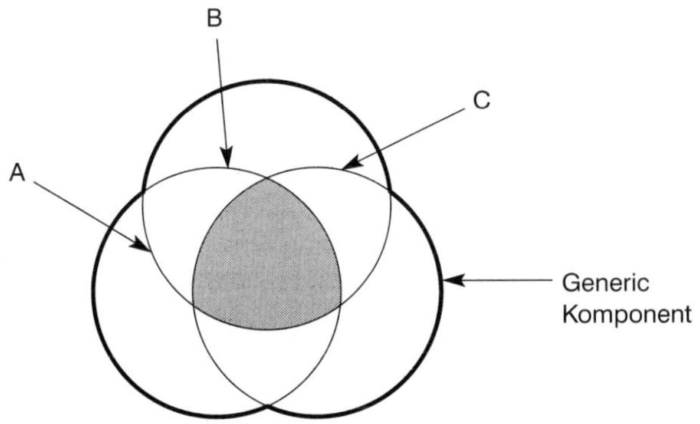

Figure 15.2 Feature overlap within generic Komponents

scenario represents totally distinct realizations of a given specification. Although this is not common it is not unreasonable. More typically, however, the different realizations of a generic Komponent share many common features, but also some differences. These differences can be quite substantial, including the presence or not of a sub-Komponent.

Like all other development concerns, KobrA handles all issues related to genericity in a recursive way according to the principle of locality. This implies a hierarchical relationship between the decision models in a framework's containment tree. Therefore, the decisions driving the selection of features in a generic Komponent's specification are influenced by the decisions in its parent's realization decision model. Similarly, the decisions driving the selection of features in a generic Komponent's realization are influenced by the decisions in its specification decision model.

This hierarchical relationship between the decision models means that choices at the root of the tree ripple down towards the leaves until a fully consistent set of features has been attained for all the Komponents in the containment tree. For this reason, unless it is the root Komponent, or a public sub-Komponent, a generic Komponent must always have a generic parent. If this were not the case, the variabilities captured by the generic component would not be visible at the root (i.e. system) level and thus would have no effect on the features of family members at the root level.

15.2 Artifacts

In this subsection we explain in detail how the Komponent-modeling and embodiment activities and artifacts described in the earlier parts of the book are enhanced when used to describe generic Komponents within a framework. The library case study developed to this point is accordingly enhanced into a framework covering four distinct variants of the library system.

15.2.1 Context realization

The purpose of a context realization in framework engineering is to represent the contexts of a set of similar systems in one generic model. This essentially corresponds to requirements engineering for a system family, which is usually known as domain analysis. The resulting products are the usual context realization models described in Chapter 7, enhanced by descriptions of the family variabilities, and a decision model describing how the variabilities related to specific family members. In addition, for the context realization, and only the context realization, a framework scope table is created to describe precisely what is covered by the framework and what lies beyond its scope.

15.2.1.1 Framework scope

KobrA's product line approach aims to maximize the return on the investment performed at the family level by devoting relatively little time to up-front analysis activities, such as exploring all facets of an application domain or analyzing in detail all subtle differences or variants of features. The analysis of the product line scope is therefore limited to the creation of a table that maps features to products. Since there are usually more features than products, the columns are used to represent products and the rows are used to represent features. When a product supports a feature, the corresponding field is checked. This kind of scoping is referred to as "product portfolio scoping" by Schmid [Sch00], and represents an economic alternative to more substantive domain scoping and asset scoping techniques that involve the analysis of much more detail. For example, asset-scoping involves the identification of assets that promise the highest benefit in the context of a product line, but this requires a significantly larger up-front effort, with a correspondingly higher risk.

Table 15.1 shows a part of the scope table for a library system framework. This embraces features derived from analyses of three real-life library systems: the Kaiserslautern City Library, the library of the University of Kaiserslautern, and the library within the Fraunhofer Institute for Experimental Software Engineering (IESE). A fourth variant, known as the basic library, has been added artificially to this product line in order to provide continuity with the single-system case study developed in Parts 2 and 3.

The bold features in the table are those supported by all library systems. At this level these can be viewed as common features, although differences may occur at a deeper level of detail. In other words, the basic structure for the classification scheme differs in some details from library to library. The reservation feature is a common feature but to illustrate how product line engineering works in KobrA, this functionality will be removed later from the basic library system. In the scope definition shown, no competitor products are included, although this is often done to illustrate the position of the planned products in the overall market.

15.2.1.2 Enterprise model

The purpose of the enterprise model is to describe the nature of the enterprise for which the system to be built is needed, what general role the system should play, and what business rules should govern its behavior. The enterprise model therefore describes the relevant aspects of the enterprise in a way that completely ignores the fact that the Komponent to be built will be realized as a computer system (see Section 7.1.1).

In framework engineering, the role of the enterprise model is thus to integrate and document the relevant aspects of the various enterprises in which members of the projected product line are intended to operate. Enterprises are typically quite different to each other. In fact, they differ far more significantly than the systems which are intended to support them. Therefore, it is not

Table 15.1 Scope definition of library system framework

	Kaiserslautern Library	University Library	IESE Library	Basic Library
Catalog				
Catalog	**X**	**X**	**X**	**X**
Monograph Catalog	**X**	**X**	**X**	
Periodical Catalog	X	X	X	
Periodical Issue Catalog			X	
Article Catalog			X	
Item Acquisition				
Acquisition on Approval		X		
Item Suggestion	X	X	X	
Firm Order	X	X		
Order Claim	X	X		
Order Cancellation	X	X		
Periodical Kardex				
Kardex		X	X	
Trial Subscription		X		
Subscription		X	X	
Subscription Claim		X	X	
Subscription Binding		X		
Routing			X	
Data Manangement				
User Management	**X**	**X**	**X**	**X**
Supplier Management	X	X	X	
Publisher Management	X	X	X	
Classification Management	**X**	**X**	**X**	**X**
Keyword Management			X	X
Descriptor Management				X
Data Exchange				
Internal Database Management			X	
External Database Management			X	
Union Catalog Interface		X		
Z39.50 Interface		X		

Table 15.1 Continued

	Kaiserslautern Library	University Library	IESE Library	Basic Library
Loan Management				
Loan	**X**	**X**	**X**	**X**
Return	**X**	**X**	**X**	**X**
Renewal	**X**	**X**	**X**	**X**
Reservation	X	X	X	(X)
Return Claim	**X**	**X**	**X**	**X**
Charges				
Loan Fee	Via annual fee			
Reservation Fee	X			
Overdue Fee	X	X		
Loss Fee	**X**	**X**	**X**	**X**
Reports/Profiles				
Account Report	**X**	**X**	**X**	**X**
User Report	**X**	**X**	**X**	**X**
DIN 1505 Report Format			X	
Accession Profiling	X	X	X	
Periodical Profiling	X	X	X	
OPAC				
OPAC	**X**	**X**	**X**	**X**
Simple OPAC Search	**X**	**X**	**X**	**X**
Extended OPAC Search		X	X	
Expert OPAC Search		X	X	
OPAC Index			X	
OPAC Reference Linking			X	
OPAC Reservations			X	

necessary to spend a great deal of effort trying to develop a generic enterprise model when the commonalities and variabilities in the various enterprises are not readily apparent. Nevertheless, it is often possible to create a simple generic enterprise model without significant effort, and when available this can be valuable input to the framework scoping activity.

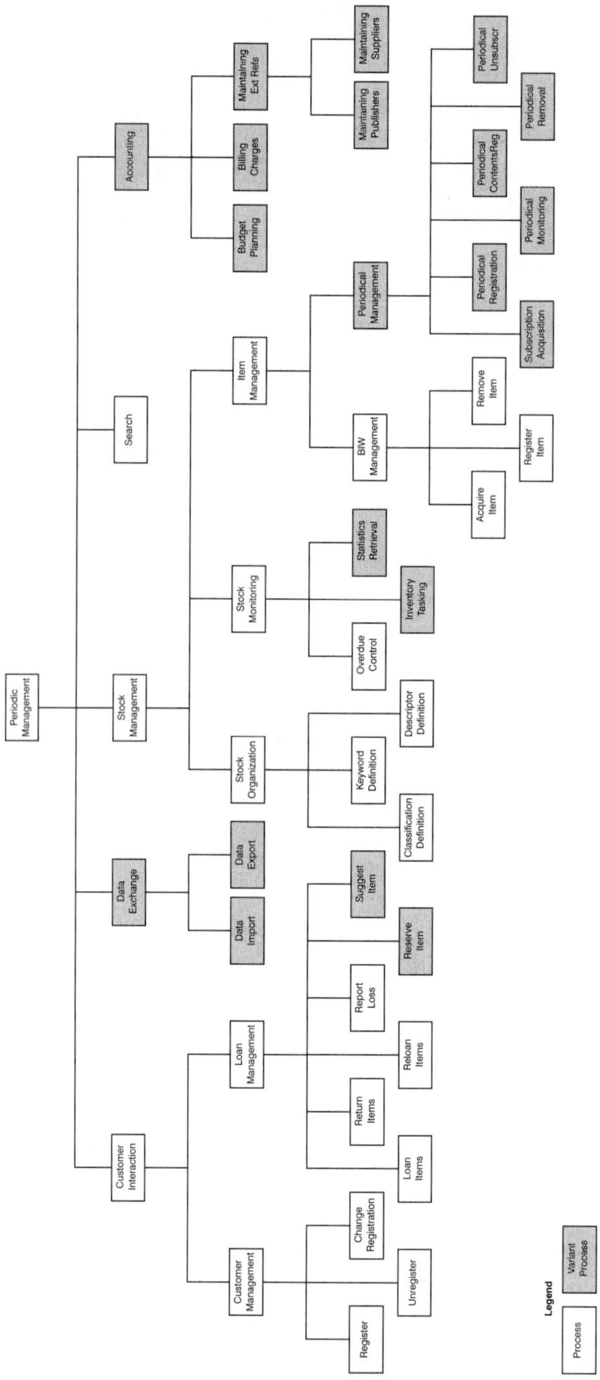

Figure 15.3
Generic enterprise process hierarchy

As in the non-product-line case, a generic enterprise model consists of a generic structural model and a generic process model. An example of the latter is illustrated in Figure 15.2. The goal of a generic business process hierarchy is to capture the processes supported by the system family. Processes not supported by all systems are represented by gray boxes. Generic enterprise models have the same form as generic structural models described below, but focus on the description of the high-level concepts of importance to the enterprise.

15.2.1.3 Structural model

The generic context realization structural model describes the classifiers that characterize in a generic way the systems' working environments and the information types that the systems need to interact with their environments. In the class diagram, basically every type of element may be variant; certain attributes, operations, classes, and relationships in the diagram may only be relevant for some of the systems in the product line.

Two styles of stereotype are used in KobrA to indicate that a feature is not possessed by every member of the product line. For model elements that otherwise would be stereotype-free, the <<variant>> stereotype is applied. In terms of Figure 15.1, it indicates that the property modeled by the element lies outside the shaded intersection that represents the features common to all family members. For model elements that already have a stereotype, for example <<x>>, stereotypes of the form <<variant x>> are used. For example, a variant Komponent would be adorned with the stereotype <<variant Komponent>>. This prevents the need for model elements to have two stereotypes.

Using these variant elements, variability within a structural diagram is made explicitly visible. However, the relationships (e.g. two classes are alternatives of each other) and dependencies between variant elements (e.g. one attribute is only needed in the case some operations are supported) are not captured in the diagram but in the decision model, as explained below.

The complete class diagram for the library system family is shown in Figure 15.4. This highlights the variant operations and variant classes in the generic LibrarySystem Komponent using the <<variant>> stereotype. Note that it is possible to have variant associations between non-variant classes. Associations that involve at least one variant class are by definition themselves variant. Therefore, to avoid diagram clutter, such relationships are not adorned with the <<variant>> stereotype since their variabilities is implicit.

In the library system product line, all classes and operations related to the managing and processing of periodicals, like monthly magazines, are <<variant>>. In fact these features, such as the operation registerNewPeriodicalIssue() and the class Periodical etc., illustrate the common "all or nothing" variability. Either a member of the product line has them all, or has none of them. It is the job of the decision model to capture these dependencies as explained in Section 15.2.1.6.

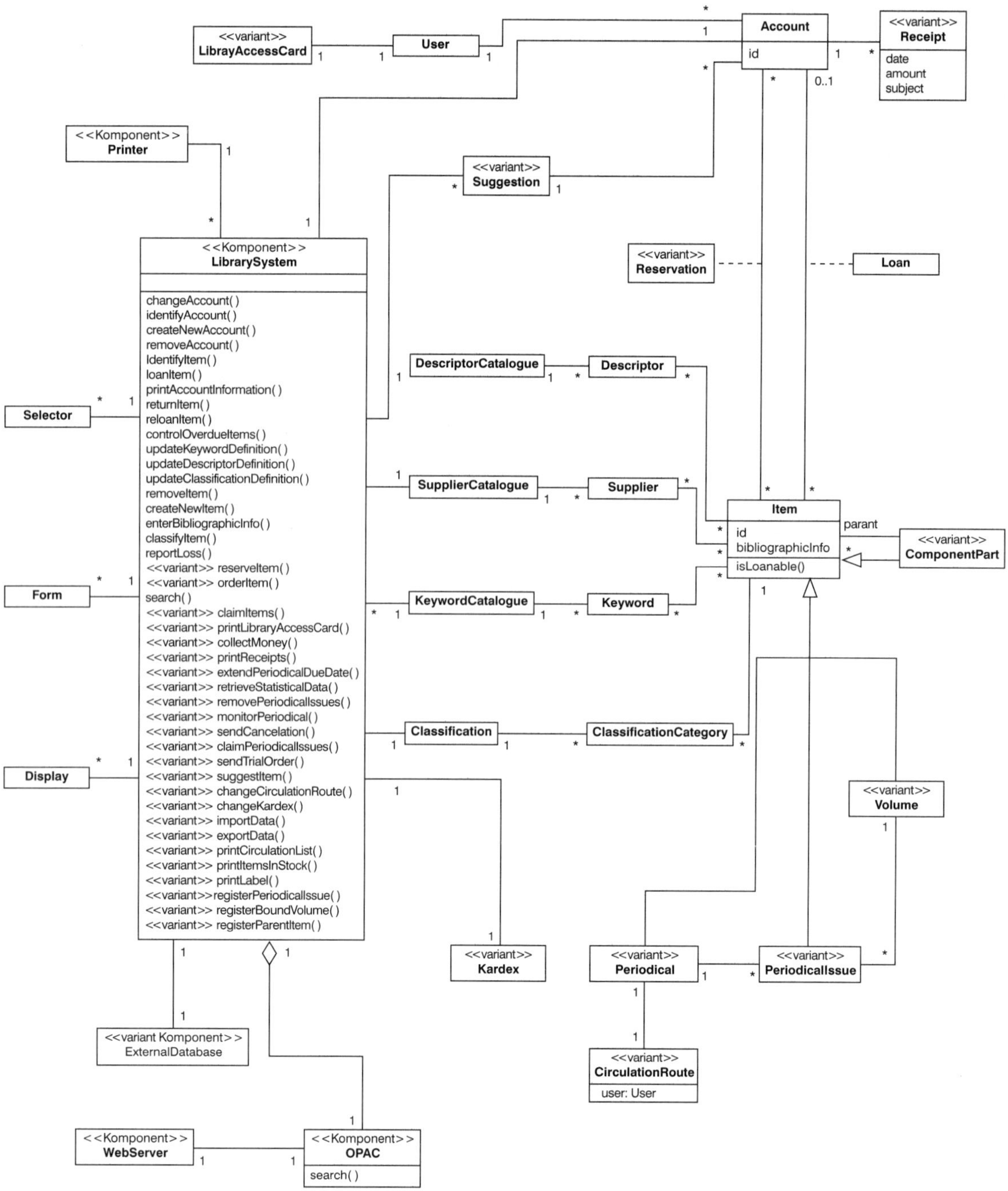

Figure 15.4

Generic context realization structural model

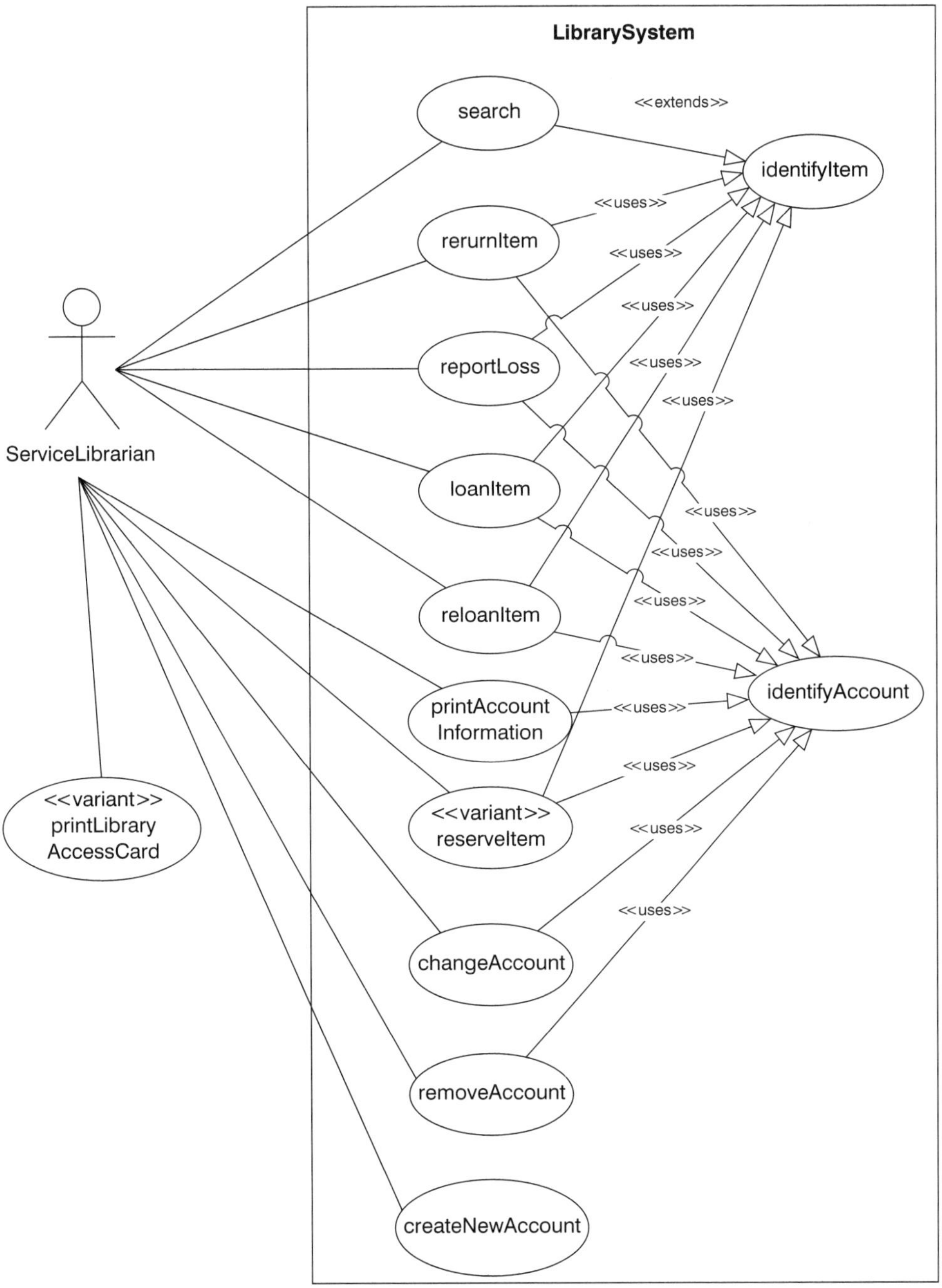

Figure 15.5
Generic use-case diagram

15.2.1.4 Activity model

The purpose of the context realization activity model in product line engineering is to generically describe the nature and the relationship of the activities that have a bearing on the operation of the systems in the product family. In general, a generic activity model can be composed of generic activity diagrams, generic activity specifications, and generic use-case diagrams. The first two of these are described below in relation to a generic Komponent realization, and thus are now discussed further below.

In the case of use-case diagrams, individual use-cases may be variant (i.e. not supported by all library systems) or <<uses>> or <<extends>> relationships between them may be variant. For example, Figure 15.5 shows the generic use-case diagram for the service librarian. From the small number of variabilities in this diagram, it is evident that the service librarian role largely represents common (i.e. core) functionality of a library system. This is as expected, since the management of customer data and the tracking of loans and returns is an important part of any library. Only the suggestion and the reservation of items are not common to all the library systems in the family. Therefore, the corresponding use-cases are adorned with the stereotype <<variant>>.

15.2.1.5 Interaction model

The context-realization interaction model ties activities and data items together and thus provides an alternative view of the way in which user tasks are realized in terms of system operations. In a product line setting these interactions may be relevant only for some of the systems in the family, that is, some of the interactions may be variant.

Figure 15.6 shows the sequence diagram for loaning an item from the library system. Because not all variants of the system require the service librarian to print out the account information after loaning an item, the invocation of the *printAccountInformation()* operation is modeled as <<variant>>.

Figure 15.6
Generic sequence diagram loanItem()

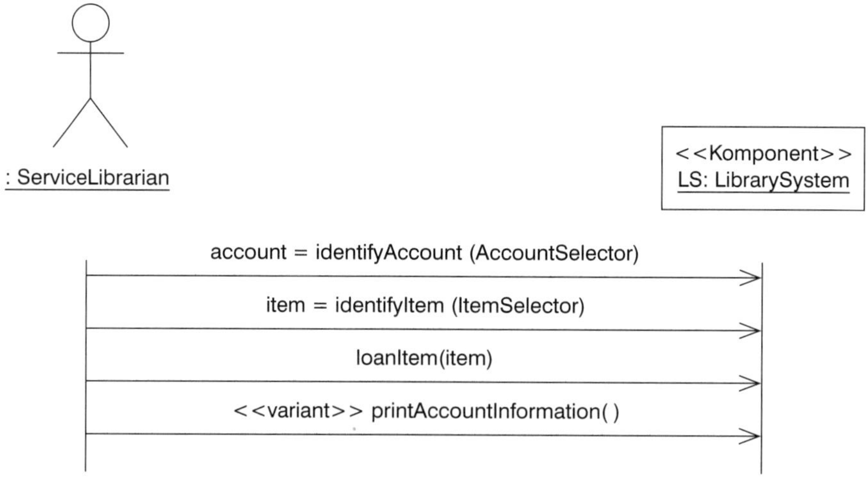

15.2.1.6 Decision model

The decision model is an auxiliary artifact of a generic context realization. Its purpose is to document all variation points within the context realization, to capture the domain knowledge behind every variant model element, and to structure the variability in a way that supports application engineering. It consists of a set of inter-related decisions at different levels of detail and abstraction.

A decision consists of:

- **ID.** Each decision has a unique identifier, which is used to reference the decision within the definitions of constraints between decisions (see below, Effect set). The ID is a hierarchical identifier that is composed of the Komponent name (here, context realization), the diagram name, and a unique number.
- **Question.** The question formulates the variability in a form that supports application engineering. The question abstracts from the model's variant elements by referring to concepts in a Komponent's environment. In the case of the context realization this is the application domain. Thus, knowledge about the application domain is an integral part of the decision model.
- **Variation point or subject.** Decisions that are directly related to a variant model element are called simple decisions. For these decisions, the variation point refers to the related variant model element. For decisions at a higher level of abstraction specific variation points are replaced by subjects. These decisions are not directly related to variant model elements, but are related to a general domain concept that causes the variability. The subject is intended to support the identification of further relationships among decisions.
- **Resolution set.** Each decision provides a set of possible resolutions that are answers to the decision's question. Optionally, the resolution that is valid for most of the systems in the application domain can be marked as the default resolution.
- **Effect set.** Each effect in the effect set corresponds to one possible resolution of the resolution set. The effects capture relationships among decisions in the decision model. In order to support application engineering, the effects describe in a precise and prescriptive way how the generic models of the context realization must be modified to match the requirements that correspond to the appropriate resolution.

At the lowest level of the decision model, there are simple decisions which are directly related to a single variant model element. These decisions should be ideally generated and maintained by a KobrA case tool (see Chapter 20).

Table 15.2 lists some of the simple decisions related to the business process hierarchy. Decision 1 corresponds to the variant process of reserving an item. It has two possible resolutions, *yes* or *no*. *Yes* means the business process *reserveItem* is supported by a library system and thus the process must be part of the process hierarchy. In this case, when an application is produced, the generic model is made specific by changing the stereotype of the process from <<variant>> to the stereotype used in component modeling and for non-variant processes. This is described in the *effect* column for a *yes* resolution of this decision. *No* means the business process "reserveItem" is not supported by a library system and thus the process must be excluded from the process hierarchy. In this case, the process is removed from the model, as described in the effect column for a "no" resolution of the decision. Analogous simple decisions for all other leaf processes exist.

The process *exchangeData* is variant because a library system may either exchange data bi-directionally with an external database, export data only, import data only, or not exchange data with an external database at all. Since it is not a leaf process, but has numerous sub-processes in the business process hierarchy, when the process *exchangeData* is removed from the process hierarchy, its leaf processes also have to be removed. This is captured in the effect entry for decision 3 by requiring the resolution of decisions 4 and 5 that concern the processes *importData* and *exportData* respectively.

Table 15.2 Decision model concerning the business process hierarchy

ID	Question	Variation point	Resolution	Effect
Business-process hierarchy				
1	Process reserveItem supported?	Process reserveItem	yes (default)	remove stereotype <<variant>>
			no	remove process reserveItem
2	Process suggestItem supported?	Process suggestItem	yes (default)	remove stereotype <<variant>>
			no	remove process suggestItem
3	Process exchangeData supported?	Process exchangeData	yes	remove stereotype <<variant>>
			no (default)	remove process exchangeData resolve decision 4: no
			resolve decision 5: no	
4	Process importData supported?	Process importData	yes	remove stereotype <<variant>>
			no	remove process importData
	...	...	...	...

Table 15.3 lists some of the simple decisions related to the structural model of the generic context realization. There are decisions concerning variant classes as well as variant operations. When a variant class is excluded its associations must be excluded too. It is also possible to have variant associations or associations with variant multiplicities, although this is not the case in the presented example. For example, in a digital library, an item could be loaned by more than one user and therefore the loan association between *Item* and *Account* would be a n:m rather than a n:1 relationship.

Table15.3 Decision model for the context realization structural model

ID	Question	Variation point	Resolution	Effect
Structural diagram				
1	Class Reservation needed?	class Reservation	yes no	remove stereotype <<variant>> remove association class Reservation remove reservation association between classes Account and Item
2	Class ExternalDatabase needed?	class External Database	yes no	remove stereotype <<variant>> remove class ExternalDatabase remove all associations of class ExternalDatabase
	...	...	...	...
11	Operation LibrarySystem. reserveItem() needed?	operation LibrarySystem. reserveItem()	yes no	remove stereotype <<variant>> remove operation LibrarySystem. reserveItem()
12	Operation LibrarySystem. suggestItem() needed?	operation LibrarySystem. suggestItem()	yes no	remove stereotype <<variant>> remove operation LibrarySystem. suggestItem()
13	Operation LibrarySystem. importData() needed?	operation LibrarySystem. importData()	yes no	remove stereotype <<variant>> remove operation LibrarySystem. importData()
	...	...	...	...

The simple decisions for the remaining diagrams are defined in the same way. Tables 15.4 and 15.5 show the simple decisions relating to the service librarian's use case diagram and the *loanItem()* sequence diagram, respectively.

Since simple decisions are at a low level of abstraction, even a small system family such as the library family can have many variant elements and thus a large number of simple decisions. Moreover, decision models consisting only of simple decisions do not add much value to the framework beyond that already provided by documenting variant elements. To address these problems, higher-level decisions can be defined that (partially) subsume several other decisions. The number of decisions that must be resolved during application engineering can then be significantly reduced and relationships and constraints between the decisions can be explicitly captured.

Table 15.4 Decision model for the use-case diagram – service librarian

ID	Question	Variation point	Resolution	Effect
Use-case diagram service librarian				
1	Use-case reserveItem supported?	use-case reserveItem	yes	remove stereotype <<variant>>
			no	remove use-case reserveItem remove association to Actor ServiceLibrarian
2	Use-case suggestItem supported	use-case suggestItem	yes	remove stereotype <<variant>>
			no	remove use-case suggestItem remove association to Actor ServiceLibrarian
	...	...	...	...

Table 15.5 Decision model for the sequence diagram – loanItem

ID	Question	Variation point	Resolution	Effect
Sequence diagram – loanItem				
1	Is message print AccountInformation sent?	message printAccount Information()	yes	remove stereotype <<variant>>
			no	remove message printAccountInformation()
	...	...	...	...

Table 15.6 illustrates some higher-level decisions that subsume several of the simple decisions related to a single diagram. In this case, the decisions concern the structural diagram only. Decision 37, for example, captures the strong relationship between the library system operations *importData(). exportData()*, and the relationship of the library system to an external database. Consequently, the effect entry for this decisions states that if the library system does not exchange data with an external database (i.e. decision 37 is resolved as *no*), the external database class is removed from the structural diagram, and the library does not offer the import and export operations. The effect is equivalent to a *no* resolution of the three decisions 2, 13, and 14 (defined in Table 15.3). Decision 37 subsumes the three decisions only partially because, in the event that data is exchanged with an external database, it is not clear whether data is imported or exported.

Decision 38, on the other hand, subsumes decisions 1 and 11 fully. When a library system supports reservations, both the class *Reservation* and the operation *reserveItem()* are needed, and vice versa.

Table 15.6 Higher-level decisions concerning the structural model

ID	Question	Subject	Resolution	Effect
Structural diagram				
37	Does the system exchange data with an external database?	Data Exchange, External Database	yes	resolve decision 2: yes
			no	resolve decision 2: no resolve decision 13: no resolve decision 14: no
38	Does the system support the reservation of items?	Reservation	yes	resolve decision 1: yes resolve decision 11: yes
			no	resolve decision 1: no resolve decision 11: no
	...	...	...	...

Because the different models in the context realization are different views of the same thing – a library system family and their contexts – it is also useful to integrate decisions defined for different diagrams into common higher-level decisions. Decisions that subsume lower-level decisions from multiple diagrams are no different from higher-level decisions of the kind just defined except that the effect entry must identify the particular diagram affected.

Table 15.7 shows higher levels that integrate decisions related to the business-process hierarchy and the structural diagram. Decision 1 in Table 15.7 (the first context-realization decision that is not related to a single diagram) subsumes decisions concerning data exchange. Decision 2, on the other hand, subsumes reservation-related decisions related to the two diagrams. For the same reasons given above, decision 1 only partially subsumes the diagram decisions, while decision 2 fully subsumes the diagram decisions concerning reservation issues.

Table 15.7 Higher-level decisions concerning multiple diagrams

ID	Question	Variation point	Resolution	Effect
Context realization				
1	Does the system exchange data with an external database?	Data Exchange, External Database	yes	resolve decision business process hierarchy 3: yes resolve decision structural diagram 2: yes
			no	resolve decision business process hierarchy 3: no resolve decision structural diagram 37: no
2	Is the reservation of items supported?	Reservation	yes	resolve decision business process hierarchy 1: yes resolve decision structural diagram 38: yes
			no	resolve decision business process hierarchy 1: no resolve decision structural diagram 38: no
	...	...	...	...

Table 15.8 Mapping features to decisions

Feature		Resolution	Effect
Data Exchange	External Database Interface	yes	resolve decision context realization 1: yes
		no (default)	resolve decision context realization 1: no
Loan Management	Reservation	yes (default)	resolve decision context realization 2: yes
		no	resolve decision context realization 2: no

The number of decision levels in the decision model is not limited and depends on the nature of the application domain and on the degree of detail already managed at the context realization level.

If possible, at the top-level decisions should be related to the features in the framework scope. Table 15.8 does this for the context-realization decisions 1 and 2. Decision 1 maps directly to the reservation feature and decision 2 is related to the external database interface feature (see also the framework scope defined in Table 15.1). In an ideal case, all decisions are fully subsumed by the features of the framework scope. In this case, application engineering is reduced to the selection of features by a customer.

If the definition of default resolutions is complete, a default application of the framework is defined. Often, the default application corresponds to a standard system that is marketed by an organization but that typically must be heavily customized to the specific needs and characteristics of its final operating environment.

15.2.2 KobrA component specification

A generic Komponent specification describes the externally visible properties and behavior of Komponents within a framework. As such, it describes the set of requirements that members of a family of Komponents are expected to satisfy. The artifacts are those described in Chapter 4, but enhanced with a description of variation points and a decision model.

15.2.2.1 Structural model

The structural model describes the nature of the classes and relationships by which the variants in the family interacts with their environment, and the parts of their structure that is visible at their interfaces. The environment of root Komponents is represented by the context realization, and in the setting of a product line, this takes the form explained in the previous section. The environments of all other Komponents, in contrast, is represented by their parent's realizations. As mentioned previously, unless a Komponent is a public sub-Komponent of its parent, it can only be generic provided that its parent realization is generic. Similarly, the specification of a root Komponent can only be generic if the context realization is generic. This is because there is no reason to support variabilities anywhere in a composition tree unless they have an effect on the properties of the overall system represented by the root.

Figure 15.7 shows a generic class diagram for the LoanManager Komponent which is a generalization of the structural model shown in Chapter 4, enhanced

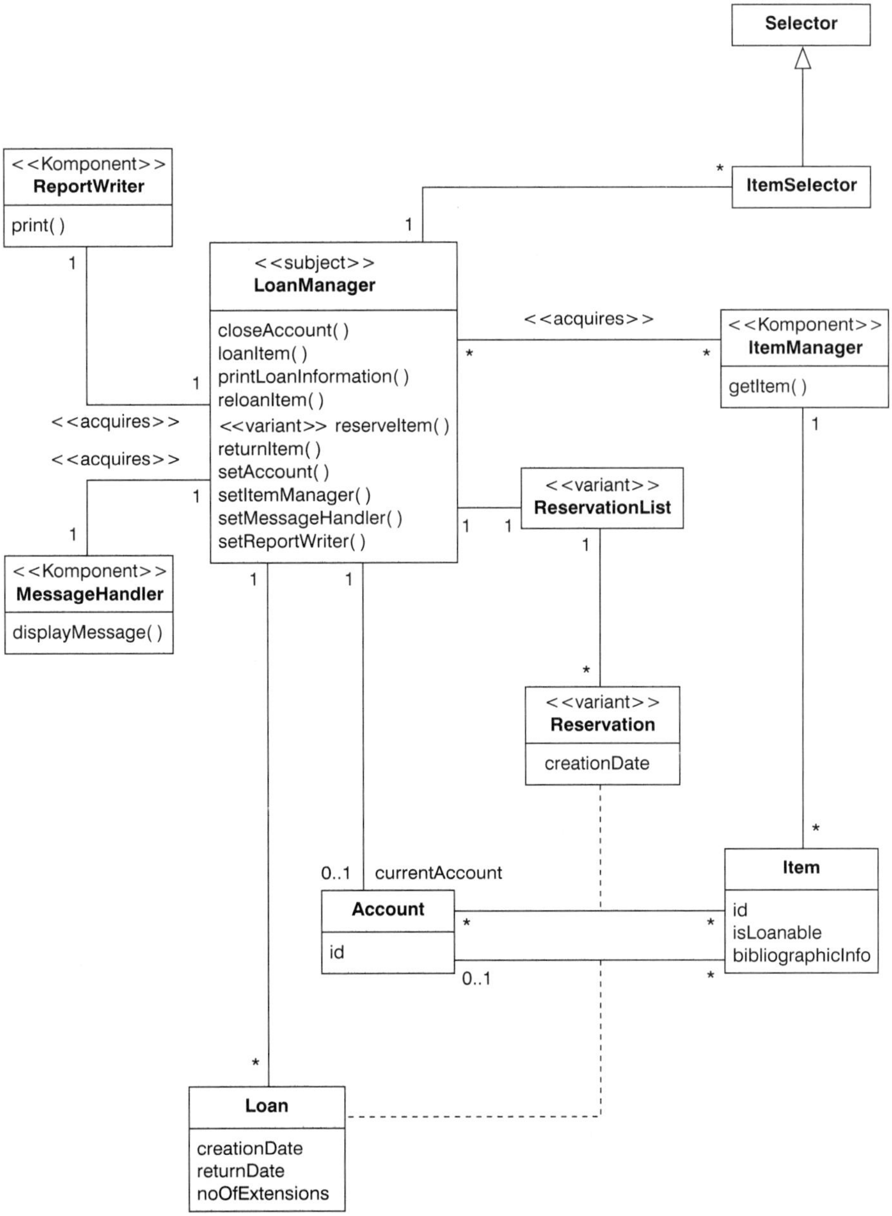

Figure 15.7
Generic structural model of the LoanManager Komponent

with variable features. Variations points are modeled exactly the same way as in the class diagram for the context realization described in the previous section. The variability of the library system family only effects the reservation capabilities of the *LoanManager* specification. The operation *reserveItem()* and the class *Reservation* are both <<variant>>.

The object diagram for the *LoanManager* is the same for all members of the library system family. This is because the run time configuration of the *LoanManager* is a commonality from an external point of view. In contrast, the internal run time configuration of *LoanManager* Komponent instances is variable as it is explained below when discussing the generic realization of the *LoanManager* Komponent.

15.2.2.2 Functional model

The functional model describes the externally visible effects of the operations supported by a Komponent in terms of a set of operation specifications. In the context of a product line, two forms of variability can be captured for a given operation:

1. Certain aspects of the operations behavior vary, but the operation is presents in all variants of the family. In this case the variant parts of the operation specification are embraced by the XML-like tags <variant> and </variant>.
2. The operation is not present in all members of the product line. In this case, the stereotype <<variant>> is placed before the name of the operation in the operation specification.

Naturally, these two are compatible, and can appear together within a given operation specification.

The generic specification of the operation *loanItem()* is described in Table 15.9. Here, the rule for deciding when an item is loanable has a variant part. Before loaning an item it is only necessary to check whether an item is reserved by another user in variants of the system in which reservation is supported.

Similarly, the specification of the *reloanItem()* operation has a variant part in the rule clause (see Table 15.10).

Apart from the variations capture in the two previous operation specifications, the only other variant part in the functional model of the *LoanManager* Komponent specification is the *reserveItem()* operation that as a whole is variant. In other words, it is not present in all variants of the system. This is documented by the <<variant>> stereotype in the name clause, as illustrated in Table 15.11.

Table 15.9 Generic specification of the operation loanItem()

Name	**loanItem()**
Description	The loan of an Item to currentAccount is registered
Receives	selector: ItemSelector
Sends	MH.displayMessage() Item item = IM.getItem(selector)
Rules	An item is loanable if it is not an item that must always stay in the library (e.g., antique books). An item is currently loanable if it is loanable and not loaned <variant> or reserved </variant> by another user.
Changes	new Loan
Assumes	Subject is in the state accountIdentified Selector selects exactly one Item
Result	item selected by selector has been obtained from the ItemManager IM by sending the message getItem(selector) if item is currently loanable a new Loan object, loan, has been created that relates item and currentAccount, and has the attribute values – creationDate = today, returnDate = today + <loanperiod> and noExtensions = 0 and, loan has been stored. if item is not currently loanable one of the following messages has been sent to the MessageHandler MH: displayMessage("AlreadyLoaned") <variant> or displayMessage("Reserved") </variant>

Table 15.10 Generic specification of the operation reloanItems()

Name	**reloanItem()**
Description	An item loaned to an Account is reloaned to the currentAccount.
Receives	selector: ItemSelector
Sends	MH.displayMessage() Item item = IM.getItem(selector)
Changes	loan
Rules	An item is not reloanable if <variant>it is reserved or </variant> the number of extensions is greater or equal to <maxExtension>.
Assumes	Subject is in the state accountIdentified. The item is loaned to currentAccount.
Result	item selected by selector has been obtained from the ItemManager IM by sending the message getItem(selector) if item is reloanable the loan containing item has the attribute values returnDate = today + <maxExtension> noExtensions = noExtensions + 1 if item is not reloanable one of the following messages has been sent to the Message Handler, MH: <variant> displayMessage("Reserved") or </variant> displayMessage("OverExtension")

15.2.2.3 Behavioral Model

The behavioral model shows the externally visible behavior of a Komponent in terms of events, operations and states. In the context of a product line, these events, operations, and states can be variable. The usual <<variant>> stereotype is then used to indicate which model elements are variant and thus are not part of every member of the product line.

Figure 15.8 shows a generic state diagram for the *LoanManager* Komponent. The reservation capability has only minimal impact on the behavior of the overall Komponent and thus the only variant part in the state diagram is the event corresponding to the invocation of the *reserveItem()* method. The fact that this does not occur in every member of the family is documented by the <<variant>> stereotype.

Table 15.11 Variant specification of the reserveItem() operation

Name	**<<variant>> reserveItem()**
Description	The reservation of an Item for the currentAccount is registered.
Receives	selector: ItemSelector
Sends	MH.displayMessage("Item cannot be reserved") Item item = IM.getItem(selector)
Changes	new Reservation
Assumes	Subject is in the state accountIdentified Selector selects exactly one Item

Result	The Item item selected by selector has been obtained from the ItemManager IM by sending the message getItem(selector)
	if item is reservable
	    a new Reservation object has been created that relates item and currentAccount, and has the attribute value creationDate = today and
	    reservation has been stored.
	if item is not reservable
	    the message displayMessage("Item cannot be reserved") has been sent to MH.

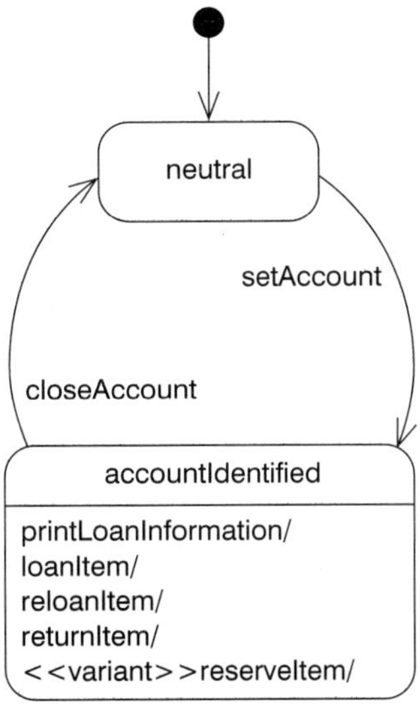

Figure 15.8 Generic state diagram of the LoanManager Komponent

This variability can also be captured in statechart tables in a similar way (see Table 15.12).

Table 15.12 Statechart table for the LoanManager Komponent

Source state	UML Transition string	Target state
neutral	set Account()	accountIdentified
accountIdentified	closeAccount()	neutral
accountIdentified	loanItem() printLoanInformation() reloanItem() <<variant>> reserveItem() returnItem()	accountIdentified

15.2.2.4 Auxiliary artifacts

Chapter 4 identifies various auxiliary development artifacts (other than the decision model) to support the main specification models discussed above. In the context of a framework all of these auxiliary artifacts are generic in the sense that they contain information related to all the variants of the Komponent for all members of the product family. However, they do not all contain definitions of variabilities.

It is clearly possible for the non-functional requirements specification to be generic because not all requirements will be the same for all variants of the Komponent. For example, the quality requirements on one variant of a system (e.g. a demo variant) will often be less stringent than those in another variant (e.g. an enterprise variant). These variable non-functional requirements can be captured in the usual way by marking them as variant in the specification, and explaining their relationships to variants of the Komponent in the decision model.

The other auxiliary artifact which will clearly often contain variabilities is the test suite. Since operations can vary in their functionality, and may not even be present in different variants of the Komponents, the test cases used to test them need to be varied accordingly. Again, the same basic strategy of marking test cases as variant, and relating them to other variabilities in the decision model, is employed.

Because the other auxiliary artifacts are concerned with development time issues (i.e. inspection and quality modeling) they do not need to contain variabilities. Indeed, part of their role is to ensure that the variabilities and decisions defined in the other models are consistent with each other. They are generic, therefore, only in the sense that they address all the features of a generic Komponent.

Table 15.13 Statechart table for the LoanManager Komponent

ID	Question	Variation point	Resolution	Effect
Structural model				
1	Class Reservation needed?	class Reservation	yes	remove stereotype <<variant>>
			no	remove association class Reservation remove reservation association between classes Account and Item
2	Class ReservationList needed?	class ReservationList	yes	remove stereotype <<variant>>
			no	remove class ReservationList remove all associations of class ReservationList
3	Operation LoanManager. reserveItem() needed?	operation LoanManager. reserveItem()	yes	remove stereotype <<variant>>
			no	remove operation LoanManager. reserveItem

Table 15.14 Decision model for functional model LoanManager Komponent

ID	Question	Variation point	Resolution	Effect
Operation specification loanItem()				
1	Variant part "or reserved by another user" needed?	Rules clause	yes	remove <<variant>> tags
			no	remove <<variant>> tags and contents
Operation specification reloanItem()				
1	Variant part "it is not reserved by another user and" needed?	Rules clause	yes	remove <<variant>> tags
			no	remove <<variant>> tags and contents
Operation specification reserveItem()				
1	Operation specification reserveItem needed?	operation specification reserveItem	yes	remove <<variant>> tag from name clause
			no	exclude operation specification reserveItem

15.2.2.5 Decision model

The decision model for a Komponent specification has the same form as that for the context realization described in 15.2.1.6. At the lowest level of abstraction it has a set of simple decisions directly related to variant elements in the specification models. Tables 15.13 to 15.14 show all the simple decisions related to the *LoanManager* specification models.

Table 15.15 Decision model for the behavioral model of the LoanManager komponent

ID	Question	Variation point	Resolution	Effect
State diagram				
1	Does reserveItem event occur in state accountIdentified?	Event reserveItem	yes	remove <<variant>> tags
			no	remove event reserveItem
Statechart Table				
1	Does reserveItem event occur in state accountIdentified?	Event reserveItem	yes	remove <<variant>> tags
			no	remove event reserveItem

Table 15.16 Top-level decision of the LoanManager specification decision model

ID	Question	Variation point	Resolution	Effect
LoanManager specification				
1	Is the reservation capability supported?	Reservation	yes	resolve decision structural model 1: yes resolve decision structural model 2: yes resolve decision structural model 3: yes resolve decision op. spec loanItem() 1: yes resolve decision op. spec reloanItem() 1: yes resolve decision op. spec reserveItem() 1: yes resolve decision state diagram 1: yes resolve decision statechart table 1: yes
			no	resolve decision structural model 1: no resolve decision structural model 2: no resolve decision structural model 3: no resolve decision op. spec loanItem() 1: no resolve decision op. spec reloanItem() 1: no resolve decision op. spec reserveItem() 1: no resolve decision state diagram 1: no resolve decision statechart table 1: no

Also, as explained for the context realization decision model, it is useful to define higher-level decisions that subsume several other decisions. In the case of the *LoanManager* Komponent, intermediate levels are not necessary because of the small number of simple decisions and their obviously strong coupling. All simple decisions can therefore be subsumed by a single higher-level decision, as illustrated in Table 15.15.

15.2.3 KobrA component realization

A generic Komponent realization describes how Komponents within a framework satisfy their specifications – in short, how they work. It describes the private designs of the members of a family of Komponents. The artifacts are those described in Chapter 5, enhanced with a description of variation points and a decision model.

15.2.3.1 Structural model

The variability in a realization structural model is managed in the same way as that in the other generic structural models described previously. Figure 15.9 shows the generic class diagram for the *LoanManager* Komponent's realization. This diagram illustrates the application of the <<variant Komponent>> stereotype mentioned in Section 15.2.1.3. The private Komponent *ReservationManager* of the *LoanManager* is marked as a <<variant Komponent>>, which means that an instance of the *LoanManger* Komponent does not in all cases contain a *ReservationManager*.

Having variant sub-Komponents means that the run time constellation of the *LoanManager* varies. This is depicted in a generic object diagram, shown in Figure 15.10.

15.2.3.2 Activity model

The activity model in a Komponent realization describes the nature and the relationship of the activities that have a bearing on the operation of the system. In the context of a product line, generic activity definitions may contain optional subactivities, alternative activities or more general alternative paths, and variant swim lanes. As in the other diagrams the stereotype <<variant>> is used to identify elements that are not part of all members of the product family. Figure 15.11 shows the generic activity diagram for *loanItem()*. Only when item reservation is supported by a library system is it necessary to check whether a user has reserved the item before it can be loaned. To capture this genericity of the *loanItem()* activity, two alternative paths are introduced, one path checking for a potential reservation, the alternative path skipping this reservation check. In the diagram this is indicated by marking decision points and/or branches from decisions points with the stereotype <<variant>>.

In some cases, the set of variant model elements is encapsulated by a swim lane. For example, the reservation capability makes the *ReservationManager* swim lane obsolete. Hence, the swim lane is marked as <<variant>>.

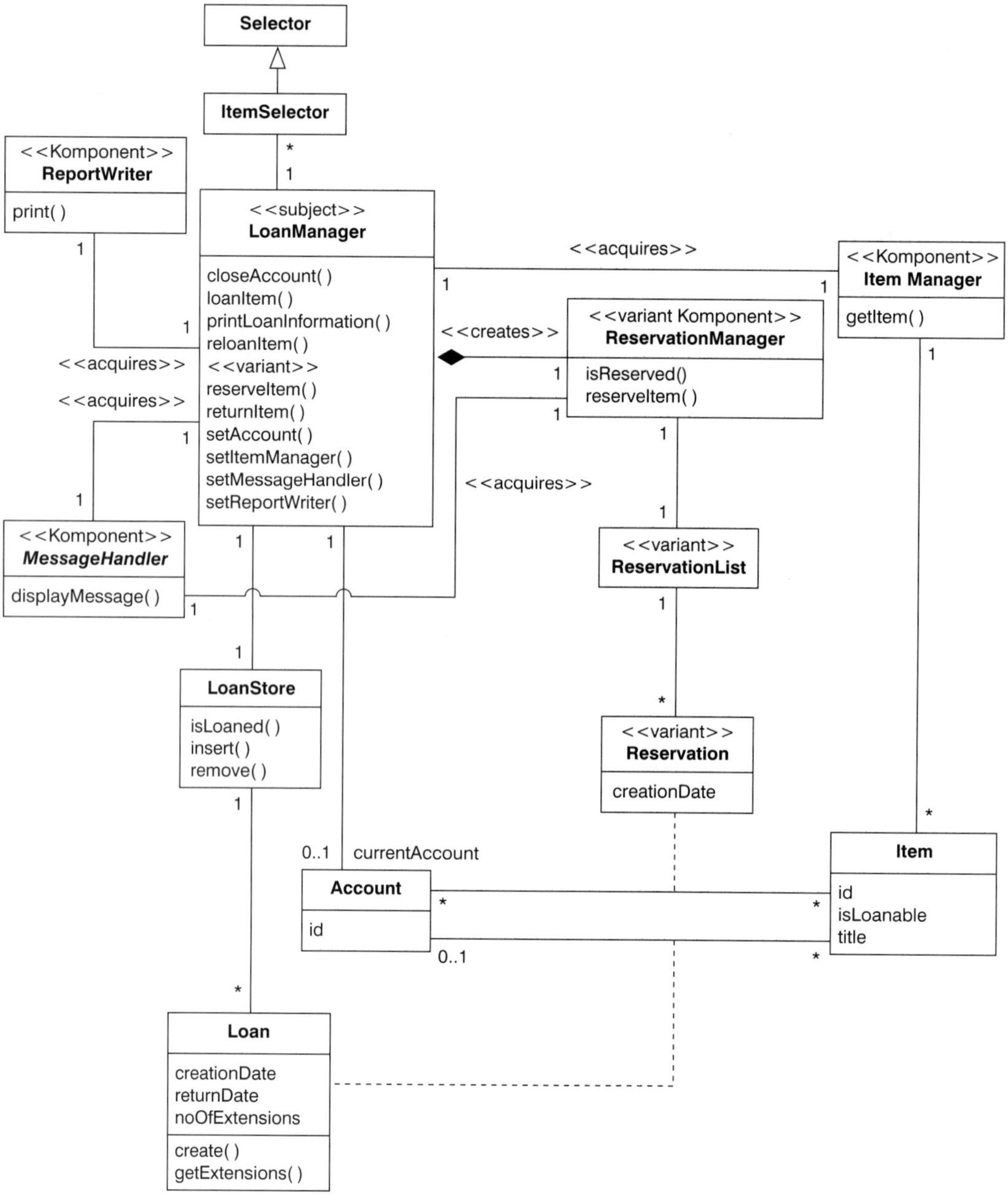

Figure 15.9
Generic structural model of the LoanManager's realization

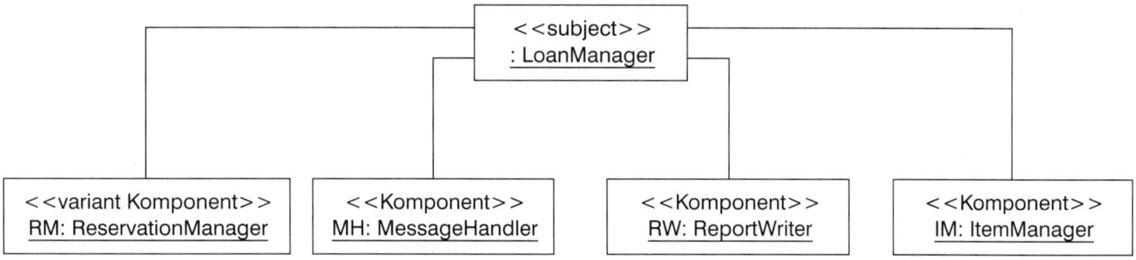

Figure 15.10
Generic object diagram of the LoanManager's realization

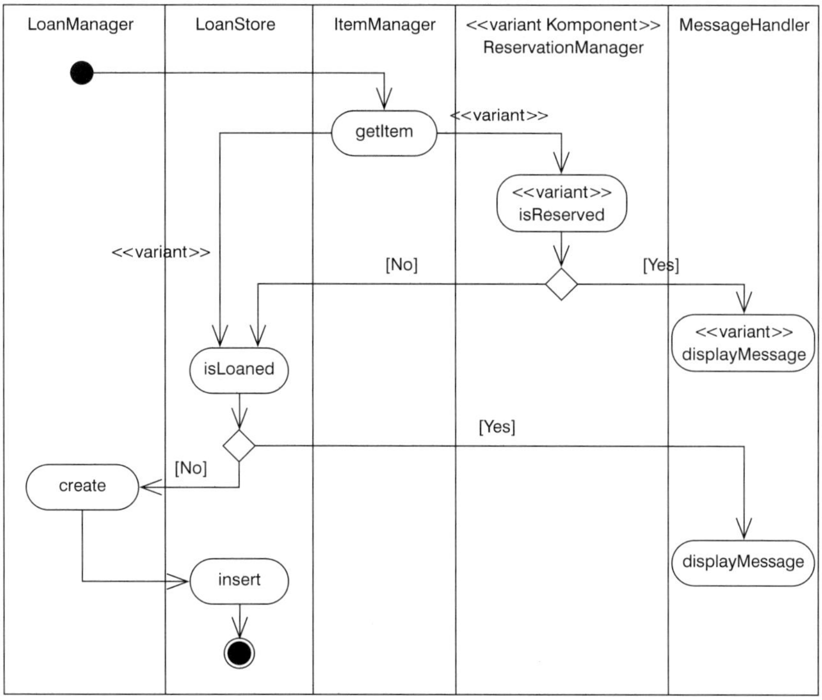

Figure 15.11
Generic activity diagram for the loanItem activity

Generic activity specifications are naturally handled in the same way as operation specifications (since the latter are a special case of the former). Thus, variant parts within activity specifications are embraced by the tags <variant> and </variant>, and activities which may be absent from some members of the product family are marked as <<variant>>.

Table 15.17 and 15.18 show the generic specifications of the intermediate activities *checkIfReloanable()* and *checkIfLoanable()* respectively. The former has no explicit variants because these are encapsulated within its subactivity, *checkIfLoanable()*.

Table 15.17 Generic specification for the checkIfReloanable activity

Name	CheckIfRelonable()
Description	Checks to see if an item can be reloaned
Receives	item: Item MH : MessageHandler
Result	If item is not loaned printMessage ("NotLoaned") has been sent to MH false has been returned else checkIfCurrentlyLoanable() and return its result

Table 15.18 Generic specification of the checkIfCurrentlyLoanable activity

Name	CheckIfCurrentlyLoanable()
Description	checks that an item is not prohibited from being loaned because <variant> it is either reserved or </variant> the MaxExtensions limit has been exceeded
Receives	item: Item, MH : MessageHandler loan: Loan
Rule	A Loan is extendable if the number of extensions is less than or equal to maxExtensions (=3)
Result	if loan containing item is not extendable printMessage ("OverExtension") has been sent to MH False has been returned <variant> else if item is reserved printMessage ("Reserved") has been sent to MH False has been returned </variant> else True has been returned

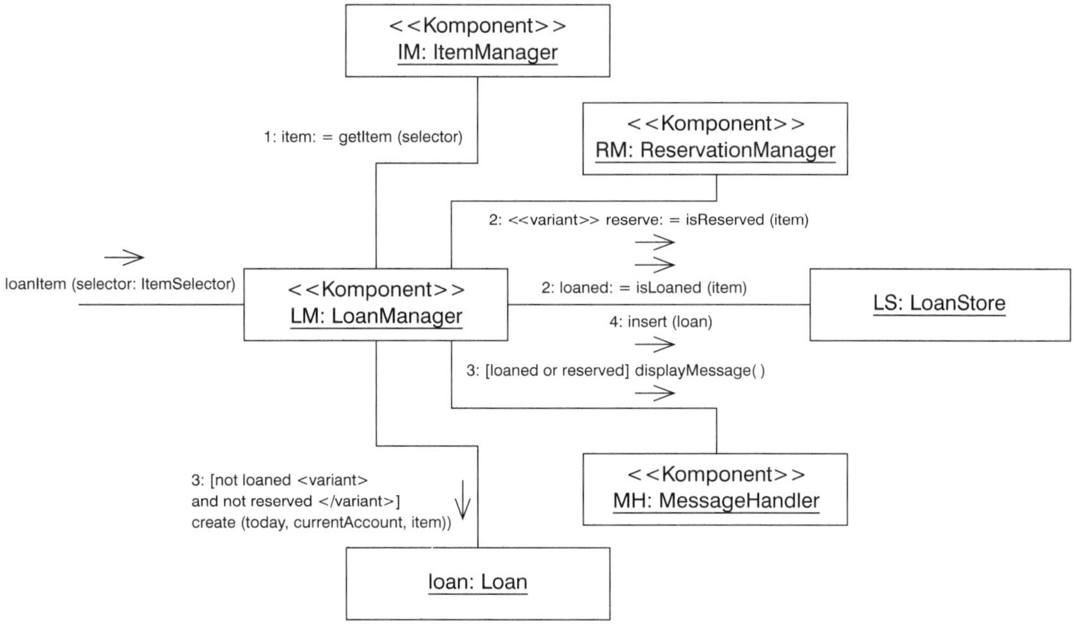

Figure 15.12
Generic collaboration diagram of the loanItem() operation

15.2.3.3 Interaction model

An interaction model describes how groups of objects interact at run time to realize an operation of the Komponent. In a product line context, these interactions can vary, so, as for the other diagram types, the stereotype <<variant>> is used to identify variant elements. Figure 15.12 shows the generic collaboration diagram for the *loanItem()* operation. As before, the *ReservationManager* Komponent is variant and thus also the check as to whether an item is reserved.

15.2.3.4 Auxiliary artifacts

A Komponent realization also has various auxiliary artifacts to support the main realization models. Essentially the same observations made above for the auxiliary specification artifacts hold for their realization counterparts. Basically, apart from the test cases, which clearly must embody variabilities related to those in the operations that they are designed to test, the quality artifacts are only generic in the sense that they cover all the features of the generic realization.

The same holds for the data dictionary. It would be possible to document all the variable and common features within the data dictionary, but since this information is already available in the other models it would unduly clutter the dictionary. It is therefore not recommended that information about variabilities be included in the data dictionary. In a framework, therefore, the data dictionary is only generic in the sense that it covers all the model elements in all variants of the Komponent.

15.2.3.5 Decision model

The decision model for a generic Komponent realization consists of decision tables of the form described in the previous section for Komponent specifications and context realizations. In the same way that the high-level decisions related to the context realization are mapped to the framework scope, the highest-level decisions related to a Komponent's realization decision model are mapped to those in the Komponent's specification. This establishes a concrete relationship between the variabilities in the specification of a Komponent and the variabilities in its realization. As is to be expected, decisions about the features to be supported by variants of the Komponent will have an impact on the way it is realized.

All the decisions related to the LoanManager Komponent's realization are concerned solely with reservation-related variability and thus can be mapped easily to the corresponding specification decisions.

15.2.4 KobrA komponent embodiment

Chapter 11 presents some useful techniques for systematically transforming the abstract models in a Komponent containment tree into machine-comprehensible source code. These basic techniques are also applicable in a product line engineering context. The main thing that distinguishes the implementation activities in a product line setting from those in a single-system setting is how to handle the variations between the members of the product family.

To understand the various options for managing the variabilities and commonalities captured in the framework it is first necessary to understand precisely what this information means. The ultimate goal of any software development enterprise is to arrive at a situation where software is running on a customer"s computer that meets expectations. This is termed a "deployed system." When the decision models in a framework define the set of features that are associated with different variants of the system, they define the minimum set of features that a deployed system conforming to that variant must contain. The set of features associated with a variant represents the minimal set of features that a user of a deployed system of that variant is entitled to expect. In terms of Figure 15.2, a user of a deployed system of variant A, for example, is entitled to expect at least the features bounded by the circle representing A.

However, there is nothing to stop a deployed system of a given variant from including, at run time, more features than those prescribed for that variant. In this sense an implementation could be considered generic. The only constraint on the range of features included in a deployed system is the non-functional requirements, which might specify minimal space usage or optimal speed. Other than this restriction there is nothing stopping a deployed system from including all the features for all the variants of the system. Thus, an extreme implementation approach is to build one implementation for the whole product line, and include all the features for all the variants in every deployed system. As long as the system has some kind of flag indicating which variant it is supposed to be at a particular point in time, the user should notice no functional difference. In terms of Figure 15.2, this would correspond to a deployed system of variant A containing all features bounded by the outer boundary (i.e. all features of all variants). At the other extreme, a deployed system of a given variant would contain only those features corresponding to that variant and no more.

15.3 Process

The framework engineering process is essentially a generalization of the Komponent-modeling and embodiment activities explained in the previous two parts of the book, but enhanced to create generic products rather than specific ones. The work involved is therefore basically identical except that variant model parts must be identified, and decision models must be constructed. These two activities, variability identification and decision modeling, are described in Sections 15.3.1 and 15.3.2. Other activities which are important to, or significantly affected by, framework engineering are discussed in the following sections, including Komponent identification, component reuse, and Komponent implementation.

15.3.1 Variability identification

The creation of Komponent models and the identification of variabilities are usually performed in parallel. Sometimes a model is first created and variant parts later added, on other occasions a variant feature is first identified and then the corresponding model feature or features subsequently adapted. The precise order in which product line issues are addressed in relation to the regular modeling activities is not important therefore. What is important is that software engineers adopt a different "product line-oriented" mind-set when thinking about the system. For variability identification, this means that whenever a new element is added to a model, a developer must think about whether the corresponding feature is always of the same form, whether it is optional or whether alternative features exist.

KobrA adopts a very simple but powerful approach for modeling variant features. This not only reduces the number of additional concepts needed to handle product lines, but also has the advantage that generic models in a framework are very similar to non-generic models for a single system. When the number of variant features becomes so large that defining them in one place would produce cluttered diagrams that are hard to understand, the KobrA approach is to split the diagram into several, alternative diagrams. The variation between members of the product family is then captured at the granularity of diagrams rather than model elements within diagrams. As with any other variability, information about which diagram corresponds to which variant of the system is documented in the decision model.

This simple but powerful set of variability modeling concepts therefore not only scales up in a straightforward way to rich product lines with large number of family members and variation points, but also facilitates the introduction of new variant parts at any point in a project. This therefore facilitates an incremental evolution to a product line from models developed for a single system. Variant features need only be integrated into the framework when they are required for a new family member.

15.3.2 Decision modeling

Decision modeling is a combination of a bottom-up and a top-down process. The former is used to model decisions in an individual specification or realization, while the latter is used to relate the decisions in different specifications and realizations within a containment hierarchy. Specifically, the decisions in a Komponent's realization should be related to those in the Komponent's specification, and the decisions in a Komponent's specification should be related to those in its parent's realization.

While modeling a Komponent's specification or realization, variabilities are identified in the models. For each variant element in a model, a simple decision is created that directly expresses the meaning of the variability in that variant element. These simple decisions represent the lowest level of abstraction in a decision model. Once the simple decisions are in place, higher-level decisions can be identified by applying the KobrA consistency rules between diagrams. For example, due to the close relationship between a Komponent's operations and the events in its state diagram, if an operation is variant the corresponding event must be variant, and vice versa.

Domain knowledge is also an important source of further dependencies and constraints leading to the identification of decisions. In general, the construction of cleanly structured and understandable decision models is a difficult task that requires domain experts with good abstraction capabilities. The final goal is to relate all decisions at one level with those of the decision model at the level above. That is, realization decision models must be related to the specification decision model, and specification decision models must be related to the parent Komponent's realization decision model. Consequently, the decisions in the context realization decision model represent the ultimate starting point for the whole decision modeling process.

15.3.3 KobrA component identification

As discussed previously, existing component-based development methods provide little guidance for finding good components. In KobrA analyzing and encapsulating the variabilities that characterize a product line can provide valuable insight into how best to consolidate functionality into good, reusable building blocks for the domain in question. In short, factoring out and encapsulating variabilities can be a good way of identifying components.

Consider the realization of the *LoanManager* Komponent, for example. The realization of this Komponent has a need for two more or less equivalent blocks of functionality: the management of loans and the management of reservations. Neither is particularly complex, and in a single system setting they could sensibly have been realized together in the form of a single sub-Komponent. However, the fact that the reservation functionality is optional, and thus not included in some variants of the product line, makes it advantageous to encapsulate this functionality in a separate, private sub-Komponent of the *LoanManager* – the *ReservationManager*. A variant of the system without this functionality therefore simply has to omit this sub-Komponent.

If the reservation management capability were encapsulated within a sub-Komponent of *LoanManager*, it would seem sensible to do the same for the loan-management functionality. However, because loan management is common to all variants of the system, there is not such a compelling reason to make it a substitutable component of the system. The only advantage in doing so would be if there was a strong likelihood of there being different realizations of the loan-management functionality, but there is not. The need to encapsulate the loan-management functionality as a separate interchangeable Komponent is therefore not as great as for the *ReservationManager*. Therefore, in this version of the library product line, reservation management is encapsulated within a distinct sub-Komponent of *LoanManager*, but loan management is not. *LoanManager* itself directly realizes it.

Variant functionality cannot always be so easily encapsulated within a single Komponent, however. In the library system product line, for example, the concept of paying for some services is a cross-cutting variability that cannot be encapsulated in a single Komponent. To control this kind of variability, variant model elements and corresponding decisions have to be included in most of the Komponents in the frameworks.

Encapsulation of variability is not the only criterion for identifying Komponents within a framework containment tree, but it is an important one. Others include the traditional "maximum cohesion, minimal coupling" criterion, the encapsulation of functionality related to dominant non-functional requirements (e.g. performance), or the existence of pre-defined external components. In general, there is no easy cookbook for identifying good Komponents in a project, and much depends on the experience and domain knowledge of the engineers. Analyzing commonalities and variabilities is an important tool in the arsenal of Komponent engineers, especially when it comes to the identification of highly reusable Komponents.

15.3.4 KobrA component embodiment

There are various points during the process of producing a deployed system that the set of features that it contains could be resolved. When the deployed system contains all the features of the entire product line, the set of features available to a user at a particular moment is obviously resolved at run time by means of some type of flag. In this case the corresponding source code (i.e. implementation) would then implement all the features of the product line, plus some mechanism for deciding between them at run time. At the other end of the spectrum, all features not required for a variant could be excluded at implementation time, so that the source code only contains the minimum set of required features. In between the two extremes, there are various other places in the typical development process at which the set of features available to a user is can be restricted:

- **Implementation time.** Certain features in the product line can be excluded from the implementation (i.e. the source code). This determines the set of variants that the source code can support.
- **Build time.** The compiler and the linker can be instructed to build a particular variant of the system. For example, preprocessor variables can be used to tag optional or variant code parts.
- **Installation time.** When a system is installed at a customer's site, certain features including the actual set of modules to be installed, can be selected. When the system is started subsequently, these initial selections will be retained, but others will be excluded.
- **Start-up time.** Depending on the nature of the implementation (i.e. whether or not it is parameterized), certain values can be chosen every time a system is started.
- **Run time.** The availability of some features can be controlled by the user while the system is running. Run-time variability ranges from appearance styles to parameters of algorithms used to perform required tasks.

There are no well-defined criteria for deciding which of these approaches to follow, and when to restrict the set of features included in deployed systems. The choice is highly dependent on the particular technology used to implement the system [CE00], and usually represents a strategic decision for the project management. To complicate things further, the optimal choice may well change over time. For example, in the early days of word-processors it was a novelty to be able to spell-check documents. At that time, people would pay for these additional features, so two variants of the same word-processor were needed, one with, and one without a spell-checker. The variability was handled during build time to deliver different products. Once spell-checkers became a standard feature, however, word-processor variants without it disappeared. Therefore, the spell-checking variability that was initially handled at development time is now handled at installation time and/or at run time.

For this reason, KobrA is not prescriptive about precisely how the different feature sets represented by different variants of the product line should be handled in the implementation, construction and deployment phases. The ability to document the features within given variants in a concrete and user-understandable way is already a step forward and users are left free to support different variants in a way that best meets their business needs.

The above discussion relates to the handling of variable features that are not shared by all members of the product line. One thing that is certain, however, is that the common features must be contained in all source code and all deployed systems, whatever variant of the product line they may be. In framework engineering, therefore, another strategic decision for the project management is when to perform the implementation of the common parts of product line, since this is not dependent on the creation of any specific applications.

15.3.4.1 Component reuse

As explained in Chapter 12, Komponent embodiment is generally much more efficient when already existing components are reused and integrated. In the context of a product line, a component could be reused either at the level of the framework, or the level of a specific application. Naturally, the advantages of reuse at the framework level are greater, because the effort is ameliorated over all applications instantiated from the framework, but the difficulty involved in achieving it are also greater.

In a framework, the initial desired specification and the final agreed specification will typically be generic, and will thus contain variabilities and related decision models. There are several ways in which this genericity can be accommodated in the reused component. In the simplest case, the reused component supports all the features defined in the agreed Komponent specification, for all the variants in the product line, but provides no means of selecting from them. As explained previously, a generic Komponent defines the minimum set of features that must be supported by the embodiment of the Komponent, but does not prohibit more features from being supported. Thus, it is perfectly possible for a reused component to support more functionality than is needed, which in the extreme case is all features for all members of the product line. The downside of this, however, is that deployed variants of the system will contain support for more features than they need.

In the ideal case, the reuse component will support the omission of features not needed for a particular variant of the product line, either because it is a Komponent whose embodiment is constructed in the typical KobrA way, or because the component is parameterized in ways that match the variabilities in the agreed specification.

16 Application engineering

The purpose of the application engineering activity is to generate a specific set of KobrA artifacts that meet the needs of a specific system. In the context of a product line it does this by instantiating the framework created within the framework engineering activity. A KobrA framework is instantiated by systematically resolving all decision models and turning the generic framework artifacts into specific application artifacts. The resulting Komponent containment tree and Komponent embodiments have the features required for one particular member of the product line.

In practice, application engineering typically also involves many of the development activities described in the previous two parts of this book. This is because it is rarely the case that a pre-defined member of a product family matches the exact needs of a customer. Usually some additional tailoring is needed to the instantiated models. Furthermore, as explained in the previous chapter, different implementation technologies have different capabilities when it comes to supporting the variabilities defined by a product line. Depending on the precise characteristics of the implementation technology, some or all of the embodiment work may need to be done at the application level.

It is also often the case that an enterprise wishes to apply KobrA outside the context of a product line just for single-system development. In such circumstances, the performance of the component modeling and embodiment activities described in the previous parts of the book can be viewed as application engineering, but without the benefit of a framework. Obviously many of the reuse advantages of product line engineering are then lost, but single-system development still has value as a way of transitioning to KobrA.

16.1 Artifacts

In KobrA, an application is almost identical to a framework. The only visible difference is that a framework may contain generic artifacts while an application may not. In the following subsections we discuss in more detail the concrete differences that distinguish framework artifacts from application artifacts.

From a practical perspective, the main thing that distinguishes a generic Komponent in a framework from a specific Komponent in an application is whether it possesses a decision model or a decision model instance. As explained in Chapter 15, generic Komponents have decision models, which describe which of the features defined by the Komponent are associated with which variants of the systems. In contrast, specific Komponents have decision resolution models that record how the decisions in the decision model were resolved for the Komponent in question. The only new kind of artifact that appears in an application Komponent tree is the decision resolution model.

16.1.1 Decision Resolution Model

Whenever a specific Komponent is instantiated from a generic Komponent the decision model is turned into a decision resolution model. The purpose of this model is to record for future reference which set of choices led to the set of features encapsulated by the specific Komponent. This is essential for proper traceability and hence maintainability of the application in relation to the evolving framework.

A decision resolution models is a table in which each row resolves one decision. The resolution of a decision consists of the decision's ID, question, variant point, and the resolution selected for a particular system.

Table 16.1 shows an example of a decision resolution model that might be created when the generic *LibrarySystem* context realization presented in Section 15.2.1 is instantiated. The choices documented in Table 16.1 represent a resolution of the decisions documented in the decision model in Table 15.3. The shown resolution excludes reservation, data exchange, and suggestion capabilities.

16.2 Process

Application engineering is generally a two-step process. First, the generic artifacts in the framework are instantiated by resolving the associated decision models. Then, in a second step the instantiated specific artifacts usually need to be adapted by the regular development techniques to meet the exact needs of the customer. The second step is necessary in practical development situations because it is rare that instantiation alone can meet all the requirements of the customer at every level of detail.

Table 16.1 Library context realization decision model instance

ID	Question	Variation point	Resolution
Structural diagram			
1	Class Reservation needed?	class Reservation	no
2	Class ExternalDatabase needed?	class External Database	no
	...	...	...
11	Operation LibrarySystem. reserveItem() needed?	operation LibrarySystem. reserveItem()	no
12	Operation LibrarySystem. suggestItem() needed?	operation LibrarySystem. suggestItem()	no
13	Operation LibrarySystem. importData() needed?	operation LibrarySystem. importData()	no
	...	...	...

This two-step process of instantiation followed by adaptation applies to all the artifacts within a framework. Since the adaptation applies exactly the same concepts and techniques used in framework engineering, just without concern for variabilities, it is not discussed further in this chapter. Instead, in the subsections that follow we focus on the instantiation activity as it applies to the various containment tree artifacts, since this is unique to application engineering.

16.2.1 Context realization instantiation

The instantiation of the context realization is the all-important first step in using a framework to create software for a specific customer. The impulse for the application of this activity occurs when a customer contacts the software-development organization and expresses an interest in a software system lying within the scope of one of the organization's existing frameworks. When such an expression of interest is received, an application engineering project is set up and the context realization instantiation is initiated.

As the first part of the instantiation process, consultants from the development organization who are experts in the application domain and other applications that have already been created from the framework work with the customer to elicit the high-level requirements for the desired application. Ideally, the consultant should lead the discussion to ensure that the information needed for the cost-effective instantiation of the candidate framework is obtained. The goal is to determine very quickly whether there is a sufficiently large overlap between the functionality offered by the candidate framework and the functionality required by the customer. This size of the overlap provides the basis for estimating the effort likely to be involved in developing the system from the framework and thus ultimately the likely costs of the project. If there is only a small overlap it may be more cost-effective to develop the system from scratch rather than try to instantiate it from the framework.

The generic context realization's decision models ultimately drive the overlap analysis. When they contain decisions that cannot be resolved directly by the customer, it is often helpful to work through several instantiation scenarios with the customer to illustrate the effects of various critical choices. This involves the creation of partially instantiated framework models to illustrate potential intermediate states of the instantiation process. In other words, the generic enterprise model, the generic structural models, the generic activity model, and the generic interaction model which, in the framework, contain several points of variation, have some of their point of variation removed so that the customers can see the effects of the related decisions graphically. After the customer has been shown the effects of the various possible resolutions of the decisions under discussion, he/she should be in a better position to select the alternative that is closest to his/her needs.

This way of eliciting requirements significantly increases the chances of reusing the framework, because only the alternatives supported by the existing framework are offered to the customer. Showing a set of alternatives that are tightly coupled to the framework helps the customer match his/her original ideas to the capabilities of the framework.

As an illustration of the results of this process, Figure 16.1 shows the context realization structural diagram of a library application instantiated according to the decision resolution model depicted by Table 16.1. All the changes described in the decision model for the selected options have been put into effect. The <<variant>> stereotypes and unnecessary classes and operations, such as *Reservation*, *ExternalDatabase*, and *reserveItem()*, have all been removed.

Once the context realization has been instantiated the specific model can be shown to the user, and if necessary the context realization activities described in Chapter 7 can be applied to further adapt the models to the customer's exact needs. Application context realization is completed when the customer accepts the specific context realization, and the resulting artifacts have been inspected to ensure that they conform to the instantiation rules as well as the regular consistency rules defined previously in the book. The decision resolution model documenting the choices made by the user can then be used as the starting point for the resolution of the specifications and realizations of the Komponents in the containment tree.

16.2.2 Specification and realization instantiation

The specific Komponent containment tree is created by the recursive instantiation of the specifications and realizations in the framework. Obviously any decisions that have been resolved at the level of the context realization must be resolved in the same way (i.e. by the same choice) wherever they appear in the Komponent containment tree. It is also likely that other decisions will arise that have not already been addressed in the instantiation of the context realization.

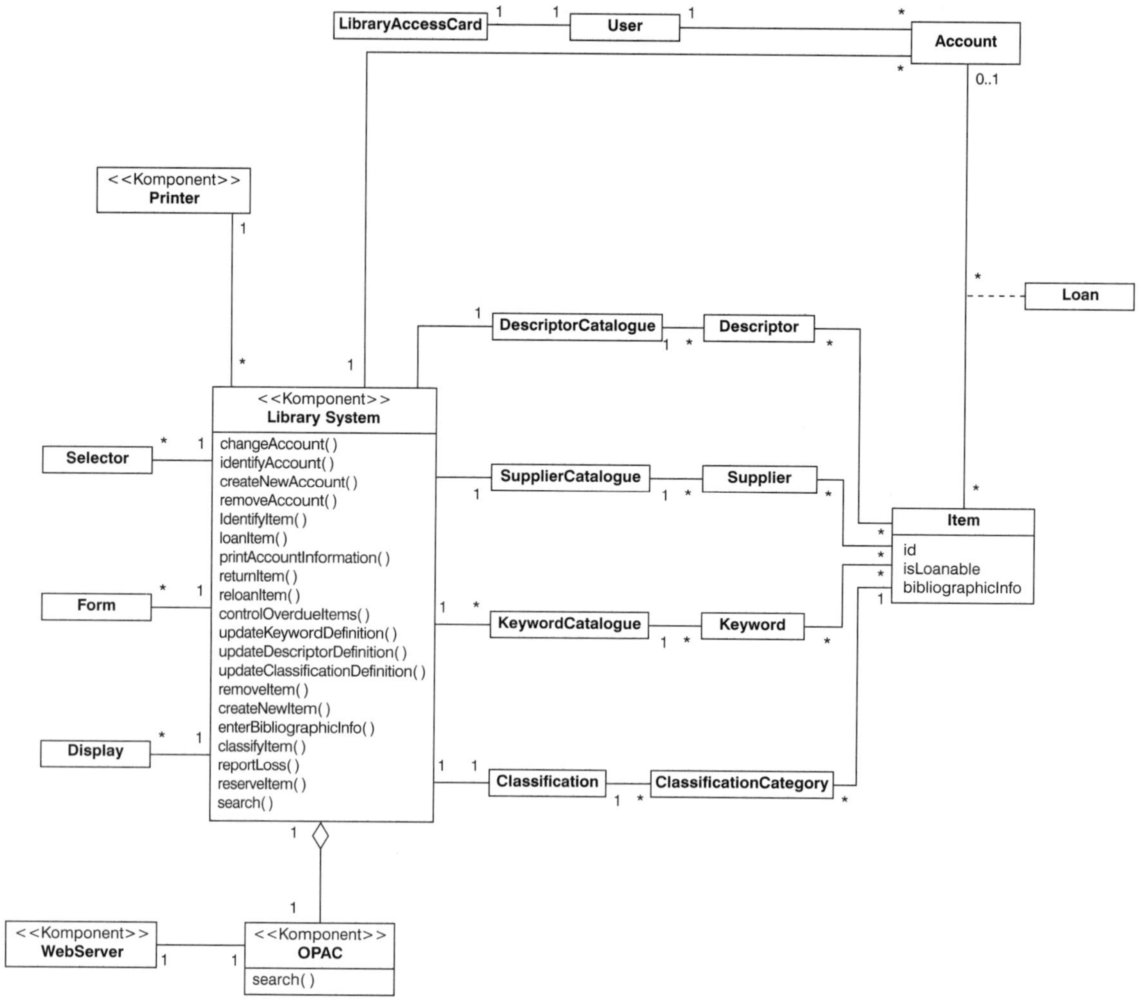

Figure 16.1
Instantiated context-realization structural diagram

If such decisions relate to the non-functional properties of the system (e.g. performance) they can be resolved in consultation with the customer. Otherwise, the help of the software engineering team may be needed to resolve the technical trade-offs implied by the various choices.

The instantiation of the containment tree is complete once all the Komponents in the tree have been successfully instantiated and all the resulting models have been inspected for adherence to the consistency rules. Note that specific Komponent models must be checked for adherence to all seven kinds of consistency rules defined in Appendix A. This adds the instantiation consistency rules to the other six rules defined previously.

As with the application context realization process, once the Komponent containment tree has been successfully instantiated it may be necessary to apply the regular Komponent modeling activities in order to adapt the application to meet new requirements. However, now an interesting dilemma arises. Should the required changes be made just at the level of the specific application, or are they general enough to be likely to be needed by other applications in the product line? This is the problem of maintenance in a product line context, and is discussed in greater detail in Chapter 17.

PART

5

Project monitoring and control

The final part of the book describes the critical supporting activities that determine the success or failure of a KobrA project and the quality of the final product.

Chapter 17 describes KobrA's model for maintenance and configuration management.

This is followed by two chapters that concentrate on quality. Chapter 18 introduces the concept of quality used in the KobrA approach and describes two of the primary techniques that are used to assure the desired levels of quality. The first is testing, and the second is an enhanced form of inspection, specialized for the needs of the component-engineering artifacts.

Chapter 19 describes quality evaluation techniques that can be applied based on the structural properties of UML models. These techniques provide valuable insight into possible error-prone areas, and also help focus the inspection effort where it is likely to have the most effect.

The final chapter looks at the issues involved in introducing KobrA into real life software development situations, and provides some recommendations for customizing the method and for transferring it into an organization.

Maintenance 17

As with most software development projects, a KobrA project involves the creation of a large number of different artifacts at various levels of granularity, abstraction, and genericity. A framework contains a tree of generic Komponents, each with a specification, a realization, and often an implementation, and each of these is composed of multiple artifacts. When the framework is instantiated to create an application, new versions of most of the framework artifacts are created and tailored to the needs of a particular customer. These are augmented by lower-level artifacts needed to create deliverable products, such as binaries, build files etc.

If keeping control of all these structural artifacts were not already complex enough, the time dimension must also be addressed. Software is subject to change perhaps more than any other engineering products, and supporting modifications over time introduces many different incarnations of each artifact. Keeping track of everything in a traceable and consistent way is therefore an enormous challenge, and can have a major impact on the success of a project. It is the job of the maintenance activity and supporting tools to manage the project artifacts and their inter-relationships, and to support the delivery of useful software products using them.

This chapter describes the maintenance aspects of KobrA. The following section first identifies the basic goals of the maintenance activities in KobrA and describes the fundamental principles involved. The section after that introduces the additional artifacts needed to support maintenance, and describes how the existing types of artifacts are affected. The final section then describes the process aspects of maintenance, and explains how they fit into the overall KobrA life-cycle.

17.1 Core principles

The underlying goal of the maintenance activity in KobrA is to ensure the mutual consistency of the artifacts in a product line in the face of constant demands for change. Requests for change come from many sources, such as the environment, the end user, the customer etc. [Smith98]. In contrast to classical software engineering approaches, the maintenance activities for a KobrA product line starts right at the beginning of the product line life-cycle. As illustrated in Figure 17.1, the KobrA method assumes a comprehensive product line development model from the initial context realization to the deployment of a deliverable.

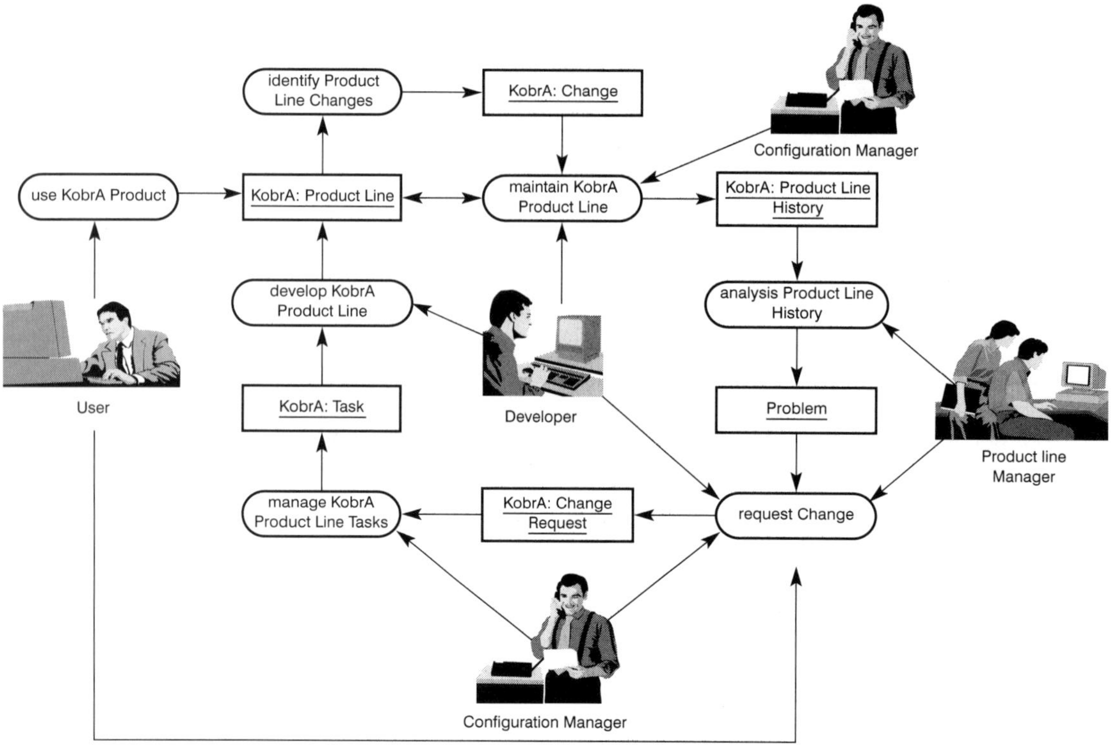

Figure 17.1
Role of maintenance

Maintenance is essentially responsible for providing the infrastructure to manage the evolution of the product line during the incremental development steps, and thus provides a closed evolutionary product line life-cycle. Three main maintenance activities complement the regular KobrA development activities outlined in the previous parts of the book:

- **Configuration management.** This activity encompasses the techniques and mechanisms involved in updating and configuring a framework and its applications (see Section 17.3.2). In essence it handles the static management of different artifact incarnations. This includes the synchronization of different versions of a framework and its applications. An orthogonal versioning approach makes it possible to build arbitrary versions of a KobrA asset for each incremental development step.
- **Change management.** This activity encompasses the techniques and methods involved in handling changes. In essence, it handles the dynamic management of the causes and effects of changes. This includes the categorization and evaluation of change requests, and the assignment of appropriate formal change operators to a change in order to support its integration into a framework and its application (see Section 17.3).
- **Maintenance planning.** This activity is concerned with constructing the infrastructure for the change and configuration management activities. It customizes the change and configuration processes according to organizational and project factors, allocates resources to the maintenance activities and defines how they should be applied.

In this chapter we focus on the configuration management and change management aspects of the KobrA approach to maintenance.

17.1.1 Framework to application coupling

A major issue for maintenance in the context of a product line is how tightly frameworks are coupled to the applications generated from them. High coupling between frameworks and their application implies that modifications in a framework can be easily reintegrated into applications, and vice versa. In contrast, low coupling implies that the integration of modifications in a framework into its applications (and vice versa) is more complex, and requires more sophisticated techniques.

The strategy for integrating modifications depends on various factors, such as the physical distance between framework and application development teams, the human resources available, the desired level of coherence between frameworks and applications, and the technical feasibility of sharing artifacts between frameworks and applications. In the ideal case, the physical distance can be neglected, human resources can be allocated as needed to both the framework and application development teams, the coherence between the framework and the applications is high, and there are no technical obstacles to sharing assets. This would guarantee high coupling between the framework and its applications. This is rarely the case in practice, however. The KobrA maintenance approach therefore assumes low coupling between frameworks and applications and hence provides the appropriate techniques for integrating framework changes into applications, and vice versa. (see 17.3.1.3 and 17.3.1.4).

17.1.2 Version orientation versus change set orientation

KobrA's maintenance approach is based on a change-oriented model rather than the pure version oriented model often provided by configuration-management approaches (see [Zel97], [Mad90]). In other words, KobrA regards a version as the result of changes applied to some artifact in the product line. The advantage of this approach is that changes can be propagated from the framework to an application, and vice versa. A pure version oriented approach does not allow changes to be easily propagated between different artifacts. With a change-oriented model, however, related changes can be grouped into change sets which can then be applied to several applications at once.

A problem with traditional change set oriented change propagation approaches is that it is hard to oversee how the changes are propagated between the different artifacts and artifacts versions. In KobrA, therefore, each change is formalized as a conceptual change operator that allows each artifact version to be reconstructed. Additional advantages of this approach are the possibility to express version rules over change operators and to control an artifact's version states independently of its revision state. This is important for controlling an artifact's state.

17.1.3 Separation of concerns

Like all the other aspects of KobrA, the maintenance approach is based on the separation of concerns. This means that the concepts are as orthogonal as possible and the approach is faithful to the principle of locality:

- **Changes and attributes.** The KobrA change-set-oriented maintenance approach allows the controlled integration of changes into a set of product line members at once whereas other approaches integrate changes into one product, or into different versions of the same product individually. KobrA maintenance strictly separates the change management and configuration management. Changes are formalized as conceptual change operators according to their impact on the assets.

- **Self-contained.** In addition a KobrA Komponent is versioned so the Komponent will be self-contained from the configuration-management point of view. That is, a Komponent knows exactly the versions of the Komponent specification, realization, and implementation that can be combined. The advantage is that the Komponent can be shipped without the versioning information about the Komponent specification and realization being stored in the framework.

- **De-coupling.** The de-coupling of the Komponent specification, realization and implementation versions from the framework version makes the Komponent versioning independent from that of the framework system or an application system. It simplifies the configuration and update mechanism of the framework and applications, because not every Komponent has to be versioned if a certain sub-Komponent version has changed. In this way, the de-coupling also reduces the number of update and configuration activities and reduces the number of Komponent versions.

- **Principle of locality.** As a consequence of the self-containment and de-coupling, the Komponent versioning activity is independent of the versioning activity of other Komponents. This means the versioning activity with respect to a Komponent is localized to the specification, realization, and implementation of that Komponent, thus preserving the principle of locality. However, a new Komponent specification requires a new Komponent version. This happens when a Komponent is explicitly versioned by the developer or when a framework is versioned. In the latter case, a recursive version process has to be started in order to version all its different assets. Nevertheless, temporary Komponent specification and realization versions can be stored and retrieved by the developer.

17.2 Artifacts

The artifacts that are managed as part of the maintenance activity are often referred to as assets in the literature. In KobrA, the maintenance assets include all the artifacts described in previous parts of the book, together with supporting management and infrastructure artifacts that are beyond the scope of this book. Additional artifacts are also needed to support the maintenance activity itself, such as changes and histories etc.

The artifacts of concern in the KobrA maintenance approach fall into four main classes – configuration items, which are subject to version control; dependencies, which relate configuration items, evolution graphs, which capture version histories; and changes, which drive the evolution of the configuration items over time.

17.2.1 Configuration items

A configuration item is any asset that is subject to version control. This can range from large-scale artifacts such as complete frameworks, to small-scale artifacts such as individual diagrams and models. Because of the principle of locality in KobrA, every configuration item is self-contained and is versioned independently of other configuration items. This is made possible in KobrA by explicitly viewing dependencies between configuration items as first-class assets in their own right. As explained in Section 17.2.2, although dependencies are not configuration items because they are not subject to version control, they play a fundamental role in the configuration-management mechanism allowing configuration items to maintain their own local configuration information.

Making dependencies first-class assets also minimizes the number of artifact types needed in the configuration mechanism. In conventional configuration-management approaches, the concept of a configuration (as opposed to a configuration item) is often used to identify a particular grouping of configuration items for a particular purpose. Configurations are therefore usually introduced as explicit objects with their own identity and life-cycle. By recognizing artifacts at all levels of granularity as configuration items (e.g. framework, applications, Komponents etc.) and by viewing the dependencies between them as objects, the explicit concept of configurations is not needed in KobrA. Specific versions of existing configuration items (e.g. framework versions, application versions) store the information that would otherwise be captured in a configuration. For example:

- A framework version has relationships to all the generic Komponent versions (including decision model versions), and Komponent embodiment versions that it contains.
- An application version has relationships to all the specific Komponent versions (including decision resolution models), and Komponent embodiment versions that it contains. In addition it has relationships to the framework from which it was instantiated.
- An implementation version has relationships to the Komponent versions which it implements, and the refinement and translation patterns that it applied.
- An executable version has relationships to the Komponent implementation version which it encodes and the build description from which it was constructed.
- A deliverable version has relationships to executable versions that it contains and to different versions of environment elements such as picture files, customization files etc. that it uses.

The precise nature of these relationships is discussed in Section 17.2.2 and is documented in the metamodel in Appendix A. The mutual consistency of the assets under version control is achieved by means of distinct feedforward and feedback configuration processes as explained in Section 17.3.1.

17.2.1.1 Version model

All of the major artifacts introduced in the previous parts of the book are viewed as configuration items subject to version control in KobrA. This includes largescale artifacts such as framework and applications, through medium-scale artifacts such as Komponents, specifications, realizations and implementations, down to relatively fine-grained artifacts such as models and diagrams. In addition, the many technology specific artifacts involved in the embodiment of a Komponent are also subject to version control, but in accordance with the version model of the technology concerned.

As illustrated in Figure 17.2, a configuration item version is captured in the form of a vector composed of four orthogonal values.

- **Revision** – a number distinguishing different semantic incarnations of the artifact over time
- **Variant** – a string distinguishing co-existing incarnations of an artifact, instantiated for the needs of particular product line family members

- **Edition** – a number distinguishing different cosmetic incarnations of the artifact over time
- **Quality state** – boolean value indicating whether the artifact is in a satisfactory quality state according to the KobrA consistency rules.

The version states (i.e. values of the version vectors) of related artifacts are also related to one another. For example, the version vector of a Komponent is related to the version vectors of its specification and realization etc.

Figure 17.3 extends the previously used cube representation of a Komponent to show how the version vector of the Komponent as a whole (the vector on the front panel) relates to the version vectors for the specification (top panel) and the realization (bottom panel). Basically, a change to the version of either the specification or realization of a Komponent requires a change to the version of the Komponent.

Each artifact is versioned in a self-contained way from the configuration management perspective. This is achieved by maintaining explicit relationships between dependent artifacts. For example, a Komponent has relationships that indicate which versions of the specification and realization it incorporates. This has the advantage that a Komponent can be shipped without containing the versioning information of the artifacts it depends on.

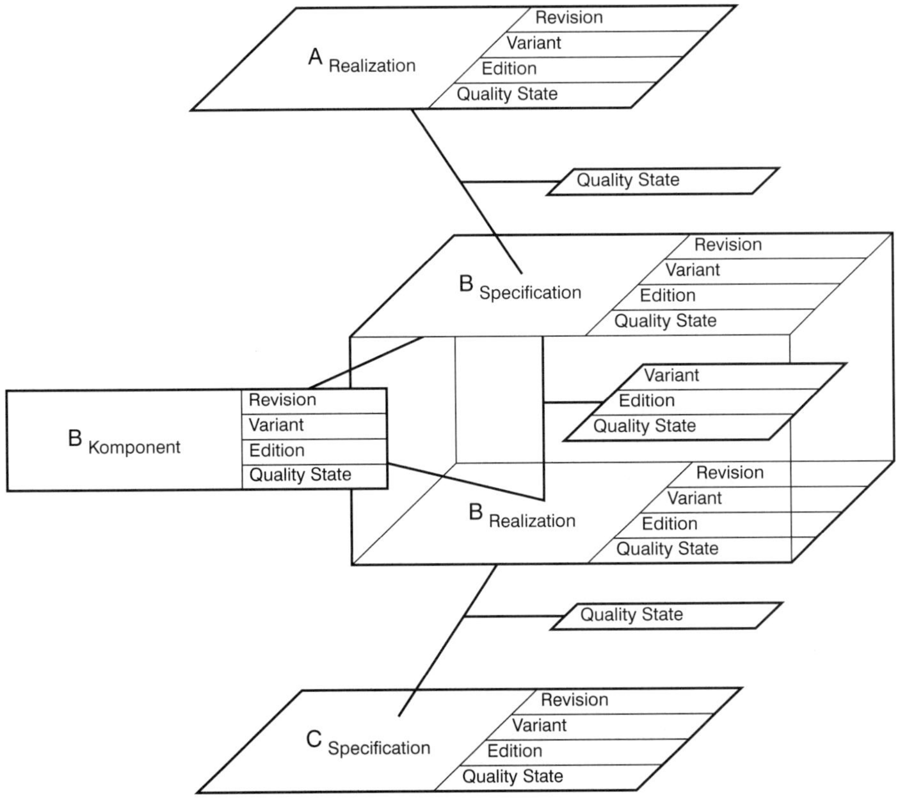

Figure 17.2
Version model

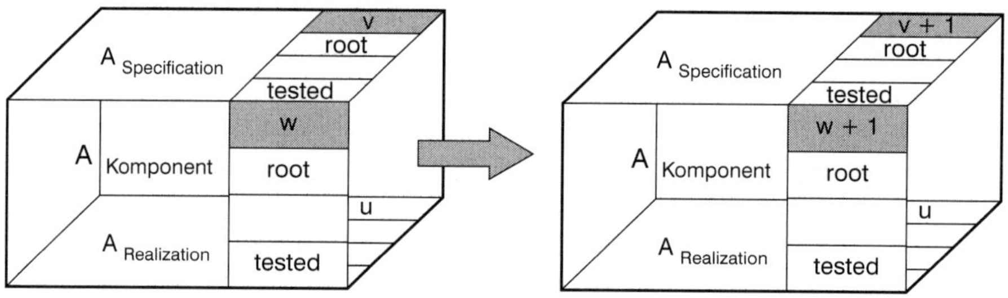

Figure 17.3
Komponent versions

17.2.1.2 Asset life-cycles

Each asset has a life-cycle that controls its behavior during the development process. This life-cycle can be represented in the form of a state machine with states for each development step and each change activity. Both are processed in parallel but are interdependent. The states are controlled by KobrA's orthogonal versioning model described above:

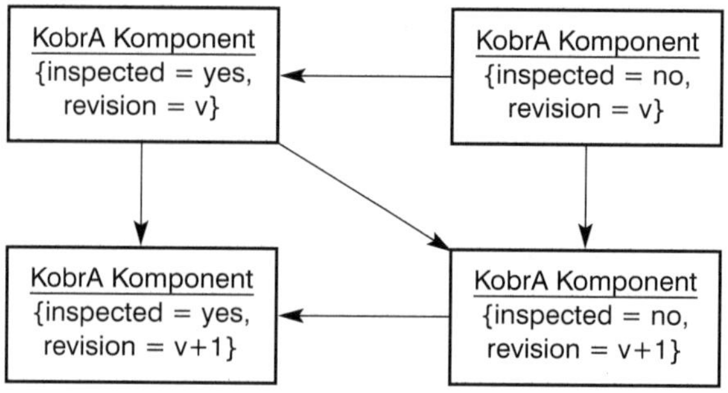

Figure 17.4
Version state model

A KobrA Komponent has its own multi-dimensional version model that describes the life-cycle according to the states and their dependencies. Figure 17.4 shows the state transition model for a sub-vector of the Komponent (revision, inspected). Inspected is one dimension of the quality state and indicates whether a Komponent has been inspected or not (see Chapter 18). The state transition fixes which state combinations are possible and how they can be transferred. For example, the KobrA Komponent revision can always be incremented and the inspected state can be transitioned simultaneously to inspected or not inspected, if the Komponent has been inspected.

Figure 17.5 illustrates the life-cycle model for a framework. The statechart shows that the establishment of a new version depends on the initial development state. If the framework is brand new, a new version is initialized. If it already existed, however, the version number will be incremented. This model is necessary to guarantee that an old version is not destroyed by change activities in the framework. Once a version is accepted, the incremental development state is changed to completed, allowing the framework development activities to be completed.

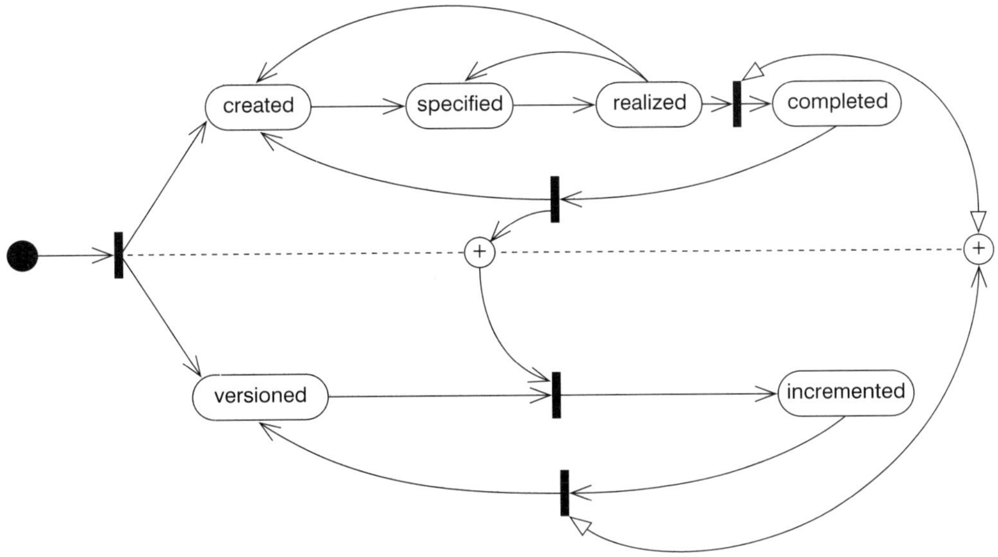

Figure 17.5
Framework life-cycle

17.2.2 Dependencies

The dependencies between assets play a key role in the KobrA maintenance approach. The artifacts developed during framework and application engineering do not themselves contain sufficient information to keep all configuration items in a consistent state. Therefore, dependencies between artifacts, as defined by the consistency rules, are viewed as assets in their own right, and capture information about the relative consistency of configuration items within a product line. In particular, the quality state of the relationships between the different asset types allows the maintenance system to automatically check consistency rules or to flag that a check needs to be performed manually.

To illustrate the idea, Figure 17.6 shows a sample of the KobrA metamodel which depicts the dependencies between the fundamental product line and Komponent concepts. The *<<contains>>* relationship between a framework and a Komponent tree captures the fact that a new version of the Komponent tree leads to a new version of the framework. Similarly, the <<contains>> relationship between a Komponent tree and a context realization captures the fact that, a new version of the context realization version leads to a new version of the Komponent tree. Other dependency relationships, such as the *<<derived>>* relationship, allow application engineers to be kept informed about changes in the application Komponent tree so that refinement, translation, and finalization activities can be started when necessary to maintain the consistency of the application.

Ideally, the artifact dependency structure is managed by the physical versioning system, with the KobrA metamodel used to control the different artifacts in the repository. The various kinds of links are classified, and activities are

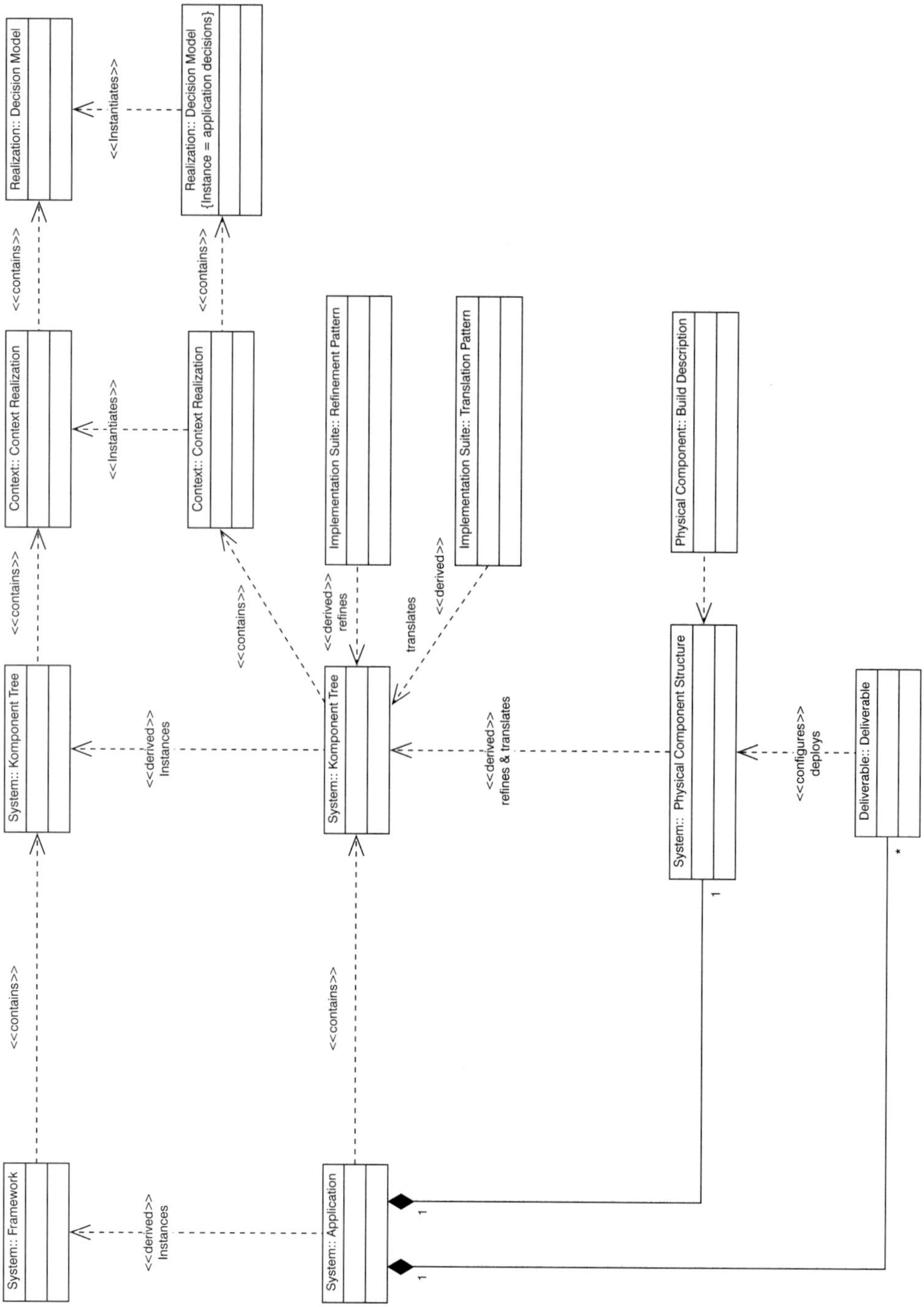

Figure 17.6
Asset dependencies

sometimes allocated to them. For example, the <<*contains*>> relationship automatically starts a versioning activity if a new version of a contained artifact is created. The consistency rules and associated trigger can be described in an abstract language like OCL. They are initially defined in the maintenance-planning activity and are subsequently implemented on top of the repository.

17.2.3 Evolution graphs

The evolution graph for a KobrA product line captures the evolution of the configuration items in the product line in the form of a directed acyclic graph. Normally, this graph will be a tree, because it is unusual for two evolution branches to merge. The graph is created for a framework and each of its applications, physical components, and deliverables. Each node of the graph denotes a version of the corresponding artifact. A branch in the graph represents an evolution step and assigns increments to the version vector for the artifact concerned. Let $\mathbf{S}_v$ denote the set of assets that belong to a version of an artifact at a given point in time, then the changes between two versions of the artifact can be described via the triple $\Delta = (\mathbf{S}^+, \mathbf{S}^-, \mathbf{T})$ where:

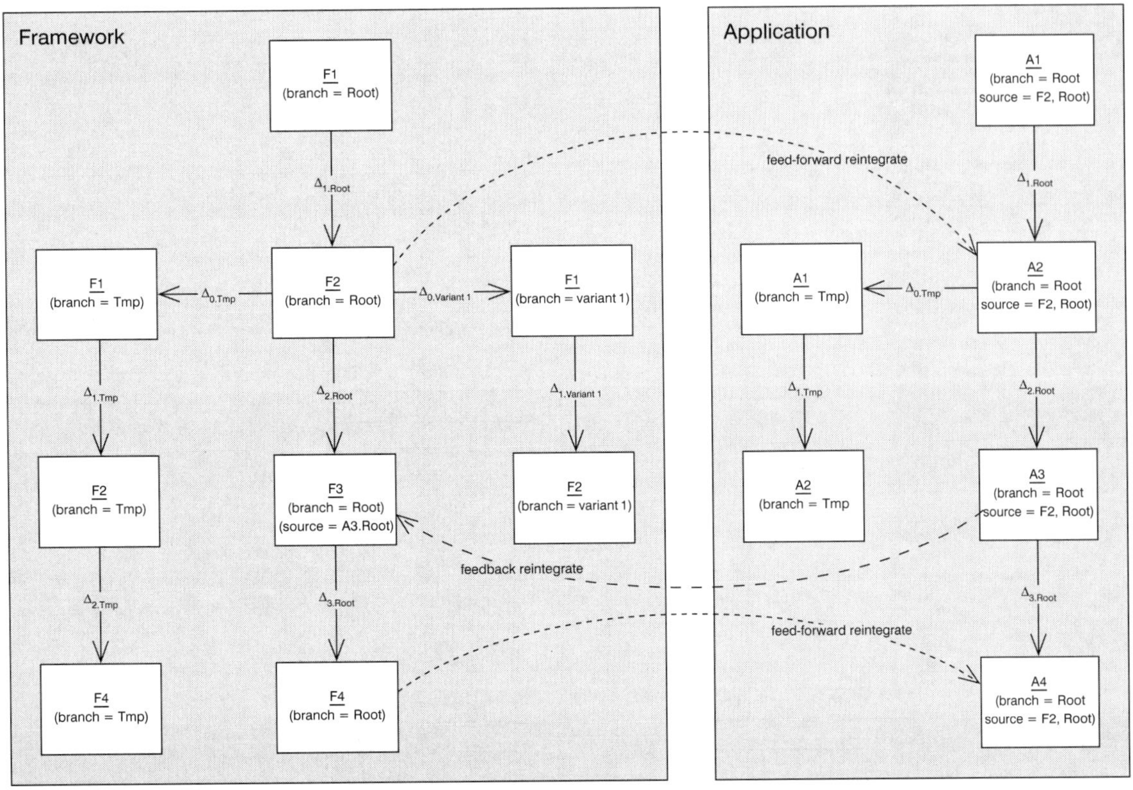

Figure 17.7
Evolution graph

- S^+ denotes the set of new assets,
- $S^- \subseteq S_{v-1}$ denotes the set of retired assets, and
- $T \subseteq (S_{v-1} \setminus S^-) \times S^+$ denotes the base transformation relation that contains the change rule how a certain $A \in S_{v-1}$ will be transformed into $A' = \sigma(A) \in S_v$.

This change description will be assigned to each branch in the evolution graph. Figure 17.7 shows an evolution graph for a framework and an application generated from it. In the graph the transitions between the different versions are augmented by a change description $\Delta = (S^+, S^-, T)$.

17.2.4 Changes

New versions of configuration items are generated as a result of changes. An important aspect of the maintenance activity is therefore the control and management of the changes that drive the evolution of the product line. In principle it is possible to accommodate changes simply by recording their

Figure 17.8
Change model

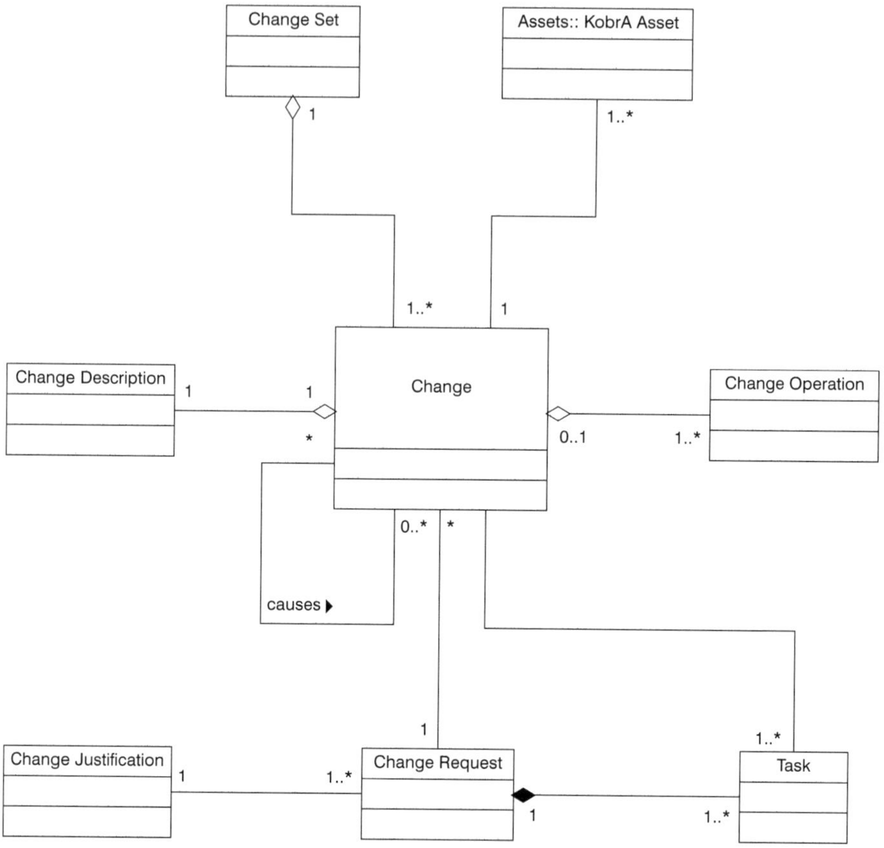

effects (i.e. by merely storing new artifact versions resulting from a change) but this greatly reduces the traceability of the maintenance processes. In KobrA, therefore, changes are treated as first-class citizens with relationships to other maintenance assets.

The KobrA change model is depicted in the metamodel fragment illustrated in Figure 17.8. This shows the various types of objects and relationships that are used to capture all the information that needs to be stored about changes.

Every change object is linked to one *change request* object that is the source of information about the rationale for the change. A change request object, in turn is linked to one *change justification* object. Each change cause object identifies a basic cause for the change request. These fall into the typical change cause categories commonly recognized in software engineering:

- **corrective changes** – diagnoses and corrects errors
- **adaptive changes** – adapts the KobrA system to changes in the environment (hardware changes, or operation system changes)
- **perfective changes** – improves the capabilities or functionality of the system (changes in the business model of a customer)
- **preventive changes** – tries to anticipate future changes and make arrangements to avoid modification of the system.

The *change justification* object defines the clear rationale and cost–benefit evaluation to justify the change. This is particularly important in large systems where a single change can affect a large number of configuration items.

A change object has a link to one *change description* object, which provides details about the nature of the change. This is important not only for maintenance, but also for other project-control activities such as quality control and planning. For example, change descriptions can help in the estimation of the costs of future changes or can be used to compare the expected and actual effort for planning purposes.

A change object also has one or more links to the assets affected by the change. This information is important for recording the impact of a change, and to identify and select the affected configuration items, dependencies and evolution graphs.

The purpose of the *change set* object is to encapsulate a set of changes made during the development of a new framework or application. This provides essential information for integrating changes into a framework or application or to exchange change information between them.

The one remaining class of object associated with changes is the *task* object. A change object has links to one or more task objects. Change operator objects capture the conceptual steps that have to be taken to enact the change, and are described in more detail in Section 17.2.4.2. Task objects group individual change operators into groups that can be explicitly allocated to individual team members. Tasks therefore provide an important intermediate grouping concept for enacting changes that may have a large impact and may take significant time to enact.

Finally, change objects can also have *cause* links to one another. Such relationships enable a trace of a sequence of consecutive changes to be recorded. This is particularly important in large product lines where the change traces can be spread over different development phases and different Komponents. Change traces can be helpful for locating defects, for example.

17.2.4.1 Change life-cycle

The life-cycle model of a change is strongly related to the asset and task life-cycle models. A change can only be enacted if the corresponding task is allocated by the product line management and is assigned to a member of the development team. Moreover, the items affected by the change can be *secured* once the effects of the change have been enacted in the whole framework or application. This is important in order to avoid inconsistencies within the product line. The transition of a task from the allocated state to the secured state will automatically trigger the transition of the change from the update-state to the completed state. This is impossible if one of the changes belonging to a task is not updated (see Figure 17.9).

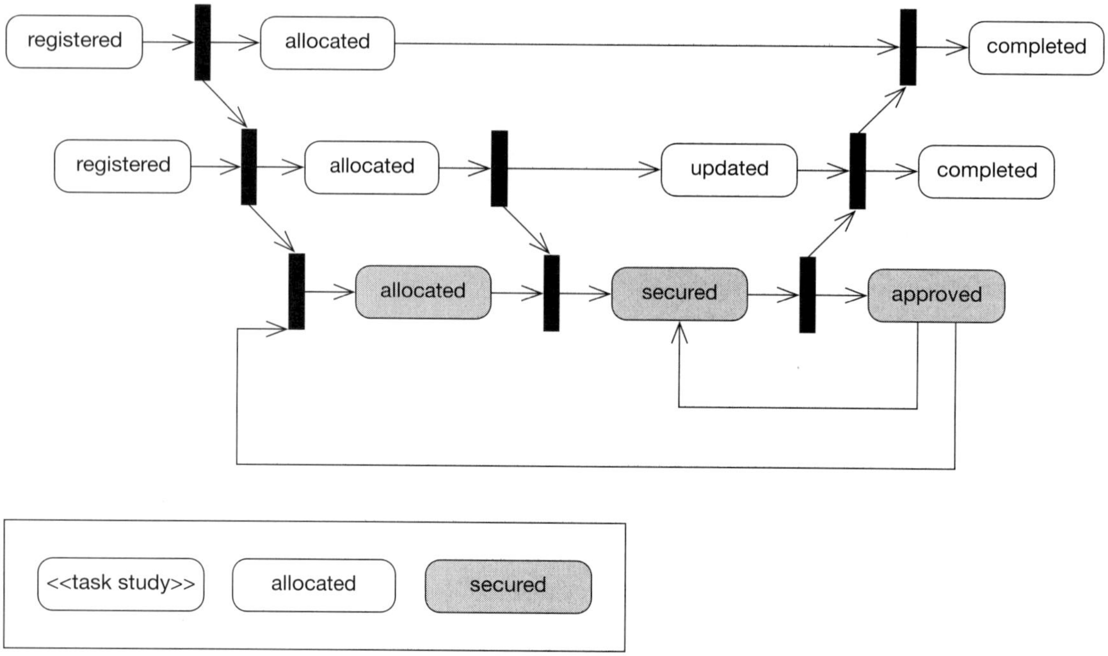

Figure 17.9
Change life-cycle model

17.2.4.2 Change operators

Change operators represent discrete conceptual changes to the configuration items in a product line. Each step in the evolution of a product line system can be decomposed into a sequence of such conceptual operators. Therefore each transition from one framework or application version to the subsequent version can be expressed as a sequence of conceptual operations. Conversely, conceptual operators can be expressed in terms of the assets.

As mentioned previously, KobrA employs a state-based approach to configuration management in which configuration items have several associated configuration states. At the top level there are essentially two states – the first representing the situation when a configuration item is stable, having been checked against all relevant consistency requirements, and the second when a newly modified item has not yet undergone these checks.

Figure 17.10 provides a schematic illustration of a stable revision of a framework containing the specifications and realizations of two Komponents, A and B. The "tick" in the corner of each specification and realization indicates that they are in a stable, validated state. The boxes above the tick describe the revision, variant, and edition values of specification and realization versions. Notice

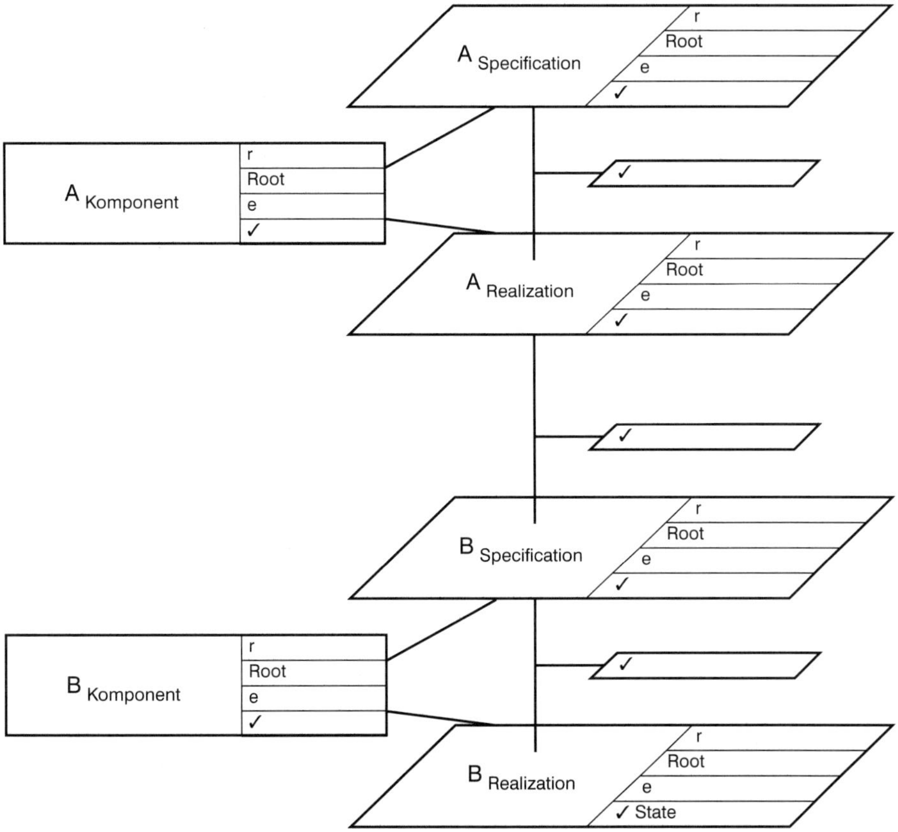

Figure 17.10
Stable framework

also that the relationships between the specifications and realizations have an associated "tick" indicating that they too are in a stable, validated state. This means that the realization **A** has been checked to be a valid realization of the specification **A** and the specification **B** has been confirmed to satisfying its contract and containment relationships with the realization **A**. Similarly, the realization B has been confirmed as a valid realization of specification **B**.

Changes to this stable framework are handled in different ways, depending on whether it is a specification or a realization that is changed. Figure 17.11 illustrates a situation in which the framework in Figure 17.10 is altered by introducing a new version of specification **A**. The figure shows that a new specification of Komponent specification $\mathbf{A}_e$ has been created, called $\mathbf{A}_{e+1}$, with some difference (or delta) ("Δ") to A_e. It also shows that the specification $\mathbf{A}_{e+1}$ is currently in an invalidated state (since it has no tick).

The effects of the introduction of the new specification **A** depend on whether the change is a *semantic change* or a *cosmetic change*. As its name implies, a semantic change creates a difference in the meaning of the specification (by changing or extending the specified requirements). A cosmetic change merely describes the existing specification in a different way, but does not change its

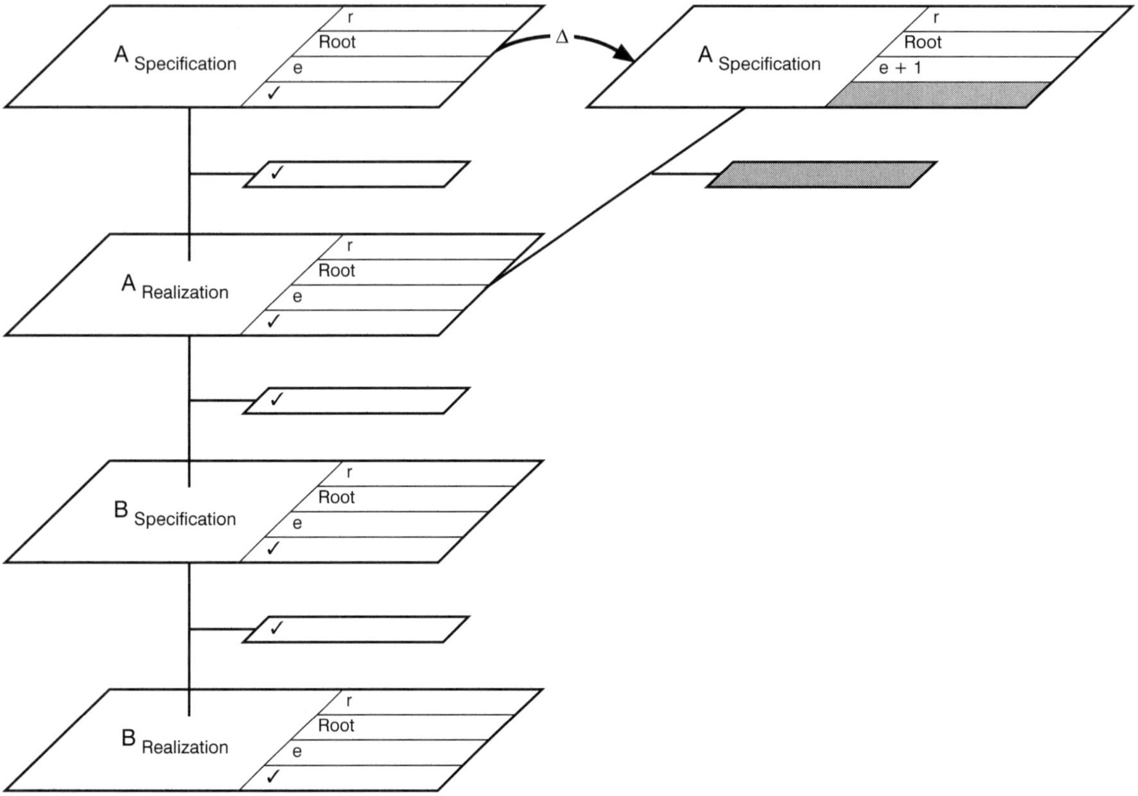

Figure 17.11
Editorial specification change

meaning. In KobrA different names are used to distinguish a version differing from its predecessor by a semantic change or a cosmetic change. A new version of an artifact created by a semantic change to the original is known as a *revision* of the original artifact, while a new version created by a cosmetic change to the original is known as an *edition* of the original artifact.

Note that at the physical level there is probably no difference. In both cases they are different versions.

The distinction between semantic and cosmetic changes is important because it determines the nature of the additional changes that must be made to the framework to bring it back to a validated state. In the example, if A_{e+1} is a new edition of A_e (i.e. based on a cosmetic change), then by definition there is (or should be) no change to the defined requirements. This means that the original realization A_e, can, and should, be a valid realization of the updated specification. Of course, this relationship also needs to be checked, so the result of an update to the specification A_e is the introduction of a new specification A_{e+1} and a new realizes relationship between specification A_{e+1} and the original realization A_e. As shown in the diagram, this relationship is also initially in the non-validated state.

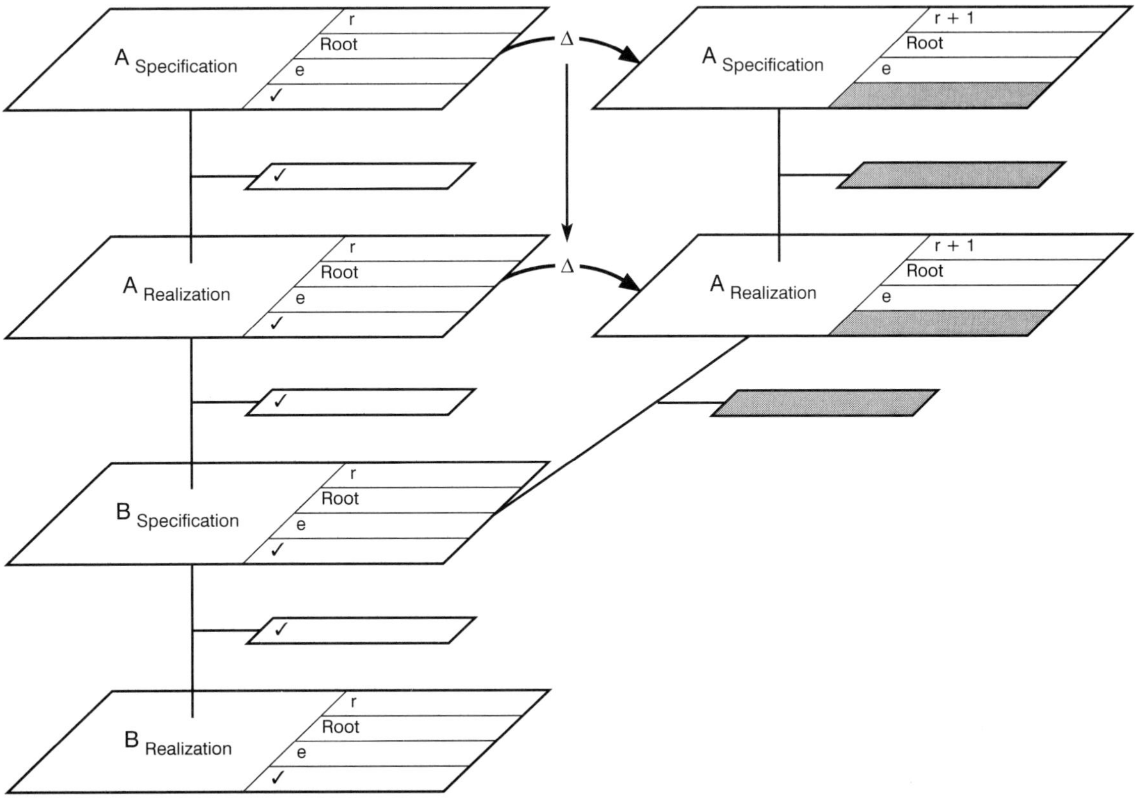

Figure 17.12
Semantic specification change

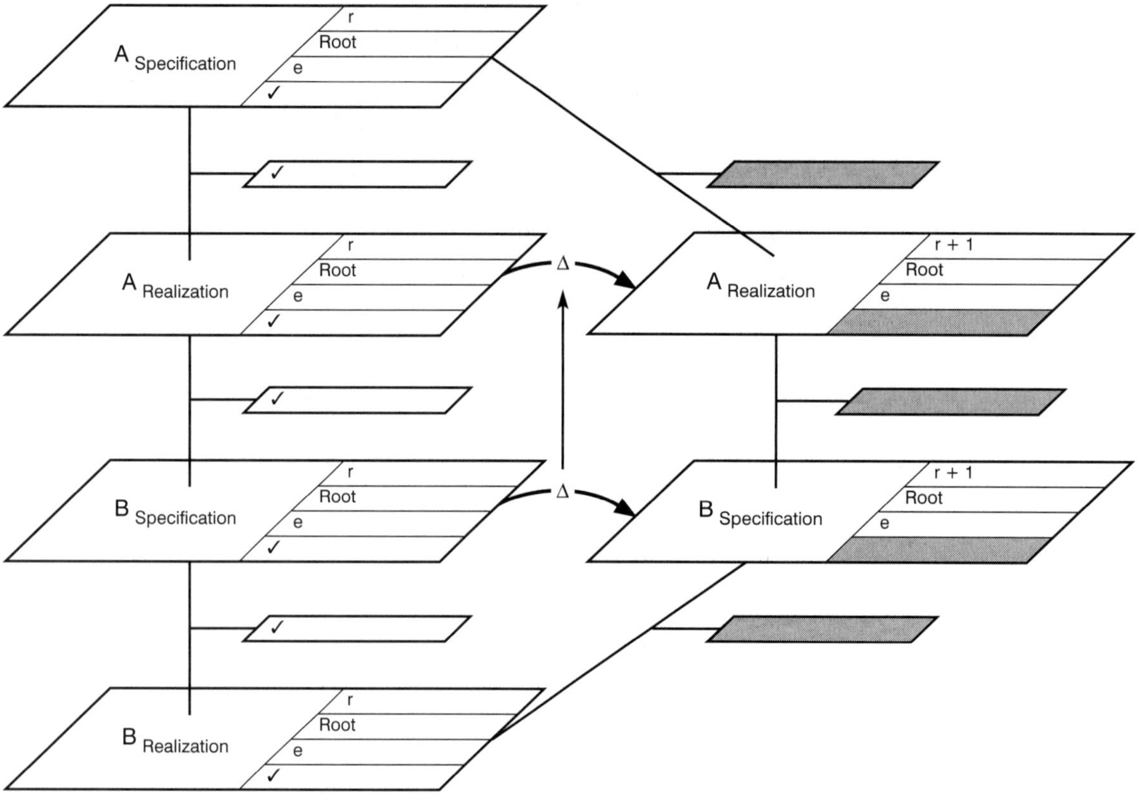

Figure 17.13
Specification change effect on parent realization

The situation would be different if specification A^{r+1} were a revision of A^r (rather than an edition). Figure 17.12 illustrates the situation that would then result. Since specification A^{r+1} now specifies different properties than A^r (because it is the result of a semantic change), it is no longer possible for the original realization A^r to realize the specification A^{r+1}. Whenever a revision is made to a specification, therefore, it is necessary to also introduce a new revision of the corresponding realization, in this case realization A^r. Whether this takes place at the same time, or at some later time, is not defined. However, the framework will not be able to re-enter a stable, validated state until the revision of the realization has been added, and the realized relationship with the revised specification has also been validated.

The consequences of adding a new realization can also be numerous. In the scenario depicted in Figure 17.13, the revised realization A^r has a clientship and containment relationship with the original specification B^r of the sub-Komponent B. This means that the semantic change to B^r causes changes in client Komponent realizations.

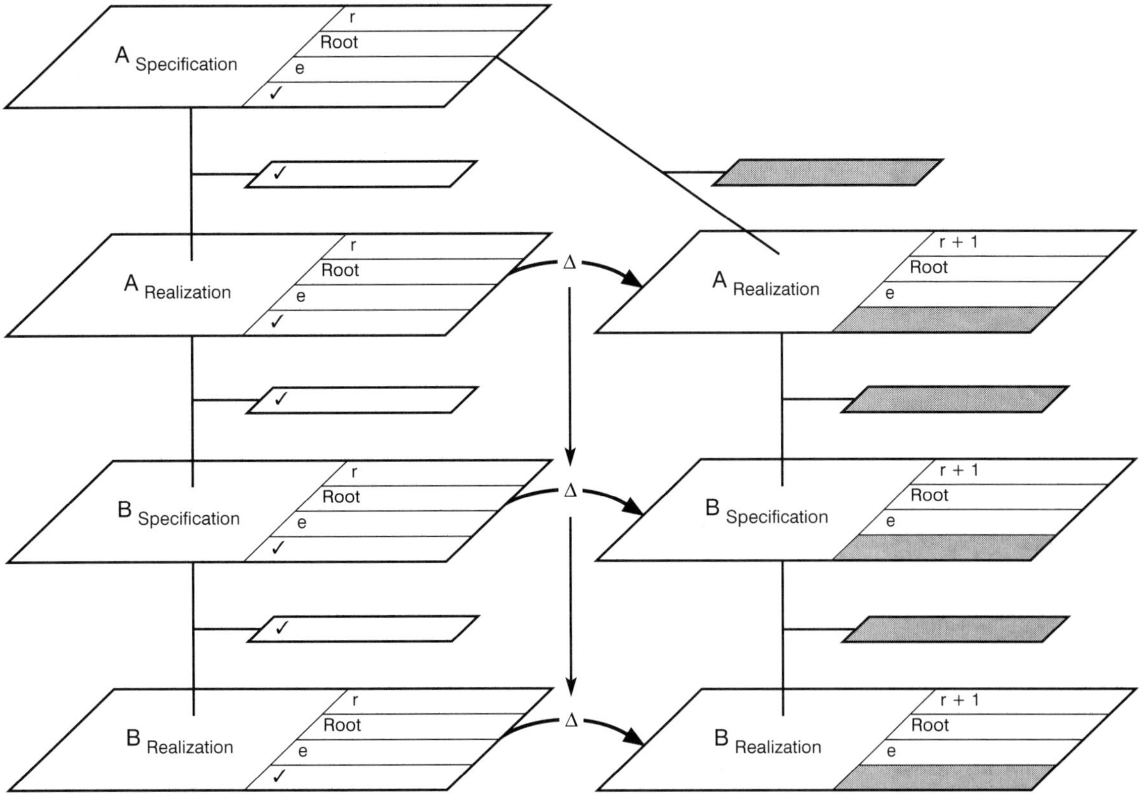

Figure 17.14
Sub-Komponent revision

However, it is also possible, of course, that the change in A^{r+1} may require a change to the specification, and hence therefore also the realization, of the sub-Komponent B. This situation is illustrated in Figure 17.14. In general, a semantic change to an entity in the framework (i.e. a revision) will have a "knock on" effect requiring new revisions to other Komponents in the tree, mainly those in its sub-tree. This is not the case with cosmetic changes (i.e. editions), however. Since these do not change the meaning of the entity, they only require the introduction (and checking) of new relationships to the old entities with which the changed Komponent is related.

The approach described above has two important properties. First, the existing entities in a framework are in no way changed by a semantic or cosmetic change to specifications and/or realizations. The effects of revisions are entirely captured by the addition of new assets (i.e. specifications, realizations, and relationships). This means old versions of a framework are always present and untouched in the project repository. Second, the relationships which tie together the different versions of specifications and realizations are organized in such a way that from any specification or realization a consistent framework can be traced out. For example, in Figure 17.12, realizations $\mathbf{A}^r$ and $\mathbf{A}^{r+1}$ are both able to include the sub-Komponent specified by $\mathbf{B}^r$ without worrying about (or even knowing about) each other. In other words, the project repository is able to store multiple framework revisions and updates, each being completely unaware of the existence of the other.

This approach satisfies the principle of locality because all configuration information about a configuration item (i.e. a specification or a realization) is stored either within the entity itself or with the relationships in which the entity participates. The configuration of a complete framework can therefore only be determined by walking through the tree of links to pick up the specification and realization revisions to which it is linked. Another advantage of this approach is that it can be (and is) applied recursively to manage versions of smaller-grained configuration items such as models and diagrams. For example, the validity states associated with the specifications and realizations in the preceding figures (indicated by the "tick"), actually represent the collective validity of all the diagrams from which they are constructed, together with the relationships between them. A revision of one of the diagrams making up a specification will thus, by definition, yield a revision of the specification. Likewise, a new edition of a diagram will result in a new edition of the specification.

17.3 Process

Two main activities drive the KobrA maintenance process: the change management activity, which deals with the creation and application of changes of the form described in Section 17.2.4; and configuration management, which deals with the consistent updating of configuration items, dependencies and evolution graphs of the form described in Sections 17.2.1, 17.2.2 and 17.2.3, respectively. In the following sections we elaborate further upon these activities.

17.3.1 Change management

The main challenge for the change management activity is to guarantee the mutual consistency of the various assets within a product line. Changes can be requested by various stakeholders in a product line, and can have various levels of impact. The change management approach must therefore provide strategies for identifying and evaluating the merits of changes, calculating the scope and impact of changes, and provide mechanisms for enacting changes in a safe and coherent way.

Change requests in the context of a product line fall into two main categories: application-specific changes requested by customers and engineers focussing on a specific application; and framework changes requested by domain experts and engineers working at the framework level. In either category, changes can be requested to satisfy new requirements, or to fix identified bugs.

The big question that has to be addressed for any change request in the context of a product line is whether to make the change at the level of the framework or the level of an individual application. Many factors influence this decision, including strategic considerations about the future of the product line. For example, customer specific change requests can only be integrated into the framework if the customer is willing to accept that the adapted application will have further new and changed features, otherwise the changes have to be integrated into the customer-specific application. This, however, has the consequence that these changes have to be reintegrated into the framework at a later stage or the customer has to accept that application specific changes may vanish in later versions.

The KobrA change management approach is designed to offer as much flexibility as possible in handling these various scenarios, while being as simple and efficient as possible. To achieve this, the change management activity explicitly identifies different strategies for the initial insertion of changes and the subsequent propagation of the changes to keep all assets in the product line consistent. In this way, the change management approach is able to ensure that each application relies on the most up-to-date framework, and vice versa.

Change management in KobrA is achieved by means of four distinct propagation integration strategies:

- **Framework change integration.** An asset is changed within a framework, and the change is propagated to other assets within the framework to assure consistency.
- **Application change integration.** An asset is changed within an application, and the change is propagated to other assets within the application to assure consistency.
- **Feedforward change reintegration.** Changes within a framework are propagated to, and reintegrated within, existing applications generated from the framework.
- **Feedback change reintegration.** Changes within an application are propagated to, and reintegrated within, the framework from which it was generated.

Over the lifetime of a product line, these different strategies will typically have to be used together to maintain the consistency of the product line. Consider, for example, the scenario depicted in Figure 17.15. Starting from the first version of the framework on the left-hand side, framework change integration is first applied to generate a new version of the framework. An application, Public Library, is then generated from this version of the framework. Changes are then made simultaneously to the application and to the framework by means of the application change integration and the framework change integration activities, respectively. Later changes in revision of 1.1 of the application are reintegrated into the framework via the feedback change-reintegration activity. Finally, feed forward change-integration is used to integrate changes in revision 1.4 of the framework into the application.

Figure 17.15 shows the four different change integration strategies for a *LibrarySystem* framework and Public *LibrarySystem* application. The example starts with a version 1.0 of the Library framework that will be changed in a version 1.1.

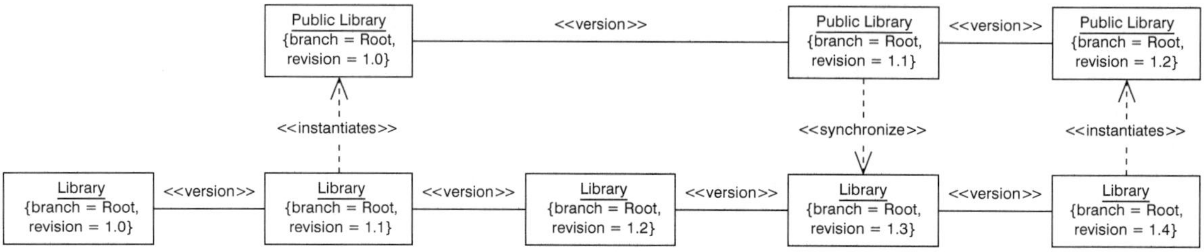

Figure 17.15
Change integration

This version is used to instantiate the version 1.0 of the Public Library that will be changed simultaneously with the library Framework (see Figure 17.18; Figure 17.19). The changes in the application will be reintegrated in the *LibrarySystem* framework that has been changed in the meantime (see Figure 17.25). After application change reintegration into the framework a new application version will be instantiated with both the framework and the application change.

The four change integration strategies share a set of common sub-activities:

- **Change identification** identifies the changes to be made within a framework or application, determines the change type, and stores the changes in a so-called change set (see Section 17.2.4).
- **Change impact calculus** analyzes the sequence of conceptual change operations associated with a change, calculates the impact of the change on the affected assets and determines further changes that will be generated.
- **Change propagation** propagates the effects through the product line framework and/or application in order to retain the mutual consistency of the product line assets.

Figure 17.16 illustrates the general pattern by which these change integration strategies and activities can be applied over the lifetime of a product line (in this case the library product line). As can be seen from this figure, an outer and an inner change cycle characterize the life-cycle. The major steps of the outer cycle, illustrated in greater detail in Figure 17.17, are always triggered by an asset change. They consist of the determination of the affected assets for the next change integration step, the propagation of further changes to related assets and the generation of tasks to instruct a developer or application engineer to carry out the change.

The inner change processes can be seen as focussing on the development and configuration management activities. These are the change, update, and version activities that will be explained in Section 17.3.2.

The following subsections elaborate upon each of the four change integration strategies in turn.

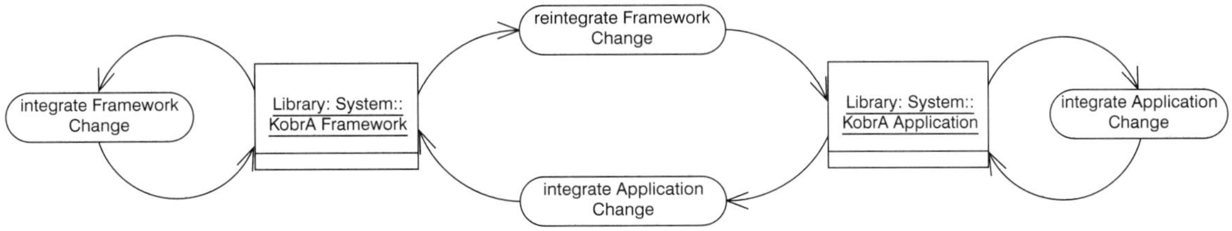

Figure 17.16
Change-management processes

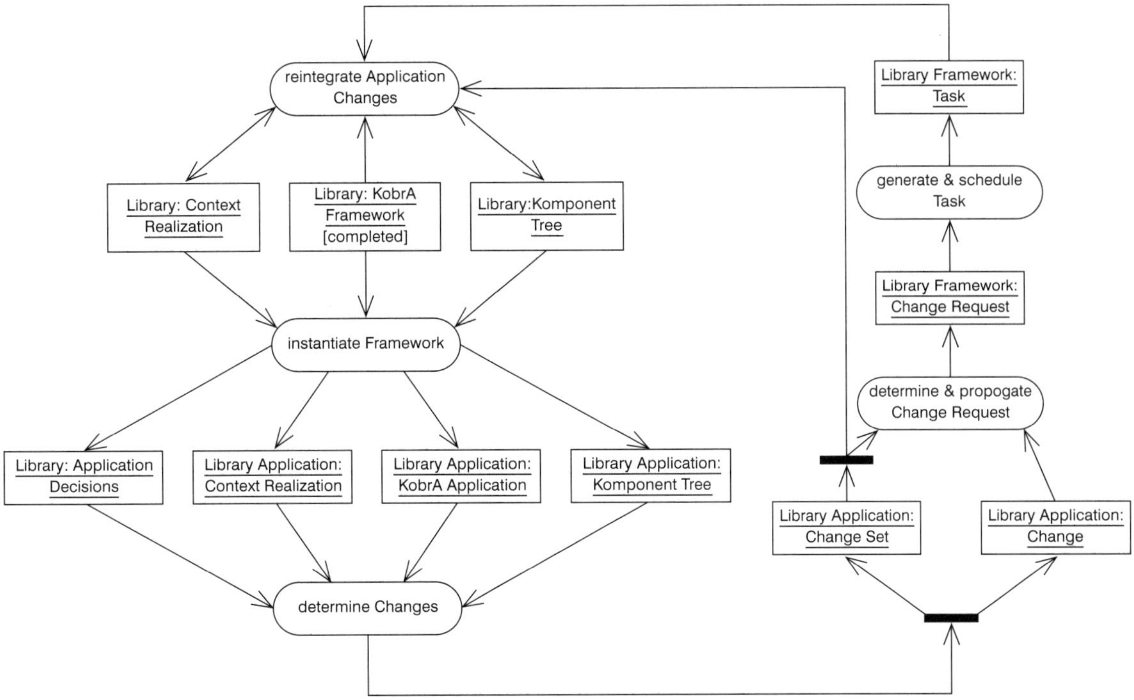

Figure 17.17
Outer feedback change process

17.3.1.1 Framework-change integration

Framework change integration introduces a change into a framework, and propagates the changes to all affected assets to ensure that the framework is in a consistent state. Five main sub-activities are involved:

1 **Change framework asset.** An asset in the framework is changed by the application of a regular development activity of the kind described in the previous parts of the book. As a result, inconsistencies may arise which require changes to other assets in the framework

2 **Identify changes.** The first step in bringing the framework back into a consistent state is to identify what changes need to be made to other parts of the framework. In general, other existing assets in the framework may need to be changed or retired, or totally new assets may need to be generated. Once identified, the required changes must be linked with the affected assets and stored in a change set. This change set is needed later to serve as an information carrier for impact analysis, updating, and the configuration process.

3 **Calculate change impact.** For each of the changes in the change set the sequence of change operations has to be determined. This sequence has the equivalent set of effects as the changes. But neither the set of conceptual operations (see Section 17.2.4.2) nor the order of their application is unique. The operation sequence is stored together will the corresponding change in the change set.

4 **Update framework.** Each of the conceptual operations in the change set can be exactly expressed in terms of the assets, base transformations, mappings, semantic operations, and constraints involved (see [Erd98]). The conceptual operations are then applied to the framework assets to generate a set of brand new and changed assets F^+, a set of retired Komponents F^- and the *base transformation T* (see Section 17.3.2.1). They are needed in the configuration process to remove retired Komponents or to verify consistency rules.

5 **Version framework.** Once the change operators have been applied, new versions of the affected assets can be established and new dependencies can be created between them. The transitive closure of the dependencies is then established according to the update rule and the mutual consistency of all the assets determined. A mutually consistent state must be established and the integration loop terminated, or the change process has to be restarted to remove inconsistencies. Once a new consistent state has been reached, a new version of the framework is established.

As an example, Figure 17.18 illustrates the class diagram for a new version of the LoanManager Komponent specification in the Library system framework. This version was generated to make it possible to store customer's loan histories. Thickened lines highlight the effects of the change at the diagram level. The example depicts the framework change from version 1.1 to 1.2 in Figure 17.15 that takes place simultaneously with the Public Library application change from version 1.0 to 1.1.

17.3.1.2 Application change integration

Framework change integration introduces a change into a framework, and propagates the changes to all affected assets to ensure that the framework is in a consistent state. Because of the similarities between frameworks and applications in KobrA, the five sub-activities involved in application change integration are very similar to those for framework change reintegration, but are applied to the specific artifacts of an application rather than the generic artifacts of the framework. The main difference, therefore, is that the application change integration must consider changes to the decision resolution model,

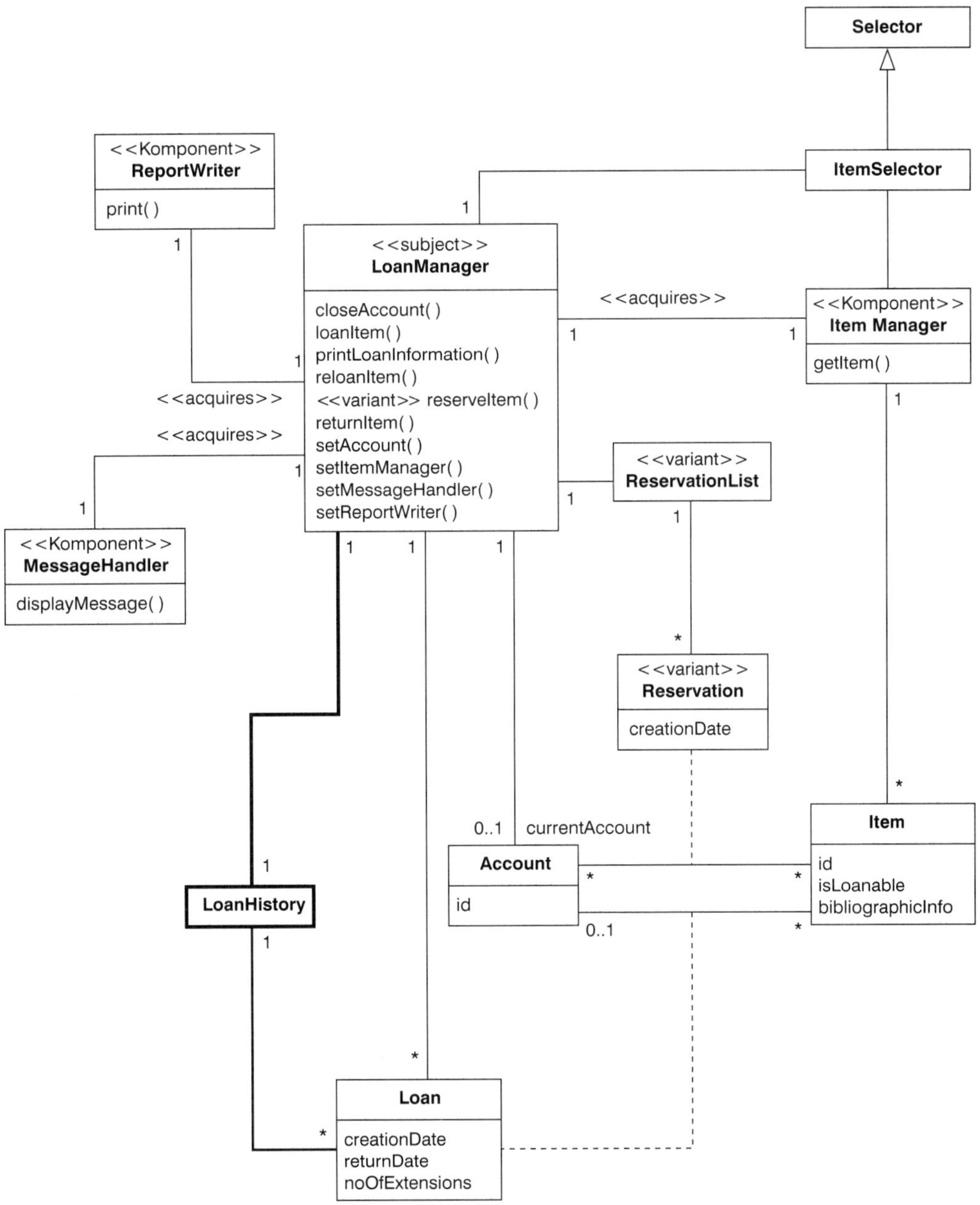

Figure 17.18
LoanManager change in the framework

while framework change integration activities must consider changes to the decisions models. Figure 17.19 illustrates the activities involved in application change integration in a graphical form.

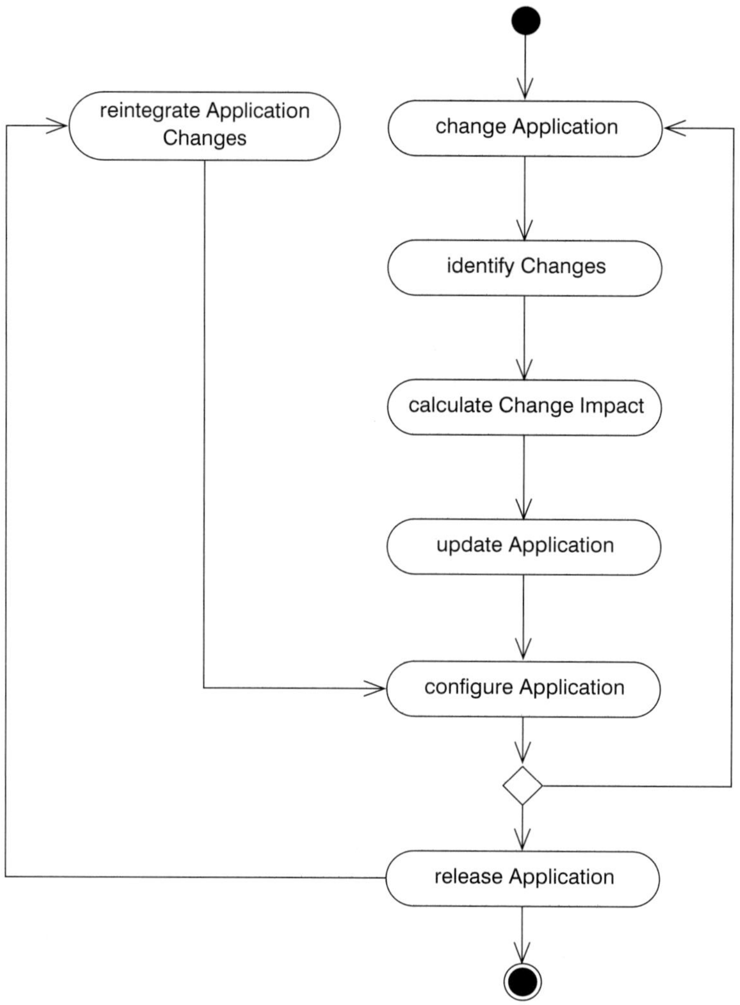

Figure 17.19
Application change integration

Figure 17.20 illustrates the class diagram for a new version of the *LoanManager* Komponent specification within a specific application in the library product line – the *Public Library* application. This version was generated in response to a customer's request for a more sophisticated account that can limit the number of loans for a particular user. Thickened lines highlight the effects of the change at the diagram level. The example depicts the application change from version 1.0 to 1.1 in Figure 17.15 that takes place simultaneously with the library framework change from version 1.1 to 1.2.

17.3.1.3 Feedforward change reintegration

Framework- and application change integration are concerned with updating the framework and applications on an individual basis. Feedforward and feedback change reintegration deal with changes at the level of the whole product line, and are concerned with bringing the latest versions of applications into a consistent state with the latest version of the framework from which they are generated.

The feedforward change reintegration process describes how a change in the framework is reintegrated into the applications derived from it. "Reintegration" is therefore concerned with propagating changes between a framework and its applications, whereas "integration" is concerned with the propagation of changes within an individual framework or application. Because a framework and its applications may evolve independently, there are two main scenarios to consider when defining the reintegration steps:

- reintegration of framework changes into an unchanged application
- reintegration of framework changes into a changed application.

In the case where the framework and the application have both evolved independently, additional changes cannot be reintegrated directly and an additional synchronization process is needed to combine both sets of changes into the reintegration version of the application. Below we consider feedforward change reintegration in the context of the two scenarios separately.

17.3.1.3.1 Reintegration of framework changes into an unchanged application

The steps involved in reintegrating framework changes into an unchanged application are as follows:

1. **Determine affected applications.** The reintegration process can only be started once the new version of the framework has been established. This means that previously changed frameworks must be versioned so that the updated application can be referenced to a stable framework version. First, all applications affected by the changes of the new framework will be determined and references to these applications will be stored. This is necessary to avoid unnecessary change activities. Nevertheless, all active applications should be linked to the newest framework version regardless of whether or not they need to be changed.
2. **Start framework change reintegration.** Strategic factors will determine which of the applications identified in the previous step should actually be brought up to date with the latest version of the framework. Some applications may actually have become inactive, or may have a relatively short predicted lifetime. It is therefore not always cost-effective to update every application the moment a new version of the framework is established. Since all the required information is stored (affected applications, change set, changed assets etc.) the application can be updated at a future point should the strategic situation change.

3 Identify framework reference version. The goal of the reintegration activity is to bring an application up to date with the latest version of the framework. However, the set of changes that are involved depend on the version of the framework from which the application was originally instantiated. In

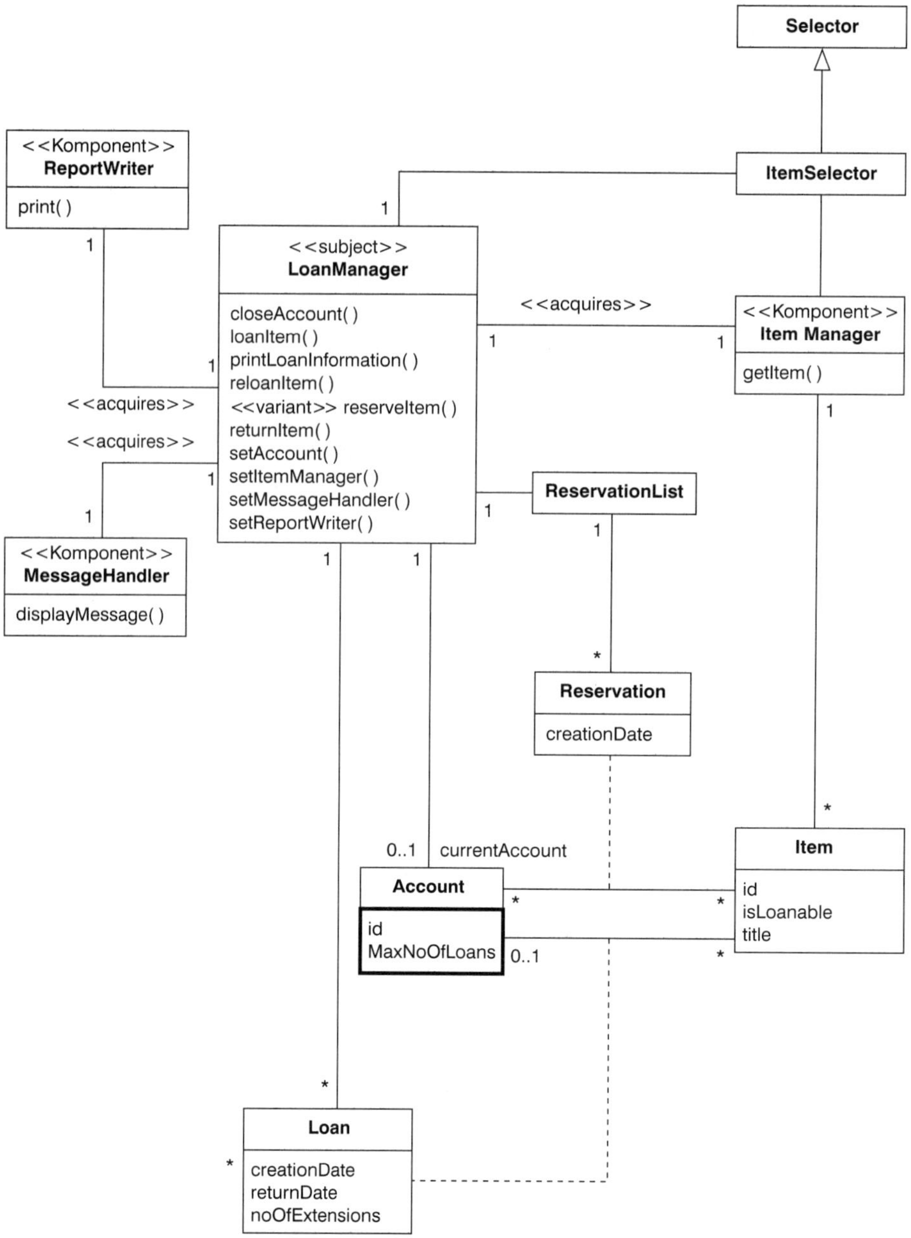

Figure 17.20
LoanManager change in the application Public Library

general, this can be any of the prior versions of the framework. The corresponding conceptual operations associated with the changes are stored in the change sets of all the framework versions from the original version to the latest version. To create the new application version a new change set has to be created that contains the cumulative set of changes, from the original to the current version of the framework, together with the corresponding conceptual operations.

4 **Create new application version.** Depending on the extent of the changes that have to be made, the new version of the application can be created in one of two ways. If the changes are substantial, and involve a significant amount of development effort it may be more cost-effective to reinstantiate the application from the latest version of the framework. Alternatively, if the changes are less significant it may be more cost-effective to create the new version by applying the identified changes directly at the level of the application. In either case the resulting application should have the same properties. The choice depends purely on the relative effort involved in reinstantiation versus application change integration.

5 **Cleanup.** Finally, the application's evolution graph must be augmented to reflect the new change set involved in moving to the new version from the previous version. In this way, the validity of the application can be verified later. Furthermore, the application will be self-contained, since it can be reconstructed from the information stored in its evolution graph.

Figure 17.21 shows an example scenario in which the feedforward change reintegration has been applied to an unchanged application. It illustrates the Library framework and the Public Library application as an evolution graph. In addition, it depicts the synchronization of both, especially the reintegration of a framework change in an unchanged Public Library. First, a Public Library was instantiated from version 1.1 of the Library Framework and changes of the Public Library made from version 1.0 to 1.1 were immediately reintegrated in the Library framework (see <<synchronize>> from Public library version 1.1 to the Library framework 1.2). After that a further change in the Library framework can be directly reintegrated in the application without a merge with meanwhile changed Public Library. Furthermore, Figure 17.21 shows the instantiation of a physical component tree that is contained in the Public Library application.

17.3.1.3.2 Reintegration of framework changes into a changed application

When framework changes have to be integrated into an application which itself evolved independently, the situation is a little more complicated. The challenge is to integrate the changes made at the framework level without losing the changes made at the application level. To do this, a temporary application is instantiated from the new version of the framework, and the changes within it are synchronized with those in the latest version of the application. The end result is a new version of the application with the synchronized change set recorded in its evolution graph.

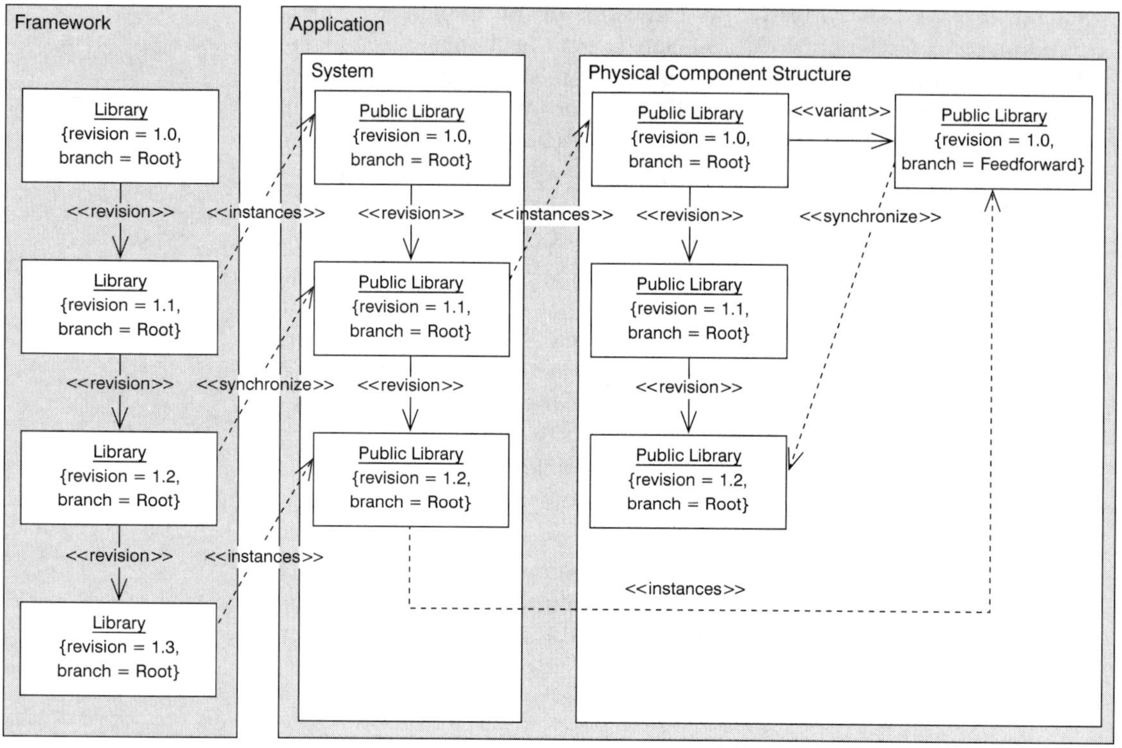

Figure 17.21
Direct reintegration of a framework change

The process for change reintegration into a changed application consist of six steps. Steps one to three are the same as for change reintegration into an unchanged application. The remaining three steps are:

4 Reinstantiate a temporary application. Once the change set and change operators capturing the difference between the original framework version and the latest version have been identified, a temporary application is instantiated with the same decision resolutions as the original version of the application. Note that there is no need to update the evolution graph for the temporary application version because this will be deleted at the end of the reintegration process.

5 Synchronize temporary application. In order to synchronize the temporary application with the latest version of the application a new application version is created to represent its predecessor. This is done to avoid the destruction of the changes contained in the latest version of the application. For example, changes from *Public Library* revision 1.0 to 1.1 would be lost if the framework would be reinstantiated as application version 1.2. The assets of the temporary and latest application versions are then merged and placed into a new application version. The details of the synchronization process are described in Section 17.3.2.4.

6 Cleanup. The cleanup process is the same as for the previous case, but in addition the temporary application version is removed from the application's evolution graph after the synchronization process has been completed.

Figure 17.22 shows an example scenario in which the feedforward change reintegration has been applied to a changed application. Like Figure 17.21 it illustrates the Library framework and the *Public Library* application evolution graph as well as the reintegration of a framework change in the *Public Library* in a changed *Public Library*. In this case, it is not possible to directly reintegrate a framework without losing the intermediate *Public Library* version 1.2. Therefore the reintegration process becomes more sophisticated. First, the change will be reintegrated in a branch of the *Public Library* version 1.0 (branch = feedforward). The predecessor of the branched Library version must be the last instantiated version that is the predecessor of version 1.1. After that the branched version will be merged with the *Public Library* version 1.1 in a new version 1.2, which will be the successor of version 1.1. The branched *Public Library* only serves as a vehicle to ship the framework changes into the *Public Library* application. In addition, Figure 17.22 shows the instantiation of a physical component tree that is contained in the *Public Library* application.

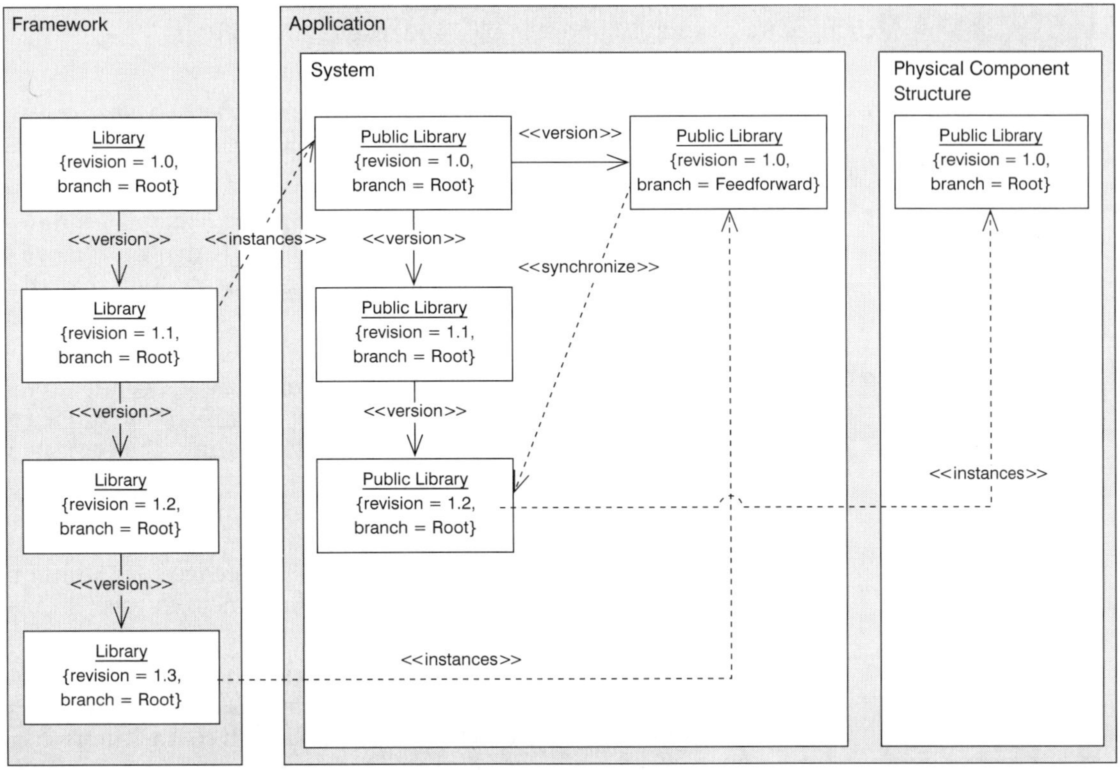

Figure 17.22
Reintegration of framework changes with application changes

17.3.1.4 Feedback change reintegration

The feedback change reintegration process describes how a change in an application is reintegrated into the framework from which it was derived. The need for feedback reintegration arises as soon as customer specific changes have been requested and integrated in an application. Unless an application is going to be retired in the near future, it is essential that application-level changes be integrated into the framework as soon as possible to minimize the number of changes that take place within other applications instantiated from the same framework. The goal of product line engineering is to keep all applications consistent with the framework, so the longer application-level changes remain unintegrated at the framework, the greater the likely divergence in the evolution of different applications, and the greater the effort involved in eventually bringing them back together.

As with feedforward reintegration, there are two main scenarios to consider when defining the reintegration steps:

- reintegration of application changes into an unchanged framework
- reintegration of application changes into a changed framework.

The decision as to when and if application changes should be propagated to the framework is a strategic one for the product line management. In any event, the reintegration should be performed as early as possible, because reintegration for changed frameworks is more complex and requires more effort.

17.3.1.4.1 Reintegration of application changes into an unchanged framework

The steps involved in reintegrating application changes into an unchanged framework are as follows:

1. **Initialization.** The initialization process is quite similar to that of feedforward framework change reintegration. The only difference is that the change set can contain additional decision resolution changes. The decision resolutions can be used to identify the place in the framework where a change in a generic Komponent or a generic context realization has to be reintegrated.
2. **Reintegrate application changes into the framework.** Once the original framework version and application versions are identified, the framework engineers have to reintegrate application asset changes into the corresponding framework assets. Maintenance engineers are supported in this task by the change sets that list the changes that have to be reintegrated. These may include changes to the existing assets, the introduction of brand new assets, or the retirement of unneeded assets. All the changes are recorded within the framework evolution change graph so that a clean framework version can be created at a future time.
3. **Reintegrate changes into the applications.** Once the changes are reintegrated into the framework, the original application has to be reinstantiated and both versions have to be checked for consistency. If so, the latest version of the application is now related to the new framework version, enabling future feedforward and feedback changes to be based on the correct

framework version. If the versions are not consistent, it could be that the application evolved while the original changes were being reintegrated into the framework. To solve such as problem, a temporary application must be reinstantiated and the version of the application has to be synchronized with the newly instantiated version. After the synchronization of the newly instantiated version, with the newest version of the application version, all the other applications which are affected by the change have to be reinstantiated.

4 **Complete framework and application history tree.** Finally, the framework's evolution graph must be augmented to reflect the new change set involved in moving to the new version from the previous version. In this way the validity of the framework can be verified later on. Furthermore, the framework will remain self-contained, since it can be reconstructed from the information stored in its evolution graph.

Figure 17.23 shows an example scenario in which the feedback change reintegration has been applied to an unchanged framework. It illustrates the Library framework and the application evolution graph as well as the reintegration of a Public Library application change into an unchanged Library framework. Therefore the changes can be directly integrated in a Library version 1.2. Unlike the reintegration of a framework change into an unchanged application that can be achieved by a simple instantiation, the framework engineer must know how to reintegrate changes from a specific application into a generic framework. In addition, Figure 17.23 shows again the instantiation of a physical component tree and the branching of the Public Library system version 1.1 with its later integration in the root of the Public Library application.

17.3.1.4.2 Reintegration of application changes into a changed framework

As with feedforward reintegration, if the application changes have to be integrated into a framework, which itself has evolved independently, the situation is a little more complicated. The challenge is to integrate the changes made at the application level without losing the changes made at the framework level. To do this, a temporary framework is created that contains the framework changes, and this will be synchronized with the latest version of the application. The changes defined in the resulting version of the application are then reintegrated into the framework.

The steps involved are basically the same as for the reintegration of application changes into an unchanged framework, with the exception that step 2 is slightly modified. Instead of one step, two new steps replace it:

2 a **Reintegrate application changes in a temporary framework.** Once the initialization activity has been completed a temporary framework is created as a variant of the original framework version that contains the application changes. Changes to assets in the application must be reflected in this temporary framework, including changes to existing assets, the introduction of new assets and the retirement of unneeded assets. Note that there is no need to update the evolution graph of the temporary framework version because this will be deleted at the end of the reintegration process.

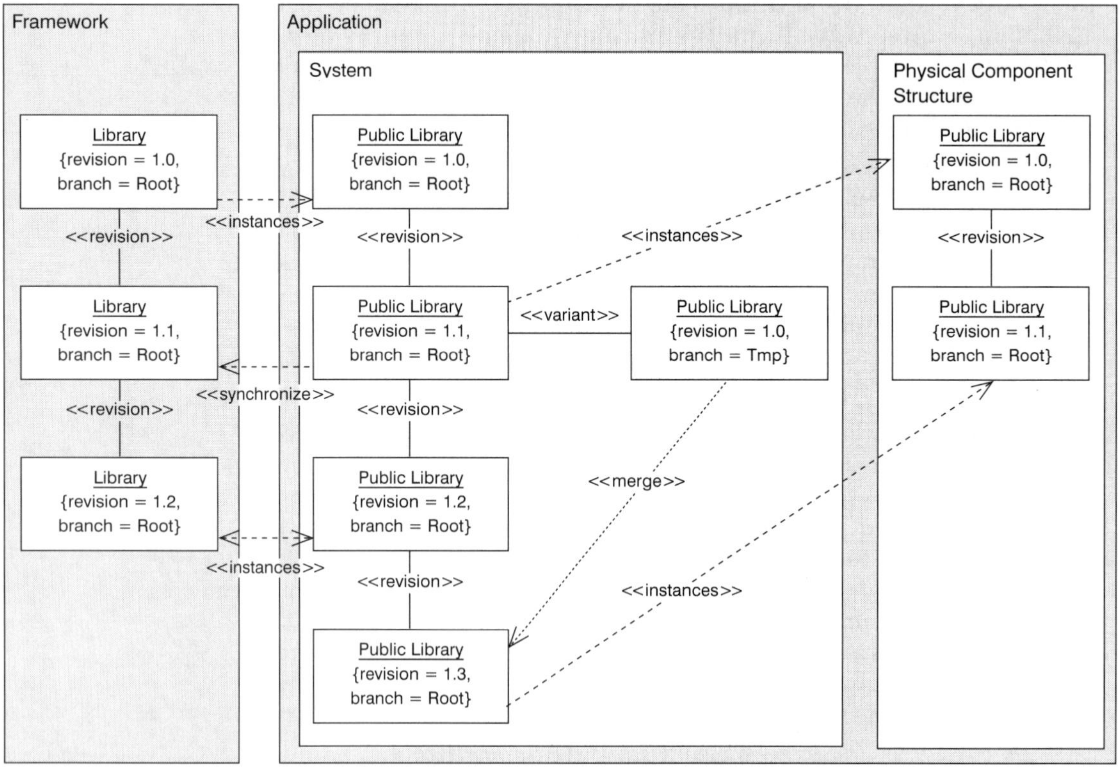

Figure 17.23
Application change reintegration into an unchanged framework

2 b Synchronize current framework with the temporary framework. In order to synchronize the temporary framework with the current framework version, a new framework revision will be created as the predecessor version corresponds to the latest version of the application. This prevents the destruction of the changes made in the latest version of the application. For example, in Figure 17.24, the changes from Library revision 1.1 to 1.2 would get lost if the framework library would be instantiated as version 1.2. The framework assets of the temporary version and the latest framework version are then merged to create the new framework version. The details of the synchronization process are described in Section 17.3.1.4.

Figure 17.24 shows an example scenario in which the feedback change reintegration has been applied to a changed framework. It illustrates again the Library framework and the Public Library application evolution graph as well as the reintegration of a Public Library application change in a changed Library framework. The process is quite similar to the reintegration of framework changes into a changed application. The difference is that a new framework version cannot be achieved by a simple instantiation; instead the framework engineer

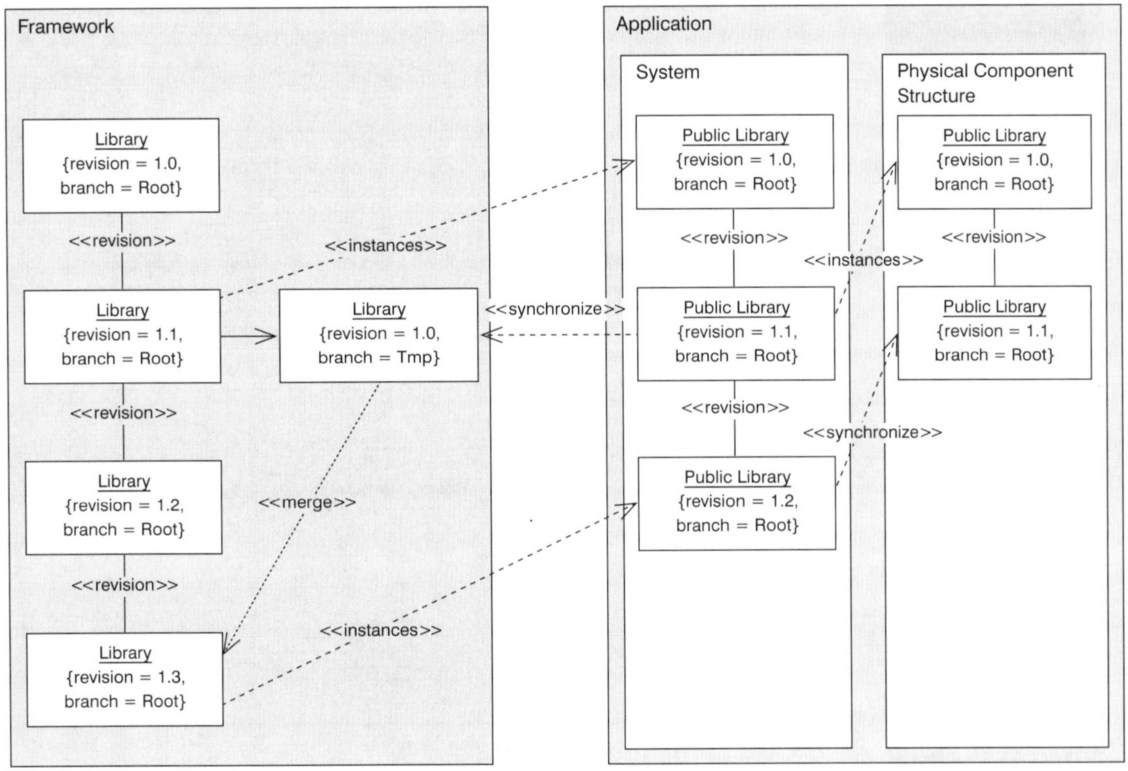

Figure 17.24
Application change reintegration into a changed framework

must know how to reintegrate changes from a specific application into a generic framework. In addition, Figure 17.24 shows again the instantiation of physical component trees.

To illustrate the principle behind these more complex change reintegration activities, imagine that the framework change shown in Figure 17.18 and the application change shown in Figure 17.20 were performed in parallel. The framework engineers have to decide whether the enhanced Account in the latest version of the Public Library is generally useful at the framework level, and whether the loan history capability is need in the Public Library.

If it is determined that the customer specific change is not relevant to the whole product line, the change has to be recorded so that a subsequent update of Public Library can be extended with this feature again. It is clear that in the latter case, the future reintegration of framework changes in the customer specific application becomes more complicated.

In general, it is more beneficial to reintegrate customer specific changes into the product line framework. This is because these features become optional rather than mandatory thanks to the generic nature of frameworks. Figure 17.25 illustrates yet another version of the Library framework in which the enhanced

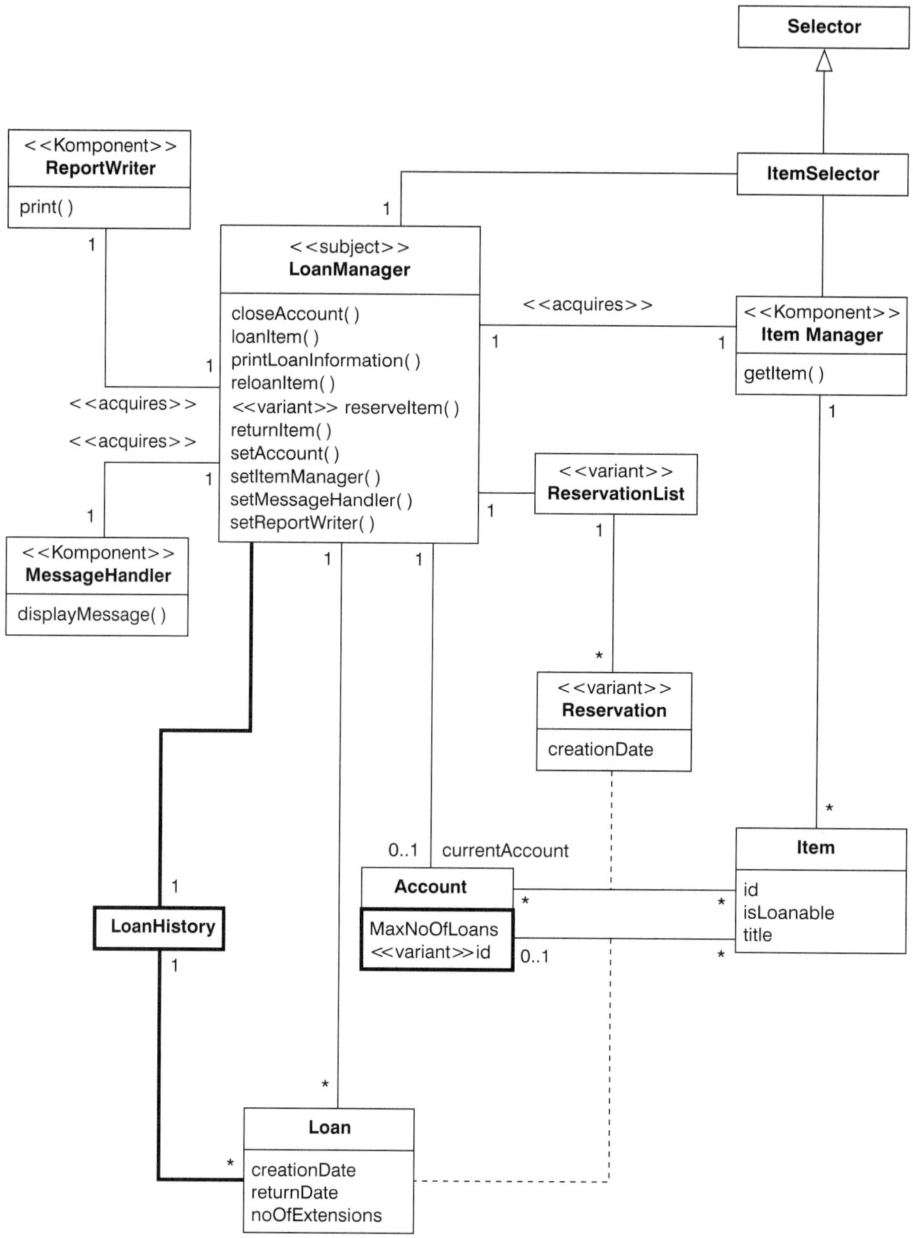

Figure 17.25
Synchronization of the framework with the application change

account features introduced into the Public Library application have been elevated to the framework level. This step corresponds to the change of the framework in Figure 17.15 from version 1.2 to 1.3.

17.3.2 Configuration management

Change management is responsible for identifying what changes need to be applied to the assets in a product line and why. Configuration management activities are responsible for how the changes are achieved. The change management activity therefore defines the technical activities needed to enact version changes in a reliable and consistent way. These activities are invoked at appropriate points during the change-management process to actually put the desired kinds of changes into effect, and to create the desired kinds of dependencies. The configuration-management activity can therefore be viewed as maintaining the members of the product line.

Seven distinct sub-activities make up the configuration management activity in KobrA:

- The **update process** propagates changes made inside the KobrA framework or a KobrA application according to conceptual change operations related to the changes.
- The **configuration process** reconciles the different asset versions by doing a transitive closure. This process is highly dependent on the type of changes carried out. As mentioned in Section 17.2.4, changes can be subdivided into semantic and editorial changes. Semantic changes such as structural level changes and functional changes (see [Mey00]) have always to be reconciled whereas editorial changes can be reintegrated without reconciliation.
- The **instantiation process** consists of an update and configuration sub-process and reinstantiates application systems in order to reintegrate framework changes into the application.
- The **synchronization process** provides a mechanism to merge two different system configurations. It consists of an update and a configuration sub-process and is applied in the change integration and reintegration processes.
- The **build process** supports the application construction process on the technical level.
- The **deployment process** configures a deliverable and prepares it for deployment to the customer.
- The **release process** provides a method to make an application system officially available for its intended use by attaching release numbers to the system assets.

17.3.2.1 Update

The **update activity** represents the primary mechanism for systematically enacting changes to the assets in a product line. It identifies all the assets that need to be changed or retired, determines what change operators are needed to affect the changes, and enacts the state changes representing a transition from one version to another.

The main steps involved in the typical update process are as follows:

1. **Calculation of S^+ and S^-.** All assets contained by the new asset versions are determined and placed into S^+. Similarly, all assets to be excluded from the new asset versions are gathered in S^-. This is achieved by using the conceptual change operations to establish which assets other than the directly changed assets are also affected.
2. **Transformation.** For all assets in $S_{v-1} \setminus S^-$ that are also in S^+ in a changed version, the necessary transformation is calculated. This is to ensure that the transition from S_{v-1} to S_v is subsequently sound and can be augmented by the *base transition relation* (see Section 17.2.3).

17.3.2.2 Configuration

The **configuration process** is responsible for generating an evolutionary graph for the product line in order to guarantee the reachability and reconstructability of consistent asset versions and dependencies. This involves the maintenance of the asset evolution graphs and allocation of a version to every change asset. For example, if the semantics of a Komponent specification have changed, a new Komponent realization version has to be created. The rules that determine when and how a new version has to be created are determined by the asset dependencies (see Section 17.2.2).

The overall configuration process consists of six sub-processes:

1. **Initialization of asset version S_v.** A new and empty set of assets S_v is created and is then subject to version control as the successor of the asset S_{v-1}.
2. **Population of S_v with additional assets.** All brand new assets $A \in S^+$ are subject to version control. Thereafter they will be added to the new version S_v and the transitive closure for the new version will be calculated.
3. **Exclusion of assets.** Initialize $S_{tmp} = S_{v-1} \setminus S^-$ with all assets of S_{v-1} that will be integrated into the new S_v. Because only assets in S_{tmp} will be integrated in S_v, the assets in S^- will be automatically excluded in S_v.
4. **Transformation of changed system assets in S_{v-1}.** For all assets in S_{tmp} version consolidation must be performed, consistency has to be checked and the transitive closure for the base transformation relation T has to be calculated. If the asset is changed, the changed asset will be confirmed as the successor version and it will be integrated in S_v. The original asset will be excluded from S_{tmp}.
5. **Transformation of unchanged assets in S_{v-1}.** The remainder of S_{tmp} will be integrated in S_v.
6. **Clean up the versioning process.** The base transformation relation $\Delta = (S^+, S^-, T)$ will be attached to the transition from system version S_{v-1} to S_v in the asset evolution graph. The temporary set S_{tmp} is then deleted.

17.3.2.3 Instantiation

The feedforward change reintegration process needs an instantiation activity for the direct integration of changes into an application. This means that the changes to the framework can be directly integrated as a successor of an application version. The instantiation process consists of three subactivities:

1. **Initialization.** This step initializes the set $\mathbf{S}^+$ of all brand new and changed assets, the set $\mathbf{S}^-$ of all retired assets, and the new asset version $\mathbf{S}_V$.
2. **Update.** After the initialization step, an adapted update process will calculate the contents of $\mathbf{S}^+$, $\mathbf{S}^-$, the *base transformation relation* **T** and the effect of the conceptual operation in **T**.
3. **Configuration.** The new A_v application will be placed under version control. The rest of the process can be adopted from the configuration process by exchanging the framework for the application.

17.3.2.4 Synchronization

The synchronization process merges two different variants of a framework or application and creates a new version. Like the instantiation process, the synchronization process consists of an initialization, an update and a configuration process. The update and configuration process merge assets from different frameworks and applications.

17.3.2.5 Build, deployment, and release activities

The deployment information defines a deliverable. The allocation of instances of the physical components to the hardware nodes is described by means of UML deployment diagrams. The deployment description defines the exact deployment of a deliverable. This includes definitions of the installation and deinstallation processes as well as the adaptation and update mechanisms (see [HHC98]).

18 Quality assurance

Quality is of importance in every software development project, but particularly so in component-based software development because of its emphasis on reuse. Like any engineering artifact, a software product is only as good as its parts, and assembling poor-quality parts is almost certain to lead to a poor end product. Only by assembling quality parts is there a reasonable likelihood of a high-quality product. The "write once – use many times" property of components gives them as much potential to spread poor quality as high quality. Thus, an enterprise that tolerates low-quality components will propagate this low quality into all the products that use them.

The need for high quality is even more critical in KobrA than in other component-based development approaches because of its emphasis on a tree based architecture. With a recursive, top-down development method, quality problems in an artifact at or near the top of a tree can have a particularly high impact on the success of a project, since they affect the nature of the work performed in lower parts of the tree. Serious errors at the root of the tree could even make the subsequent work performed lower down worthless. KobrA therefore places a premium on systematic quality control in all parts of the development process.

The umbrella activity that embraces the various quality checking, measurement, and improvement activities in a KobrA project is called quality assurance. Its primary goal is to minimize the initial occurrence of software defects and to support the detection, analysis, and removal of those defects[49] that do occur as soon as possible.

In this chapter we first describe general issues related to software quality. We define what we mean by software quality and how it is related to non-functional requirements and quality assurance techniques. We then elaborate

49. A defect can be considered any deviation from an expected quality attribute.

upon testing and software inspection as the primary techniques for achieving high quality in a KobrA development project.

18.1 Attaining quality

Quality can mean different things to different people. An end user, for example, might expect a high-quality system to support expected tasks in a fast, simple and reliable manner. To a system administrator, a high-quality software system might be one that is easy to install and adapt to new hardware configurations, and has no side effects in relation to other installed software. For the developers, a key quality attribute of the system could be that it is easy to adapt to changing requirements.

To accommodate these various interpretations of the concept of quality, and to identify concrete practices from these abstract ideas, an effective quality-assurance program requires:

1. a precise definition of what quality is and how it manifests itself in different kinds of artifacts
2. a precise description of what quality aspects are important for different kinds of artifacts, and what quality levels are required
3. systematic methods for judging quality and improving it to the required levels
4. a plan for putting the previous three together.

18.1.1 What is quality?

Ideally it is desirable to control the quality of all the entities involved in the full software life-cycle, including processes and resources[50] as well as the development artifacts such as models, documents and code. Controlling the quality of processes and resources, however, is beyond the scope of the KobrA method. The quality-assurance activities in KobrA are only concerned with the quality of the development products.

A *quality attribute* (or attribute for short) of a development artifact is any feature or property of the artifact. A *quality measure* is an empirically objective assignment of a number (or symbol) to an artifact to characterize a specific attribute. *Measurement* relates quality attributes and measures. Formally, measurement can be defined as the process by which numbers or symbols are assigned to attributes of entities in the real world.

Standard definitions for a comprehensive set of quality attributes that have been found to be relevant and useful in the context of software products are provided by the ISO 9162 standard [ISO/IEC FCD 9126]. At its core, this standard consists of the definitions of six software quality attributes, which are further refined into quality sub-characteristics. Table 18.1 gives an overview of these six quality attributes and sub-characteristics, which are sometimes called the non-functional requirements, or, informally, the "-ilities" of the system.

50. Resources can be concrete, such as development tools, or abstract, such as development methods.

Table 18.1 ISO 9126 Quality attributes and sub-characteristics

Quality attribute	Sub-characteristic
Functionality	Suitability Accuracy Inter-operability Security
Reliability	Maturity Fault tolerance Recoverability
Usability	Understandability Learnability Operability Attractiveness
Efficiency	Time behavior Resource utilization
Maintainability	Analyzability Changeability Stability Testability
Portability	Adaptability Installability Co-existence Replaceability

Two further qualities that are often deemed important, but are not explicitly included in the above taxonomy are:

- **reuseability** – reusing a (part of a) system requires a change to the system (part) so that it fits the new context. Reusability is therefore subsumed by maintainability and portability. Consequently, techniques for supporting maintainability/portability are also suitable for reusability.
- **cost** – although not often considered a quality attribute in the usual sense, cost is of course an attribute of the highest importance. Many cost- and risk-analysis techniques exist (e.g. Function Points in various flavors, the COCOMO model [BAB00], etc.) and measurement of structural properties, especially size, is often one of their main inputs.

18.1.2 Non-functional requirements specification

In an ideal world, every artifact would have 100 per cent quality, but this is clearly not possible in practical development projects. Quality must compete for resources with other success factors (e.g. functionality, time-to-market etc.). Therefore, a pre-requisite for a successful quality program is a concrete description of what quality levels are acceptable under what circumstances.

This information is usually contained in the *non-functional requirements specification*. This is a common software engineering artifact that describes all the requirements that are not directly concerned with the functionality of the system. In other words, the non-functional requirements define how, or how well, the functional requirements are to be implemented. In general, they can include all kinds of requirements, such as "the system shall be implemented in Java," but quality requirements are usually the largest part. Like all requirement statements, the only constraint on non-functional requirements is that they be unambiguous and testable. In other words, it should be possible to determine unambiguously whether a completed artifact meets the specified requirement.

The non-functional requirements specification is an auxiliary artifact of a Komponent specification. Like all other aspects of the software documentation, non-functional requirements are organized according to the principle of locality. Non-functional requirements should therefore be oriented towards specific Komponents, wherever possible, and should be contained in the non-functional requirements document for that Komponent. As explained in Chapter 4, every Komponent can have (but need not have) a non-functional requirements document.

The non-functional requirements specification for a Komponent can make statements about (i.e. place constraints upon) the specification, realization, and implementation of the Komponent. For example, if the Komponent must be implemented in Java, this would be stated in the non-functional requirements specification. In addition, because a Komponent logically owns its children, the non-functional requirements specification of a Komponent can also place constraints on its children. For example, "the Komponent and all its children shall be implemented in Java." This prevents the need to replicate the same requirement multiple times for Komponents within a sub-tree. In particular, non-functional requirements that pertain to the whole systems can be placed in the non-functional requirements specification for the root or system Komponent.

KobrA is not prescriptive about precisely what form the non-functional requirements should take. Nor is it prescriptive about how any textual functional requirements should be written if they are used to augment the KobrA specification artifacts. In general it is advisable to use a mature standard such as ISO 9126 [ISO/IEC FCD 9126].

18.1.3 Quality assurance techniques

Quality attributes that cannot be measured are of no value. To be useful, therefore, every quality attribute must have at least one associated analysis technique that can be used to assure or assess it. Table 18.2 extends Table 18.1 with a column of quality-assurance techniques that are suitable for validating the different kinds of quality attributes. Although this list of techniques is by no means comprehensive, it can already be seen from this table that there are quite a variety of techniques. Some are more generally applicable than others.

Table 18.2 Quality attributes and quality-assurance techniques

Quality attribute	Suitable techniques
Functionality	**Inspections** (Completeness, consistency, correctness of the system parts). **Testing** Scenario-based evaluation: attack scenarios (for security), deployment scenarios (interoperability).
Reliability	**Inspections** **Testing** Measurement of **structural properties.** Defect modeling, reliability growth models, etc,
Usability	Prototyping, Mockups, User Lab, Interviews, Cognitive Science
Efficiency	Scenario-based evaluation: (load analysis based on operational profiles), stress test, etc. Theoretical computer science (optimal algorithms)
Maintainability	Measurement of **structural properties**. Scenario-based evaluation (change scenarios) Variability modeling
Portability	Scenario-based evaluation (change scenarios) Variability modeling

A prescriptive method for applying each of these techniques in the context of component-based and product line development would take up a whole book in its own right. Therefore, KobrA explicitly addresses only three of these technologies: testing, inspection, and the measurement of structural properties. The first two are addressed because they are not only the most mature and well-known quality assurance techniques, but they also focus on one of the most-critical quality attributes, that is, functional accuracy (or correctness). Testing and inspection also happen to be complementary in the context of KobrA and are therefore described in Sections 18.2 and 18.3 respectively.

The measurement of structural properties is addressed explicitly because it is of particular importance in model-driven methods like KobrA where there is a large potential time lag between the initial specification of a Komponent and the creation of an executable incarnation. Along with inspections, structural measurement is invaluable for obtaining early indicators of the quality of an artifact. This is the topic of the next chapter.

18.1.4 Quality policy specification

Based on the three selected quality-assurance techniques, a general quality program that accompanies the KobrA software-development project can be organized around the following main steps:

1. Identify the external quality attribute(s) that are of importance. The ISO 9126 [ISO/IEC FCD 9126] quality attributes in Table 18.1 can serve as a starting point.
2. For each quality attribute, identify suitable analysis techniques. Table 18.2 provides some pointers of techniques for specific quality attributes.
3. Implement the selected analysis techniques during the development project. Depending on the analysis techniques, these activities occur in certain phases (e.g. early during context realization, or after implementation), or accompany the whole process (e.g. regular inspection of artifacts). The final result is a set of quality measures as prescribed by the analysis technique.
4. Apply the quality model (if any) that is part of the analysis technique. This is a prescription of how the obtained quality measures are to be processed or interpreted. Quality models are discussed further in the next chapter.
5. Take action. Either the quality is found satisfactory according to the requirements specified in the non-functional requirements specification and nothing needs to be done, or certain action needs to be taken, such as revising the system and reevaluating its quality.

Table 18.3 shows examples of possible quality programs based on some of the quality attributes mentioned in Table 18.1 for some selected techniques mentioned in Table 18.2.

The KobrA artifact that defines the quality program for a Komponent is the quality policy. For the quality attributes identified in the non-functional requirements specification, the quality policy defines:

1. what measures will be used as indicators of these attributes
2. what techniques will be used to determine these measures
3. what quality models (if any) should be used to process and/or interpret the measures
4. the policy for reacting to the interpreted measures (what actions are necessary and when).

Table 18.3 Examples of quality programs

Step	Example 1	Example 2	Example 3
1 Quality attribute	Maintainability	Reliability	Reliability
2 Selected analysis technique	SAAM (Software Architecture Analysis Method) [BCK98]	Reliability Growth Modeling (RGM) [Musa98]	Measurement of structural properties (see Chapter 19)
3 Apply technique, obtain quality measures	Identify a set of likely future change scenarios. Assess their impact on the current system architecture (top levels of the KobrA Tree) and estimate the required change effort.	Develop an operational profile for the system, then design and execute system test cases according to this profile. Record failures and the time at which they occurred.	Perform static measurement of structural Komponent properties (e.g. coupling, complexity) on Komponent models.
4 Apply quality model	In each Komponent, take sum of estimated change effort over all scenarios	Calculate current failure intensity and project when failure intensity goal is reached.	Calculate predicted fault-proneness of Komponents from their structural properties.
5 Take action	For example, redesign Komponents that may be subject to a large number of changes	Continue testing and remove faults until reliability goal is reached.	Perform intensive inspections on Komponents with high predicted fault-proneness; consider revision of their design

In other words, the quality policy defines *how* the quality requirements defined in the non-functional requirements specification are to be realized. It is therefore to the non-functional requirements what the realization is to a specification (i.e. a way of putting it into effect).

Like all other artifacts the quality plan is organized according to the principle of locality. Therefore, every Komponent can have a quality policy defining a quality program for its specification, realization and implementation(s). To avoid unnecessary repetition of general quality policies, like non-functional requirements specifications, a quality plan can define policies for sub-Komponents as well. This means that system-wide quality policies can be defined in the system (or root) Komponent's quality policy specification. As with the non-functional requirements specifications, if a Komponent does not have a quality policy, the quality policy for its parent will be adopted.

18.1.5 Quality documentation

Every quality-assurance technique involves some kind of examination or execution of an artifact to gather data. Carrying out the quality program defined in the quality policy therefore invariably involves the creation of many different kinds of artifacts and data sets. In order to provide a convenient way of referring to these, the term "quality documentation" is used to refer collectively to all the artifacts involved in the enactment of the quality policy.

Typical examples of quality artifacts and data sets embraced by the quality documentation include:

- **test suites** – the test cases and oracle used to exercise the system
- **test results** – the results of running and analyzing the test cases
- **quality measures** – the values generated by examining the structural properties of development artifacts
- **defect descriptions** – the defects detected during an inspection.

It is important to note that these artifacts and data sets can be generated at all points in a development project. For example, KobrA requires inspections to be performed as soon as a critical mass of inspectable artifacts is available, so the inspection data is created soon after a Komponent specification or realization is finalized. KobrA also recommends that test cases be generated along with the regular specification and realization models. The common property that determines whether an artifact or data set forms part of the quality documentation is whether it is generated as a result of the enactment of the quality policy.

18.2 Software testing

As indicated in Table 18.2, software testing can be used to measure various different quality attributes, but is most widely used for the detection of defects. This is the purpose for which it is used in KobrA. In general, testing involves three main tasks:

1. test suite definition
2. test execution
3. test result analysis.

The main limitation of testing is that it can only be performed within or after the implementation activity. This is because only executable artifacts can be tested. As a consequence, the time span between the realization of a Komponent and the availability of the test results can be potentially quite long. In the extreme case, it is possible to fully define all the specifications and realizations in a containment tree before any executable incarnations of the Komponents are created. Quality problems may therefore be uncovered by testing long after they have impacted the development of other Komponents in the containment tree. It is for this reason that the recommended development sequence in KobrA is to create and test executable incarnations of Komponents as soon as possible (see Chapter 13).

Like all incremental development approaches, in order to test early increments it is necessary to create "stubs" or "proxies" that simulate the essential features of increments to be developed later. In the context of KobrA, early increments are the Komponents near the root of the tree, and the later increments are those towards the leaves of a tree. Essential, a non-leaf Komponent needs "stubs" of its sub-Komponents in order to be tested.

There is, however, one part of the testing process that does not have to wait until any implementation work has been performed. This is the test suite generation activity. Because the specification provides comprehensive details of the functionality of Komponent operations, and the realization provides comprehensive details of the realization of Komponent operations, it is possible to define the test cases during the Komponent specification and realization activities. Numerous studies have shown that the definition and subsequent inspection of test cases as part of the artifact development process can contribute significantly towards the identification of defects, not only in the test cases, but also in the artifacts themselves.

Because KobrA focuses on the front-end development activities akin to analysis and design, in the remainder of this section we consider the first sub-activity of testing – test suite generation. The test execution and analysis activities should be performed according to a standard object-oriented testing process such as that described by [BG00] or [Bin99].

18.2.1 Testing artifacts

For a given Komponent operation, a test case consists of two basic elements:

1. input and initial state
2. output and final state.

The basic principle is that for a given set of input values/objects and a given starting state, the output values/objects and the final state should be uniquely determined.

The key to successful test design in the context of limited resources is to pick the optimal set of test cases. Various strategies have been defined for picking test cases, depending on whether only the specification of a Komponent is visible, or whether the realization is visible as well. Test cases designed only on the basis of the specification are said to lead to a functional test, while those based on the realization are said to yield a structural test. Test cases that consider the state of a Komponent or parts of it in addition to functional and/or structural criteria lead to state-based test cases.

The set of test cases used to test an operation or a Komponent is known as a test suite. In the following, we present some example test cases for each of the different approaches.

18.2.1.1 Functional test cases

Table 18.4 presents example functional test cases to check the operation specification of *loanItem()* described in Chapter 4.

Table 18.4 Examples of functional test cases

Test case ID	Testing pre-conditions	Test case input	Expected result
1	(from assumes clause) subject is in the state accountIdentified selector selects exactly one Item	Item is not loanable (antique book).	No new Loan object, loan, has been created. Message ("Item cannot be borrowed")
2	(from assumes clause) subject is in the state accountIdentified selector selects exactly one Item.	Item is loanable, but not currently loanable (i.e. reserved by another user).	No new Loan object, loan, has been created. Message ("Item cannot be borrowed").
3	(from assumes clause) subject is in the state accountIdentified selector selects exactly one Item.	Item is currently loanable, current system date + 4 weeks is a normal working day.	A new Loan object, loan, has been created that relates item and account, and has the attribute values creationDate = today, returnDate = today + 4 weeks and no Extensions = 0 and, loan has been stored.
4	(from assumes clause) subject is in the state accountIdentified selector selects exactly one Item.	Item is currently loanable, current system date + 4 weeks is a holiday.	A new Loan object, loan, has been created that relates item and account, and has the attribute values creationDate = today, returnDate = first working day after today + 4 weeks (lack in specification!) and no Extensions = 0 and, loan has been stored.
5	(from assumes clause) subject is in the state accountIdentified selector selects exactly one Item.	Item is currently loanable as it is actually loaned by the current user (fault in specification – rules!), current system date + 4 weeks is a normal working day.	No new Loan object, loan, has been created. Message ("Item cannot be borrowed"). This test case contradicts the specification as the definition of currently loanable holds true! (instead of false). The item is not loaned by another user as it is loaned by myself!
...			

Table 18.5 Examples of structural test cases

Test case ID	Testing pre-conditions	Test case input	Expected result
1	(from Collaboration Diagram) loanItem(selector: Itemselector)	Path1: ->getItem->IsReserved-><>yes->sendMessage	(from Collaboration Diagram): [reserved] displayMessage()
2	(from Collaboration Diagram) loanItem(selector: Itemselector)	Path2: ->getItem->IsReserved-><no>->isLoaned-><yes>->sendMessage	(from Collaboration Diagram): [loaned] displayMessage()
3	(from Collaboration Diagram) loanItem(selector: Itemselector)	Path3: ->getItem->IsReserved-><no>->isLoaned-><no>->createLoan->Insert	(from Collaboration Diagram): [not loaned and not reserved)] create (today, currentAccount, item); [not loaned and not reserved)] insert (loan)
...			

18.2.1.2 Structural test cases

Table 18.5 presents example structural test cases to check some of the transitions of the collaboration diagram for loanItem in Chapter 5.

18.2.1.3 State-based test cases

Table 18.6 presents example test cases to check the states of the loanItem() operation. The test cases are based on the state diagram in Chapter 4.

18.2.2 Testing process

In this subsection we briefly describe the approach for performing functional, structural, and state-based testing in the context of a KobrA project. Since KobrA supports inheritance in the guise of Komponent specification, we also address the question of how this should be handled in the testing process.

18.2.2.1 Functional, structural, and state-based testing

Testing in the context of a KobrA project is based on the most prominent testing approaches, that is, functional, structural, and state-based testing (as a specific instantiation of functional testing). The definition of functional and state-based test cases is supported by a Komponent's specification, while the definition of structural tests is supported by a Komponent's realization.

Table 18.6 State-based test cases

Test case ID	Testing pre-conditions	Test case input	Expected result
	Starting state	**Event**	
1	unconfigured		neutral
2	neutral	setAccount	accountIdentified
3	accountIdentified	printAccountInformation	accountIdentified
4	accountIdentified	loanItem	accountIdentified
5	accountIdentified	reloanItems	accountIdentified
6	accountIdentified	returnItem	accountIdentified
7	accountIdentified	reserveItem	accountIdentified
8	accountIdentified	closeAccount	neutral

Because of the large combination of input/state combinations, achieving complete functional testing is usually impractical. Common techniques for addressing this problem are the equivalence-class partitioning of input and output parameters and boundary value analysis. An equivalence class is a set of input or output values which have the property that if any value in the set is processed/displayed correctly, all others will be processed/displayed correctly. Equivalence-class partitioning is applied in the following manner.

1 *Analysis*

For each input or output value:

a Identify valid ranges, valid data formats, a valid set of data, a valid number of data, and valid conditions.

b Identify invalid ranges, invalid data formats, data not in the valid set, an invalid number of data values, and valid conditions that are not fulfilled.

2 *Test-case specification*

For the equivalence classes of the input/output values:

a Give each equivalence class a unique identity.

b If there are still valid equivalence classes unmarked, determine a test case that covers them to the greatest extent possible. Mark identities as completed and repeat step a.

c If there are still invalid equivalence classes unmarked, determine a test case that covers them to the greatest extent possible. Mark identities as completed and repeat step b.

d For each test case, select test data in the "middle" of each equivalence class. Identify the expected result.

3 Add the test case specification for boundaries:

   For each equivalence class add test cases so that additional test data are:

   a beneath the boundary

   b above the boundary.

Of course, the use of equivalence classes only takes advantage of input and output parameters and does not consider state information. Hence, state-based testing approaches [Bin99] complement this basic approach.

For the realization, tests are constructed from a structural perspective. In structural testing, the implementation details of a Komponent are a key source of information to set up test cases. Coverage criteria are defined to indicate the completeness of the testing. For example, one may set up test cases to ensure that every branch of an operation is exercised at least once. Other useful coverage criteria are loop coverage, and relational coverage [Bei90], [Mar95].

The basic process for structural testing can be outlined as follows:

1 Identify the coverage criteria for the Komponent/operation.

2 Select test values to ensure full coverage of the criteria.

3 Determine the results for these inputs.

Unfortunately, the principle of information hiding used in an object-oriented context limits the possibilities for structural testing. The reason for this is that operations local to a Komponent are invisible to the outside and thus difficult to test according to structural criteria. Another important aspect to consider is that the testing results depend on the state of the Komponent. This requires that the Komponent be in a well-defined state before test execution starts, which is sometimes difficult to achieve.

Functional and structural testing both focus on correct input/output mappings. However, the specification and realization of a Komponent must both also be faithful to the constraints of the specified states. Hence, the states of a Komponent or operation provide another basis for the design and execution of tests, in addition to structural and functional criteria. To set up test cases and test sequences for state-based testing, one can take advantage of the state diagrams in a Komponent's behavioral model. These should be complete, consistent, and correct before they are used to create a test suite (software inspection described later in this chapter allows the scrutiny of these criteria).

Several strategies have been described for state-based testing (see, for example, N+-Strategy described in [Bin99]). Although full coverage of the published material on this subject is beyond the scope of this chapter, a useful process for state-based testing can be outlined as follows:

1 Identify the transitions in the state-diagram.

2 Determine a test suite that consists of at least one test sequence for each transition.

3 Determine the expected results for the test suite.

Using state-based testing in addition to functional and structural testing adds to the quality of a Komponent.

In summary, the ideal test strategy in KobrA is to use equivalence classes and state-based testing criteria to develop test cases from a Komponent's specification, and to use branch coverage and data-flow-based state-transition testing to develop test cases from a Komponent's realization.

18.2.2.2 Testing and generalization

The presence of generalization complicates functional, structural, and state-based testing. Therefore, when a Komponent is specialized from another Komponent in a KobrA containment tree, we suggest the following testing strategies as an extension to the approaches described above [Mar95]:

1. If the specialized Komponent modifies functionality of the general Komponent, the test cases for the specialized Komponent are inherited from the general Komponent and changed accordingly.
2. If the specialized Komponent adds functionality to the general Komponent, all test cases of the general Komponent can be applied to the specialized Komponent and new test cases need to be added.

These principles address some of the issues involved with inheritance. In both cases, the generalization relationship between the Komponents leads to a similar generalization relationship between the corresponding test cases.

A more detailed explanation of how to approach testing in the context of object-oriented development can be found in [Mar95] and [Bin99].

18.3 Software inspection

Software inspection is the primary quality assurance method in KobrA. The basic idea is that the completion of every major development artifact should be followed by its systematic and rigorous inspection. An inspection involves activities in which qualified personnel determine whether software artifacts are of sufficient quality for subsequent development phases. It is important not to confuse inspections with informal reviews and artifact cross-checks which can be expected to occur from time to time. These are primarily extensions to the development process that help developers keep abreast of consistency rules while they are working on the creation of an artifact. Inspection, on the other hand, is a more formal[51] process that is conducted when the artifact reaches maturity and stability.

Inspections are particularly valuable in KobrA because they can be performed immediately after the completion of a software artifact. In contrast with testing, software inspections can uncover between 50 and 90 percent of the defects in a software artifact [BEFL98] before they are propagated to other dependent

51. In the organizational rather than mathematical sense.

artifacts. This significantly decreases the cost of removing defects before they have had an impact on other artifacts. Inspection is thus the key ingredient in (near) zero defect development methods such as Cleanroom. In the Cleanroom approach, systematic inspection is woven into every step of the development method. This explains why software inspection has become one of the most cost-effective defect-detection techniques since its introduction in 1976 [Fag76].

Software inspections can be performed on all kinds of software artifacts, at all levels of granularity and abstraction. Not surprisingly, therefore, many different variants of the approach have emerged over the years [LD00]. In the following sections we explain the key principles of the KobrA inspection approach, and then describe the main inspection artifacts and process steps.

18.3.1 Organization technique

The first thing to decide when organizing an inspection is to determine precisely *what* is going to be inspected. In the early days of inspection methods, the decision was fairly easy because there were relatively few kinds of logical artifacts (e.g., systems, functions) from which software was assembled, and these were described in terms of relatively few physical artifacts (e.g. requirements specifications, design documents, and source code). In today's modern object-oriented methods, however, there are many more logical concepts making up a software system (e.g. packages, components, classes, objects, or methods), and these are described in terms of many more kinds of physical artifacts (e.g. class diagrams, statechart diagrams, source code etc.). Moreover, these logical and physical artifacts tend to be related in arbitrary ways. A physical artifact (e.g. class diagram) will often describe aspects of many logical artifacts (e.g. classes), and a logical artifact (e.g. method) can appear in many physical artifacts (e.g. class diagram, collaboration diagram, statechart diagram etc.).

There are two basic strategies for handling this situation when selecting one or more artifacts to serve as the subject of a particular inspection. One approach, which has been called the *document-centric* approach [LA99], is to select specific documents to serve as the focus of inspections. For example, an inspection may be organized to identify defects in a class diagram. Another inspection may be organized to identify the defects in a statechart diagram, and so on. This approach has the advantage that selecting artifacts for inspection is relatively easy. However, the disadvantage is that inspectors will often lack crucial pieces of information to judge the quality (e.g. correctness) of an item. Under such circumstances inspectors tend to jump to unwarranted conclusions which ultimately reduce the cost-effectiveness of the inspection. The other strategy, which has been called the *architecture-centric* strategy, is to select the documents to be examined within a particular inspection on the basis of one or more logical elements of the system, rather than physical artifacts. For example, an inspection may be organized to identify defects in a class. Another inspection may be organized to find defects in a package, and so on. The term "architecture-centric" is used to characterize this approach because the collection of logical artifacts is commonly viewed as characterizing a system's architecture [LA99] [Lai00].

In general, the architecture-centric approach leads to more documents needing to be examined, but less information within them. Thus, the architecture-centric and document-centric approaches represent different strategies for grouping the physical documentation into manageable inspection units. The advantage of the architecture-centric approach is that it typically provides inspectors with the optimal set of information needed to judge the presence or absence of defects. However, the disadvantage is that the identification of which parts of which documents should be inspected is much more complicated. Fortunately for KobrA, these disadvantages with the architecture-centric approach are greatly reduced, for two reasons.

First, the principle of locality means that information is already organized in an "architecture-centric" way[52] because every physical artifact (e.g. UML diagram) is focused on the description of one and only one logical artifact. When such a logical artifact is chosen as the subject of an inspection, therefore, it is clear what set of physical artifacts (e.g. UML diagrams) needs to be inspected, and what information within them needs to be examined (all, if not all, of the information). The only caveat to this rule is that when a Komponent is inspected, information about its server Komponents is also required.

Second, KobrA's emphasis on minimal concepts means that there really are only two candidate artifacts that can serve as the subject of an inspection – Komponents and activities. The only other candidate, the simple class, is almost invariably too small.

KobrA, therefore, adopts the architecture-centric approach to inspection organization in which specific inspections focus on the examination of specific Komponents or activities (including operations). Whether a Komponent or an activity is used as the subject of an inspection is a matter of scale. Komponents at or near the root of the tree are often too large to be inspected in one inspection. The operations of such Komponents then usually represent the optimal level of granularity. Nearer the leaves of the tree whole Komponent are more likely to be of the right scale, and inspecting individual operations is likely to involve too much overhead.

18.3.2 Reading technique

Having dealt with the question of what to inspect, the next question is how to perform the inspection. The part of an inspection in which documents and models are actually examined is typically called the *defect detection* or *reading phase*. The question, therefore, is what reading technique should be followed during an inspection to achieve an optimal return on investment.

Early inspection methods used what have come to be known as *ad hoc* reading techniques. As its name implies, this kind of reading technique provides no explicit guidance on how to examine the documents. Not surprisingly, therefore, *ad hoc* techniques soon gave way to *checklist-based* reading techniques [GG97] which provide guidance on what to look for in the form of a list of items to check. This technique naturally represented a big step forward over *ad hoc* techniques, but still left inspectors to work out how to check the specified

52. Where the architecture is represented by the Komponent containment tree.

items. The next major step forward came in the form of *scenario-based* reading techniques [Bas97]. These improve upon checklist-based techniques by also providing guidance on how the required checks should be achieved.

KobrA advocates a scenario-based reading technique for two reasons. First, it generally leads to the best results, and, second, it is complementary to the architecture-centric organization approach. In fact, it would be difficult to follow an architecture-centric inspection approach without scenario-based reading. As mentioned above, one of the main problems with the architecture-centric approach is identifying which segments of which documents need to be examined to check a given logical artifact. With a checklist-based reading technique the inspector is given no help in this respect. However, the key difference between scenario-based and checklist-based reading techniques is precisely the fact that inspectors are given advice on where to look to get the information needed to carry out specific checks. In this way, scenario-based reading complements the architecture-centric inspection approach because inspectors need advice about what fragments of what documents to inspect.

The particular form of scenario-based reading adopted in KobrA is known as *perspective-based* reading [Bas97] [Lai00]. This is an approach that divides the reading work into slices according to critical stakeholders' perspectives on the subject of the inspection. The rationale for this approach is that different inspectors reading from different perspectives tend to catch different classes of defects. Empirical evidence suggests that for a given level of inspection effort, the combination of multiple perspectives typically detect more defects than a single, monolithic reading session from one perspective. The reading technique provides guidance for inspection from a particular perspective in the form of a scenario.

Choosing the right set of perspectives is critical for achieving these benefits, however. The inspection perspectives used in KobrA are:

- **Komponent modeler.** The Komponent modeler is responsible for generating Komponent specification and realization models within a Komponent containment tree.
- **Product-line engineer.** The product line engineer is concerned with instantiating the generic framework to develop applications for specific customers.
- **Implementer.** The implementer is responsible for generating the models and artifacts making a Komponent implementation.
- **Tester.** The tester is responsible for testing Komponents. This involves the definition of test case and analyzing the test results.

It is important to realize that these represent roles rather than actual people. When inspecting large artifacts different people may play these roles, but for smaller artifacts one person may play all the roles.

18.3.3 Inspection artifacts

The key artifacts involved in the inspection activity are the scenarios that guide the inspectors during the reading process, and the defect collection forms that are used to record the results of the inspection.

Table 18.7 Inspection scenarios

	Komponent	Operation
Specification	Komponent modeler Product-line engineer Tester	Komponent modeler Product-line engineer Tester
Realization	Komponent modeler Product-line engineer Tester	Komponent modeler Product-line engineer Tester
Implementation	Implementer Tester	Implementer Tester

18.3.3.1 Inspection scenarios

When the three main forms of description (specification, realization and implementation) and the two main logical artifacts (Komponents, operations) are combined with the four perspectives identified above, there are 16 meaningful inspection scenarios in KobrA. These are summarized in Table 18.7. The rows in this table identify the description levels and the columns identify the possible logical artifacts.[53] The cell entries list the perspectives (or roles) that make sense for each combination of description level and logical artifact.

Assume you are inspecting a Komponent specification from the perspective of a Komponent modeler. The main concern of a Komponent modeler is to ensure the consistency of the various descriptions of the Komponent in the specification models (i.e., the structural, behavioral, and functional models). High quality therefore corresponds to conformance to the relevant well-formedness rules. The development products that are of relevance are all primary specification models plus the class diagram of the encapsulating realization and the collaboration diagrams of client Komponents.

Consider the class diagram of the subject Komponent:

1 Do any classes in the diagram without the <<Komponent>> stereotype have any operations listed? If so list these as defects.

Consider now also the object diagram of the subject Komponent:

2 Is every object in the object diagram an instance of a class in the class diagram? If not list those that are not as defects.

Consider now also the operation specifications:

1 Does the subject Komponent (i.e. the class marked with the stereotype <<subject>>) have any operations? If so, go through each operation specification in turn and write down its name. Is every operation listed in the operation compartment of the subject Komponent, and vice versa? If not list all the discrepancies as defects.

2 Go through each operation specification and write down every message appearing in the sends clause as well as the class of the instance to which the message is sent. Turn now to the class diagram. Is there an operation corresponding to each message in the operation compartment of each acquired Komponents (i.e. classes marked with the stereotype <<Komponent>> and connected to the subject Komponent with an association stereotype <<acquired>>) and vice versa. If not list the discrepancies as defects.

Figure 18.1 Komponent specification inspection scenario from the Komponent modeler's perspective

53. In the sense of architecture-centric inspection explained in Section 18.3.1.

Figure 18.2
Defect detection form

Inspector: ____________________
Defect detection effort (minutes): ____________________

The following defects where found during the defect detection phase

No.	Document	Location of the defect	Defect description	Defect class and severity
1.				
2.				
3.				
4.				
5.				
6.				

Each scenario identified in Table 18.7 has an associated textual description, which guides the inspector through the enactment of that inspection scenario. Figures 18.1 shows the first part of the textual description for the top left-hand scenario in Table 18.7, namely, the scenario for inspecting the specification of a Komponent specification from the perspective of a Komponent modeler.

Based on the example in Figure 18.1, it is straightforward to set up the scenarios for the other perspectives identified in Table 18.7. A more precise scenario development process is outlined in [Lai00].

18.3.3.2 Defect collection forms

During the inspection, participants need to fill out forms of the kind shown in Figure 18.2 and 18.3 to collect the relevant measurement data.

Figures 18.4 shows an example of the defect detection data that was generated by an inspection of the specification of the LoanManager Komponent by combining the results of the various perspectives.

18.3.4 Inspection process

An inspection in KobrA consists of four essential tasks. These are inspection planning, defect detection, defect collection, and correction:

1 Inspection planning (inspection organizer)

a The logical artifact for inspection is selected and the required set of physical artifacts (e.g. models) is identified. The appropriate set of inspection scenarios is also determined. The entry criteria for the set of physical artifacts (e.g. models) are also checked.

b The inspection participants are selected (i.e. the inspection moderator and the inspectors) and their assignment to roles and perspectives is determined. The meeting for defect collection is scheduled and the inspection materials are sent to the inspection participants.

2 Defect detection (inspector)

a The relevant reading scenarios are applied to the subject artifact (i.e., perform the activities and answer the stated questions). Detected candidate defects are documented on individual defect detection forms.

b After completion of individual reading, individual defect detection forms are completed (e.g., consumed effort is added).

Artifact information

Version	Name	Location (node, directory, filename) if not administrated in version control system	Remark

Inspection dates

Inspection Information

Total defect detection effort: ________

Date of the meeting: ________

Place of the meeting: ________

Moderator name: ______________

Total meeting effort ________

Number of meeting participants ________

Inspectors

Name	Department	Preparation effort + remark

The following defects were accepted or found in the defect detection phase

No.	Document	Location of the defect	Defect Description	Defect severity	Found in meeting (Yes/No)
1.					
2.					
...					

Figure 18.3 Defect collection form

Figure 18.4
Completed defect detection form

The following defects were accepted or found in the defect detection phase

No.	Document	Location of the defect	Defect description	Defect severity	Found in meeting (Yes/No)
1	Context Specification	Class diagram	No possibility to store the reference of the Message handler Komponent.	High	No
2	Context Specification	State diagram	The state diagram indicates that the loan item operation can be executed even in the state "empty." This is inconsistent with the assumes clause.	Medium	No
...					

3 Defect collection (moderator, inspector, author)

a In a group meeting, the validity of each candidate defect is collectively judged, the accepted defects are collated in the defect collection form and each defect is assigned a severity level of high, medium, or low.

b The moderator should ensure that the meeting is performed in a positive atmosphere. Inspectors should use the meeting to look for defects not detected throughout the defect detection activity and get to know possible sources of defects. Authors should try to understand the defects and learn to avoid them in the future.

c Once all defects have been discussed, it is determined whether the artifact subsequently needs to be reinspected.

d The defect collection form is completed (e.g. consumed effort is added).

4 Defect correction (author)

a The quality policy is applied to determine how the defects should be corrected.

The literature on software inspection identifies two additional tasks: an overview meeting after the planning activity, and a follow-up activity after correcting defects. More details about these optional phases can be found in [Fag76] or [GG93]. A more detailed description of inspection methodological variations can be found in [LD00].

19 Quality modeling

Another approach that can provide insight into the quality of development artifacts, before executable incarnations have been generated, is quality modeling based on measurement of structural properties. The structural properties of a piece of software (e.g. the coupling of one class to another class, or the internal complexity of a class) can serve as indicators of important software quality attributes. For instance, highly complex classes are more likely to contain faults, and therefore may negatively impact on system reliability. A class that is strongly coupled to other classes is more difficult to modify, which may negatively affect the maintainability of the system.

Quality modeling based on the structural properties of UML artifacts is particularly valuable because, like inspections, it can provide early feedback on the quality of Komponents. As early as the completion of the context realization activity, the resulting UML models can be analyzed to identify potential quality problems and help focus effort spent on the other quality activities, in particular inspections. The KobrA method is especially sensitive to quality problems at the top of the Komponent tree, because errors there can affect much of the later development work (i.e. the sub-Komponents). Other advantages of quality modeling based on the measurement of structural properties are that the measurement process is objective, does not rely on subjective judgment, and can be automated. Once a suitable tool is in place, measurement of even large designs can be accomplished fast and cheaply.

In this chapter we describe how certain structural properties such as coupling, complexity, and size can be measured for KobrA artifacts, (i.e. how these quality measures are defined), and then explain how these quality measures can be used for decision making. This is achieved by means of a quality model that defines how structural measures are related to externally meaningful quality attributes. The key ideas behind quality models are described in the following subsection. Sections 19.2 and 19.3 then explain how these can be put to use in KobrA to help improve the quality of the final product.

19.1 What is a quality model?

A fundamental distinction for product quality attributes is between internal and external attributes [FP96]. An *internal attribute* of a product can be measured in terms of the product itself. Examples of internal product attributes are the structural properties coupling, cohesion, complexity, or size. All information that is needed to quantify the internal attribute is available from a representation of the product. Therefore, these attributes are measurable during and after creation of the product.

External attributes of products can only be measured with respect to how the product relates to its environment. External attributes are properties or features of the product that are externally visible (hence the name). Consider, for example, the ISO9126 quality attributes reliability and maintainability. The former is visible to the user of the system if the system does not perform as expected; the latter is visible to the development organization, which finds it increasingly costly to adapt the system to changing requirements. To measure external attributes *directly* requires additional information besides the product itself. For instance, maintainability can be measured directly only when the product undergoes maintenance, for example, in terms of the person hours spent to modify the product, maybe for a specified period of time, or for a specified functional enhancement. Reliability can be measured in terms of the mean time to failure of the operational product. Thus, external attributes can be measured directly only some time after the product has been created.

Internal attributes, on the other hand, describe no externally visible quality of a product, and therefore have no inherent meaning in themselves. For example, during the operation of a system, the users will not notice whether the Komponents of the system are loosely coupled (an internal quality of the system).

Since internal attributes are not inherently meaningful, why should we care about them at all? Because we assume that internal attributes of a product have a causal impact on its external attributes. For instance, systems with loosely coupled Komponents (internal attribute = coupling) are hypothesized to be more maintainable (impacted external attribute maintainability) than highly coupled Komponents. Komponents of high structural complexity are hypothesized to be more fault-prone (external attribute = reliability) than low-complexity Komponents. In other words, measures of internal attributes are indicators of external attributes. Measuring internal attributes is a means of *indirectly* measuring external attributes.

Making the link from measurement of internal attributes of an artifact to what the implications are for the external attributes of the artifact is precisely the purpose of a quality model. It provides an interpretation of the readily available internal attribute measurement data in terms of the external quality the artifact is later likely to exhibit.

Note that the quality model only makes a prediction or statement about the external quality of an artifact, it does not define what actions should be taken in response to certain results. Deciding what to do based on the output of a quality model is the job of the quality policy.

19.2 Measurement of structural properties

The internal quality attributes of relevance in KobrA are structural properties of the UML artifacts making up Komponent specifications and realizations. The specific structural properties of interest are coupling, complexity, and size. These attributes are well understood, and have been shown empirically to be indicators of various external qualities in OO design (see, e.g., [BWDP00] [BWLI99] [CDK98] [NQ98] [LH93]).

To support the definition or selection of coupling, complexity, and size measures that are likely indicators of the selected qualities at hand, we provide a "construction plan" for coupling, complexity, and size measurement.

19.2.1 Size measurement

Size measures are counts of elements in the UML design (the more elements, the larger the design) and are good candidates for developing cost-estimation models for designing, implementing, inspecting, and testing a Komponent. Such estimates can be used as input for effort-planning purposes and the allocation of personnel.

The definition of a size measure involves the resolution of three decisions which should be driven by the overall measurement goal:

1. the entities of which the size is to be measured (e.g. the size of a Komponent, the size of a class)
2. the elements of which these entities are composed, and which contribute to its size
3. how to quantify the size of the entity in terms of the size or number of its constituting elements.

Table 19.1 summarizes the possible choices for decisions (1) and (2) in the above list. The left column states the entities for which size can be measured, the right column indicates what elements make up each entity, and thus contribute to its size.

Table 19.1 Building blocks for size measures

Entity	Countable elements within
Komponent	Sub-Komponents Classes Objects States
Classes or Objects	Operations Attributes
Operation	Parameters passed to the method

In KobrA, Komponents and classes are likely to be the most useful choice of entity to measure the size of. Investigating size at a finer level also requires investigating external quality at that level, which may be impractical (for instance, collecting and planning effort on a per-operation basis).

When defining size measures, we do not recommend counting entities of different types in one measure (e.g. defining the size of a class as the number of its operations plus the number of its attributes). This would be implicitly assuming that the entities have equal impact on the quality (e.g. an operation adds, on average, the same amount to the required development effort as an attribute) – something that may not necessarily be true and needs to be justified explicitly. Rather, we suggest measuring the entities separately, and using the separate measures as building blocks for quality models.

19.2.1.1 Size quantification

For the quantification of the size of the selected type of entity there are two options. The size of a "higher-level" entity can be measured in terms of:

1. **the number** of a certain type of countable elements within it. Examples are:
   - the size of a Komponent measured in terms of the number of classes within the Komponent
   - the size of a class measured in terms of the number of operations it contains.
2. **the sum of the size** of certain countable elements, taken over all such elements. Examples are:
   - the size of a Komponent measured in terms of the number of operations of the classes it contains
   - the size of a class measured in terms of the number of parameters passed to its methods.

In general, the first option results in rougher, more coarse-grained size measurement than the second option. Which option to choose largely depends on how stable the number of counted elements is at the time in the life cycle when the measurement is performed.

As an example, consider the *LoanManager* Komponent. Based on its specification class diagram (Figure 4.2), the size of the LoanManager Komponent can be measured as follows:

- Komponents (decision 1) consist of classes (decision 2) and we just count their number (decision 3).
  Resulting size measurement: 8 classes (note that acquired classes/ Komponents external to *LoanManager* are not counted)
- Komponents (decision 1) consist of classes (decision 2), and we sum up their size (decision 3). As a measure of the size of classes (decision 1) we take the number (decision 3) of the operations they contain (decision 2). Resulting size measurement for Komponent *LoanManager*: 10 operations (since only operations for one class, *LoanManager* itself, have been identified at this stage).

- Similarly, we can measure the size of the Komponent in terms of the number of attributes of its classes.
  Resulting size measurement: 8 attributes.

Considered in isolation, these numbers give us little, if any, useful information. However, when compared to one another they can help allocate resources. For example, measuring the size of all Komponents in the library system containment tree allows us to make relative statements about the size of these Komponents (e.g., "*LoanManager* is twice as large as *MessageHandler*, and three times as large as the *ReportWriter* Komponent"). This could be used as input to the allocation of people to the realization and implementation of these Komponents.

More importantly, if empirical data is available on the cost of implementing a class, an operation or an attribute at this particular level of specification in the KobrA tree, the size measures can be used to predict the effort needed to realize, implement, verify, and test the Komponent, and thus support project planning and management.

19.2.2 Coupling measurement

Coupling is the degree to which various entities in the software are connected. These connections cause dependencies between the entities that have an impact on various external system quality attributes. For instance:

- the system maintainability (a modification of an entity may require modifications to its connected entities)
- testability (a fault in one entity may cause a failure in a completely different, connected entity)
- fault-proneness (dependencies usually define a client–server relationship, the developer needs to know exactly how to use the service).

Therefore, a common design principle is to minimize coupling. KobrA supports the enforcement of this principle by suggesting ways to quantify coupling, so that excessive violations of the principle can be spotted, and, if necessary, removed.

Defining a coupling measure involves five decisions (this is a variation of the coupling measurement framework presented in [BDW99]):

1. relevant mechanism by which the entities are coupled
2. the direction of coupling (import or export coupling)
3. the granularity of the measure
4. direct and/or indirect connections to count (yes/no decision)
5. special relationships that should exist between connected entities.

These are explained in more detail below. For each decision, we have a number of options from which to select. This selection should be driven by the measurement goal we have in mind (the development phase at which the measurement is applied, and the external qualities we wish to control).

19.2.2.1 *Types of coupling connections*

Choosing a type of connection implies choosing the mechanism that constitutes coupling between two entities. Table 19.2 below summarizes the possible types of connectors, that is, links between a client and a server item[54] (operation, class, Komponent).

Table 19.2 Building blocks for coupling measures

Item 1 (Client)	Item 2 (Server)	Connector	Found in ...
Class of Komponent A	Class of Komponent B	Association	Class and object diagrams, code
Class of Komponent A	Class of Komponent B	Aggregation	Class and object diagrams, code
Operation O of class C	Operation P of class	O invokes P	Activity and interaction diagrams, operation specification, code
Operation O of class C	Attribute A of class D	O references A	Operation specification, code
Operation O of class C	Class D	O receives by parameter, or returns, and instance of type D	Class and interaction diagrams, operation specification, code

Different types of connections have different strengths. For instance, according to information hiding principles, directly using an attribute of another class is worse (i.e. stronger) than invoking an operation of another class. Therefore, we recommend to avoid incorporating more than one type of connection into a single measure. Rather, separate measures for each type of connection should be used.

19.2.2.2 *Direction of coupling*

If the connector is directed, we can decide whether to count import or export coupling:

- Import coupling analyzes the entities in their role as clients (users) of other attributes, methods, or classes.
- Export coupling analyzes the entities in their role as servers to other attributes, methods, or classes.

54. Most, but not all connectors are directed and thus impose client and server roles on the connected items. For undirected connectors, the connected items are just two equal peers.

High import coupling indicates that the Komponent depends strongly on other Komponents. Import coupling may therefore be relevant in conjunction with the following external attributes:

- **Understandability.** To understand the Komponent, we must know about the services it uses.
- **Fault-proneness.** For similar reasons as understandability – if we incorrectly use an external service because we misunderstand it, we are likely to introduce errors.
- **Maintainability.** Low understandability and high fault-proneness result in low maintainability.

High export coupling means that the Komponent is used a lot by other Komponents. Export coupling may be relevant in conjunction with the following external attributes:

- **Criticality.** Any defects in an entity with high export coupling are more likely to propagate to other parts of the system. Such defects are more difficult to isolate. In that respect, entities with high export coupling are particularly critical. An export coupling measure could therefore be used to select entities that should undergo special (effective but costly) verification or validation processes.
- **Testability.** A class with high export coupling can be difficult to test. If defects need to propagate to other parts of the system to cause failures there, they may not be detected when testing the class in isolation.

19.2.2.3 Granularity of the measure

The granularity of a measure is the level of detail at which information is gathered, and is determined by two factors, the domain of the measure and how the connections are counted:

- **The domain of the measure.** This relates to the choice of entity for which the coupling is to be determined. Possible entities are
  - Komponents
  - classes
  - objects
  - operations.

  The choice here is influenced by the selections made for decisions 1 and 2 – the finest possible granularity of the measure is either the client or server item of the chosen type of connection. For example, if we decide to measure import coupling and invocations of operations, the finest possible granularity of the measure is the operation.

  However, for some measurement goals we may want to choose a coarser domain for the measure than the finest possible. For instance, if the measurement goal is to characterize understandability, reusability etc. of a class, a measure defined at the class level is appropriate. This can be accomplished by taking into account all connections starting from or ending in the operations or attributes of the class.

- **The counting of connections.** This addresses the question of whether we want to count multiple connections between identical items separately or not. For example, if an operation O invokes the same operation Q several times (e.g. in different collaboration diagrams), should we count each invocation individually, or not? This will depend on how stable information on the precise number of invocations is at the time the measurement is to be taken. If this number is likely to change when the Komponent is refined in the future, then there is no point in counting the precise number of invocations. Just distinguishing the fact that an invocations occurs but not how often would then be more honest.

  For some external quality attributes, we may not want to count multiple connections individually. For example, assume we want to characterize understandability of a fully implemented Komponent. The hypothesis is that the more operations it invokes from classes of external Komponents, the lower the understandability. To test this hypothesis, we only count the number of operations used by the Komponent, even though its source code is available and the precise number of invocations of each operation is fixed. According to the hypothesis, it is the number of different operations that impacts understandability, not how often each operation is used.

19.2.2.4 Direct and indirect connections

This is a yes/no decision as to whether the coupling measure should only count direct connections between elements, or also indirect ones. For example, if an operation O invokes an operation P, which in turn invokes an operation Q, we can say that O indirectly invokes Q. Operations O and Q are indirectly connected.

Indirect connections can be relevant when estimating the effort for run-time activities such as testing and debugging, or to estimate the impact of a modification to a Komponent on the system (ripple effects).

Direct connections are sufficient for the analysis of understandability – to understand a Komponent, we need to know the functionality of the services directly used by it. We do not need to know how these services are implemented, and therefore, what other services these services need (at least, this is true if the services are well documented).

19.2.2.5 Special relationships

In some cases, we may only want to count coupling between items that have a special relationship, for example:

- counting only coupling between classes with an inheritance relationship (or no such relationship) between them

- stability of connected elements (e.g. count only coupling connections to Komponents that are generic).

19.2.2.6 Examples of coupling measures

Table 19.3 defines some coupling measures (one per row) by specifying for each decision (all but the first column) how the decision is addressed for the measures.

Table 19.3 Example coupling measures

Measure	Dec. 1: Type of connection	Dec. 2: Import or export coupling	Dec. 3a: Granularity	Dec. 3b: Multiple connections	Dec. 4: Direct/ indirect	Dec. 5: Special relationships
1	Operation invocation	Import coupling	Komponent	Count each invocation individually	Direct connections only	No special relationships
2	Operation invocation	Import coupling	Komponent	Do not count multiple invocations of same operation	Direct connections only	No special relationships
3	Operation invocation	Export coupling	Class	Count each invocation individually	Direct connections only	No special relationships
4	Operation invocation	Import coupling	Operation	Count each invocation individually	Count also indirect connections	No special relationships
5	Association	n.a. (connector not directed)	Class	Count each association individually	Direct connections only	No special relationships

Based on the collaboration diagrams of the operations *loanItem()*, *reserveItem()*, of the *LoanManager* (Chapter 5), measures 1 to 3 have the following values for Komponent *LoanManager*:

- **measure 1** – 6 invocations of operations of other Komponents (2x *getItem()*, 1x *isReserved()*, 1x *reserveItem()*, 2x *displayMessage()*). Note that invocations to the *LoanStore* and *Loan* objects are not counted as we are here only interested in operation calls that leave the boundary of the subject Komponent *LoanManager*.

- **measure 2** – 4 different operations invoked (*getItem(), isReserved(), reserveItem(), displayMessage()*). Multiple invocations of the same operation are ignored here.
- **measure 3** – 2 invocations of operations of class *LoanStore* (1x *isLoaned()* and 1x *insert()*). Our focus here is coupling of an individual class, and we count operation invocations from outside the class (not necessarily outside its Komponent).

Based on the activity diagram for operation *loanItem()* of the *LoanManager*, measure 4 counts 6 operations that can be invoked directly or indirectly in response to an invocation of operation *loanItem()*.

Similarly, based on the realization class diagram of the *LoanManager* Komponent, measure 5 calculates the following values for the classes of the *LoanManager*:

Class	Value
Account	3
Item	3
ItemSelector	1
Loan	1
LoanStore	2
LoanManager	6
Reservation	1
ReservationList	2
Selector	0

Note that this assesses coupling at the class level and we do not distinguish whether the coupling link leaves the boundary of the Komponent.

Again, considered in isolation the meaningfulness of these measures is not obvious. However, using them in combination with one of the quality models presented in Section 19.3 allows useful interpretations to be attached to them, for example, calculate predicted fault-proneness after implementation of the classes, etc.

19.2.3 Complexity measurement

Whereas coupling counts connections between different entities (e.g. connections between two distinct Komponents), complexity counts connections between elements within an entity (connections between elements within a Komponent).

The decisions to make when defining complexity measures are therefore similar to the coupling measurement decisions:

1 Relevant mechanism by which the entities are connected.

The coupling connectors in Table 19.2 between different entities can also occur within the same entity (e.g., operation invocations within a Komponent), and thus are also relevant for the definition of complexity measures. In addition, state transitions and the relationship between states and operations contribute to the complexity of a Komponent. Table 19.4 summarizes the relevant intra-artifact connections for complexity measures.

Table 19.4 Building blocks for complexity measures

Client Item	Server Item	Connector
Class C of Komponent A	Class D of Komponent A	Association
Class C of Komponent A	Class D of Komponent A	Aggregation
Operation O of class/object C	Operation P of class/object C	O invokes P
State S in Komponent A	State T in Komponent A	State Transition from S to T
Operation O of class C	State S	O features state S as part of its assumes or results clause

2 The granularity of the measure:

  a domain of the measure (depending on the intended application of the measurement, we can measure the complexity of Komponents, classes, objects, operations).

  b counting multiple connections between the same element or not (the same guidelines as for coupling measures concerning the stability of the connections apply).

3 Direct and/or indirect connections to count (yes/no decision).

4 Special relationships that should exist between connected entities.

As an example of a complexity measure for a Komponent (decision 2 – domain of the measure), we can count the number of associations (decision 1) between classes of the Komponent, counting multiple associations between two classes separately (decision 2 – multiple connections), without regard for special relationships. Applying this measure to the realization class diagram of the *LoanManager* Komponent, there are 7 associations between classes of the Komponent.

Another example of a complexity measure for a Komponent (decision 2 – domain of the measure) is the count of the number of state transitions between states (decision 1). Multiple transitions between two states (in the same direction) are not possible, so we have no option for decision 2b, and decision 4 is also not applicable (this could become relevant, for example, when certain state transitions are possible only for certain variants of the Komponent). Applied to the statechart diagram of the *LoanManager* specification, there are two state transitions.

19.3 Example quality models

The previous subsection identified several internal quality attributes that can be measured directly from a Komponent's UML diagrams. However, except when compared to one another to get a feel for the relative sizes of Komponents, these internal attributes are naturally of little direct value. In this section we show how the application of different quality models allows the above measures to be used to assess certain quality attributes of a system. In particular, we show how such models can be used to identify potential risk areas/trouble spots in the system design. Such risk areas can then be made the focus of inspection and testing activities, thus increasing the efficiency of these activities.

19.3.1 Prediction models

Predictive models aim to predict the future quality of the system from internal system quality attributes that are measurable at present. This is achieved by empirically exploring the relationship between internal and external quality from systems developed in the past, and then applying the findings to new systems.

In the following we describe how to build and use a prediction model for class fault-proneness from the structural properties of the class. Figure 19.1 depicts the steps involved in building the quality model.

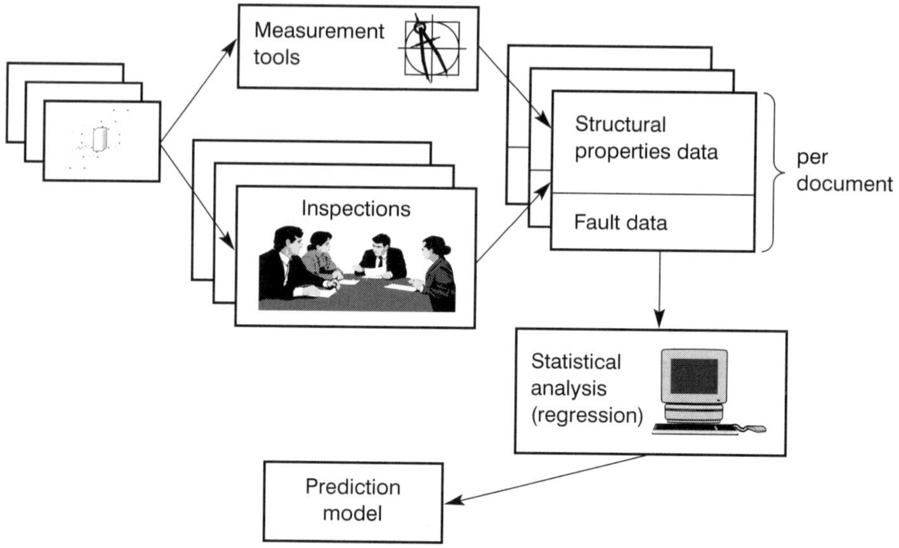

Figure 19.1
Building a prediction model

The starting point is a set of documents (e.g. Komponent specifications, realizations, implementations) that have been created in the past. A measurement tool is then applied to automatically obtain the structural properties from these documents. Inspections are also performed on the documents and the faults recorded.

A set of documents is now available with well-known structural properties on the one hand, and fault data from inspections on the other. A statistical analysis (classi-

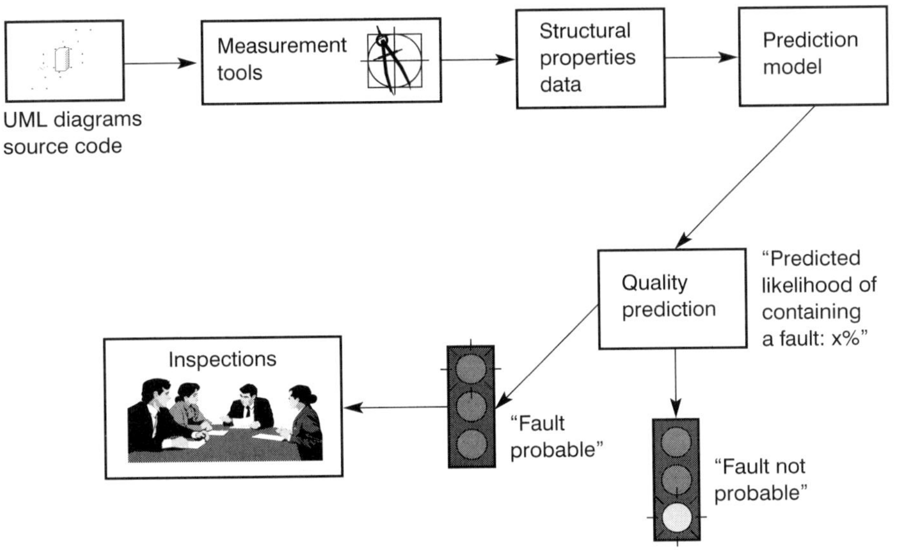

Figure 19.2 Using the prediction model

fication or regression analysis) is then performed to identify relationships between fault data and structural properties. The result of this analysis is a prediction model in the form of a regression equation that computes a predicted fault-proneness (or predicted number of faults) from the structural properties of a document.

Once such a prediction model is in place, we can use it to make predictions for new documents. This procedure is depicted in Figure 19.2.

The starting point is a new document (Komponent specification, realization, implementation), to which we again apply our measurement tool to obtain data on the structural properties (e.g. data on a number of coupling and complexity measures). This is an automatable process that is cheaply and quickly performed.

This data is then fed into the prediction model. This model uses what it learned about the relation between coupling, complexity, and faults, to calculate, for instance, a predicted fault-proneness for the document, that is, the probability that a fault is found in the document upon inspection.[55]

The output of the prediction model is useful for decision-making as defined by the quality policy. For instance, if the predicted fault-proneness is low, say, below 25 per cent, it is not necessary to inspect the document but possible to proceed immediately to the next activity (realization, next lower-level specification, implementation, or test). However, if the predicted fault-proneness is above 25 per cent, this can be the trigger for the document to be inspected. This is determined by the quality policy. Thus, inspection effort can be focussed on the system parts that are more likely to contain faults.

Prediction models for other system qualities can be built in the same way. Examples include models to predict the effort needed to implement and test a

55. What kind of prediction the model makes depends on the regression analysis used, which in turn depends on the type of (fault) data which is collected.

Komponent from the size of the Komponent specification, or the effort needed to maintain a Komponent from its coupling and/or complexity.

The advantage of using such a prediction model is that it provides a mapping from non-interpretable internal quality data (coupling = 5, ...) to the easily interpreted external quality data (fault-proneness: 20 per cent). The result of the prediction model is an absolute, quantitative statement about the external quality of the system expressed in the same unit in which the external quality is measured, such as "the predicted effort required to implement Komponent X is 105 person hours," or "the predicted number of faults of Komponent Y is 4," and so on. These are very concrete statements that can easily be interpreted by developers.

However, a disadvantage of the approach is that the data requirements are high, and the activities involved in building and using a prediction model require expertise in statistics.

It is important to understand the limitations of such prediction models. First and foremost, the predictions of such models are of a statistical nature. As such, they come attached with a prediction error. This is because the external attribute we are investigating (e.g. fault-proneness) is usually influenced by many factors and the prediction model only captures those that are easily obtainable. For instance, class fault-proneness is likely to be affected not only by the structural complexity of the class, but also by the experience of the person developing the document, the available time (e.g. if there was time pressure), and many more factors.[56] Since these factors cannot all be taken into account, the model explains only a part of the variation of the external quality. As a consequence, it is clear that such models cannot be used to certify the reliability of a system, for instance.

Despite these limitations, however, empirical evidence shows that highly accurate prediction models can be built from structural properties [BWDP00] [BWLI99] [CDK98] [LH93] [NQ98], and these are good at highlighting potential trouble spots in the system and supporting project planning and steering.

19.3.2 Quality benchmarks

The idea of this approach is to compare the structural properties of the Komponent to be evaluated to those of previous systems that are "known to be good" and have stood the test of time.

To this end, the measurement values for selected size, coupling, and complexity measures obtained from previously developed, successful, Komponents are stored in a database. If a new or modified Komponent is to be evaluated, the same measurements are applied to this Komponent. Then, the *distribution* of each size, coupling, and complexity measure is compared to the distribution of these measures stored in the database.

As an example, Figure 19.3 shows a (fictitious) distribution of the number of operations invoked from within a Komponent. The vertical axis indicates, for each value on the horizontal axis (number of invoked operations), the percentage of Komponents that have that particular value. In the example, the distribution of the new Komponent follows closely the distribution of

56. It is possible to quantify such "soft" factors and use them along with the measures of structural properties as additional variables in the prediction model.

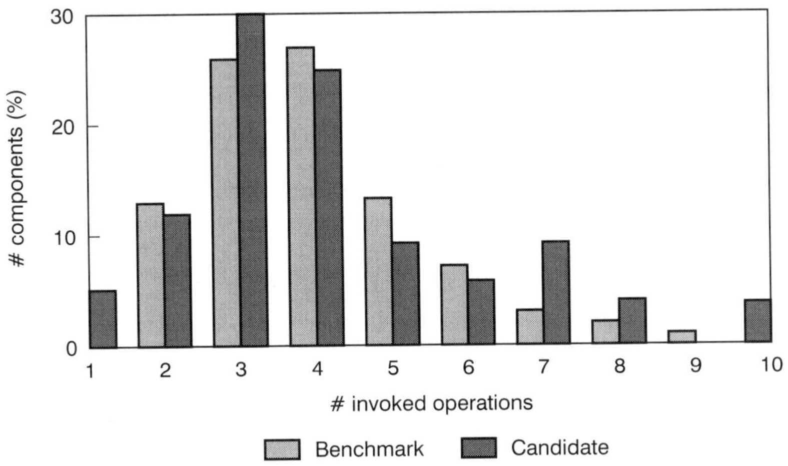

Figure 19.3
Quality benchmark example: number of invoked operations

the benchmark, except for values 7 and 8, which occur more frequently. Such deviations from expected distributions pinpoint potential risk areas in the system. These areas could then be inspected to verify whether such a deviation is justified, or if a redesign of the system should be recommended.

The example also illustrates why looking at distributions of measures provides more information than simply defining thresholds for each measure that are not to be exceeded. Though the number of invoked operations for the new system is not exceedingly high (the benchmark suggests 10 as an upper value), there are more than an average number of Komponents with 7 and 8 invoked operations. This may point to a problem.

Unlike the previous approach of building predictive models, this approach does not require external quality data. No absolute statement about the quality of the Komponent in terms of the external system quality can be made. A database of past "known to be good" Komponents has to be acquired first before this approach can be used in an operational manner.

Such a quality benchmark can then be used to:

- Provide some guidance for design decisions or design documentation decisions. As a simple example, a quality benchmark for the size of class, object, or interaction diagram provides objective criteria as to whether such a diagram is too big and should be split.
- Define exit criteria for a design before it can enter the next phase. For instance, deviations from the quality benchmark must either be explicitly justified, or else the design reworked.
- Identify potential trouble spots in the system that should be the focus of inspections and testing activities.

19.3.3 Simple ranking models

This approach uses structural property measures to select potentially critical areas in a design that should be inspected. Although it does not require the measurement of external quality data, it is important that the measures used are indicators of qualities of interest (see Section 19.3.1 for how to do this).

The steps are as follows:

1. Apply the design measurement to the Komponents.
2. Independently for each measure, rank the Komponents in decreasing order, and assign the rank to each Komponent. The Komponent with the lowest measurement is assigned rank 1, the Komponent with the next lowest measurement is assigned rank 2, and so on. The Komponent with the highest measurement value is assigned rank n, given that there are n classes.
3. For each Komponent, take the average rank of that Komponent for all measures.
4. Sort the Komponents in decreasing order of their average rank.

An example application of the procedure is shown in Table 19.5. The first column indicates the name of a Komponent, the next three columns show the ranks of each Komponent for three selected measures. The last column is the average of these ranks, the rows are sorted in decreasing order of this average.

The Komponents near the top of the table are then potentially critical, because they display less desirable structural properties than Komponents near the bottom. We could thus select Komponents starting from the top of the table for inspections or testing, until the allocated resources for inspection/testing are depleted.

The advantage of this approach is that it is immediately applicable – neither external quality data nor a database of past "known to be good" systems is required. Its disadvantage is that no absolute statement about the quality of the software is made. Also, we cannot make a statement about the quality of an individual entity (Komponent or class) considered in isolation. Only a relative ranking of an entire set of classes/Komponents is possible. Also, in creating such a ranking, we are actually losing information: the measures used are defined on an interval or even ratio scale, but we use them only on an ordinal scale (see [FP96] for a discussion of measurement scales in a software engineering context). This approach is therefore less powerful than the previous two. For this approach to be useful it must be ensured that the used measures indeed are related to the external quality in mind (e.g. to select classes for testing/inspection, the measures used should be reliable indicators of class fault-proneness, something which can only be demonstrated by empirical investigation).

Table 19.5 Example of a simple ranking model

Komponent	Rank measure 1	Rank measure 2	Rank measure 3	Average rank
Komponent2	24	3.5	12	14,67
Komponent18	9	16	14	13
Komponent7	17	3.5	7	10,67
...	...			

Applying KobrA 20

The cost effective, and thus successful, development of software involves establishing a careful balance between the expected benefits of an activity and the likely costs. As a result, many activities that are beneficial from a theoretical perspective (given unlimited resources) may not make sense for a specific organization under specific circumstances (e.g. limited resources or market pressures). General purpose methods such as KobrA must, by definition, describe activities and products in a sufficiently general way to cover the majority of situations in which the method is potentially applied.

A major question that therefore arises in the use of a general-purpose method such as KobrA is the customization of the method to the particular needs of a specific organization and the transfer of the customized method into that organization. Customization addresses the technical and methodical aspects of how to adapt the KobrA method to make it fit to a specific organization. Transfer, on the other hand, is concerned with the organizational and management aspects of preparing an organization to use the customized KobrA method to its best advantage. This includes training project members, project management, and tool support.

This chapter describes how to instantiate and use KobrA in the context of an industrial software development project. The first subsection reviews some of the general characteristics of KobrA and observes how these help or hinder the migration to KobrA. This is followed by two separate subsections discussing the problems of customization and transfer, respectively.

20.1 General characteristics of KobrA

The KobrA method is built upon a number of proven software engineering technologies, especially components, object-orientation, frameworks, patterns, product lines, Cleanroom, quality modeling, testing, and inspections. When used together, these offer a powerful vehicle for the development, application, and maintenance of software according to engineering principles. However, few, if any, organizations can move to a new development approach overnight. KobrA therefore packages these ingredients in a modular, flexible, and scalable way to support the incremental introduction of the method.

The three main characteristics of KobrA responsible for its inherent modularity, flexibility, and scalability are its strict separation of concerns, its focussed prescriptiveness, and its uniform application of basic concepts.

- **Separation of concerns.** KobrA provides a clean and explicit separation of the three most fundamental concerns in component-based product line engineering:
  - **containment** – the containment and nesting of components and component instances
  - **genericity** – the usage of generic components to cover members of a product line
  - **abstraction** – the description of components at different levels of abstraction to provide design level and implementation level representations. This dimension can be further subdivided into concerns for refinement and translation.

  As illustrated in Figure 2.1, these dimensions are completely orthogonal to one another, both with respect to products and activities. This enables a clear separation of the dimensions, thereby allowing an incremental introduction of the method. An organization can first try out the component engineering aspect of the method, (as described in Part 2) while still using large portions of their previous implementation approach. Later, once they are comfortable with the component-oriented aspects of the method, the organization can start to introduce product line considerations (i.e., moving from application engineering to framework engineering along the genericity dimension) by gradually adding variable features to the components (as described in Part 3). At any point during this process, the organization can also change their implementation approach to that recommended by KobrA (or any other for that matter), which can include the introduction of physical components at the implementation level.

  Even the introduction of the component paradigm can take place in steps. A company can first introduce KobrA's modeling approach at the system level (as a requirements engineering approach) but develop the internals of the system using their existing methods. Once familiarity with the modeling techniques has been attained, the internals of the system can be componentized and modeled according to the KobrA approach. Initially, however, the containment tree only needs to have two levels (the root and its children) corresponding to a policy of global visibility for all components (see Section 10.3.1). Only later need the concept of private composition/containment and deeper trees be introduced. As a final step, advanced users can introduce the concepts of public containment, public composition, and specialization.

The component-reuse mechanism of KobrA (Chapter 12) also plays an important part in KobrA itself to be introduced incrementally. This is because it enables existing systems, or parts of them, to be integrated into KobrA containment hierarchies as black-box units. Organizations using KobrA, therefore, have a choice as to which parts of their system they wish to reimplement in a component-oriented way, and which parts they simply wish to "wrap up" and disguise as a component. Since it is often impractical, if not impossible, to reimplement all parts of existing enterprise systems in a new paradigm, this capability is essential for the incremental introduction of the method.

- **Focussed prescriptiveness.** A critical judgment for any general purpose method is how prescriptive it should be. In other words, to what extent should the method define what *must* be done as opposed to what *may* be done? KobrA's solution for this issue is to tailor the prescriptiveness of the method to the different paradigms and phases in the life-cycle. As illustrated in Figure 2.21, the method is most prescriptive in relation to framework engineering, and gradually reduces its prescriptiveness towards the lower-level implementation, building, and deployment aspects of the software life-cycle. In contrast, the prescriptiveness of implementation technologies likely to be used in conjunction with KobrA increases in the opposite direction. Most contemporary component and middleware technologies give a lot of advice on how to develop code and binary executables, but relatively little on how to support the evolution of a system based on a model-driven architecture.

  This is important, because it concentrates KobrA's help for developers where it is most needed – in connection with the most novel paradigms (component and product line paradigms), and the most challenging early stages of the software life-cycle – and allows developers to integrate their tried and trusted implementation technologies with KobrA's component and product line modeling approaches. This characteristic allows KobrA to dovetail relatively easily with the existing development practices of many organizations, and to provide help where there is most likely to be a gap.

- **Uniform concepts and activities.** Another important aspect of KobrA is its minimal set of concepts and its uniform application of these concepts to the widest possible levels of granularity and abstraction. As well as decreasing the learning effort for users of the method, this characteristic is valuable because it means that the method is applicable in basically the same way to small and large systems alike, without quantum changes in development concepts, products, or activities. The development activities applicable at a given point in a project are primarily determined by the properties of the artifact under development rather than its location in the architecture or its size.

20.2 Customization

Customization involves the adaptation of a method to better fit the needs of a specific project. Ideally, during customization users of a method should be given explicit methodical support to help them adapt the method to the context in which it will be applied. This means that the method should provide support for being changed in terms of its processes and products.

Many factors influence a KobrA project. They can be divided into four categories:

- **The application domain for which a framework is developed.** Different application domains demand different methodical support. An example is real-time support for applications in the telecommunication domain.
- **The organizational context in which the project is performed.** This includes factors like the availability of tools and expertise of non-project members.
- **The structure of the project.** This category includes, for example, available resources in terms of personnel, time, and money, as well as factors concerned with the experience of the project staff.
- **The goals of the project.** For a KobrA project the main factor in this category is the scope of the developed framework, that is the number of applications covered by the framework. This scope can range from a single product to a complete software product line. The schedule for incremental development of a product line is also covered here.

There are numerous factors that influence how KobrA processes and products should be modified to fit a given project situation, some of which might contradict each other. Therefore, the goal of customizing KobrA is to modify its processes and products so that the application of the method is optimized.

In the software process modeling community, the problem of tailoring processes to the specific context in which they will be used in order maximize their value is well known [BR86]. The basic idea of these process-centered approaches is that modifying the processes modifies the products that are consumed and produced by those processes. None of the existing approaches start by modifying the products and then subsequently modify the processes accordingly. However, even though processes and products are (or should be) closely related, there is no way for such process-centered customization approaches to ensure that the products produced by a modified process are still consistent and complete. The fundamental philosophy in KobrA is that the product drives the process, not the other way round. Therefore, customization of KobrA is centered on the products, with the process being adjusted subsequently to best meet the needs of the customized artifacts.

20.2.1 Framework and component genericity

Genericity is one of the best ways of modeling variability. Variability can range from single-system options that can be controlled by each end user to major differences among systems in the same application domain built for diverse target environments. As described in Part 3, these different situations cannot in practice be accommodated using one single technique across the complete spectrum of variability. Single-system variability is typically handled in an *ad hoc* manner, whereas for product lines, domain engineering or product line methods are used. However, these methods typically introduce product lines in one single step and thus require a significant and risky up-front investment. Additionally, they are not very detailed about the relationship of product line activities and the development activities typically used for single system development.

The concept of KobrA is to first introduce the KobrA method with a focus on single system development to allow an organization to become familiar with the KobrA models and component engineering process. Afterwards, the product line concepts are added for some of the Komponents. Typically, as a first step the product line concepts of KobrA are applied to a medium-sized Komponent, which is well understood and which has few dependencies on other Komponents. This Komponent is modeled in a generic way and then deployed using the concepts of application engineering. When this is successful, the product line concepts are extended to other Komponents, step by step.

20.2.2 Context realization adaptation

The context realization activity provides the entry point for a KobrA project. As part of this activity, the requirements for a KobrA framework are elicited and modeled. As mentioned in Chapter 7, the nature of requirements and their optimal representation depends to a large extent on the application domain. The context realization models and activities, described in Chapter 7, are tailored to the Enterprise Resource Planning (ERP) domain and are, therefore, centered on business processes. As a result they may not be optimal for other application domains such as embedded or real-time systems. Different abstractions and concepts tend to be used in different application domains, so the requirements documents, domain models, or in the case of KobrA, context realizations, that are developed to describe them should ideally use these different concepts and notations.

The abstractions and notations used in different application domains are usually well known and familiar to the organizations that use them. Therefore, when a KobrA framework is being developed, idiosyncrasies in the concepts and abstractions used in the domain should become apparent and be deployed as early as the context realization activity. Existing domain analysis approaches are well suited for this purpose, and can be used as the front end to the KobrA context realization activity. Examples are the Feature-Oriented Domain Analysis (FODA) [KCN90], the commonality and variability analysis that is part of the Family-oriented Abstraction, Specification, and Translation process (FAST) [WL99], and the Customizable Domain Analysis (CDA) [BMW99].

The only concrete requirement that a KobrA context realization must fulfill is that the final product set should include the regular Komponent realization artifacts. Since these are common object-oriented models described using the UML, there is a reasonable chance that they will be part of the product set produced by a foreign domain-analysis method in any case. If not, missing products or mappings among products must be introduced so that the context realization complies with the requirements of a Komponent realization.

In the case of organizations that have previously developed single systems in isolation and wish to move towards the development of product lines using KobrA, the established requirements documents must be augmented with information about variations in the requirements for the different product line members. This can be achieved using the techniques of variability and decision modeling described in Chapter 15.

20.2.3 Komponent modeling adaptation

The KobrA method defines a comprehensive set of models to capture all the possible characteristics of Komponents. However, not all components will require all the models to be developed all the time. A Komponent that does not exhibit interesting external states, for example, does not need a behavioral model, while a Komponent with very simple operations does not need an activity model. This choice can be made on a Komponent-by-Komponent basis and is only a function of the properties of the Komponent, not its absolute position in the containment hierarchy. This ability to select which models to use based on the Komponent's characteristic significantly reduces the amount of modeling work that is required in practice.

20.2.4 Implementation, building, and deployment

As explained above, the KobrA method is progressively less prescriptive about the implementation, building, and deployment activities in the software-development process. This is to enable organizations to marry the KobrA front-end technology with their preferred and established implementation technologies. Therefore, an important customization step, which must be performed before a project can be completed, is the integration of the already installed implementation technologies with KobrA.

As described in Chapter 11, KobrA's recommended approach to implementation is based on systematically applying refinement and translation pattern oriented towards a particular implementation profile. Adapting the implementation activity to a different implementation technology may therefore involve the definition of a new implementation profile, the definition of new transformation patterns, the adaptation of existing patterns, and the deletion of patterns targeted to different languages or patterns that may invalidate the prevailing non-functional requirements. The adaptation of existing patterns may also be used to incorporate feedback on single patterns gathered in their practical application. If no translation patterns are available for the company's implementation language (e.g. for Java or Ada), these could be created.

The creation of new implementation profiles and transformation patterns follows the guidelines described in [Bun01] with respect to the chosen target language. In general, the goal is to define a pattern set containing at least one translation pattern, targeted at the used programming language, for each used modeling construct. Another important factor in pattern adaptation is their optimization concerning non-functional requirements which means checking the impact a pattern will have, as well as fine-tuning patterns to better fulfill the prevailing requirements. In summary, this activity can be performed at different levels of detail. Consequently, the effort for adaptation may grow quickly, especially if KobrA is applied for the first time within an organization. However, each adaptation enhances and improves the set of available patterns.

Another issue in the overall customization activity are the tools used for implementing the different UML models, including code-generators, compilers, and linkers. The chain of tools applied in a project may also require adaptation to KobrA. First, translation patterns have to reflect the special use of implementation constructs imposed by a specific compiler. Second, it may not be possible to fully support the KobrA way of implementing models based on the prevailing non-functional requirements, since this requires a fully flexible generation mechanism. Therefore a strategy for using transformation patterns with a specific case-tool is needed [Bun01].

20.3 Transfer

As mentioned above, transfer is concerned with preparing an organization to use the customized KobrA method to its best advantage. In this subsection, we concentrate on the issues involved in planning such a transfer, training for the target method users, and tool support.

20.3.1 Transfer planning

When a new technology is introduced into an organization, it is important to take into account the technical and non-technical aspects of the technology. For example, technologies that aim to reduce development effort or foster reuse are often anticipated as job-killers. Alternatively, technologies that aim to monitor work practices and gather project statistics are often viewed as intrusive and "big-brother" like. These social aspects of technological changes are often not taken into account, resulting in frustrated employees and puzzled management. Migration to a new method therefore has to be accompanied by a coaching and training program that not only introduces the new technology, but also explains why it is introduced.

Another issue that influences the transfer planning are the aspects of KobrA that are new to the organization. This depends to a large extent on the experience and background of the company. Companies already familiar with component technology and product line concepts will have the least problem learning KobrA's way of applying these ideas. However, companies that have no prior experience in these paradigms face a more significant training challenge.

The quality improvement paradigm (QIP) [BCR94b] is a method that enables the establishment of a continuous improvement program. It is based on the ideas of a step-wise controlled improvement and experimental software engineering. QIP consists of the following steps:

1. Analyze and characterize the current situation.
2. Set goals on the basis of the characterization.
3. Choose a process that allows the achievement of the goals in the current situation.
4. Perform the processes and feed back data on goal achievement.
5. Analyze the data at the end of each specific project.
6. Package the experience to enable its later reuse.

QIP is usually accompanied by the Goal/Question/Metric Paradigm (GQM) [BCR94a] to measure the achievements against the project goals.

Planning the introduction of KobrA should begin with a rating of the importance of its different technological ingredients. These should then be introduced in a step-wise manner following the QIP. Of course, the method can be introduced in one major step, but this is usually risky and impractical.

If a product line is to be incorporated with the initial introduction of KobrA, the products that comprise the product line must be determined. This can be achieved using scoping approaches of the kind commonly used in product line engineering [Sch00]. An important aspect is the scheduling of the development of different product line members: are they needed all together at once or is it possible to develop them one at a time? Based on the main technological ingredients and the schedule for the individual product line members, resource planning can be performed and a detailed project plan can be established.

20.3.2 Training

Everybody that comes into contact with a new technology like KobrA should be trained appropriately. Training courses and seminars can be used to teach people about the ideas and concepts of the method and the ingredients that are considered most important. Follow-up workshops can also help provide practical experience.

Ideally, a pilot study with a small team should also be performed to evaluate whether KobrA matches the needs of an organization. A method described in a book might seem appealing to the people planning its introduction, but only a trial application can really establish whether it is appropriate for the company from both an organizational and technical perspective.

Training should always be based on the known problems and issues in an organization and point out how KobrA can help deal with those. The prerequisites for the training and for the KobrA method must be made clear. It is essential that the future users and their trainers are aware of this. The training for other new technologies introduced along with KobrA must be coordinated with the KobrA training. This is supported by the modular nature of the KobrA ingredients.

20.3.3 Tool support

Tool support is critical for a method like KobrA. Two main categories of tool support are required, particularly when applying KobrA in an industrial context: CASE tool support for UML diagram development, and configuration management support for keeping track of the many artifacts developed during a KobrA project, especially when applied in a product line context.

Because no CASE tools provide explicit support for KobrA, help in diagram development can only be attained today by adapting and using existing tools. Unfortunately, however, the current generation of UML CASE tools typically requires users to follow a particular process, with a built-in set of well-formedness rules and constraints. Many of their modeling features are unavailable unless these often undocumented rules are followed. This is particularly true for the features that distinguish CASE tools from pure drawing tools, such as the provision of syntactic, semantic, and consistency checks within and among models. Unfortunately, most of the tools currently available are not sufficiently customizable to the needs of a new method such as KobrA.

Nevertheless, since KobrA uses standard UML models, it can in principle be applied using most existing UML CASE tools. New stereotypes such as the stereotype <<variant>> that are needed within the generic KobrA models can usually be defined. Moreover, textual models such as the operation specifications in the functional model are usually supported through some kind of user-defined reports. If not, a standard word processor must be used to augment the CASE tool.

The main problem for CASE tools is supporting the different types of presentation elements required in a Komponent containment tree. Several different views of a given model element are usually required showing different details and characteristics of the element at different points in the containment tree. For example, the LoanManager Komponent, which is represented by a single class, is presented in:

- the LibrarySystem realization as a Komponent showing all external visible operations and attributes
- the LoanManager specification as the subject showing all external visible operations and attributes
- the LoanManager realization as the subject showing all operations and attributes including operations and attributes only visible inside the LoanManager Komponent.

The visibility issues can sometimes be addressed by tailoring the visibility options for operations and attributes, but the change of a stereotype in different diagrams is usually not supported in the current generation of CASE tools.

Apart from the syntactic representation of variant elements using stereotypes, no existing CASE tool has the capacity to support the product line engineering aspects of KobrA. The automated application of decisions models to create applications from a framework can at present only be supported by an external tool that transforms the UML models accordingly. At the Fraunhofer IESE, a prototypical COM component has been developed that demonstrates this idea [Fle00].

The other category of tool support required in industrial applications of KobrA is support for configuration management. This is best based on a repository capable of providing full life-cycle support for the various artifacts defined by the method. As part of the KobrA project, configuration management support for KobrA artifacts was implemented on Softlab's Enabler Repository, and a prototype Komponent manager was developed to facilitate the visual browsing of the artifacts in a KobrA framework.

APPENDIX

A Metamodel

This appendix summarizes in the form of a metamodel the concepts, rules, and constraints that underpin the KobrA method. The elements in the metamodel are organized into a hierarchy of packages, each focussing on a key ingredient of the method. The presentation of the concepts encapsulated by a package consists of two parts:

- **overview** – a short description of the purpose of the package accompanied by a graphical overview of its elements in the form of a UML class diagram
- **abstract syntax** – a textual definition of each element in the package.

In addition, for those packages primarily concerned with the UML modeling artifacts (Komponent, Specialization, Realization, Implementation, System, Physical Component, Context Realization) a set of well-formedness rules is defined. These define constraints on how the modeling elements in the package should be applied, and thus provide the foundation for the prescriptive and systematic nature of the method. They can either be checked automatically (via tool support, if available) or manually (by inspection). Seven distinct types of constraints can be distinguished:

1. **Intra-diagram rules** ensure that individual diagrams are complete and well formed.
2. **Inter-diagram rules** ensure that the diagrams within a specification, realization or implementation are consistent with each other.
3. **Realization rules** ensure that a Komponent's realization represents a faithful representation of its specification.
4. **Clientship rules** ensure that a client and server both fulfill their contract.
5. **Containment rules** ensure that a Komponent's relationships with other Komponents are consistent with its location in the containment tree.

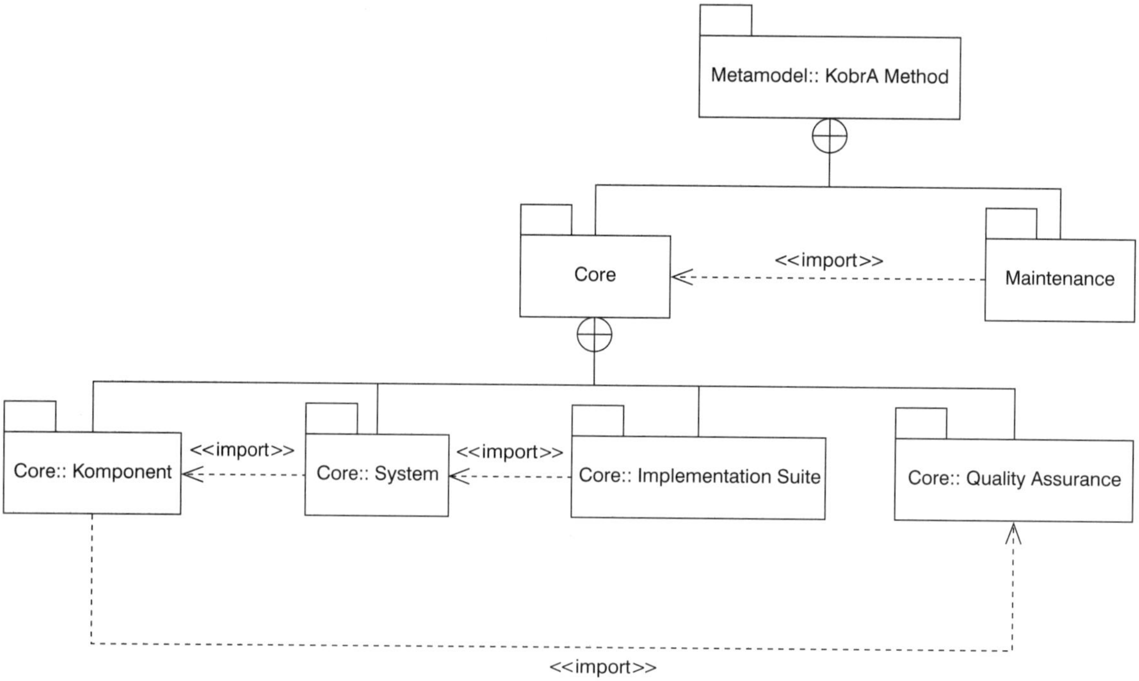

Figure A.1
Method packages

6 **Specialization rules** ensure that a Komponent specialization conforms to the class from which it was specialized.

7 **Instantiation rules** ensure that a specific Komponent is consistent with the generic Komponent from which it is instantiated.

The high-level package structure of the metamodel is illustrated in Figure A.1. The overall KobrA method contains two main sub-packages, Core and Maintenance, which both contain further sub-packages. The Core package encapsulates the fundamental elements in the method, while the Maintenance package describes the configuration management and change management concepts used to control the life-cycle of the core elements.

A.1 Core

The Core package contains four sub-packages, three of which are organized in terms of further sub-packages. The Komponent package defines the fundamental elements associated with the logical component model of KobrA. This is accompanied by a Quality Assurance package, which contains all the elements needed to ensure the optimal quality of Komponents. The System package defines the elements needed to define bounded, self-contained systems from Komponents. This is accompanied by the Implementation Suite package, which defines the technical elements needed to generate deliverable artifacts.

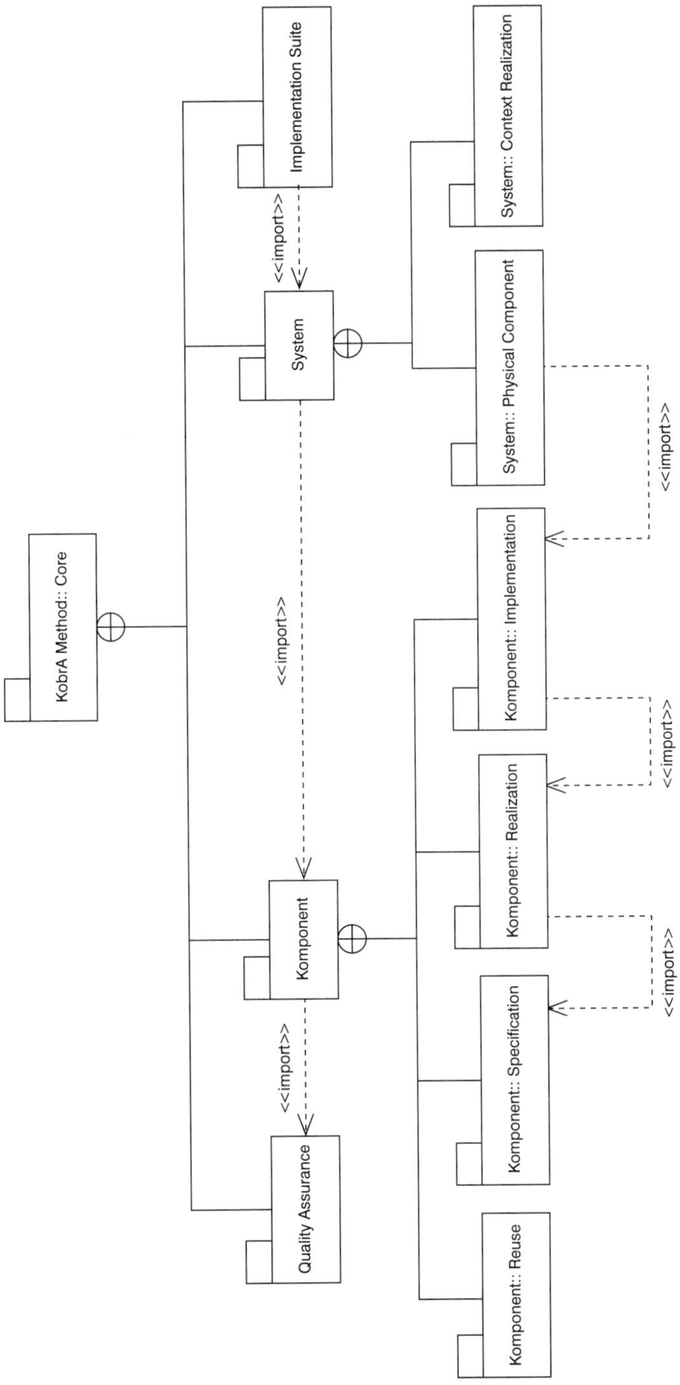

Figure A.2
Core packages

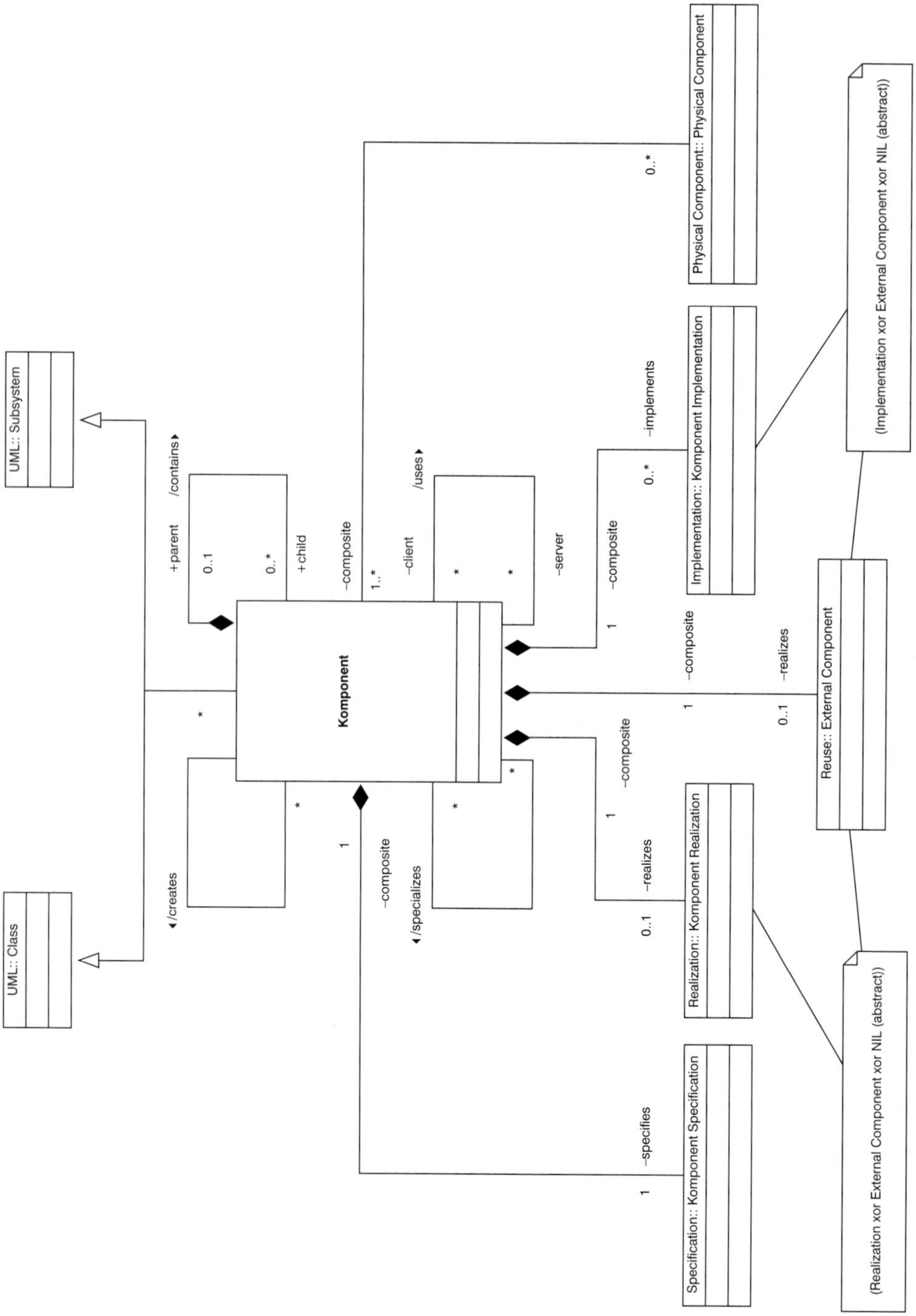

Figure A.3
Static Komponent structure

A.1.1 Komponent

A.1.1.1 Overview

The Komponent sub-package (see Figure A.3) defines the basic structure of Komponents within the framework and application engineering processes. Elements specified in the Komponent sub-packages include Komponent Realization, Komponent Specification, Komponent Implementation and External Component. Komponent is derived from UML Class and UML Subsystem, which provide their respective class and package facets. Komponent also has a link to the Physical Component element.

A1.1.2 Abstract syntax

Komponent
A logical unit of behavior with a separately defined specification and realization. A Komponent has the properties of a class in that it defines the features and behavior of Komponent instances instantiated from it, and the properties of a package in that it acts as a container for other elements, including possibly sub-Komponents.

Komponent Specification
A Komponent Specification describes the externally visible properties of a Komponent and can be viewed as the interface of a Komponent (see Chapter 4). It consists of a Structural Model, a Behavioral Model, a Functional Model, and either a Decision Model or a Decision Resolution Model. A Komponent has exactly one Komponent Specification.

Komponent Realization
A Komponent Realization describes how the private design of a Komponent fulfills the requirements of the Komponent specification (see Chapter 5). It comprises a Structural Model, an Activity Model, an Interaction Model, a Specific Data Dictionary, and either a Decision Resolution Model or a Decision Model. A Komponent can have zero or one Komponent Realizations.

External Component
An External Component can be a KobrA Komponent created in another KobrA Framework or a COTS component. It can have a Komponent Specification, a Conformance Map, a List of Adaptations, a Semantic Map, and a link to a reused component. This can be another Komponent or a foreign component.

Physical Component
A Physical Component describes the physical structure for a set of logical Komponents so that development tools (generally a compiler) can treat them in order to produce executable binaries (see Chapter 11 and Chapter 13). A Physical Component contains a Binary and a Construction Model.

Creates
The Creates relationship defines which Komponents create instances of other Komponents. The creation of a Komponent instance by another Komponent can be shown in a Komponent Specification Object Diagram. The properties of this relationship are inherited from the UML Usage relationship (stereotyped as <<create>>) by virtue of both the Class and Subsystem characteristics of a Komponent.

Uses
The Uses or clientship relationship defines a client–server relationship between Komponents. It is captured by the equivalent Komponent Specialization and Komponent Realization client–server relationship. The properties of this relationship are inherited from the UML Usage relationship by virtue of both the Class characteristic (stereotyped as <<call>>, <<create>>, <<instantiate>>, or <<send>>) and the Subsystem characteristic (stereotyped as <<import>> or <<access>>) of a Komponent.

Contains
The Contains relationship defines the parent–child relationship between two Komponents. The name space of the child Komponent is extended with the name space of the parent Komponent. A Komponent can have an arbitrary number of child Komponents. The properties of this relationship are inherited from the UML name space mechanism by virtue of the Subsystem characteristic of a Komponent.

Specializes
The Specializes relationship describes the generalization–specialization relationship between two Komponents. The properties of this relationship are inherited from the UML generalization mechanism by virtue of both the Class and Subsystem characteristic of a Komponent.

A1.1.3 Well-formedness rules

A.1.1.3.1 Realization rules

Specification structural model ↔ Realization structural model

- The realization structural model should be a refinement and extension of the specification structural model. In essence, this means that every entity appearing in the specification structural model must appear in the realization structural model, although possibly in a refined form (i.e. same information, but perhaps in a more detailed form).

Specification functional model ↔ Realization interaction model

- The input parameters in a collaboration diagram must match those defined in the corresponding operation specification (i.e. entities in the receives clause).
- The output parameters in a collaboration diagram must match those defined in the corresponding operation specification (i.e. entities in the returns clause).
- The effects defined in an operation specification must be attained by the corresponding collaboration diagram or hierarchy of collaboration diagrams.

- Every object accessed or changed in an operation specification must receive a message in the corresponding collaboration diagram or hierarchy of collaboration diagrams.

Specification functional model ↔ Realization activity model

- The input parameters in an activity diagram (if any) must match those defined in the corresponding operation specification (i.e. entities in the receives clause).
- The output parameters in an activity diagram (if any) must match those defined in the corresponding operation specification (i.e. entities in the returns clause).
- The effects defined in an operation specification must be attained by the corresponding activity diagram or hierarchy of activity diagrams.
- Every object accessed or changed in an operation specification must receive a message in the corresponding activity diagram or hierarchy of activity diagrams.

A.1.1.3.2 Clientship rules

Realization structural model ↔ Specification structural model

- A Komponent's realization class diagrams must contain classes representing each of the Komponent's server Komponents.
- The class representing a server Komponent in the realization class diagrams of one of its clients must only contain the operations called by that client.
- The realization class diagrams of a Komponent must contain all the elements that it needs to call its servers' methods. Thus, a Komponent's realization class diagrams must contain all the necessary elements from its servers' specification class diagrams.

Realization interaction model ↔ Specification structural model

- For every message sent to an instance of a server Komponent in a collaboration diagram, the server Komponent must have a matching operation.
- For every parameter type of every message sent to an instance of a server Komponent in a collaboration diagram there must be a corresponding model element in the server Komponent specification class diagrams.

Realization interaction model ↔ Specification functional model

- The parameter profile of every message sent to an instance of a server Komponent in a collaboration diagram must match the parameter profile of the corresponding operation specification of the server Komponent.

- The intended effect (post-condition) of a message sent to an instance of a server Komponent in a collaboration diagram must match the results clause of the corresponding operation specification of that Komponent.
- The pre-condition of a message sent to an instance of a server Komponent in a collaboration diagram must match the assumes clause of the corresponding operation specification of that operation.

Realization interaction model ↔ Specification behavioral model

- The message invocation sequencing allowed by the collaboration diagrams of a Komponent's client must be consistent with the message sequencing allowed by the Komponent's behavioral model.

A.1.1.3.3 Containment rules

- The elements in the structural model of a sub-Komponent must be a subset of those in the realization structural model of its super-Komponent.
- For server sub-Komponents (i.e. sub-Komponents which are servers of their parents) the features presented for the sub-Komponent class in the parent's realization class diagram must be a subset of the Komponent's features.
- For sub-Komponents whose only client is their parent, the features presented for the sub-Komponent class in the parent's realization must be the same as those presented in the sub-Komponent's specification.

A.1.1.3.4 Specialization rules

Generalization structural model ↔ Specialization structural model

- The pattern roles names in a specialization class diagram must match the class names in the generalization class diagram.
- Every class in the general class diagram must appear at least once in the specialization class diagram, but can appear multiple times.
- The class diagrams of a specialized specification must not have conditions.

Generalization functional model ↔ Specialization functional model

- The assumes clause of an inherited operation specification in a specialized specification must be the same as, or stronger than, that of the corresponding operation specification in the general specification.
- The result clause of an inherited operation specification in a specialized specification must be the same as, or stronger than, that of the corresponding operation specification in the general specification.

Generalization behavioral model ↔ Specialization behavioral model

- The states, events and transitions defined in the behavioral model of a specialized specification must be a superset of those inherited from the general specification.

A.1.2 Specification

A.1.2.1 Overview

The Komponent Specification sub-package (see Figure A.4) defines the basic structure of Komponent Specifications developed within the framework and application engineering processes. Instances specified in the Komponent Specification sub-package include the Functional Model, the Behavioral Model, a Structural Model, either a Decision Model or a Decision Resolution Model, a Non-Functional Requirements Specification, a Quality Policy, and Quality Documentation. Komponent Specification also references Realization.

A1.2.2 Abstract syntax

Structural Model
The Structural Model describes the nature of the classes, the relationships by which the Komponent interacts with its environment, and the structure of the Komponent interface (see Section 4.1.1). It consists of one or more UML Class Diagrams and zero or more UML Object Diagrams. Each Komponent Specification has exactly one Structural Model.

Functional Model
The Functional Model describes the externally visible effects of the operations supplied by the Komponent (see Section 4.1.2). It consists of an arbitrary number of Operation Specifications. Each Komponent Specification has exactly one Functional Model.

Behavioral Model
The Behavioral Model shows how the Komponent behaves in response to external stimuli (see Section 4.1.3). It consists of zero or one Statechart Tables and an arbitrary number of UML Statechart Diagrams. Each Komponent Specification has zero or one Behavioral Models.

Decision Model
The Decision Model describes the variabilities among the products in a product line. It contains the variation points, their resolution space, and their effects on the Structural Model, Behavioral Model and Functional Model. Each Komponent Specification has zero or one Decision Models.

Decision Resolution Model
The Decision Resolution Model describes decisions that resolve the variabilities in the generic framework artifacts in order to make them specific for a certain

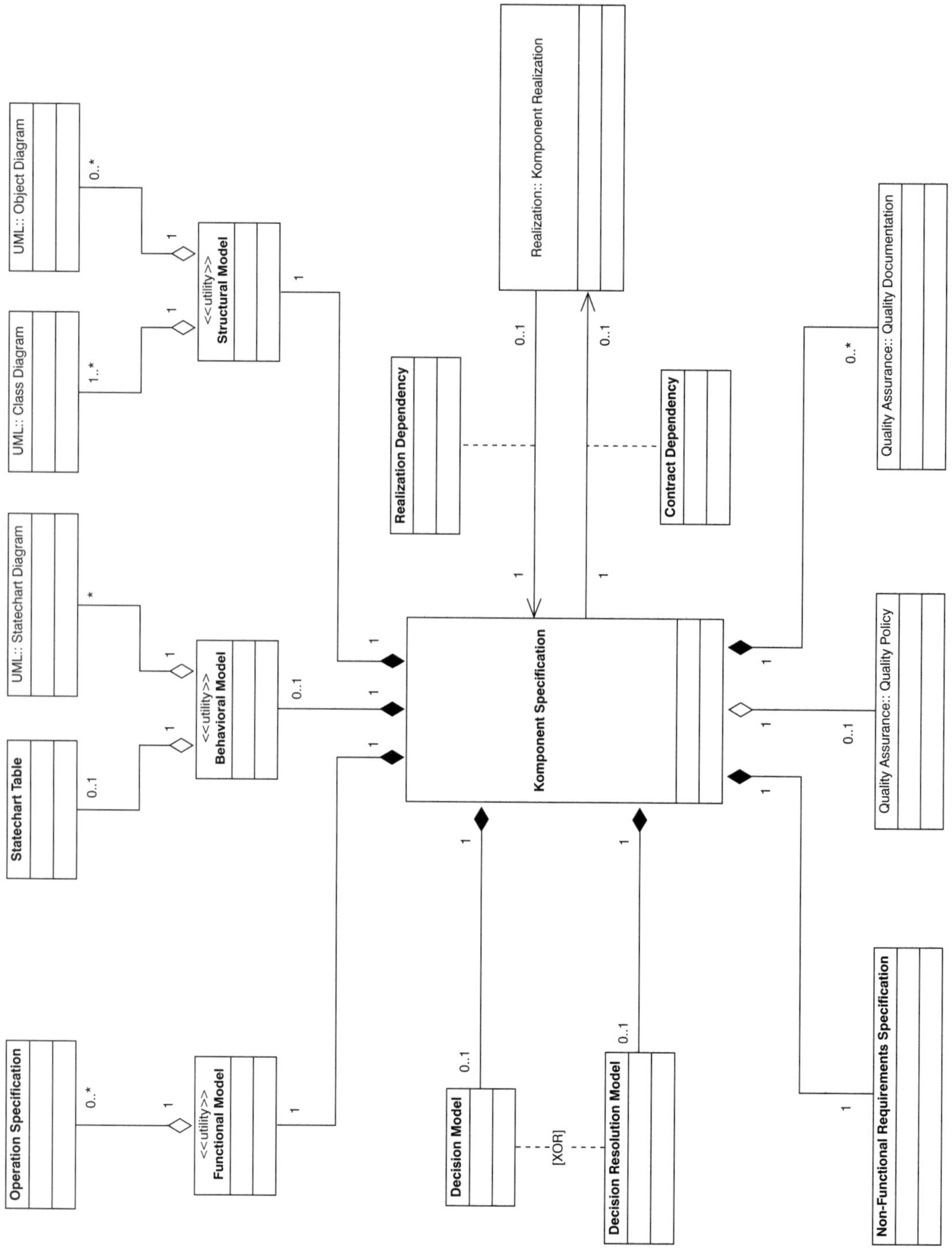

Figure A.4
Static Komponent Specification structure

application. Therefore it contains the variation points and their resolutions. The Structural Model, Behavioral Model and Functional Model are made specific according to these resolutions. Each Komponent Specification has zero or one Decision Resolution Models. A Decision Resolution Model will exist if and only if a Decision Model exists for the generic Komponent from which the specific Komponent was instantiated.

Non-Functional Requirements Specification
The Non-Functional Requirements Specification describes the quality levels that are acceptable under specified circumstances. It consists of information about the quality level for the reliability, usability, efficiency, maintainability, portability, and functionality issues like completeness, consistency, or correctness of system parts.

Quality Policy
The Quality Policy contains the artifacts and concepts needed to execute the quality assurance program. The artifacts involved include the defect detection forms, the defect collection form, the inspection scenarios, the quality models and the measurement plan.

Quality Documentation
The Quality Documentation contains the artifacts generated by the execution of the quality assurance program in accordance with the Quality Policy. This includes defect descriptions, test suite, test results, and the quality measures.

Realization Dependency
The Realization Dependency captures the relationship between a Komponent Specification and the Komponent Realization that realizes it. It contains information about the compatibility of the Komponent Specification and Komponent Realization. A Komponent Specification can have zero or one Realization Dependencies, and thus can have zero or one Komponent Realizations.

Contract Dependency
The Contract Dependency (or clientship dependency) captures the relationship between a Komponent Specification and a Komponent Realization that uses it (i.e. is a client of it). It contains information about the compatibility between the Komponent Specification and Komponent Realization. For each Komponent Specification version a new Contract Dependency is created.

A.1.2.3 Well-formedness rules

A.1.2.3.1 Intra-diagram rules

Class diagrams

- Only Komponents can have an operation compartment.
- If the subject Komponent has an operation compartment, it must list all the operations of the Komponent (i.e. all or nothing).

- If an acquired Komponent has an operation compartment it should only list the operations that the subject Komponent invokes (listed in the sends clauses of its operation specifications).

Statechart diagrams/tables

- The outgoing transitions of each state in each statechart diagram/table must be disjoint, that is, the behavioral model must be deterministic.

Operation specifications

- There must be one operation specification for every operation belonging to the Komponent under specification.
- Assumes and result clauses must be Boolean expressions.
- Terms in sends, receives, returns, and changes clauses can only use items that are mentioned in the assumes and result clause.

A.1.2.3.2 Inter-diagram rules

Object diagrams ↔ Class diagrams

- Object diagrams can only define instances of entities defined in the class diagrams of the same Komponent Specification.

Statechart tables ↔ Statechart diagrams

- If a statechart table supplements a statechart diagram the names of states must be consistent, i.e. the names used in the table must occur in the statechart diagram of the same Komponent Specification, but not necessarily vice versa.

Structural model ↔ Behavioral model

- Attributes referred to in a statechart diagram must appear in the class representing the Komponent and must match in name and type.
- Conditions within a statechart diagram may be defined only in terms of local entities or entities defined in the class diagrams (e.g. attributes).
- Sub-Komponents referred to in statechart diagrams/tables (i.e. for public composition) must appear in one of the object diagrams.

Class diagrams ↔ Operation specifications

These two models must basically agree on the nature of the entities that exist in order for operations to do their work. In particular, this means:

- All classes, attributes and relationships referred to in an operation specification must be in the class diagram. In the case of associations, this means that operation specifications can only create links for associations and aggregation relationships defined in one of the class diagrams.
- Every class in the class diagrams should appear in at least one operation specification.

- Every attribute appearing in the class diagram should appear in at least one operation specification.
- Every operation listed in the operation compartment of an acquired Komponent must be involved in at least one operation specification.

Object diagrams ↔ Operation specifications

- The names of instances used in the operation specifications must match those in the object diagram.

Operation specifications ↔ Statechart diagrams/tables

These two models must basically agree on the states, events and operations of the Komponent. In particular, this means:

- Operation specifications may only mention states and events appearing in one of the statechart diagrams/tables.
- Statechart tables and diagrams may only mention operations specified in the functional model.

A.1.3 Realization

The Komponent Realization sub-package (see Figure A.5) defines the structure of the Komponent Realizations developed during the framework and application engineering processes. Elements specified in the Komponent Realization sub-package include an Interaction Model, an Activity Model, a Structural Model, either a Decision Model or a Decision Resolution Model, a Quality Policy, a Data Dictionary and Quality Documentation. Furthermore it has references to a Komponent Specification and to Komponent Implementations.

A.1.3.1 Overview

A.1.3.2 Abstract syntax

Structural Model

The Structural Model describes the nature of the classes and relationships from which the Komponent is realized and the architecture of the Komponent (see Section 5.1.1). It consists of one or more UML Class Diagrams and zero or more UML Object Diagrams. Each Komponent Realization has exactly one Structural Model.

Interaction Model

The Interaction Model describes the control flow of an algorithm used to realize an operation in the form of a flowchart-oriented graph (see Section 5.1.3). It consists of one or more UML Interaction Diagrams. Each Komponent Realization has zero or one Interaction Models.

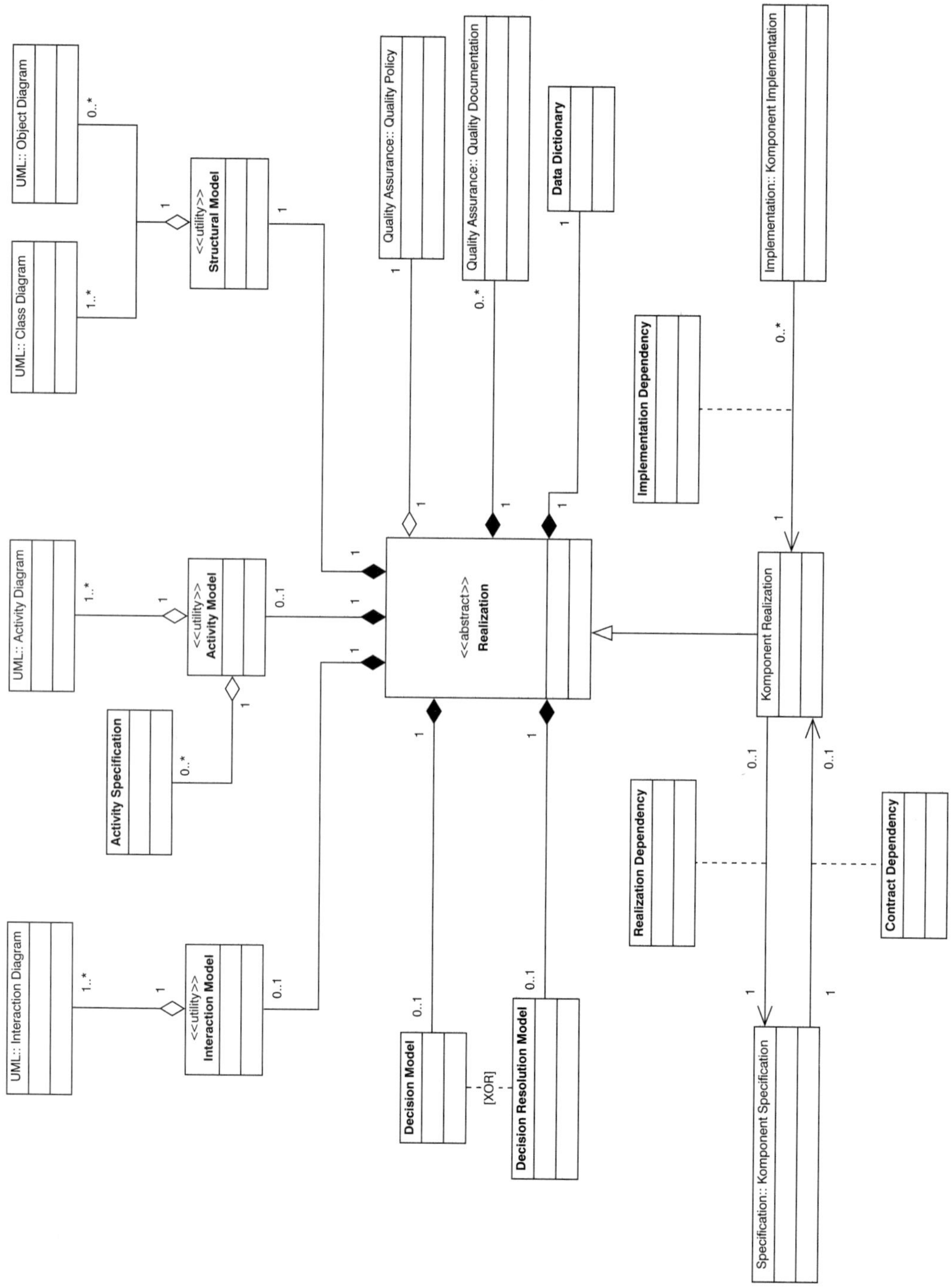

Figure A.5
Static Komponent realization structure

Activity Model

The Activity Model describes the algorithms by which the operations of a Komponent are realized (see Section 5.1.2). It consists of one or more UML Activity Diagrams and zero or more Activity Specifications. Each Komponent Realization has zero or one Activity Models.

Decision Model

The Decision Model describes the variabilities among the products in a product line. It contains the variation points, their resolution space and their effects on the Interaction Model, Activity Model and Functional Model. Each Komponent Realization has zero or one Decision Models.

Decision Resolution Model

The Decision Resolution Model describes decisions that resolve the variabilities in the generic framework artifacts in order to make them specific for a certain application. It contains the variation points and their resolutions. The Structural Model, Interaction Model and Activity Model are made specific according to these resolutions. Each Komponent Realization has zero or one Decision Resolution Models. A Decision Resolution Model will exist if and only if a Decision Model exists for the generic Komponent from which the specific Komponent was instantiated.

Quality Policy

The Quality Policy contains the artifacts and concepts needed to execute the quality assurance program. The artifacts involved include the defect detection forms, the defect collection form, the inspection scenarios, the quality models and the measurement plan. Normally they will be shared by more than one Komponent.

Quality Documentation

The Quality Documentation contains the artifacts generated by the execution of the quality assurance program in accordance with the Quality Policy. This includes defect descriptions, test suite, test results, and the quality measures.

Implementation Dependency

The Implementation Dependency defines which Komponent Implementation a Komponent Realization is implemented by. It contains information about the compatibility between the Komponent Realization and the Komponent Implementation. For each new Komponent Realization version a new Implementation Dependency will be created.

A.1.3.3 Well-formedness rules

A.1.3.3.1 Intra-diagram rules

Class diagrams

- If the class representing the Komponent under specification has any operations, it must list all the operations of the Komponent (i.e. all or nothing).

Activity diagrams

- Activity diagrams with only free activities must have no swim lanes.
- Activities diagrams with associated activities must have swim lanes.

Interaction diagrams

- Each interaction diagram must contain an instance of the subject Komponent.
- Each interaction diagram must have one source-free message, and the target of this message must be the instance of the subject Komponent.
- The source-free message must not have a sequence number.

A.1.3.3.2 Inter-diagram consistency rules

Class diagrams ↔ Object diagrams

- Object diagrams can only define instances of entities defined in the class diagrams.

Class diagrams ↔ Activity diagrams

- Objects named in swim lanes must be instances of classes in the class diagrams.

Object diagrams ↔ Activity diagrams

- The names of instances in swim lanes must match the names of the corresponding instances in the object diagrams.

Class diagrams ↔ Interaction diagrams

These two models must basically agree on the nature of the Komponents that exist in order for operations to do their work. In particular, this means:

- Instances in an interaction diagram must be instances of classes appearing in the class diagrams.
- Each class in the class diagrams must either be a type of a parameter of a message exchanged in an interaction diagram, or must be an instance appearing in an interaction diagram.
- Every message received by an instance in an interaction graph must appear as a method of the instance's class in the class diagram.

Object diagrams ↔ Interaction diagrams

- Every instance in the object diagrams must appear in at least one interaction diagram.
- The names of instances in an interaction diagram must match those in the object diagram.

Activity specifications ↔ Class diagrams

- All classes, attributes, and relationships referred to in an activity specification must be in the class diagram.

Activity specifications ↔ Object diagrams

- Instance names used in an activity specification must match those in the object diagram.

Activity diagrams ↔ Activity specifications

- For a given activity, the sub-activities and algorithms defined in its activity diagram must match those defined in its specification.

Interaction diagrams ↔ Activity diagrams
These two models must also agree on the nature of the objects that exist in order for activities to do their work. In particular, this means:

- The algorithm defined in an activity diagram must be the same as that in the equivalent interaction diagram if one exists, and vice versa.
- The sub-activities defined in an activity diagram must match the messages in the equivalent interaction diagram if one exists, and vice versa.

A.1.4 Implementation

A1.4.1 Overview

The Komponent Implementation sub-package (see Figure A.6) describes the elements which make up the implementation of a Komponent. Elements specified in the Komponent Implementation sub-package include Structural Model, Implementation Model, and Refinement/Translation Patterns.

A.1.4.2 Abstract syntax

Pseudo-Code
Pseudo-Code is a container for the algorithmic description of the methods belonging to and within a Komponent (see Section 11.4.2).

Structural Model
The Structural Model contains one or more UML Class Diagrams, which describe the static structure of a Komponent's implementation at an abstraction level close to source code.

Refinement Pattern
A Refinement Pattern defines a conceptual mapping between UML Model Elements, with associated Name, Context, Description, and Trace information.

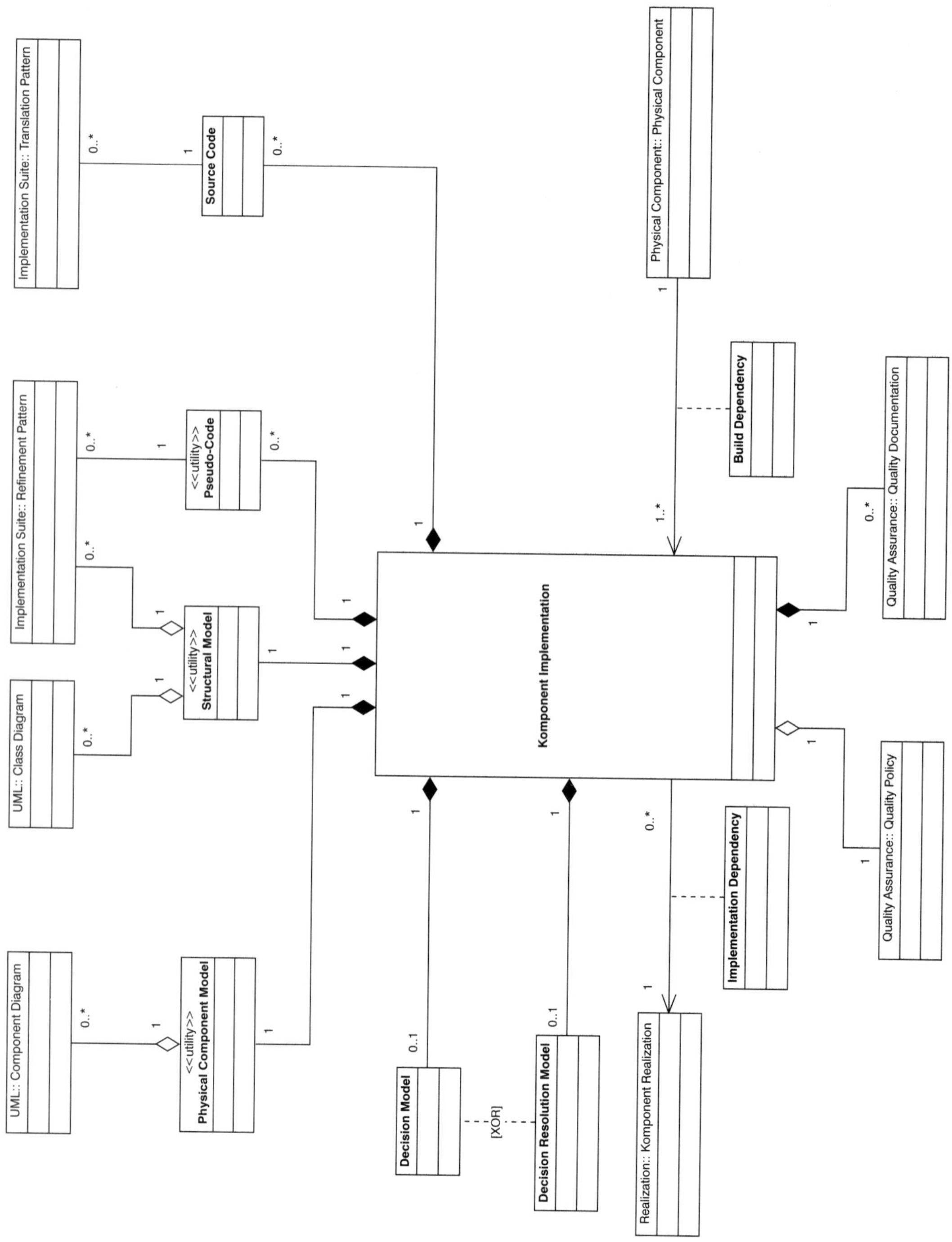

Figure A.6
Static Komponent Implementation structure

Translation Pattern
A Translation Pattern defines a conceptual mapping between a UML Model Element and a corresponding set of elements in a target implementation technology. It has associated Name, Context, Description, and Trace information.

Source Code
Source Code provides a tool-comprehensible representation of a Komponent which can be automatically compiled into one or more executables.

Implementation Dependency
The Implementation Dependency defines which Komponent Implementation a Komponent Realization is implemented by. It contains information about the compatibility between the Komponent Realization and the Komponent Implementation. For each new Komponent Realization version a new Implementation Dependency will be created.

Komponent Implementation
The Komponent Implementation contains a Structural Model, a Physical Component Model, Pseudo-Code, Source Code, optionally a Decision Model or a Decision Resolution Model, a Quality Policy and Quality Documentation. It is also associated to a Komponent Realization.

Build Dependency
The Build Dependency defines the association between a Physical Component and the implementation defining it. The build dependency identifies the implementation used to actually build the component.

Test Suite
The Test Suite of a Komponent Implementation contains all test cases, as well as relevant additional information (e.g., stubs) which is needed for successfully testing a Komponent against its requirements.

A.1.5 Reuse

A.1.5.1 Overview

The Reuse package (see Figure A.7) describes the entities related to the incorporation of External Components within a KobrA project. This supports the underlying principle of component-based development (i.e. reusing components multiple times) and defines a precise set of entities to facilitate the incorporation of predefined Komponents within a Komponent Tree or Physical Components Structure.

A.1.5.2 Abstract syntax

Conformance Map
A Conformance Map specifies how the native-offered interface of a foreign component is mapped to a KobrA representation of the interface. Basically it contains a table which describes for each specified operation an appropriate sequence of method calls to the foreign components. This information can be used during the construction activity to define a Wrapper, which implements this access.

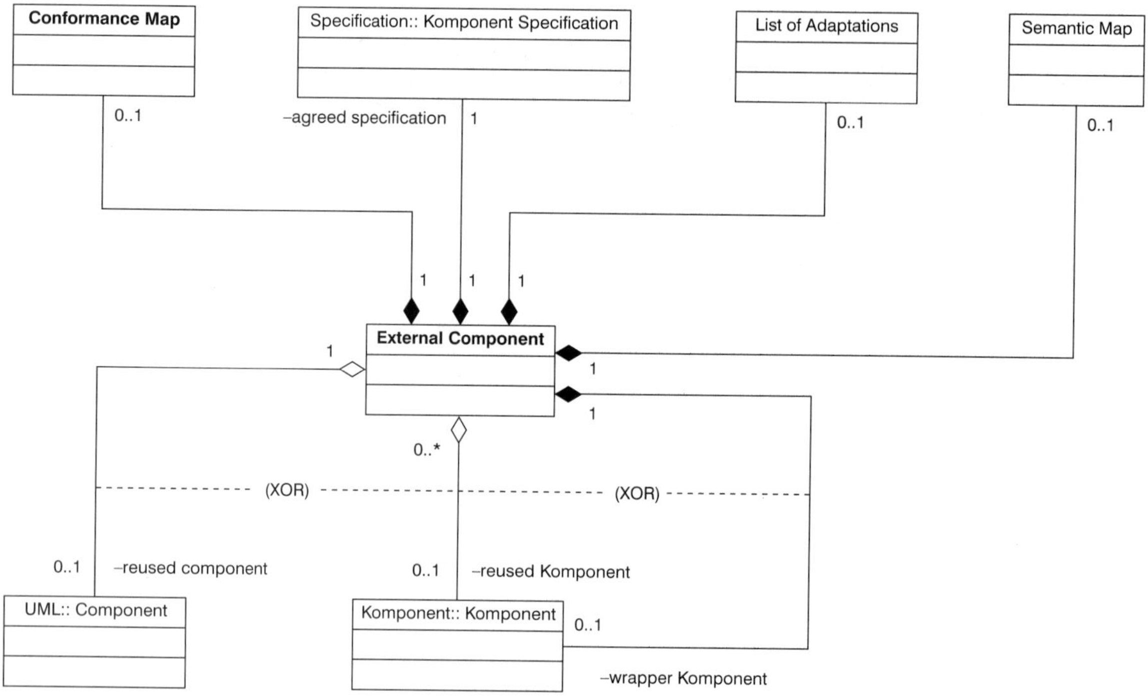

Figure A.7
Static Reuse structure

List of Adaptations
A List of Adaptations describes how the External Component has to be modified to be used within the final system. Typically it lists a series of issues which have to be solved by the manufacturer of the component prior to its incorporation into a KobrA Framework or Application.

External Component
An External Component can be a Komponent created in another KobrA Framework or a COTS component. It can have a Komponent Specification, a Conformance Map, a List of Adaptations, a Semantic Map, and a link to a reused component. This can be another Komponent or a foreign component.

A.1.6 System

A.1.6.1 Overview

The System package (see Figure A.8) is one of the most fundamental sub-packages of the metamodel. It defines the basic structure of a KobrA System (i.e., Application and Framework) as well as the core constructs involved in a KobrA development project. Elements specified in the System sub-package include Komponent Tree, Physical Components Structure, and Deliverable.

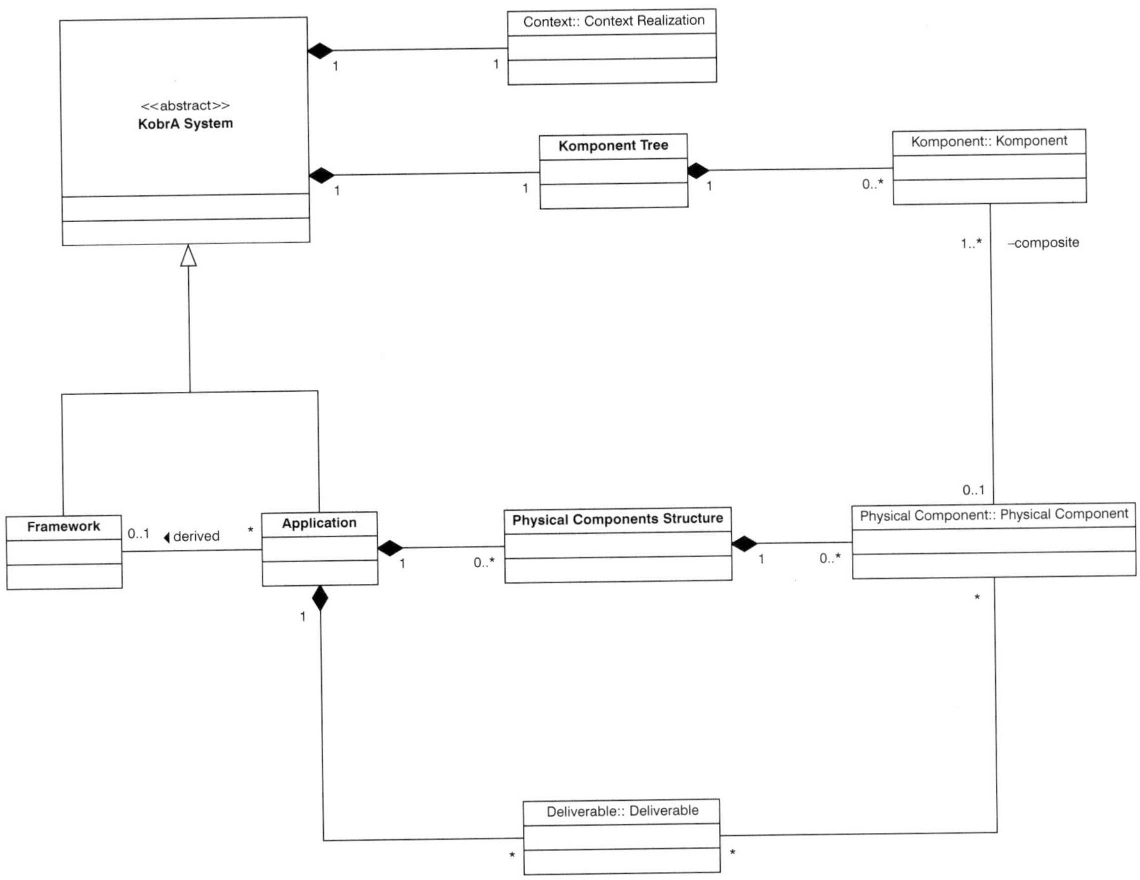

Figure A.8
Static System structure

A.1.6.2 Abstract syntax

KobrA System

A KobrA System is the product of a KobrA development project. It consists of a Context Realization, describing the environment of the system under development and a Komponent Tree which describes the logical entities within the system. There are two distinct incarnations of a KobrA System: a Framework which provides a generic description at the level of a product line, and an Application, which provides a specific description for one member of the product line.

Framework
A Framework is a generic description of the software elements making up a family of Applications. The Framework has Decision Models and can be instantiated to create Applications. This is achieved by resolving all decisions contained within the Decision Models.

Application
An Application is a single, specific system, created by instantiating a Framework. It contains a Physical Components Structure which describes the physical (i.e. runnable) components of the system as well as a Deliverable which represents the final delivered artifacts (i.e. the executable, final system).

Komponent Tree
The Komponent Tree is the major artifact developed during a KobrA development project. It defines a logical containment tree embracing all Komponents within the system.

Physical Components Structure
This Physical Components Structure shows the dependencies between the Physical Components. It also encapsulates the Source Code needed for the creation of an executable representation of individual Physical Components.

A.1.7 Physical Component

A.1.7.1 Overview

The Physical Component sub-package (see Figure A.9) defines the basic structure of a Physical Component. Elements specified in the Physical Component sub-package include Construction Model and Binary.

A.1.7.2 Abstract syntax

Main
The Main construct is part of a component's Source Code. It is basically a program file that is needed in programming languages such as C++ to represent the boundary of a Physical Component.

Komponent Source
The Komponent Source represents the translation of the logical UML models to code in a specific language.

Wrapper
A Wrapper is a Source Code description, which maps a modeled interface to that of an already existing component. Consequently, there is a maximum of one Wrapper per Source Code.

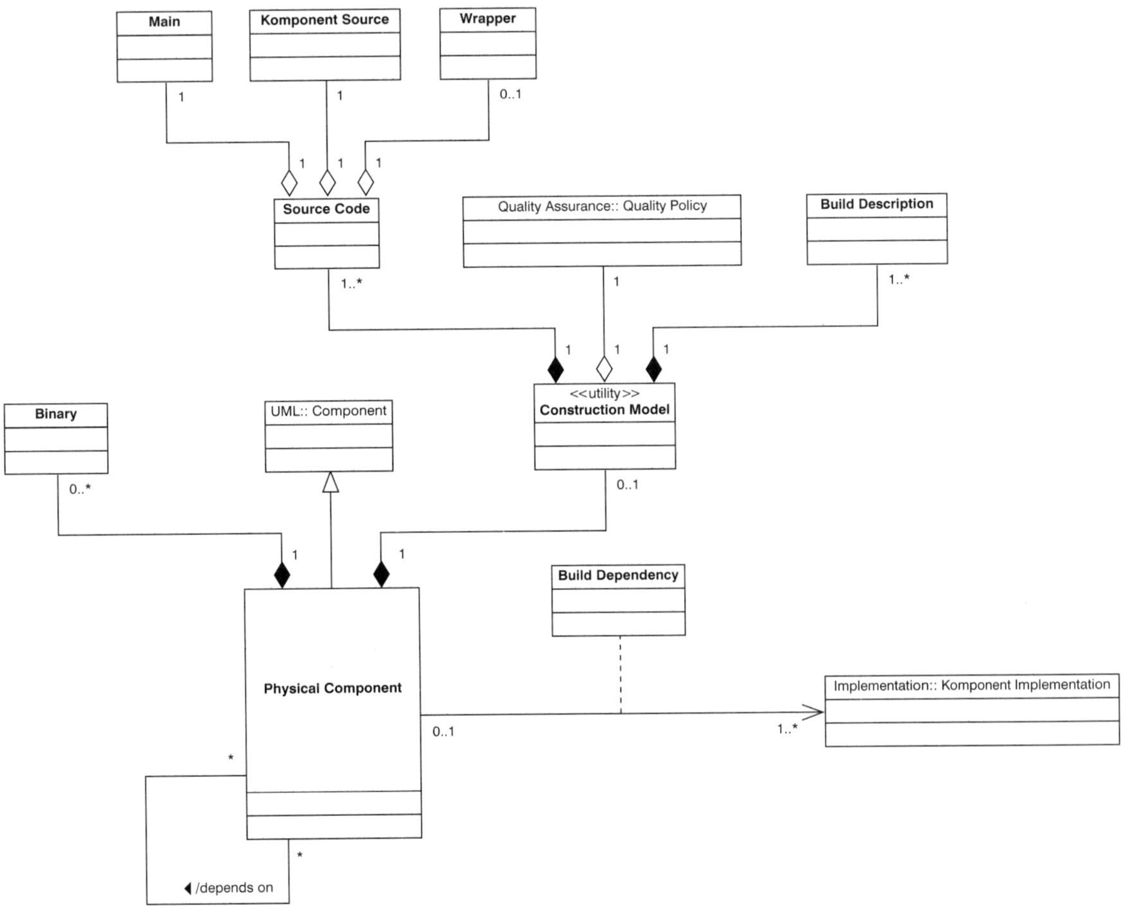

Figure A.9
Static Physical Component structure

Source Code

The Source Code represents a compilable description of a Physical Component. It consists of the Komponent's Source, a Wrapper (optional), and the Main routine needed to initiate execution.

Build Description

A Build Description can be compared with a Makefile in standard system development. It describes the places where individual files can be found, the dependencies between these files, required libraries, and compiler options. It is used during compilation to create an executable image of a Physical Component.

Binary

A Binary is an executable image of a Physical Component created by compiling the associated Source Code. Depending on the number of Source Code representations there may be many binaries for a given Physical Component.

Construction Model
The Construction Model is a logical description of all entities needed for building an executable image of a Physical Component. It consists of a Build Description and the Source Code.

Physical Component
A Physical Component is a separately deployable and executable unit. It is represented by an executable Binary and a Construction Model. Physical Components can depend on other Physical Components (i.e. components which are needed/called at run time).

Build Dependency
A Build Dependency identifies the Komponent Implementations, which are actually used to build the Physical Component.

Depends On
The Depends On relationship describes the dependencies among software components, including source code components, binary code components, and executable components. The properties of this relationship are inherited from the UML Package definition.

A.1.8 Context Realization

A.1.8.1 Overview

The Context Realization sub-package (see Figure A.10) defines the basic structure of a Context Realization. Elements specified in the Context Realization sub-package include all model elements of a Realization plus the Enterprise Model that comprises an Enterprise Concept Diagram and an Enterprise Process Diagram.

A.1.8.2 Abstract syntax

Enterprise Model
The Enterprise Model is a container for one or more Enterprise Concept Diagrams and Enterprise Process Diagrams (see Chapter 7).

Enterprise Concept Diagram
The Enterprise Concept Diagram is a UML Class Diagram, which describes the fundamental enterprise concepts.

Enterprise Process Diagram
The Enterprise Process Diagram is a UML Class Diagram, which shows the nesting hierarchy of enterprise activities.

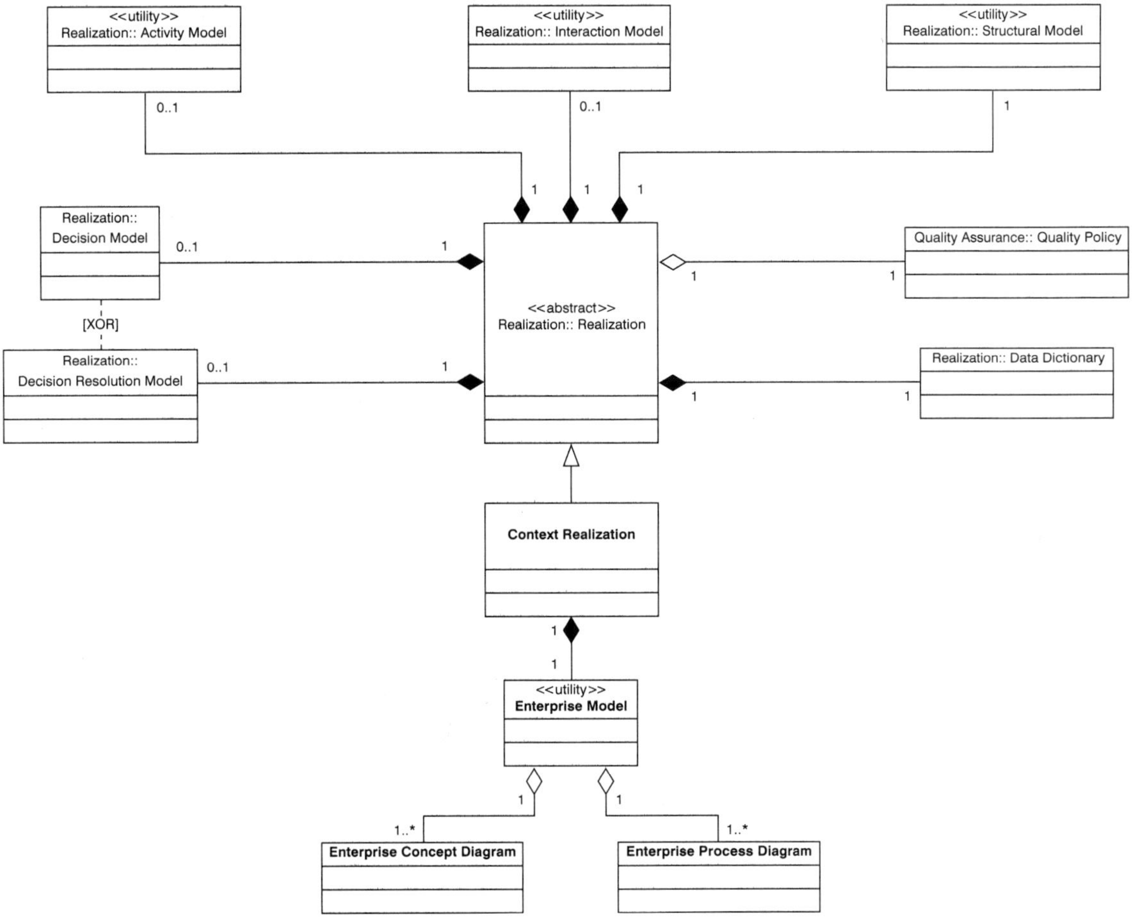

Figure A.10
Static Context Realization Structure

A.1.9 Quality Assurance

A.1.9.1 Overview

The Quality Assurance sub-package (see Figure A.11) defines the basic structure of the quality assurance artifacts developed and used during the framework and application engineering processes. The main elements specified in the Quality Assurance sub-package include the Quality Policy and Quality Documentation. A Quality Policy references zero or more Defect Detection Forms, Defect Collection Forms, Inspections Scenarios and Measurement Plans. The Quality Documentation comprises zero or more Defect Descriptions, Test Results, Test Suites, and Quality Measures.

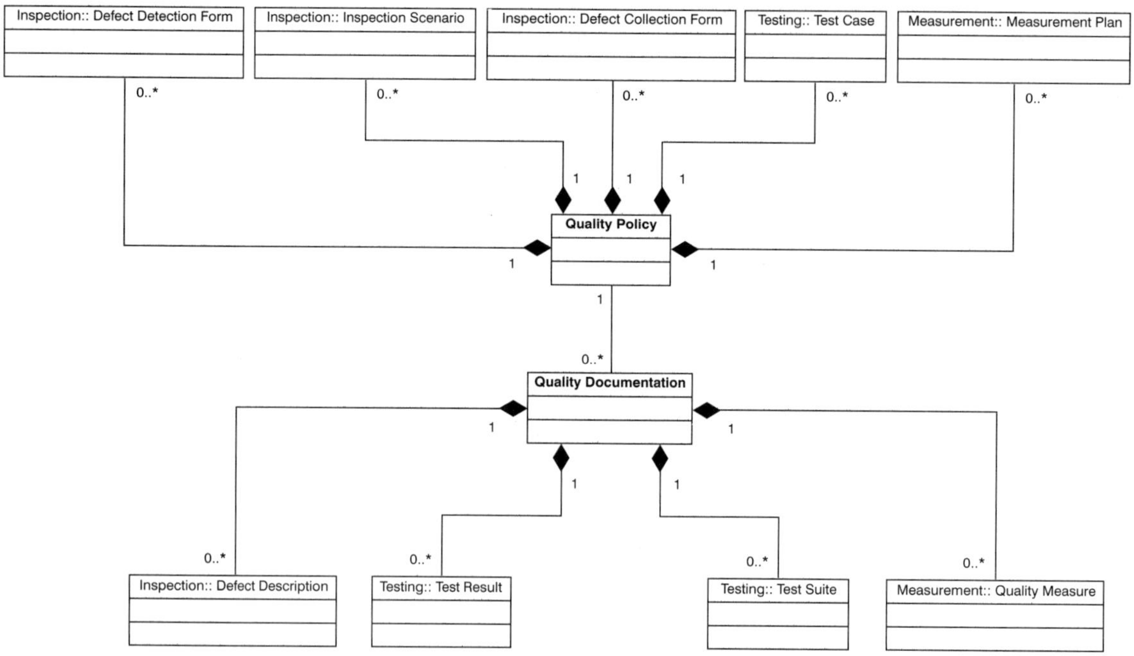

Figure A.11
Static Quality Assurance structure

A.1.9.2 Abstract syntax

Quality Policy
The Quality Policy contains the artifacts and concepts needed to execute the quality assurance program. The artifacts involved include the Defect Detection Forms, the Defect Collection Forms, the Inspection Scenarios, the Quality Models and the Measurement Plan.

Quality Documentation
The Quality Documentation contains the artifacts generated by the execution of the quality assurance program in accordance with the Quality Policy. This includes Defect Descriptions, the Test Suites, Test Results, and the Quality Measures.

A.1.10 Inspection

A.1.10.1 Overview

The Inspection sub-package (see Figure A.12) defines the basic structure of the Inspection artifacts developed and used within the framework and application engineering processes. Elements specified in this sub-package include an Inspection Scenario, a Defect Detection Form, a Defect Collection Form, and zero or more Defect Descriptions.

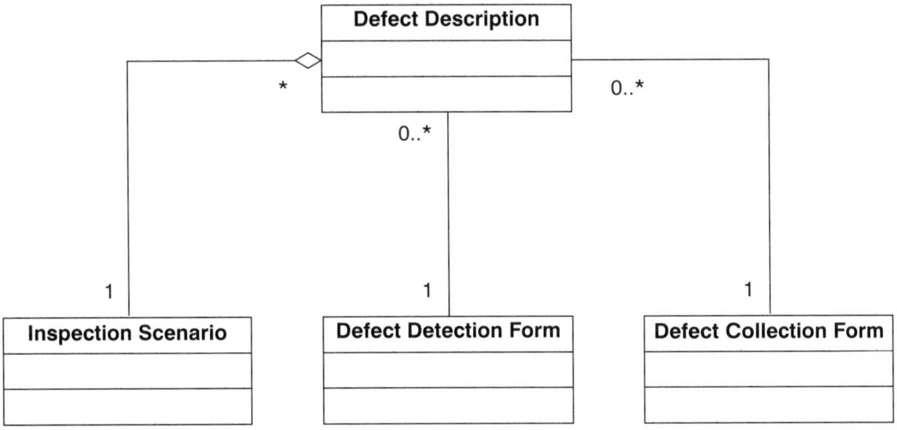

Figure A.12
Static Inspection structure

A.1.10.2 Abstract Syntax

Defect Description
The Defect Description contains the defect data and provides a collection of mechanisms needed to detect, collect and describe these defects. It refers to exactly one Defect Detection Form and one Defect Collection Form, and contains an Inspection Scenario. The Defect Description describes defects according to the Defect Detection Form and Defect Collection Form.

Inspection Scenario
An Inspections Scenario describes how to detect defects, collect defects, and enact corrective steps.

Defect Detection Form
A Defect Detection Form provides a schema to describe defect data and is referenced by zero or more Defect Descriptions.

Defect Collection Form
A Defect Collection Form describes a schema to collect the relevant defect data and it is referenced by zero or more Defect Descriptions.

A.1.11 Measurement

A.1.11.1 Overview

The Measurement sub-package (see Figure A.13) defines the basic structure of Measurement artifacts developed and used within the framework and application engineering processes. Elements specified in this sub-package include a Measurement Plan, zero or more Quality Models, Measurement Procedures, and Quality Measures.

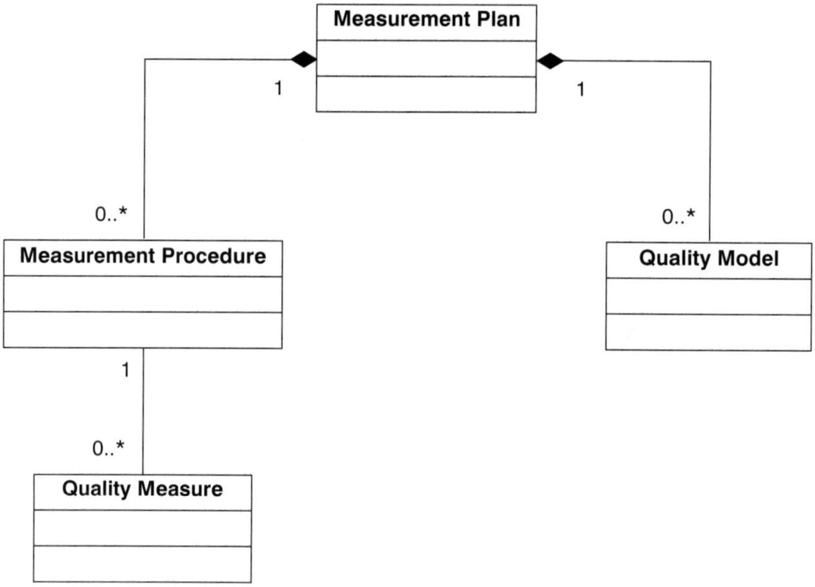

Figure A.13 Static Measurement structure

A.1.11.2 Abstract syntax

Measurement Plan
The Measurement Plan describes how to execute the Measurement Procedures and to apply the Quality Models. It contains zero or more Quality Models and Measurement Procedures.

Quality Model
The Quality Model describes the external attributes of the KobrA artifacts in order to predict their external quality.

Measurement Procedure
The Measurement Procedure describes how to attain the required Quality Measures to predict the external quality of a KobrA artifact. It refers to zero or more Quality Measures.

Quality Measures
The Quality Measures contains the results of the quality measurement activity and refers to exactly one Measurement Procedure.

A.1.12 Testing

A.1.12.1 Overview

The Testing sub-package (see Figure A.14) defines the basic structure of the Testing artifacts developed and used during the framework and application-engineering processes. Elements specified in this package include the Test Suite, the Test Cases, and the Test Results. Three specializations of Test Case are defined: Functional Test Cases, State-Based Test Cases and Structural Test Cases.

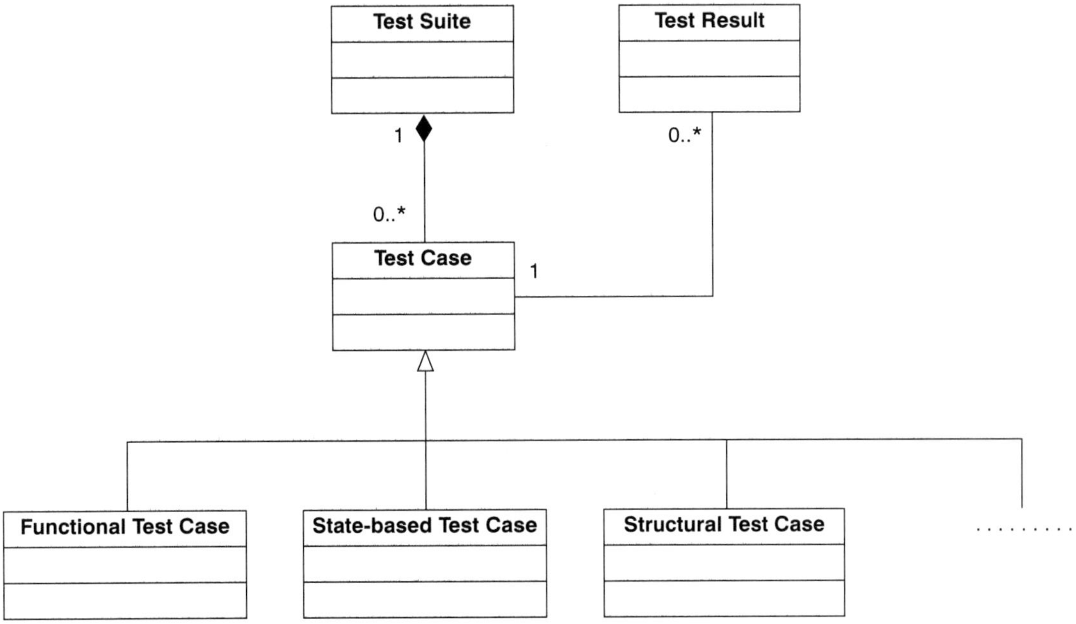

Figure A.14
Static Testing structure

A.1.12.2 Abstract syntax

Test Suite
The Test Suite contains zero or more Test Cases and describes them on the basis of the Komponent's operations in the Komponent Specifications and Komponent Realizations.

Test Result
A Test Result is the outcome of the test execution activity and is input for test result analysis. It is associated with exactly one Test Case.

Test Case
The Test Case describes test scenarios and is contained in exactly one Test Suite.

A.1.13 Implementation Suite

A.1.13.1 Overview

The Implementation Suite sub-package (see Figure A.15) defines a tailored set of modeling concepts together with rules for the implementation level at which refinement ends and translation is performed in the Komponent implementation activity according to the UML profile mechanism. Elements specified in this package include the Implementation Profile, the Normal Object Form, the Implemenation Suite, the Refinement Pattern, and the Translation Pattern.

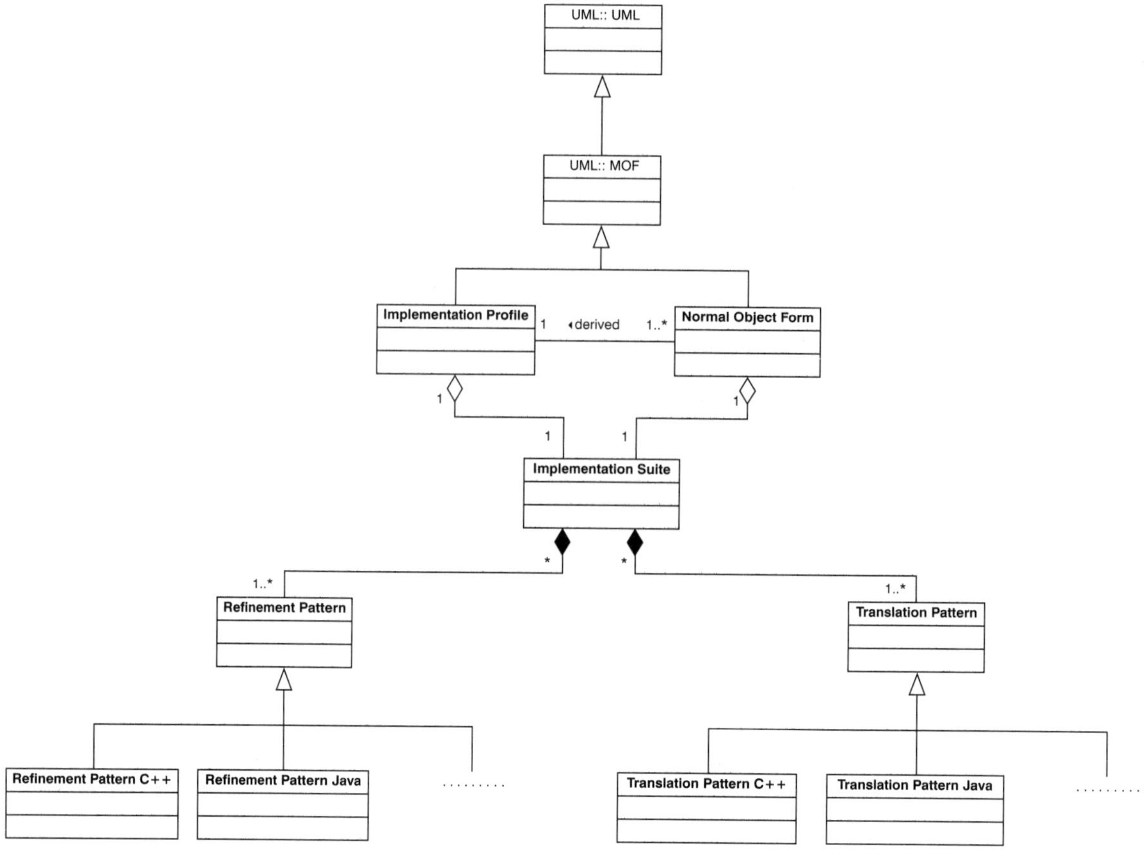

Figure A.15
Static Implementation Suite structure

A.1.13.2 Abstract syntax

Normal Object Form
The Normal Object Form (NOF) is a pre-defined UML profile that captures the core concepts of object-oriented programming (see Section 11.2.1). It is an

extension of the UML core metamodel which defines new model elements and constraints on the use of the new and old model elements, using the UML's built-in extension mechanisms.

Implementation Profile
An Implementation Profile is a profile of the UML tailored towards the creation of a tool-comprehensible (i.e. implementation level) structural model of a Komponent. It therefore contains elements semantically close to the constructs of a programming language or other implementation vehicle.

Implementation Suite
The Implementation Suite is a container for the company-specific Translation and Refinement Patterns. It belongs to exactly one Implemetation Profile and can contain one or more Translation and Refinement Patterns. It has to be constructed during the company-specific customization of the KobrA method.

Refinement Pattern
A Refinement Pattern defines a refinement mapping from UML regular modeling elements to NOF-UML elements (see Section 11.2.1 and 11.2.2). The available Refinement Patterns are collected in the Implementation Suite.

Translation Pattern
A Translation Pattern defines a refinement mapping from NOF-UML elements to the Source Code (see Section 11.2.1 and 11.2.2). The available Translation Patterns are collected in the Implementation Suite.

A.2 Maintenance

The Maintenance package (see Figure A.16) contains three sub-packages, which define the elements associated with the KobrA maintenance activities. The Asset sub-package characterizes KobrA development artifacts in the form used by the configuration and change-management activities. The Change sub-package describes the nature of the changes that drive the maintenance process, and how they relate to the assets. Finally, the Deliverable sub-package describes the concepts used to generate configurations of assets that can be delivered to customers.

A.2.1 Assets

A.2.1.1 Overview

The Asset sub-package (see Figure A.17) defines the basic structure of Assets developed and used within the framework and application engineering processes. Elements specified in this package include the Realization Dependency and the Contract Dependency between Komponent Realization, and the Komponent Specification, the Implementation Dependency between the Komponent Implementation and Komponent Realization, and the Build Dependency between Komponent Implementation and Komponent

Realization. These associations are needed to describe the dependencies between the different Komponent-related artifacts.

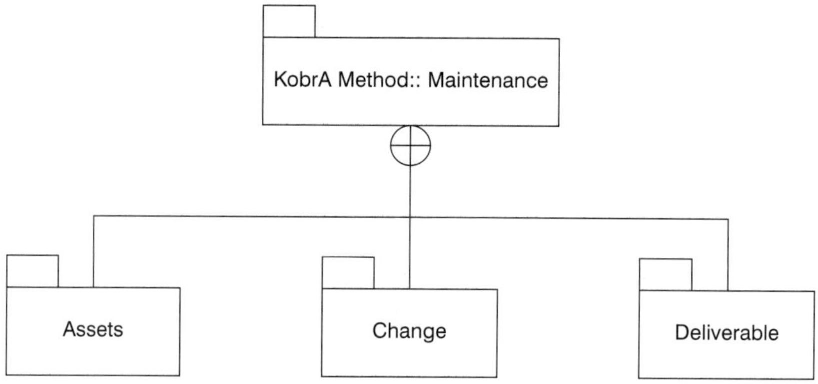

Figure A.16 Maintenance package structure

A.2.1.2 Abstract syntax

Realization Dependency
The Realization Dependency associates a Komponent Realization with exactly one Komponent Specification, thereby defining the Komponent Realization that realizes the Komponent Specification and fixing the state of the association for quality assurance. For each new version of a Komponent Realization or Komponent Specification a new Realization Dependency is created so the state of the different versions can be checked independently.

Contract Dependency
The Contract Dependency associates a Komponent Specification with exactly one Komponent Realization, thereby establishing a client–server relationship between Komponents on the Komponent Specification and Realization level and fixing the state of the association for quality assurance. For each new version of a Komponent Specification or Realization a new Contract Dependency is created so the state of the different versions can be checked independently.

Implementation Dependency
The Implementation Dependency associates a Komponent Implementation with exactly one Komponent Realization and it describes which Komponent Realization is implemented by a Komponent Implementation. A Komponent Realization can be implemented by zero or more Komponent Implementations. The Implementation Dependency fixes the state of the association for quality assurance. For each new version of the Komponent Realization or Implementation a new Implementation Dependency is created so the state of the different versions can be checked independently.

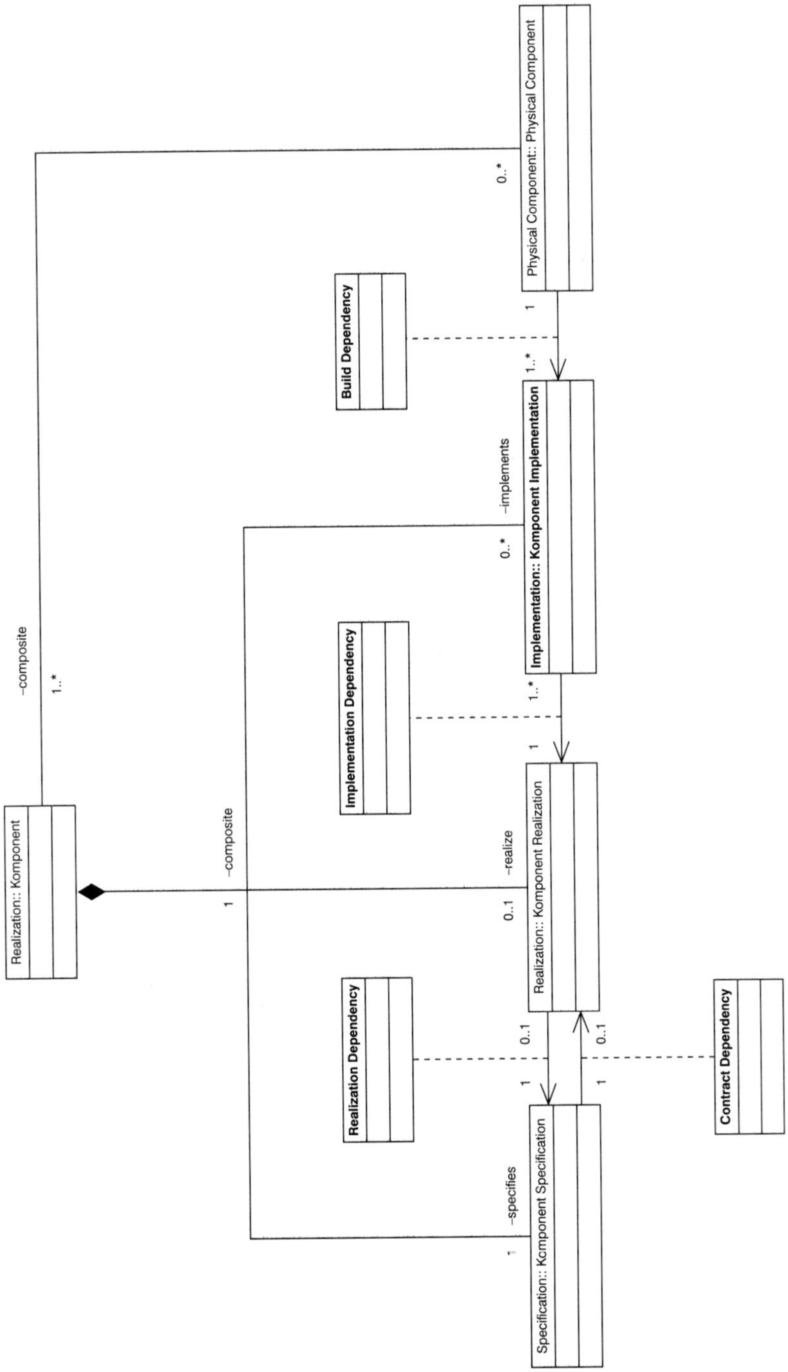

Figure A.17
Static Asset structure

Build Dependency
The Build Dependency associates a Physical Component with one or more Komponent Implementations, thereby capturing the build relationship between Komponent Implementations and Physical Components and fixing the build-state of the associations for quality assurance. For each new version of a Komponent Implementation or Physical Component a new Build Dependency is created so the state of the different versions can be checked independently.

A.2.2 Change

A2.2.1 Overview

The Change sub-package (see Figure A.18) defines the basic structure of Changes used to drive the framework and application engineering processes. Elements specified in the Change sub-package include Change, Change Requests, Change Information, Change Justification, the Change Operator, Tasks, and Change Set.

A.2.2.2 Abstract syntax

Change
The Change associates a Change Request with the Assets that need to be modified, and serves as the container for the Change Operators that will be applied to the Asset. A Change can result from all development activities of the framework engineering and application engineering processes in conjunction with the change management process.

Change Request
A Change Request can be created by a developer or a customer and captures the rationale for a Change. A Change Request is therefore associated with one or more Change Justifications. In addition, it is a container for the tasks that will be initiated by the Change Request.

Change Description
A Change Description provides details about the nature of a Change. Each change activity within the framework and application engineering processes generates a Change Description, which is referenced by exactly one Change.

Change Justification
A Change Justification defines the rationale and cost–benefit evaluation that justifies a Change. A Change Justification is associated with one or more Change Requests.

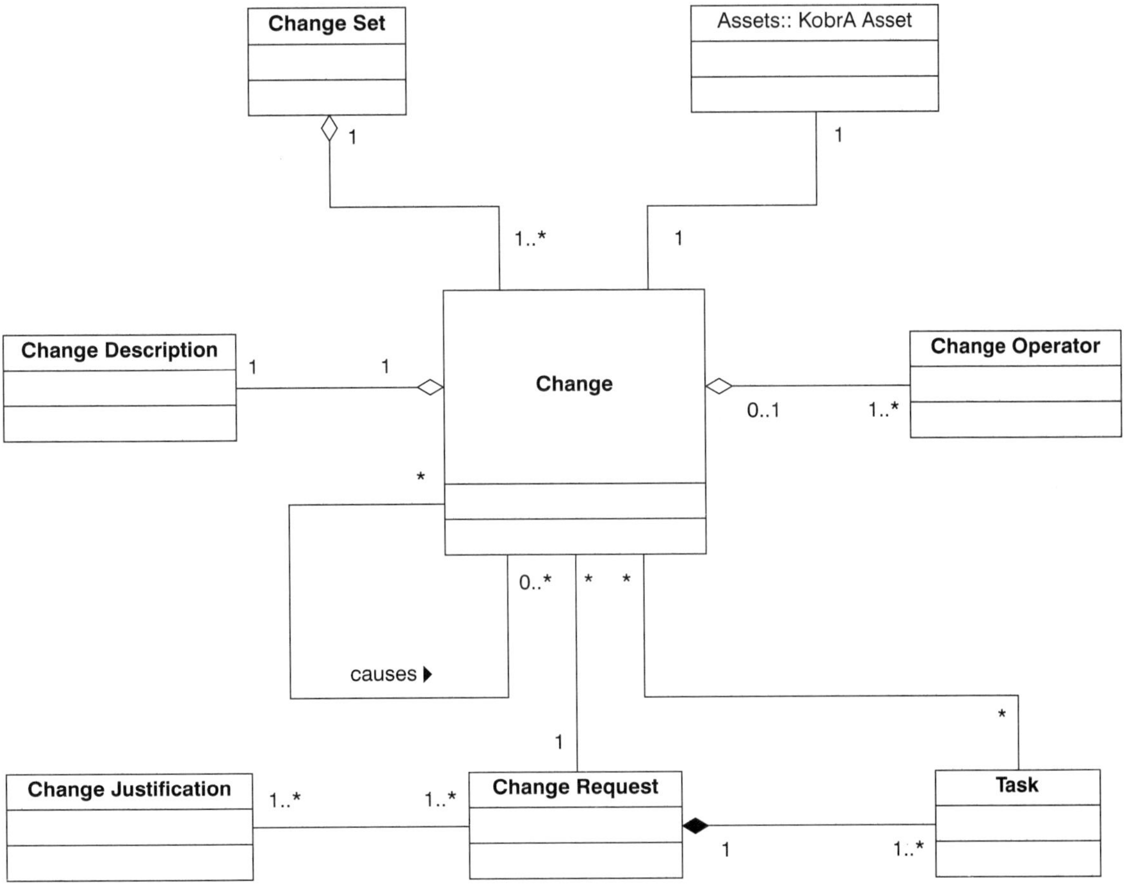

Figure A.18
Static Change structure

Change Operator

A Change Operator represents a discrete conceptual change to a configuration item in a product line. A Change encapsulates all the different conceptual operators that are needed to perform the associated modifications.

Change Set

A Change Set encapsulates a set of Changes made during an activity so that they can be transferred between different Framework and Application versions. Each Change is therefore contained in a Change Set to facilitate this transition.

Task

Each Change Request is divided into a set of Tasks that can be better distributed and allocated among developers. Each Change Request therefore contains one or more Tasks.

A.2.3 Deliverable

A.2.3.1 Overview

The Deliverable sub-package (see Figure A.19) defines the basic structure of a Deliverable produced by the application engineering process. Elements specified in this sub-package include Deliverable, Release Description, and Deliverable Information, which in turn contains one or more Deployment Descriptions and Deployment Diagrams.

Figure A.19
Static Deliverable structure

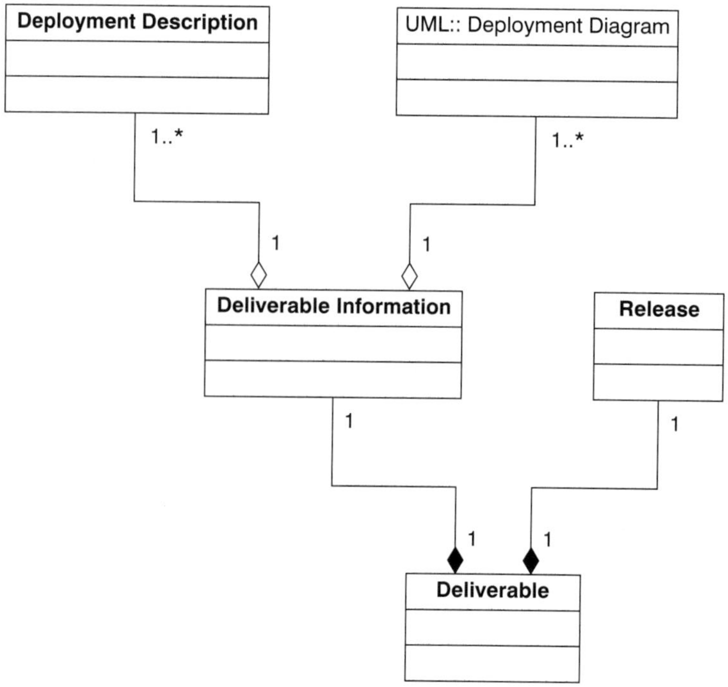

A.2.3.2 Abstract syntax

Deployment Description
The Deployment Description describes how a certain Deliverable will be deployed to customers. It explains how the Deliverable will be distributed and installed or updated on the customer systems.

Deployment Diagram
The Deployment Diagram provides a graphical depiction of the installation of a system's components using a UML Deployment Diagram.

Release

The Release contains information about the binary releases within the Physical Components belonging to a Deliverable. A Deliverable contains exactly one Release.

Deliverable

The Deliverable contains and describes the end product that will be delivered to a customer and contains Deliverable Information that comprises one or more Deployment Descriptions and Deployment Diagrams.

APPENDIX

B Process

This appendix complements the informal, natural language description of the KobrA process provided in the main part of the book with a concise, semi-formal process model. The process metamodel and the notation used for the process model are introduced in Section B.1, while the process model itself is presented in Section B.2.

In general, the complete description of a software engineering method involves four major components [FH99]:

- **Work products.** The work products of the KobrA method are described in the respective chapters of this book. What kind of products are involved and how they interrelate is described by the KobrA metamodel in Appendix A.
- **Modeling language.** The KobrA method makes extensive use of the UML [OMG00a], supplemented by (structured) natural language.
- **Techniques.** The techniques to be applied when dealing with certain products are described in the respective chapters of this book.
- **Modeling process.** The KobrA method's modeling process is partly described in the respective chapters of this book. A compact overview is provided in this appendix.

An electronic version of the process description – an Electronic Process Guide (EPG) – is available on the KobrA website [www.iese.fhg.de/KobrA/book].

B.1 Process metamodel and notation

The process model used to describe the KobrA method's modeling process is based on a small set of very simple concepts. These concepts are introduced in Section B.1.1 in terms of the process metamodel.

Since the whole book is focused on component-based product line engineering with UML, we have chosen the UML for representing the process as well. Section A.1.2 explains the notation used for the process model.

B.1.1 Process metamodel

The process metamodel used to describe the KobrA method's modeling process is shown in Figure B.1. The following simple and intuitive process concepts are used:

- **Activity.** Something that needs to be done. An activity may be performed by one person alone, by a team, or by a complete organization. The term **process** is used for large and complex activities that are themselves decomposed into many lower-level activities, forming a conceptual unit.
- **Activity containment.** Activities can be decomposed into smaller activities. There is no special term for atomic activities.
- **Artifact.** Something that is used or produced during the process, for example, a document, a UML model, or a source code file. The types of artifacts involved in a KobrA project are defined in Appendix A.
- **Artifact containment.** Artifacts can be decomposed into smaller artifacts. These relationships are also defined in Appendix A.
- **Product flow.** Activities can consume, modify, or produce artifacts. An artifact is usually produced by at most one activity, but it may be consumed (used without changing it) or modified (used and changed) by an arbitrary number of activities. (If an artifact is produced by an activity and this activity is part of a more complex activity, the artifact may appear as a product of the complex activity as well.)

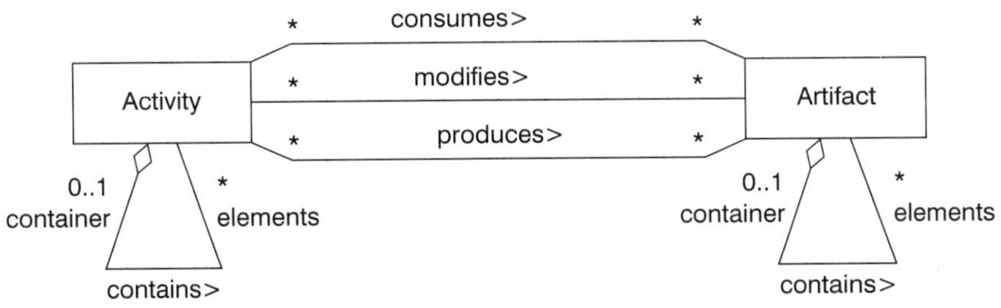

Figure B.1
Process metamodel

It should be noted that the process metamodel only captures the principal ingredients of a process, that is, its separate parts and the logical dependencies between them. This is called the static structure of the process. The static process structure helps in understanding the process, but it does not prescribe a specific order in which activities are to be performed. However, logical dependencies between activities impose certain constraints that need to be considered by a project manager who is planning process performance. The static process structure can be understood as a set of process patterns that are applied in an architecture-centric way.

The dynamic aspects involved in applying the KobrA method are discussed in Chapter 13 and Chapter 20.

Table B.1 Stereotypes and notation for process models

Stereotype	UML Activity Diagram Element	Notation
Activity	Action State	
Artifact	Object, Object in State	
Product Flow	Object Flow	Dashed Arrow

B.1.2 Notation

The KobrA method's modeling process is presented textually and graphically:

- Activities and their decomposition form a hierarchy, which is presented as a textual list with appropriate indentation (see Section B.2.1). The same applies for artifacts and their decomposition (see Section B.2.2).
- Product flow between activities (and processes) is depicted graphically using UML activity diagrams (see Section B.2.3).[57]
- Table B.1 lists the stereotypes that have been defined in the process metamodel and their associated extended notation.

B.1.3 Background

The process metamodel introduced in Section B.1.1 is a simplified version of the one presented in [WB97]. It is also compatible with both the OPEN process metamodel [FH99] and the concepts underlying the Rational Unified Process (RUP) [Kru00]. Table B.2 provides a mapping between the different terminologies.

Given the KobrA method's use of a standard modeling notation (the UML) it would obviously have been desirable to have followed a standard approach to software process modeling as well. This was not possible at the time of writing, however.

Object Management Group (OMG) is in the process of soliciting proposals for specification of a Software Process Engineering metamodel or UML Profile for Software Process Engineering. The longer-term aim of this request is to provide

Table B.2 Process terminology translation table (based on [OMG01])

KobrA	OPEN [GHY97]	RUP [Kru00]	SPEM [OMG01]	ISO/IEC 12207 [ISO12207]
Process	Activity	Discipline	Discipline	Process
Activity	Task, Subtask	Activity, Step	Activity, Step	Task
Artifact	Deliverable	Artifact	Work product, Information element	Product

57. The diagrams in this chapter were drawn with SPEARMINT™, a software process modeling tool developed by Fraunhofer IESE. More information on SPEARMINT™ can be found in [BHK99] and at [www.iese.fhg.de/Spearmint_EPG/].

inter-operability and repository facilities in the software-process-engineering field. More than two years after this process has been initiated, several companies (IBM, Rational Software, Fujitsu/DMR, and others) have finally combined their proprietary ideas into one single proposal called the Software Process Engineering Metamodel (SPEM) [OMG01]. The SPEM is intended to be useful for describing concrete software development processes, or families of related software development processes. The SPEM adopts an object-oriented approach to modeling, and employs the UML notation. Consequently, it is defined as a UML profile.

However, the SPEM is currently just in the proposal stage and at the time of writing has not been voted on. In the absence of a recognized standard, therefore, we have adopted a simple but powerful approach to describing the KobrA method's process.

B.2 Process model

The KobrA method defines a process for component-based product line engineering that integrates aspects of multiple software engineering disciplines: object-oriented modeling, component-based development, product line engineering, and quality assurance. The process description tries to make this explicit wherever possible.[58]

The technical activities of the product line engineering process are accompanied by managerial activities related to project management, configuration and change management, and environment management. Together, these four top-level processes (Figure B.2) form a portfolio similar to the core workflows of the Rational Unified Process.

Of these top-level processes, the product line engineering process is described in more detail in the subsequent sections:

- Section B.2.1 gives an overview of all activities that make up the product line engineering process at different levels of abstraction.
- Section B.2.2 gives an overview of all artifacts that are used or produced in the product line engineering process at different levels of abstraction.
- Section B.2.3 describes all product flows among activities in the product line engineering process at different levels of abstraction.

B.2.1 Activities

KobrA's product line engineering process consists of two primary sub-processes: the framework engineering process and the application engineering process. Each is further decomposed into a set of activities and sub-activities. The motivation for, and details of, these activities are described extensively in the main part of the book. This section provides a compact overview of all activities in one place. Figure B.3 shows the complete decomposition structure of the product line engineering process.

58. Extreme Programming (XP) [Bec00] is a good counter-example: XP is a software development process based on the combination of 12 so-called practices. However, since there is no comprehensive description of how these practices are performed together it is unclear what the overall process looks like.

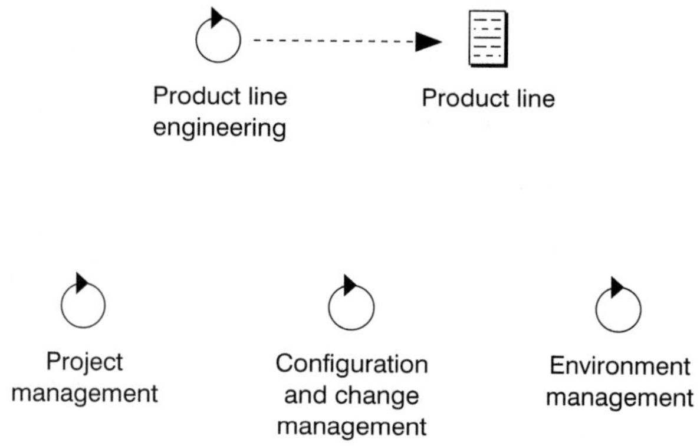

Figure B.2
Top-level processes of the KobrA method

Figure B.3
Product line engineering process: activities

Product line engineering

Framework engineering

- **Context realization**
  - Enterprise modeling
  - Structural modeling
  - Usage modeling
  - Interaction modeling
  - Variability identification
  - Decision modeling
  - Komponent identification
- **Komponent specification**
  - Structural modeling
  - Functional modeling
  - Behavioral modeling
  - Variability identification
  - Decision modeling
- **Komponent realization**
  - Structural modeling
  - Activity modeling
  - Interaction modeling
  - Variability identification
  - Decision modeling
  - Komponent identification
  - Containment tree refactoring
- **Komponent implementation**
  - Flattening
  - Refinement
  - Translation
  - Finalization
  - Variability identification
  - Decision modeling
- **Component reuse**
  - Component selection
  - Negotiation
  - Component adaptation
  - Containment tree adaptation
- **Inspection**
  - Inspection planning
  - Defect detection
  - Defect collection
  - Enactment of corrective steps
- **Measurement of structural properties**
  - Measurement planning

- Quality measurement
- Quality analysis

- **Testing**
  - Test suite definition
  - Test execution
  - Test result analysis

Application engineering

- **Application context realization**
  - Instantiation
  - Enterprise modeling
  - Structural modeling
  - Usage modeling
  - Interaction modeling
- **Application Komponent specification**
  - Instantiation
  - Structural modeling
  - Functional modeling
  - Behavioral modeling
- **Application Komponent realization**
  - Instantiation
  - Structural modeling
  - Activity modeling
  - Interaction modeling
- **Application Komponent implementation**
  - Instantiation
  - Flattening
  - Refinement
  - Translation
  - Finalization
- **Construction**
- **Component building**
- **Releasing**
- **Inspection**
  - Inspection planning
  - Defect detection
  - Defect collection
  - Enactment of corrective steps
- **Measurement of structural properties**
  - Measurement planning
  - Quality measurement
  - Quality analysis
- **Testing**
  - Test suite definition
  - Test execution
  - Test result analysis

Figure B.3 continued

B.2.2 Artifacts

KobrA's product line engineering process makes use of many artifacts: Komponents, models, UML diagrams, textual descriptions, etc. The KobrA meta-model in Appendix A defines precisely which types of artifacts exist, and how they are related to one another.

In the process model describing the product line engineering process, instances of the artifact types are used to illustrate the context of, and the interdependencies between, activities (see product flow graphs in the subsequent sections). Figure B.4 gives an overview of all artifacts that are used in the process model.

Figure B.4
Product line engineering process: artifacts

Product line

Framework

- **Generic context realization**
  - Generic enterprise model
  - Generic structural model
  - Generic activity model
  - Generic interaction model
  - Generic data dictionary
  - Quality policy
  - Quality documentation
  - Decision model
- **Generic containment tree**
  - Generic Komponent
    - Generic Komponent specification
    - Generic structural model
    - Generic behavioral model
    - Generic functional model
    - Generic non-functional requirements specification
    - Quality policy
    - Quality documentation
    - Decision model
  - Generic Komponent specification (abstract)
  - Generic Komponent realization
    - Generic structural model
    - Generic activity model
    - Generic interaction model
    - Generic data dictionary
    - Quality policy
    - Quality documentation
    - Decision model
  - Generic Komponent implementation
    - Generic physical component model
    - Generic structural model
    - Generic pseudo-code
    - Generic source code
    - Quality policy
    - Quality documentation
    - Decision model
  - External component
    - Reuse candidate (intermediate)
    - Conformance map
    - Reused component
    - Agreed specification
    - List of adaptations
    - Semantic map
    - Wrapper Komponent
- **Quality policy**
  - Subject artifact
  - Inspection scenario
  - Defect detection form
  - Defect collection form
  - Measurement plan
    - Measurement procedure
    - Quality model
- **Quality documentation**
  - Quality measure
  - Test suite
  - Test results

Application

- **Specific context realization**
  - Specific enterprise model
  - Specific structural model
  - Specific activity model
  - Specific interaction model

Figure B.4
continued

- Specific data dictionary
- Quality policy
- Quality documentation
- Decision resolution model

- **Specific containment tree**
  Specific Komponent
  - Specific Komponent specification
    - Specific structural model
    - Specific behavioral model
    - Specific functional model
    - Specific non-functional requirements specification
    - Quality policy
    - Quality documentation
    - Decision resolution model
  - Specific Komponent specification (abstract)
  - Specific Komponent realization
    - Specific structural model
    - Specific activity model
    - Specific interaction model
    - Specific data dictionary
    - Quality policy
    - Quality documentation
    - Decision resolution model
  - Specific Komponent implementation
    - Specific physical component model
    - Specific structural model
    - Specific pseudo-code
    - Specific source code
    - Quality policy
    - Quality documentation
    - Decision resolution model

External component
  - Reuse candidate (intermediate)
  - Conformance map
  - Reused component
  - Agreed specification
  - List of adaptations
  - Semantic map
  - Wrapper Komponent

- **Physical components structure**
  Physical component
  - Construction model
    - Source code
    - Build description
    - Quality policy
    - Quality documentation
  - Binary

- **Deliverable**
  - Release
  - Deployment information

- **Quality policy**
  - Subject artifact
  - Inspection scenario
  - Defect detection form
  - Defect collection form
    - Measurement plan
      - Measurement procedure
      - Quality model

- **Quality documentation**
  - Quality measure
  - Test suite
  - Test results

B.2.3 Product flow

Activities depend on each other in terms of artifacts they exchange, and in terms of tasks that need to be performed in a particular sequence. The latter is determined by project management and is documented in a project plan. Product flow dependencies guide the project manager in defining a meaningful task order.

In this section, the product flow within the product line engineering process is described in more detail. For every activity including the process itself, the product flow between its constituent activities is shown. If the constituent activities are themselves decomposed, the product flow between their respective constituent activities is shown, again. This is repeated until the lowest level of granularity is reached.

B.2.3.1 Product line engineering

The product line engineering process is the most important top-level process of the KobrA method (see Figure B.2). It consists of the following sub-processes:

- framework engineering
- application engineering.

They are described separately in Section B.2.3.2 and Section B.2.3.3. The product flow between them is shown in Figure B.5.

Figure B.5
Product line engineering

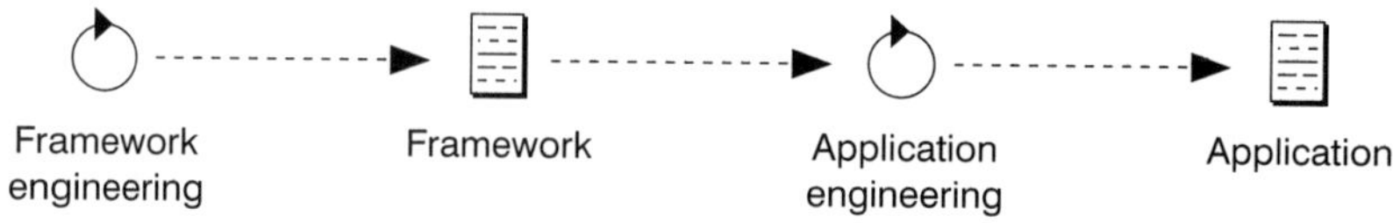

It is important to note that Figure B.5 does not imply sequential execution. As explained in Section B.1.1, the process model only defines the static structure of the process, not its dynamic performance. Figure B.5 states that application engineering needs a framework as input and can thus only be performed if such a framework is at least partly available. However, framework engineering can continue in parallel with application engineering, or can be re-entered later to make changes to the framework.

The performance of the process must be managed with concepts like product life cycle, phases, and iterations, which are tightly related to project management. These concepts are out of the scope of the process model and of this appendix. They are discussed in Chapter 13 and Chapter 20.

B.2.3.2 Framework engineering

The framework engineering process is part of the product line engineering process (see Figure B.5). It consists of the following activities:

- context realization (Section B.2.3.2.1)
- Komponent specification (Section B.2.3.2.2)

- Komponent realization (Section B.2.3.2.3)
- Komponent implementation (Section B.2.3.2.4)
- component reuse (Section B.2.3.2.5)
- inspection (Section B.2.3.2.5)
- measurement of structural properties (Section B.2.3.2.7)
- testing (Section B.2.3.2.8)

All activities are further decomposed to show their internal structure. They are described separately in the respective sections. The product flow between them is shown in the Figure B.6:

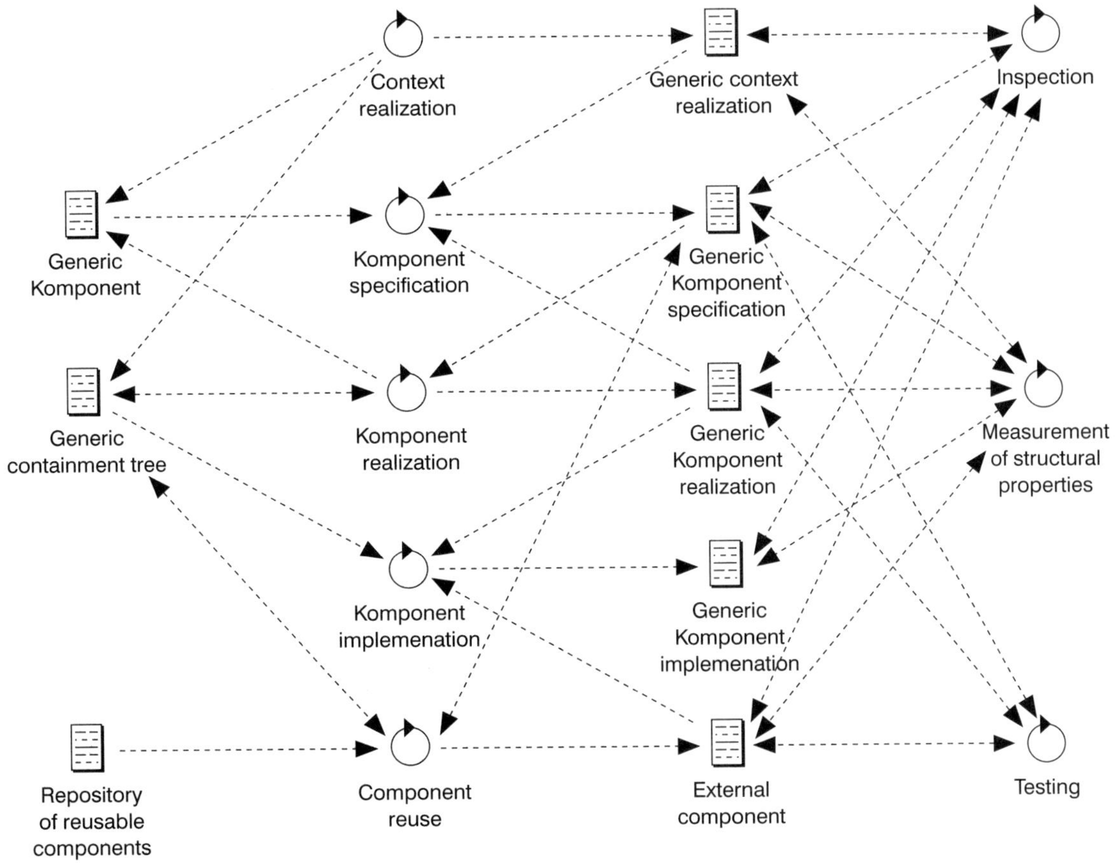

Figure B.6
Framework engineering

B.2.3.2.1 Context realization

The context realization activity is part of the framework engineering process (see Figure B.6). It consists of the following activities:

- enterprise modeling

- structural modeling
- usage modeling
- interaction modeling
- variability identification
- decision modeling
- Komponent identification.

The product flow between them is shown in Figure B.7.

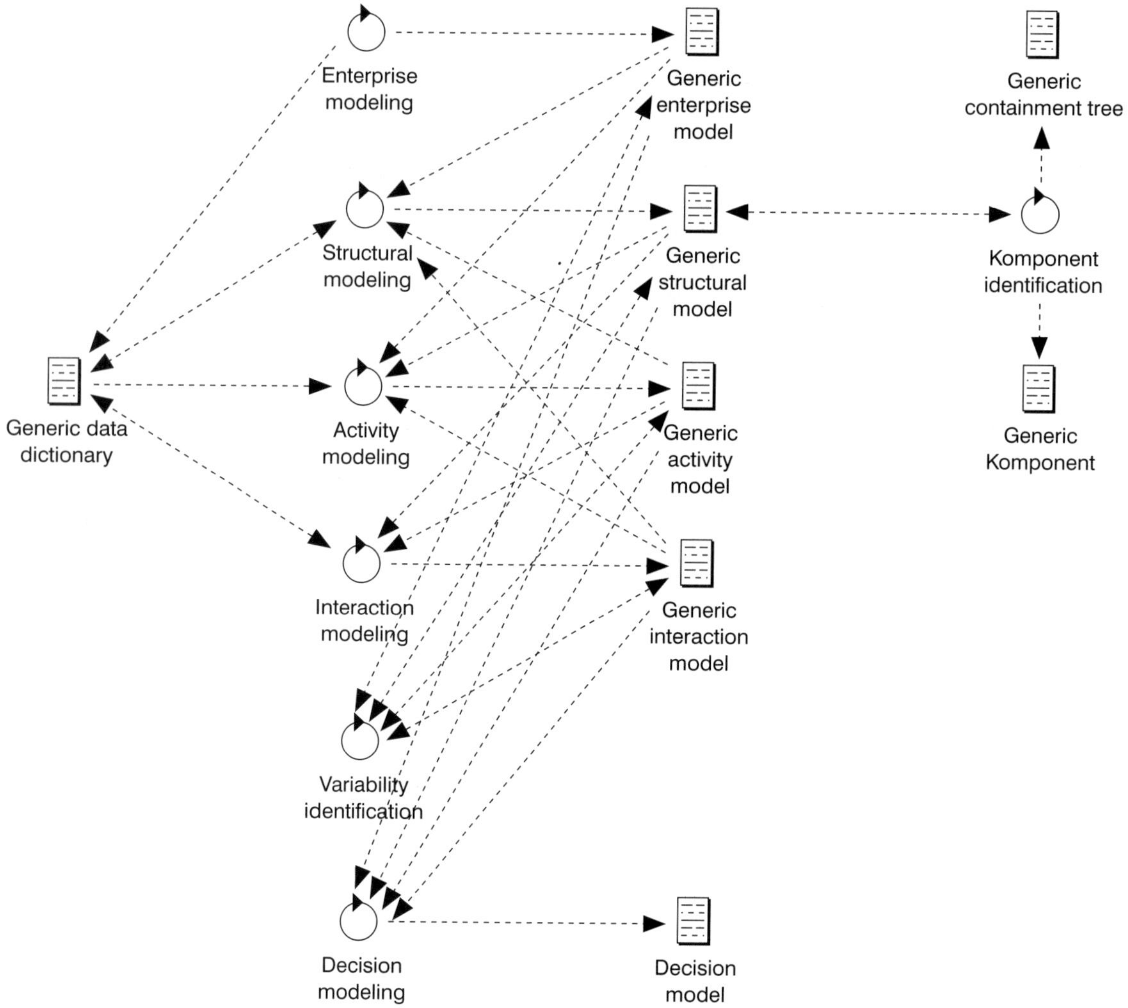

Figure B.7
Context realization

The context realization activity is described in Chapter 7 and Chapter 15.

B.2.3.2.2 Komponent specification

The Komponent specification activity is part of the framework engineering process (see Figure B.6). It consists of the following activities:

- structural modeling
- functional modeling
- behavioral modeling
- variability identification
- decision modeling.

The product flow between them is shown in Figure B.8:

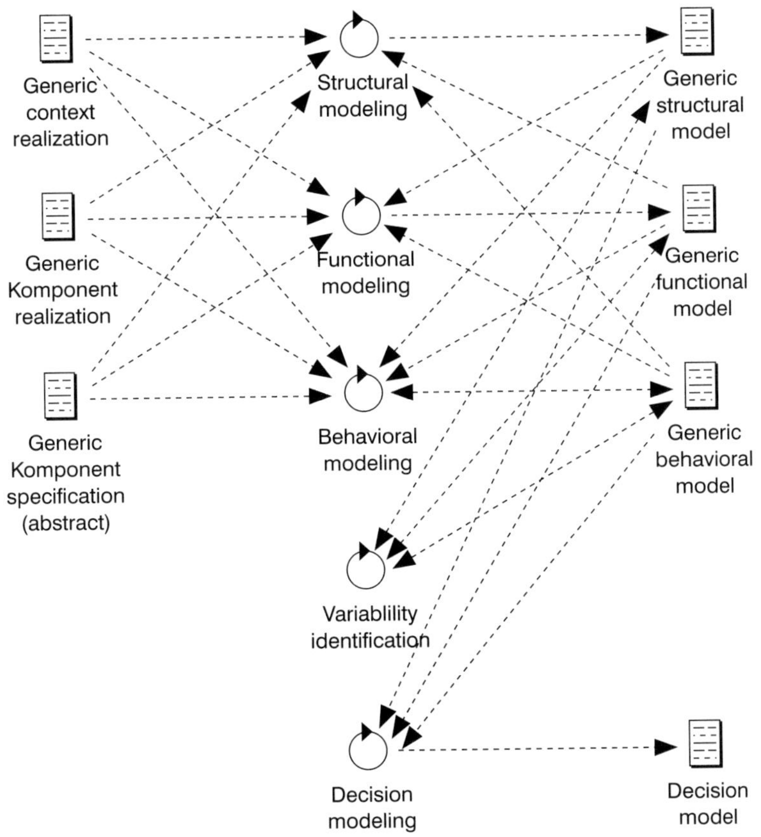

Figure B.8 Komponent specification

The Komponent specification activity is described in Chapter 4, Chapter 9, and Chapter 15.

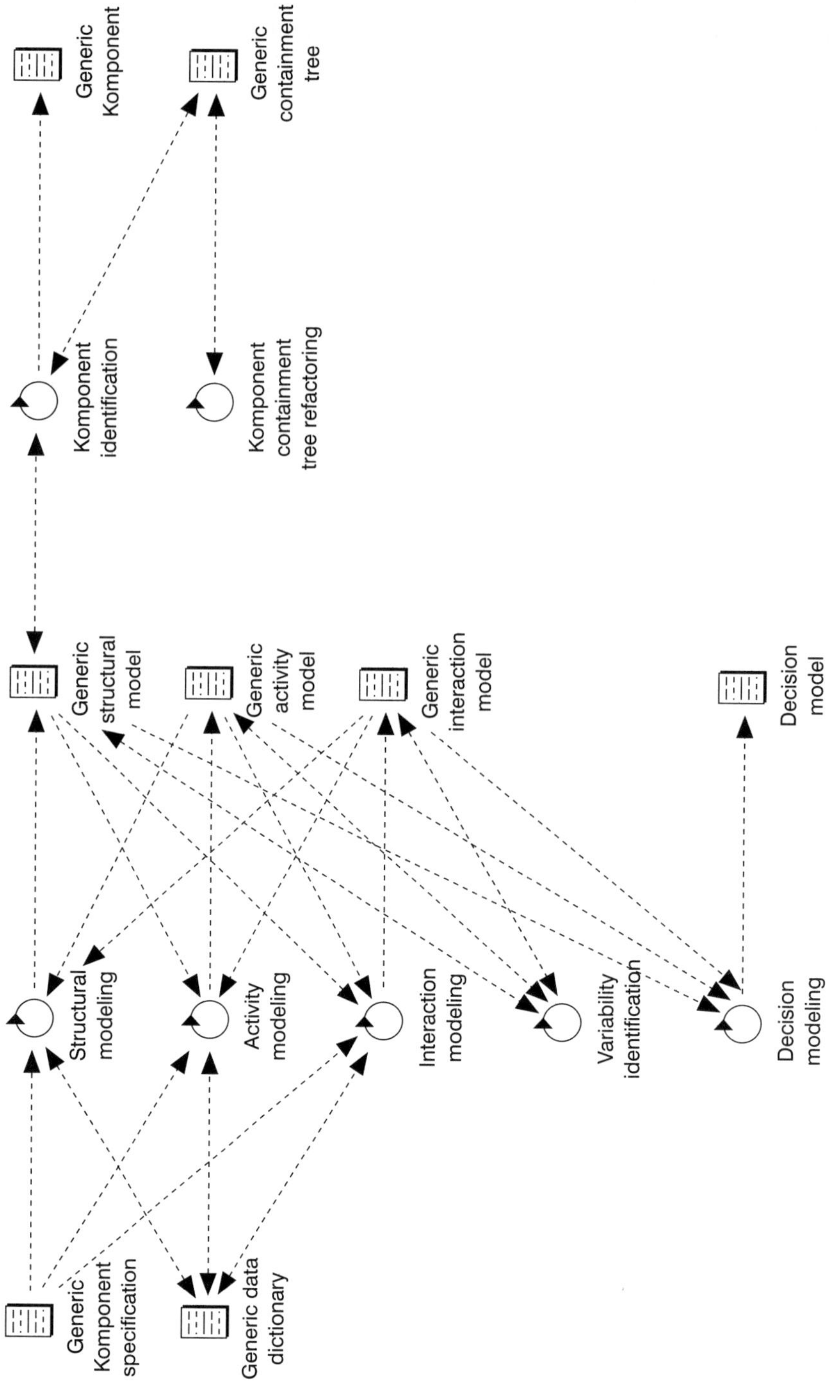

Figure B.9
Komponent realization

B.2.3.2.3 Komponent realization

The Komponent realization activity is part of the framework engineering process (see Figure B.6). It consists of the following activities:

- structural modeling
- activity modeling
- interaction modeling
- variability identification
- decision modeling
- Komponent identification
- containment tree refactoring.

The product flow between them is shown in Figure B.9. The Komponent realization activity is described in Chapter 5 and Chapter 15.

B.2.3.2.4 Komponent implementation

The Komponent implementation activity is part of the framework engineering process (see Figure B.6). It consists of the following activities:

- flattening
- refinement
- translation
- finalization
- variability identification
- decision modeling.

The product flow between them is shown in Figure B.10. The Komponent implementation activity is described in Chapter 11 and Chapter 15.

B.2.3.2.5 Component reuse

The component reuse activity is part of the framework engineering process (see Figure B.6). It consists of the following activities:

- component selection
- negotiation
- component adaptation
- containment tree adaptation.

The product flow between them is shown in Figure B.11. The component reuse activity is described in Chapter 12.

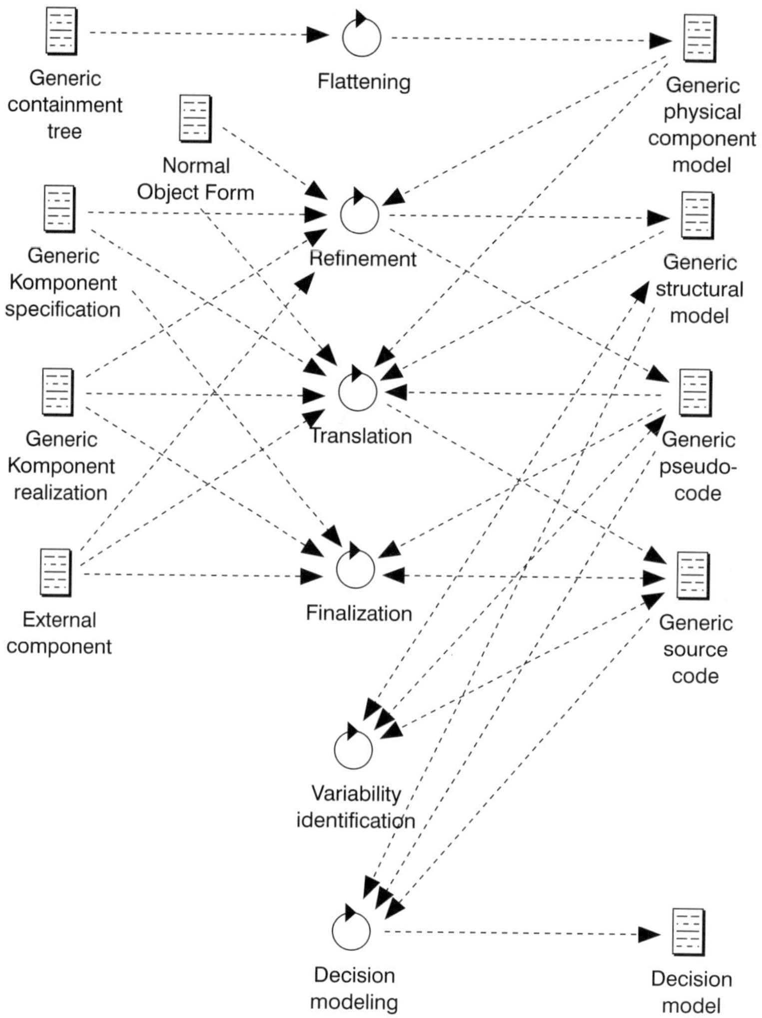

Figure B.10
Komponent implementation

B.2.3.2.6 Inspection

The inspection activity is part of the framework engineering process (see Figure B.6). It consists of the following activities:

- inspection planning
- defect detection
- defect collection
- enactment of corrective steps.

The product flow between them is shown in Figure B.12. The inspection technique is described in Chapter 18.

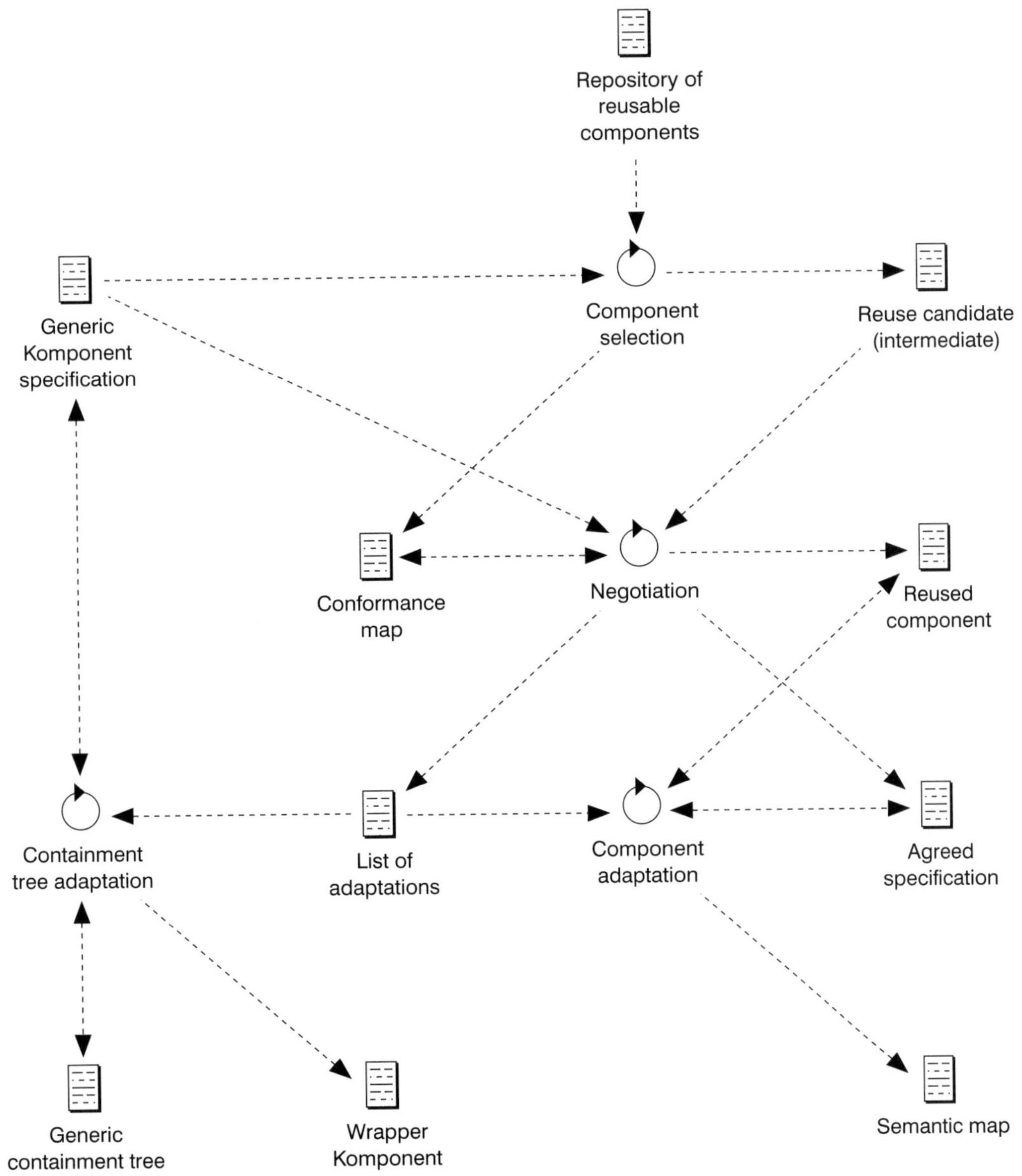

Figure B.11
Component reuse

B.2.3.2.7 Measurement of structural properties

The measurement of structural properties activity is part of the framework engineering process (see Figure B.6). It consists of the following activities:

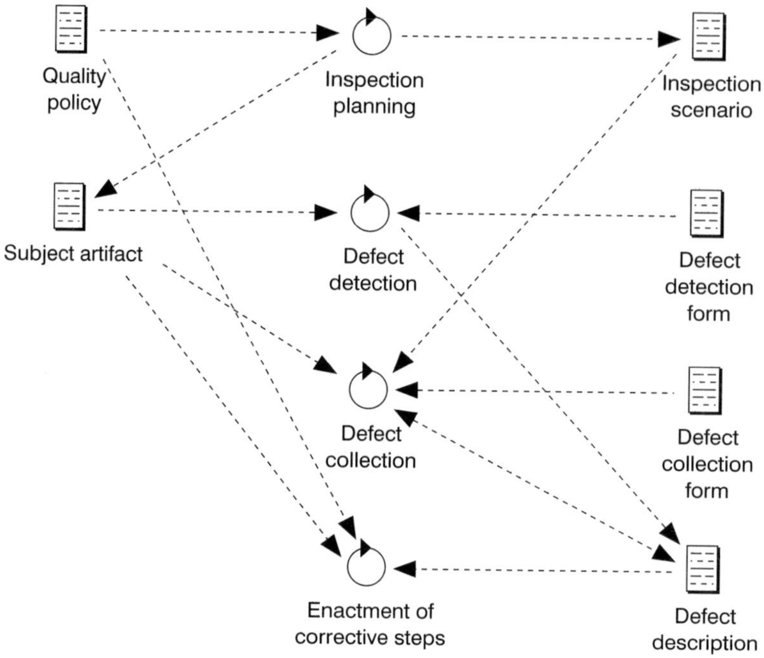

Figure B.12
Inspection

- measurement planning
- quality measurement
- quality analysis.

The product flow between them is shown in Figure B.13. The measurement of structural properties technique is described in Chapter 19.

The reader should be aware that measurement planning is usually performed at the beginning of the project. It might also be embedded into an organizational improvement program as well.

B.2.3.2.8 Testing

The testing activity is part of the framework engineering process (see Figure B.6). It consists of the following activities:

- test suite definition
- test execution
- test result analysis.

The product flow between them is shown in Figure B.14. The testing technique is described in Chapter 18.

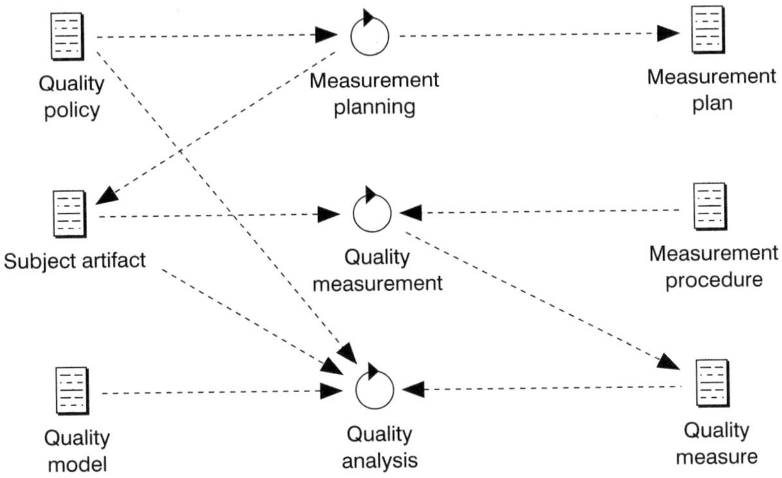

Figure B.13
Measurement of structural properties

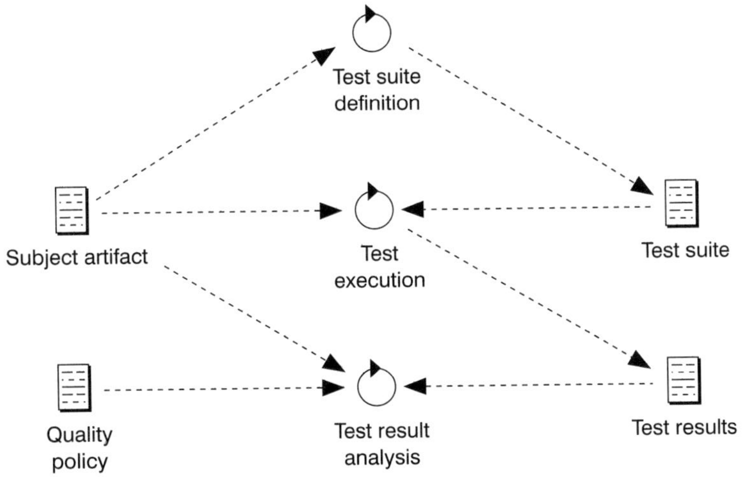

Figure B.14
Testing

The reader should be aware that test suite definition is usually performed as part of framework engineering, whereas test execution and test result analysis can only be performed as part of application engineering, or specifically, after construction and component building.

B.2.3.3 Application engineering

The application engineering process is part of the product line engineering process (see Figure B.5). It consists of the following activities:

- application context realization (Section B.2.3.3.1)
- application Komponent specification (Section B.2.3.3.2)

- application Komponent realization (Section B.2.3.3.3)
- application Komponent implementation (Section B.2.3.3.4)
- construction
- component building
- releasing
- inspection (Section B.2.3.3.5)
- measurement of structural properties (Section B.2.3.3.6)
- testing (Section B.2.3.3.7).

Most activities are further decomposed to show their internal structure. They are described separately in the respective sections. The product flow between them is shown in Figure B.15.

The KobrA method is very prescriptive about modeling Komponents, but it leaves freedom for those activities at the "back end" of the development process. Consequently, the construction activity, the component building activity, and the releasing activity have deliberately not been refined. The engineer is free to apply any technology for these activities that may help or already be in use.

In fact, all activities up to application Komponent implementation take place in a Komponent-centered environment. Such an environment should provide specific support for dealing with Komponents.[59] The remaining activities – construction, component building, and releasing – take place in a different environment that is centered around physical components instead. This might be a collection of command-line tools working on a file system (e.g. Emacs, Java compiler, and make), or a CASE tool working on a repository (e.g. VisualAge).

B.2.3.3.1 Application context realization

The application context realization activity is part of the application engineering process (see Figure B.15). It consists of the following activities:

- instantiation
- enterprise modeling
- structural modeling
- usage modeling
- interaction modeling.

The product flow among them is shown in Figure B.16. In Figure B.16, the left part deals with instantiating the framework, whereas the right part deals with extending the application according to customer-specific requirements. The application context realization activity is described in Chapter 16.

59. Please refer to the KobrA website at [www.iese.fhg.de/KobrA/book/] for more information on such tools.

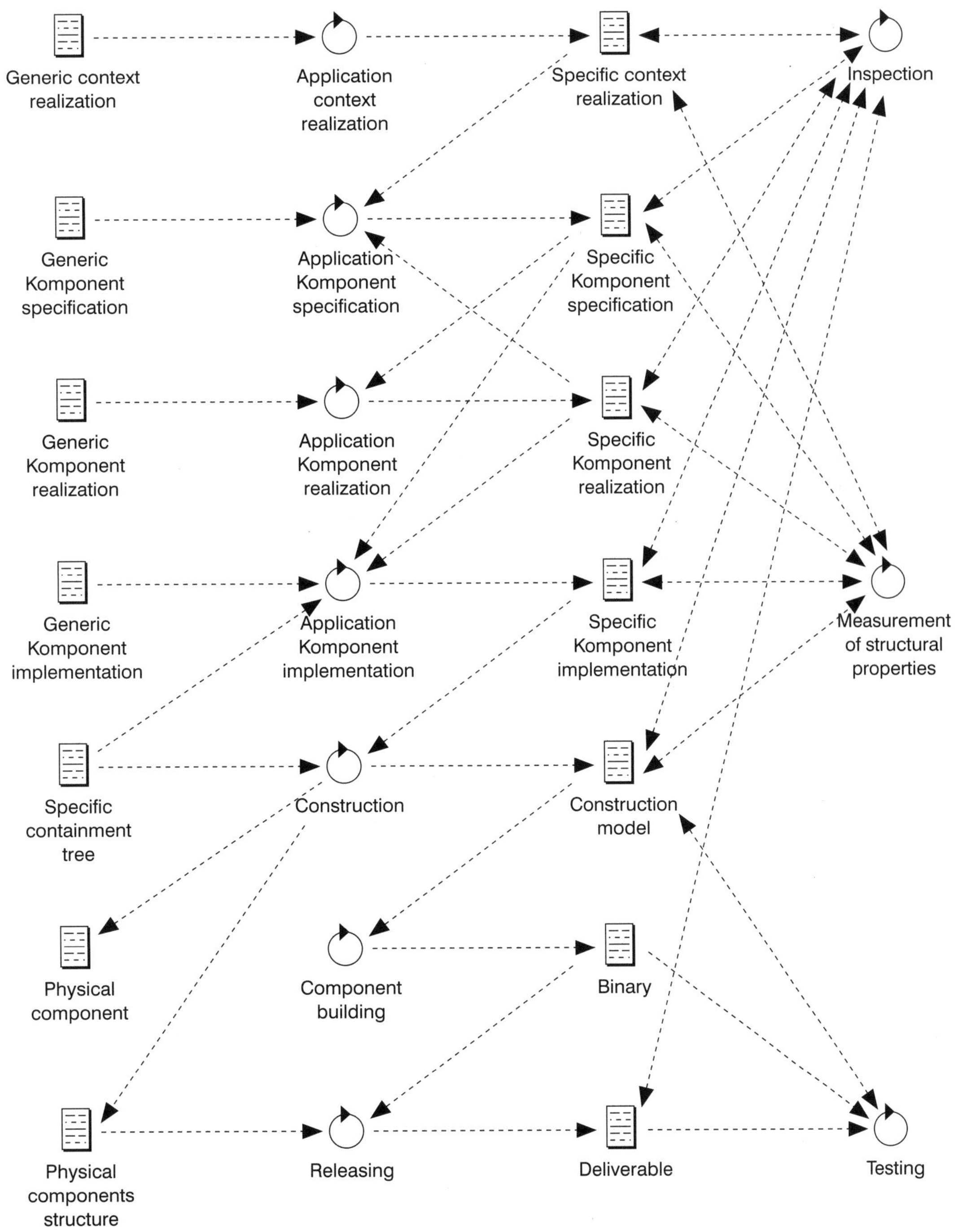

Figure B.15
Application engineering

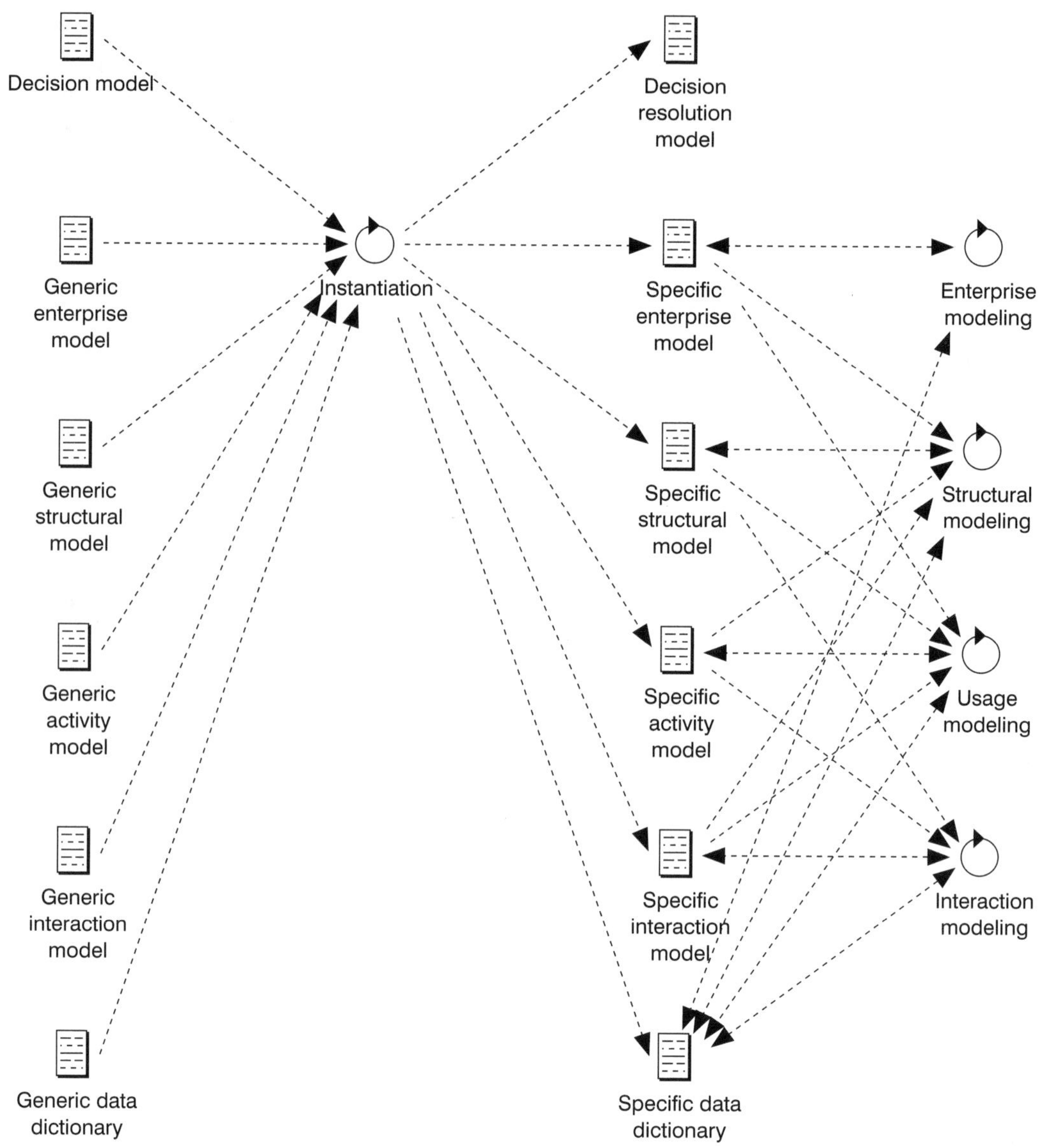

Figure B.16
Application context realization

B.2.3.3.2 Application Komponent specification

The application Komponent specification activity is part of the application engineering process (see Figure B.15). It consists of the following activities:

- instantiation
- structural modeling

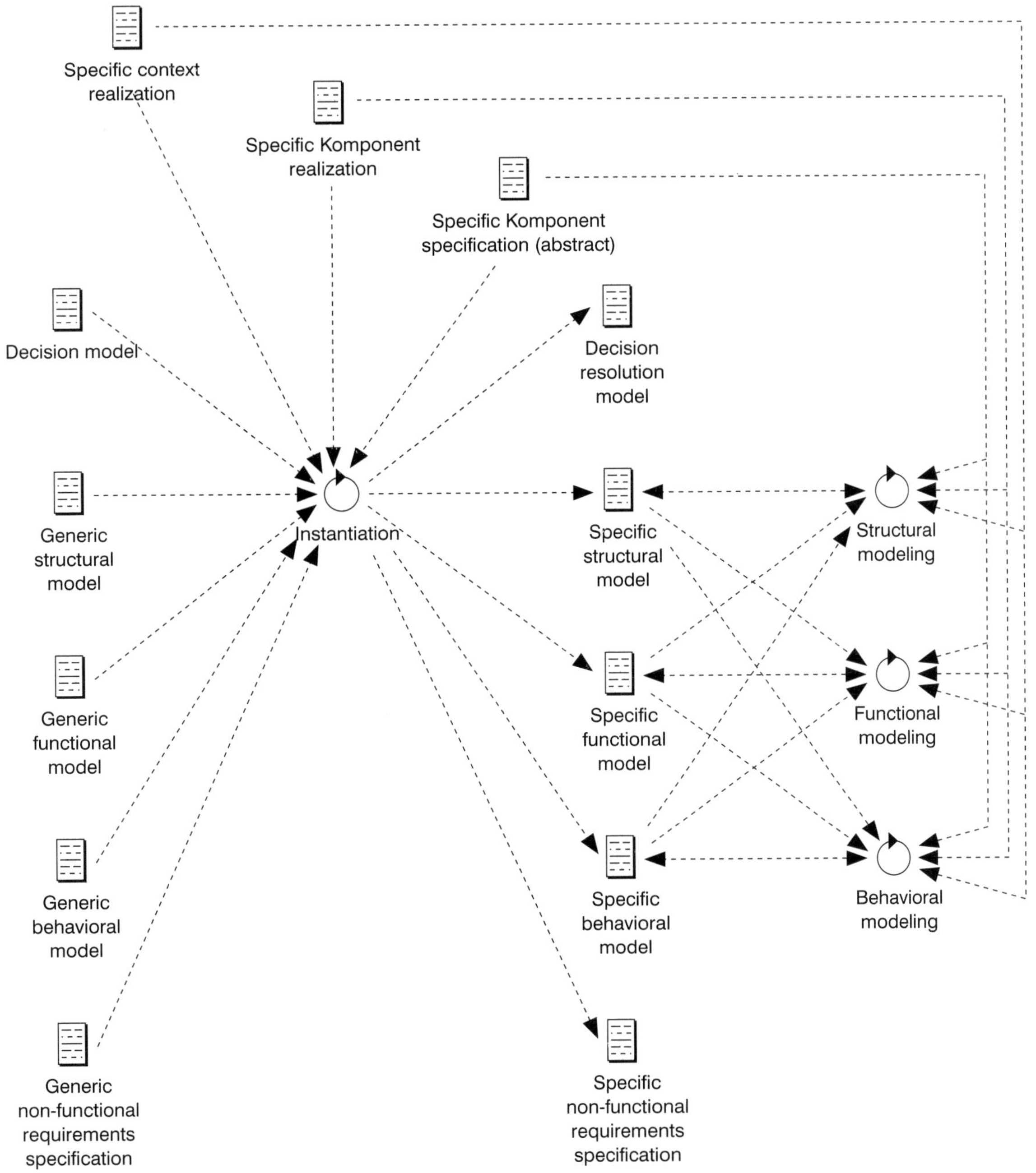

Figure B.17
Application Komponent specification

- functional modeling
- behavioral modeling.

The product flow between them is shown in Figure B.17. In Figure B.17, the left part deals with instantiating the framework, whereas the right part deals with extending the application according to customer-specific requirements. The application Komponent specification activity is described in Chapter 16.

B.2.3.3.3 Application Komponent realization

The application Komponent realization activity is part of the application engineering process (see Figure B.15). It consists of the following activities:

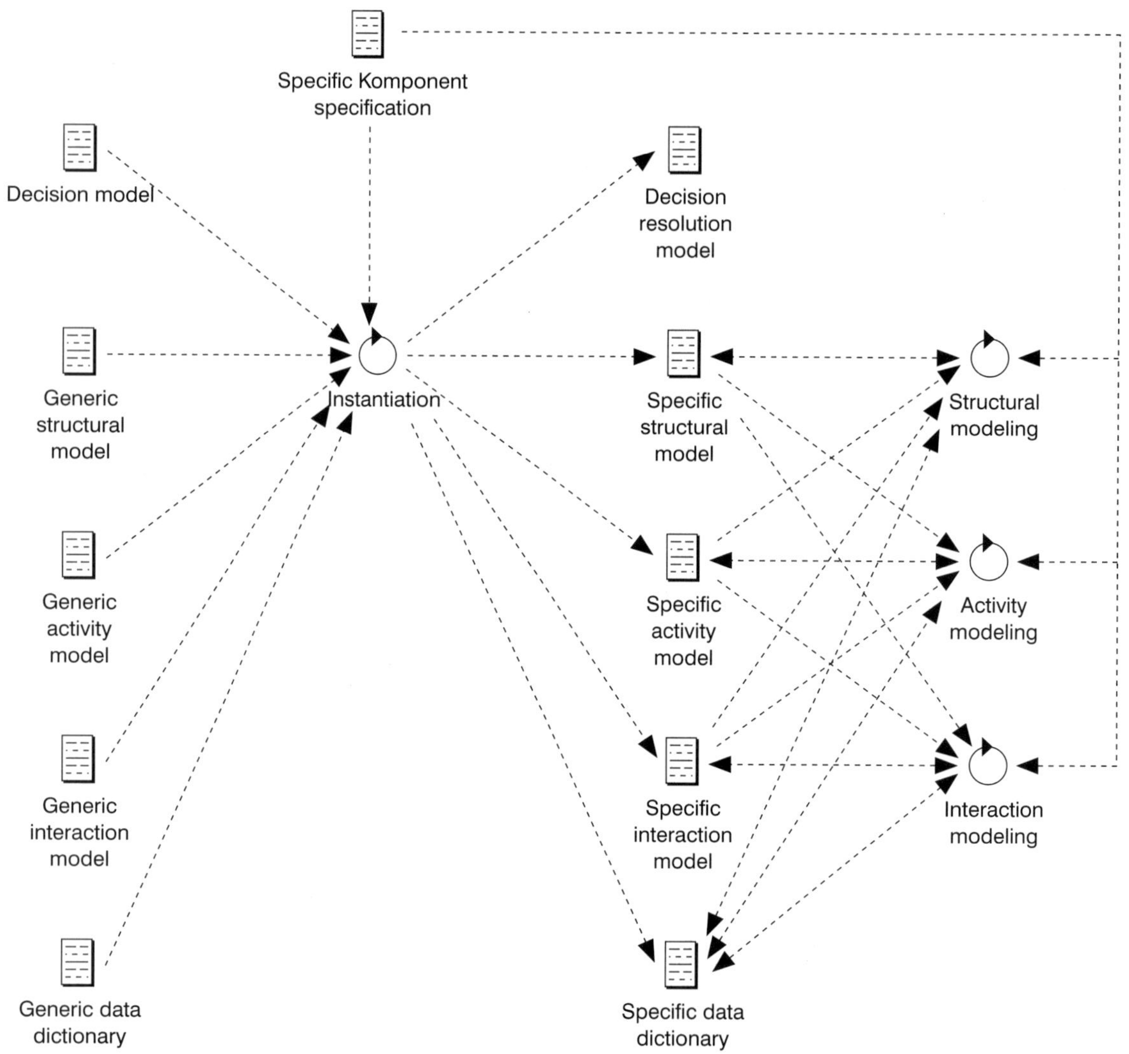

Figure B.18
Application Komponent realization

- instantiation
- structural modeling
- activity modeling
- interaction modeling.

The product flow among them is shown in Figure B.18. In Figure B.18, the left part deals with instantiating the framework, whereas the right part deals with extending the application according to customer-specific requirements. The application Komponent realization activity is described in Chapter 16.

B.2.3.3.4 Application Komponent implementation

The application Komponent implementation activity is part of the application engineering process (see Figure B.15). It consists of the following activities:

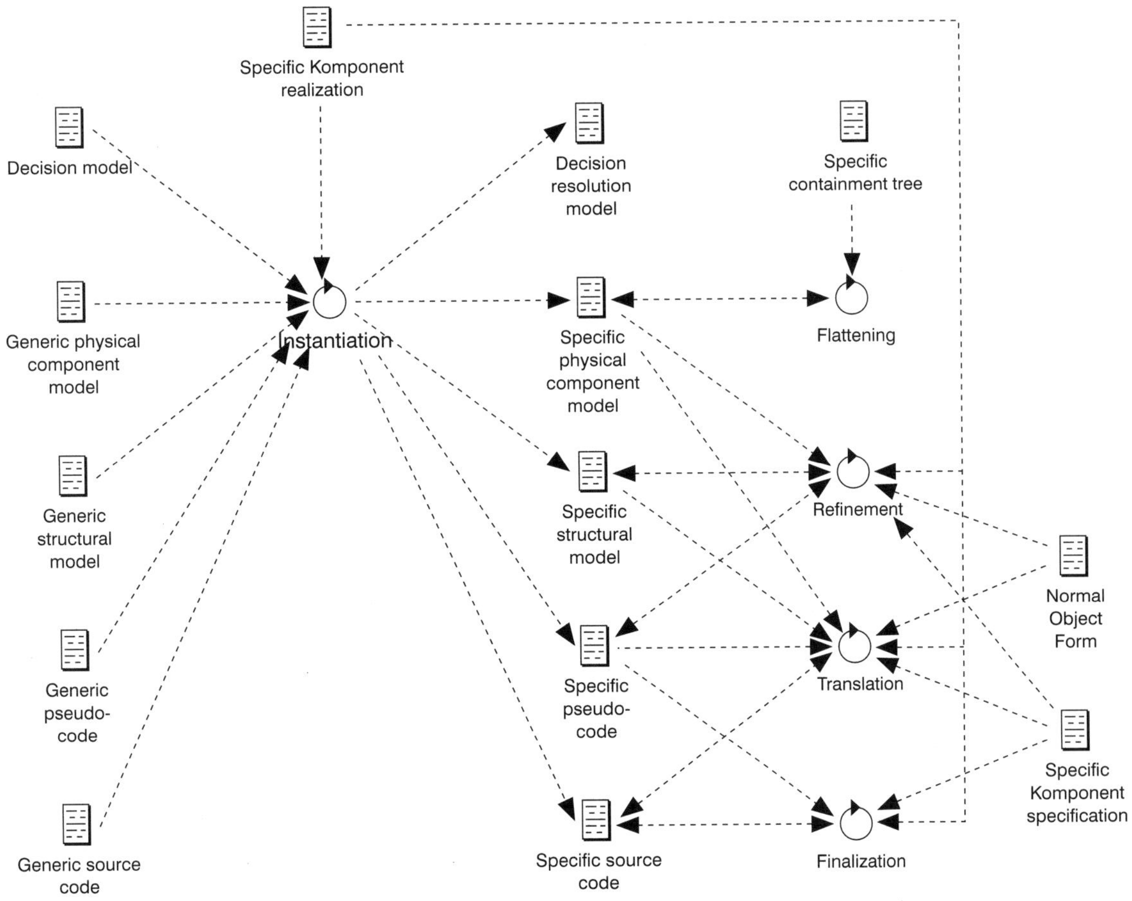

Figure B.19
Application Komponent implementation

- instantiation
- flattening
- refinement
- translation
- finalization.

The product flow between them is shown in Figure B.19. In Figure B.19, the left part deals with instantiating the framework, whereas the right part deals with extending the application according to customer-specific requirements. The application Komponent implementation activity is described in Chapter 16.

B.2.3.3.5 Inspection

The inspection activity is part of the application engineering process (see Figure B.15). It consists of the following activities:

- inspection planning
- defect detection
- defect collection
- enactment of corrective steps.

Figure B.20
Inspection

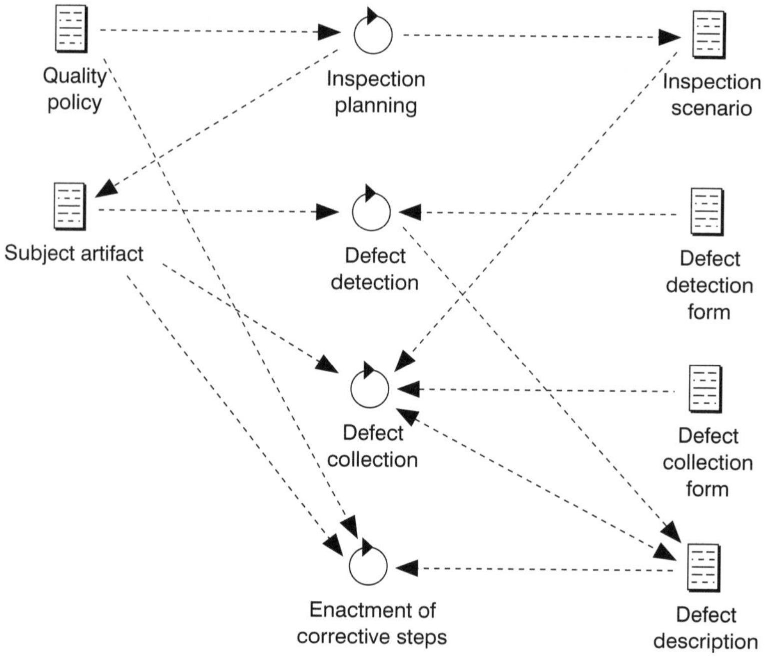

The product flow among them is shown in Figure B.20. The inspection technique is described in Chapter 18.

B.2.3.3.6 Measurement of structural properties

The measurement of structural properties activity is part of the application engineering process (see Figure B.15). It consists of the following activities:

- measurement planning
- quality measurement
- quality analysis.

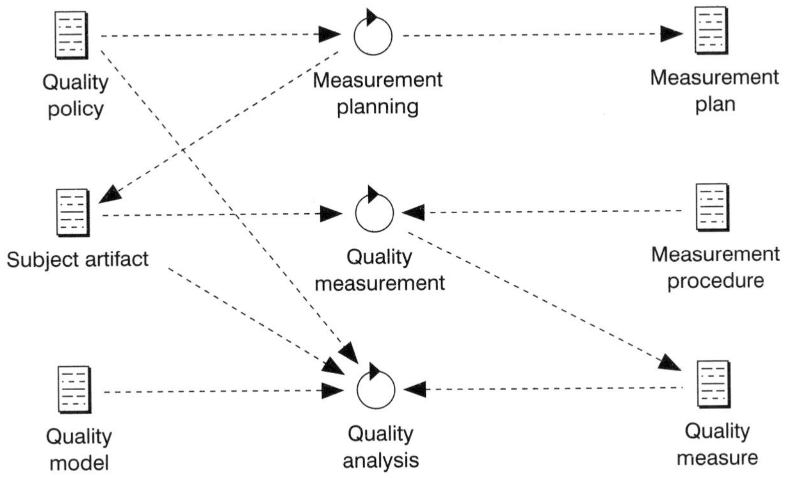

Figure B.21 Measurement of structural properties

The product flow between them is shown in Figure B.21. The measurement of structural properties technique is described in Chapter 19.

The reader should be aware that measurement planning is usually performed at the beginning of the project. It might also be embedded into an organizational improvement program.

B.2.3.3.7 Testing

The testing activity is part of the application engineering process (see Figure B.15). It consists of the following activities:

- test suite definition
- test execution
- test result analysis.

The product flow between them is shown in Figure B.22. The testing technique is described in Chapter 18.

The reader should be aware, that test suite definition is usually performed as part of framework engineering, whereas test execution and test result analysis can only be performed as part of application engineering, or specifically, after construction and component building.

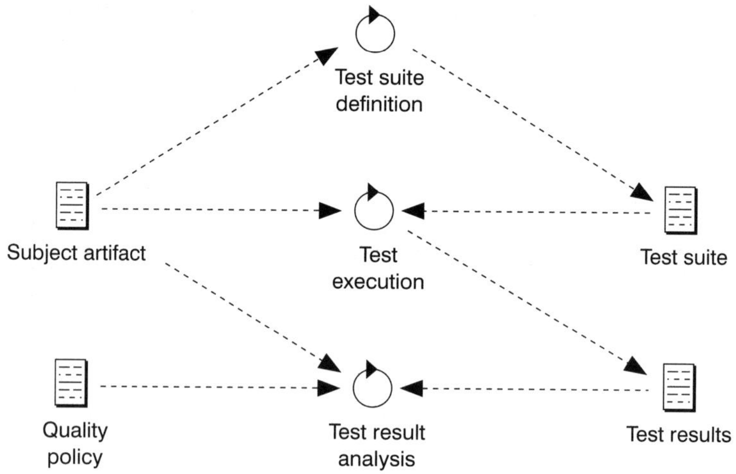

Figure B.22
Testing

Glossary

Acquired server. A server of a Komponent which the Komponent itself does not create but receives access to during the execution of the system. This means that the server is passed to the Komponent as the parameter of an operation invocation.

Activity model. A part of a Komponent realization which describes in terms of UML activity diagrams and activity specifications the algorithms by which the operations of a Komponent are realized.

Application. A single, specific system, usually created by instantiating a framework. In KobrA, this includes a specific context realization, a containment tree with specific Komponents, Komponent implementations, and executable physical components.

Behavioral model. A part of a Komponent specification which describes in terms of statechart diagrams and/or statechart tables how a Komponent behaves in response to external stimuli. A Komponent's behavioral model describes the events it recognizes, the externally visible states it exhibits, and the possible transitions between these states.

Client. A client of a component instance X is any other component instance that invokes X's methods. A client of a component Y (as a type) is any other component with instances that are clients of Y's instances.

Clientship. The clientship (or uses) relationship defines a client–server relationship between components or component instances. It indicates that an instance of one component (the client) uses, or invokes, the services of instances of another component (the server). Clientship is an asymmetric relationship requiring the client to have visibility of the server but not the server to have visibility of the client.

Component. The development-time description of a unit of behavior and composition with contractually specified interfaces and context dependencies only. It therefore represents a type from which component instances can be instantiated at run-time. A Komponent is one particular kind of component.

Component instance. A run-time unit of behavior and composition with contractually specified interfaces and context dependencies only. A Komponent instance is one particular kind of component instance.

Composition. A run-time dependency defining the composite/part relationship between Komponent instances. It is a form of aggregation which implies a strong composite/part relationship. In particular, the part is private to the composite, and its lifetime is bound by that of the composite.

Conformance. A relationship between two Komponents which guarantees that instances of one Komponent will be able to satisfy all the requests that can be satisfied by instances of the other Komponent. Instances of the former can therefore be substituted for instances of the latter.

Conformance map. Specifies how the native offered interface of a foreign component is mapped to a KobrA representation of the interface. Basically it contains a table, which describes for each specified operation an appropriate sequence of method calls to the foreign component.

Containment. A development-time dependency defining the parent–child relationship between Komponents. Its effect is to extend the name space of the sub-Komponent with the name space of the Komponent. A Komponent can have an arbitrary number of sub-Komponents. The properties of this relationship are specialized from the UML ownership relationship by virtue of the package characteristics of a Komponent.

Containment tree. A tree structure of nested Komponents, created by the recursive application of the specification and realization activities.

Contract Dependency. See *Clientship.*

Context Realization. Describes the properties of the environment of the Komponent at the root of a containment tree. Since the root Komponent represents the system under development, the environment of the root Komponent corresponds to the context of the system. The context realization includes all the models of a normal Komponent realization, as well as additional enterprise models to capture the abstract business for which the system is needed.

Creation. Links a Komponent instance to the Komponent instance that created it. It is therefore also reflected in the development-time description of Komponents. The properties of this relationship are specialized from the UML usage relationship (stereotyped as <<creates>>) by virtue of both the class and package characteristics of a Komponent.

Created server. A server of a Komponent which the Komponent itself creates.

Creation tree. A run-time tree traced out by the creation relationships within a system of Komponent instances.

Data dictionary. A table which contains a description of every model element involved in the description of a Komponent. Each row of the table corresponds to one element, and contains the name of the element, the type of the element, and a short description of its role in the model.

Decision model. Characterizes a Komponent as being generic, and describes the groups of features possessed by the Komponent for different members of a product line. It identifies the variation points and the corresponding resolution space in terms of the effects of each resolution option on the other models in the Komponent specification, Komponent realization, or Komponent implementation.

Decision resolution model. Characterizes a Komponent as being specific, and describes the choices that were made to resolve the variabilities in a generic Komponent in order to make it specific for a certain application. It therefore describes the resolution of the variation points in the product line for a specific application.

Defect. Can be characterized as any product anomaly representing a deviation from the required quality properties. Examples include correctness, reliability, maintainability, or testability. An error usually carries the connotation of human fallability in requirements analysis, design, coding, or test planning, and leads to a fault in a program that may or may not cause a failure.

Enterprise concept diagram. Captures the actors and data entities of importance in a business process.

Enterprise model. Part of a context realization, and describes the concepts and activities of importance in a business process. It consists of an enterprise-concept diagram and an enterprise-process diagram.

Enterprise process diagram. Captures the hierarchy of activities and tasks of importance in a business process.

External component. A component that is outside a given containment tree. It can either be a Komponent created within another containment tree or a foreign component.

Framework. A generic description of the software artifacts making up a family of applications. In KobrA, this includes a generic context realization, a containment tree with generic Komponents, and also possibly Komponent implementations.

Foreign component. A non-KobrA component.

Functional Model. A part of a Komponent specification which describes the externally visible effects of the operations supplied by the Komponent in terms of textual-operation specifications. Each operation has one operation specification.

Generalization. The generalization/specialization relationship between two Komponents resembles sub-typing inheritance in object-oriented programming languages. It indicates that one Komponent (the specialized Komponent) has at least all the properties of another Komponent (the general Komponent) and thus conforms to it.

Generic Komponent. A Komponent that embraces the features needed for all members of a product line and contains points of variation corresponding to the differences between these members. The points of variation, and possible resolutions thereof, are captured in a decision model.

Implementation. A Komponent implementation provides a language-level description of how a Komponent works. It consists of a structural model, a physical component model, source code, and pseudo-code.

Implementation dependency. Represents the relationship between a Komponent realization and a Komponent implementation, and contains information about the consistency of the two.

Implementation profile. A UML profile containing elements which are semantically close to the constructs of an implementation language such as a programming language. An implementation profile defines the constructs which are the target of refinement patterns and the source of translation patterns.

Inspection. Defines a rigorous analysis technique which encompasses a rigorous process, well-defined roles, and the measurement of inspection data. Inspections cover requirements documents, design, code, user documentation, and even test plans and test cases.

Inspection scenario. Provides prescriptive instructions for detecting and collecting defects from the perspective of a particular stakeholder in a development project.

Interaction model. A part of a Komponent realization which describes how the Komponent interacts with its server Komponents in order to realize its operations. It consists of one or more UML interaction diagrams.

Komponent. Short form of "KobrA component."

KobrA component. A logical component documented according to the KobrA approach. In particular, a Komponent has a specification, describing what the Komponent does, a realization, defining how it does it, and an implementation, describing how it is implemented using language-level constructs and physical components.

Komponent realization. Describes how the private design of a Komponent fulfills its specification. It comprises a structural model, an activity model, an interaction model, a data dictionary, quality documentation, and either a decision model or a decision-resolution model, depending on whether it is part of a framework or an application respectively.

Komponent specification. Describes the externally visible properties of a Komponent and can be viewed as defining its interface. It consists of a structural model, a behavioral model, a functional model, a non-functional requirements specification, quality documentation, and either a decision model or a decision resolution model depending on whether it is part of a framework or an application, respectively.

Logical component. A logical unit of behavior which has the properties of a class in that it defines the features and behavior of its instances, and the properties of a sub-system in that it acts as a container for other elements, including possibly logical subcomponents.

Measure. An empirical objective assignment of a number (or symbol) to a software entity (product, process, resource) to characterize a specific attribute.

Measurement. The process by which numbers or symbols are assigned to quality attributes of software entities in the real world.

Normal Object Form (NOF). A predefined implementation profile which contains elements corresponding to the core features of object-oriented programming languages, and thus can be simply translated into any mainstream object-oriented language.

Non-functional requirements specification. Describes the constraints on how the functional requirements should be satisfied, including the quality levels that are acceptable under various circumstances.

Operation specification. A table which provides a declarative description of a Komponent operation. Of 11 possible clauses, the assumes and result clauses are the most important. The assumes clause defines the pre-conditions that must be true for the operation to be guaranteed to work properly, while the result clause defines the post-conditions that become true if the operation is invoked with a satisfied assumes clause.

Physical component. Represents a modular, deployable and replacable part of a system that encapsulates implementation and exposes a set of interfaces. Physical components provide a mechanism for the embodiment of logical components such as KobrA components. A given physical component may encapsulate one or more logical components.

Principle of encapsulation. The principle that the description of what a component does should be separated from how it does it.

Principle of locality. The principle that all development-time artifacts, including UML diagrams, should be oriented towards the description of one component or activity, and should only contain local information from the perspective of that entity.

Principle of parsimony. The principle that every descriptive artifact should have just enough information – no more and no less. This is also known as "Occam's Razor."

Principle of uniformity. The principle that every behavior-rich entity type should be treated as a component, and that every component should be treated the same, irrespective of its position in the containment hierarchy.

Pseudo-code. An abstract algorithmic description of a Komponent's operations.

Quality. The totality of features or properties of an entity that bear on its ability to satisfy stated and implied needs.

Quality attribute. Any feature or property of a software product, process, or resource that bears on its ability to satisfy stated and implied needs.

Quality documentation. The container for the artifacts generated by the execution of the quality assurance program in accordance with the quality policy. This includes defect descriptions, test suites, test results, and structural measures.

Quality model. A formula and/or set of instructions for how the obtained quality measures are to be interpreted to draw conclusions about the quality of a software entity (product, process, resource).

Quality policy. Defines the artifacts and concepts needed to execute the quality assurance program. The artifacts involved include defect detection forms, defect collection forms, inspection scenarios, quality models, and the measurement plan.

Reading technique. Defines a series of steps for the individual analysis of software documents in the context of an inspection to achieve the understanding necessary for detecting defects.

Realization dependency. Represents the relationship between a Komponent's specification and realization. It contains information about the consistency of the Komponent specification and Komponent realization.

Refinement pattern. Defines a conceptual mapping between UML model elements, with associated name, context, description and trace information. The target of the mapping is the set of constructs defined by a suitable implementation profile.

Required interface. The set of services provided by an acquired server of a Komponent. The Komponent instance uses these services in order to realize its own services. A given Komponent instance can have multiple required interfaces.

Semantic map. Describes a mapping between the specification of a desired Komponent and the KobrA specification of the interface offered by a foreign component. It contains no functional code, but simply defines name mappings and feature correspondences.

Server. A server of a component instance X is any other component instance whose methods X invokes. A server of a component Y (as a type) is any other component with instances that are servers of Y's instances.

Source code. The source code for a Komponent is a tool-comprehensible representation of how the Komponent works, written in a specific programming language or other implementation notation.

Specific Komponent. A Komponent that has no points of variation, and is thus tailored to the needs of a particular application. When instantiated from a generic Komponent, a specific Komponent possesses a decision resolution model to describe how the points of variation were resolved.

Structural model. Describes classes and the nature of their attributes, operations and relationships. Structural models are used in three places in KobrA: in Komponent specifications to define the externally visible structural and informational properties of a Komponent; in Komponent realizations and context realizations to define the structural and informational properties; and in Komponent implementations to provide a detailed description of a Komponent's implementation in terms of language-level constructs.

Supplied interface. Services offered by a Komponent to its clients in the form of a set of operations.

Testing. The process of executing a component or parts of it for the purpose of detecting defects or showing the reliability of a component.

Translation pattern. Defines a mapping between a UML model element and a corresponding set of elements in a target implementation language. It has associated name, context, description, and trace information.

References

[ABJ+98] Atkinson, C., Bunse, C., Jungmayr, S. and Kamsties, E., "Systematic Object-Oriented Development," in *Proceedings of the GI-Workshop "Anwendung von objektorientierten Entwicklungsstrategien und deren Unterstützung durch Vorgehensmodelle,"* Frankfurt, Germany, 1998.

[ABM00] Atkinson, C., Bayer, J. and Muthig, D., "Component-Based Product Line Development: The KobrA Approach," lst International Software Product Line Conference, Pittsburgh, August 2000.

[AF98] Allen, P. and Frost, S., *Component-Based Development for Enterprise Systems, Applying the SELECT Perspective*, Cambridge University Press/SIGS, Cambridge, 1998.

[AI95] Atkinson, C. and Izygon, M., *ION a Notation for the Graphical Depiction of Object-Oriented Programs*, University of Houston – Clear Lake, Technical Report, 1995.

[AIS77] Alexander, C., Ishikawa, S., Silverstein, M., Jacobson, M., Fiksdahl-King, I. and Angel, S., *A Pattern Language*, Oxford University Press, 1977.

[AKZ96] Awad, M., Kuusela, J. and Zieger, J., *Object-Oriented Technology for Real-Time Systems: A Practical Approach using OMT and Fusion*, Prentice Hall, 1996.

[Alex79] Alexander, C., *The Timeless Way of Building*, Oxford University Press, 1979.

[Allen00] Allen, P., *Realizing e-Business with Components*, Addison-Wesley, 2000.

[Atk97] Atkinson, C., "Adapting the Fusion Process to Support the UML," Object Magazine, 1997.

[BA99a] Bunse, C. and Atkinson, C., "Improving Quality in Object-Oriented Software: Systematic Refinement and Translation of Models to Code," in *Proceedings of the 12th International Conference on Software & Systems Engineering and their Applications* (*ICSSEA '99*), Paris (France), 1999.

[BA99b] Bunse, C. and Atkinson, C., "The Normal Object Form: Bridging the Gap from Models to Code," in *Proceedings of the 2nd International Conference on the Unified Modeling Language* (*UML '99*), Fort Collins (USA), 1999.

[BAB00] Boehm, B., Abts, C., Brown, W., Chulani, S., Bradford C., Kendall, H. Madachy, R., Reifer, D. J. and Steece, B., *Software Cost Estimation with COCOMO II*, Prentice Hall PTR, 2000.

[Bas97] Basili, V. R., "Evolving and Packaging Reading Technologies," *Journal of Systems and Software*, 38(1), July 1997.

[Bay99] Bays, M. E., *Software Release Methodology*, Prentice Hall, Upper Saddle River, NJ, 1999.

[BFK99] Bayer, J., Flege, O., Knauber, P., Laqua, R., Muthig, D., Schmid. K., Widen, T. and Debaud, J.,-M., "PuLSE. A Methodology to Develop Software Product Lines," in *Proceedings of the Symposium on Software Reuse* (*SSR '99*), May 1999.

[BBC99] Bär, H., Bauer, M., Ciupke, U., Demeyer, S., Ducasse, S., Lanza, M., Marinescu, R., Nebbe, R., Nierstrasz, O., Przybilski, M., Richner, T., Rieger, M., Riva, C., Sassen, A.-M., Schuiz, B., Steyart, P., Tichelaar, S. and Weisbrod, J. *The FAMOOS Object-Oriented Reengineering Handbook*. Available at http://www.iam.unibe.ch/~famoos/handbook/

[BC87] Beck, K. and Cunningham, W., "Using Pattern Languages for Object-Oriented Programs," Technical Report No. CR–87–43, Computer Research Laboratory, Tektronix, Inc., September 1987.

[BCK98] Bass, L., Clements, P. and Kazman, R., *Software Architecture in Practice*. Addison-Wesley, 1998.

[BCR94a] Basili, V.R., Caldiera, G. and Rombach, H. D., "Goal Question Metric Paradigm," in J. J. Marciniak, ed., *Encyclopedia of Software Engineering*, vol. 1, pp. 528–532. John Wiley & Sons, 1994.

[BCR94b] Basili, V. R., Caldiera, G. and Rombach, H.D., "Experience Factory," in J. J. Marciniak, ed., *Encyclopedia of Software Engineering*, vol. 1, pp. 469–476. John Wiley & Sons, 1994.

[BDW99] Briand, L., Daly, J. and Wüst, J., "A Unified Framework for Coupling Measurement in Object-Oriented Systems," IEEE *Transactions on Software Engineering* 25(1), 91–122, 1999.

[BEFL98] Briand, L., El-Emam, K., Fussbroich, T. and Laitenberger, O., "Using Simulation to Build Inspection Efficiency Benchmarks for Development Projects," in *Proceedings of the 20th International Conference on Software Engineering*, pp. 340–349. IEEE Computer Society Press, 1998.

[Bei90] Beizer, B., *Software Testing Techniques*, 2nd edn., Van Nostrand Reinhold, New York, 1990.

[BFV96] Budinsky, F. J., Finnie, M. A., Vlissides, J. M. and Yu, P. S. "Automatic Code Generation from Design Patterns," in *Object Technology*, 35(2), 1996.

[BG00] Bashir, I. and Goel, A. L., *Testing Object-Oriented Software. Life Cycle Solutions*, Springer, New York, 2000.

[BHK99] Becker-Kornstaedt, U., Hamann, D., Kempkens, R., Rösch, P., Verlage, M., Webby, R. and Zettel, J., "Support for the Process Engineer. The Spearmint

Approach to Software Process Definition and Process Guidance," in *11th International Conference on Advanced Information Systems Engineering (CAiSE 99)*, LNCS 1626, pp. 119–133. Springer, 1999.

[Bin99] Binder, R., *Testing Object-Oriented Systems*, Addison-Wesley, 1999.

[BMG01] Bayer, J. Muthig, D. and Göpfert, B., "The Library System Product Line – A KobrA Case Study," Technical Report IESE-Report No. 024.01/E, Fraunhofer Institute for Experimental Software Engineering, Kaiserslautern, Germany, 2001.

[BMR96] Buschmann, F., Meunier, Fl., Rohnert, K, Sommerlad, P. and Stal, M., *A System of Patterns. Pattern-Oriented Software Architecture*, Wiley & Sons, Chichester, 1996.

[BMW99] Bayer, J., Muthig, D. and Widen, T., "Customizable Domain Analysis," in *1st International Symposium on Generative and Component-Based Software-Engineering (GCSE '99)*, Erfurt, Germany, September 1999. Lecture Notes in Computer Science 1799. Springer-Verlag, Heidelberg, Germany, 2000.

[Boe88] Boehm, B. W., "A Spiral Model of Software Development and Enhancement," IEEE *Computer* 21(5), May, 1988.

[Bo87] Booch, G., *Software Components with Ada: Structures, Tools and Subsystems*, 1987.

[Bo94] Booch, G., *Object-Oriented Analysis and Design with Applications*, Benjamin Cummings, 1994.

[BR86] Basili, V. R. and Rombach, D., "Tailoring the Software Process to Project Goals and Environments," Technical Report, University of Maryland, 1986.

[Bun01] Bunse, C., *Pattern-Based Refinement and Translation of Object-Oriented Models to Code*, Fraunhofer IRB Verlag, 2001.

[BWDP00] Briand, L., Wüst, J., Daly, J. and Porter, V., "A Comprehensive Empirical Validation of Product Measures for Object-Oriented Systems," *Journal of Systems and Software* 51, pp. 245–273, 2000.

[BWLI99] Briand, L.,Wüst, J., Lounis, H. and Ikonomovski, S., "Investigating Quality Factors in Object-Oriented Designs: an Industrial Case Study," in *Proceedings of the 21st International Conference on Software Engineering (ICSE 99)*, 345–354, Los Angeles, USA, 1999.

[CE00] Czarnecki, K. and Eisenecker, U. W., *Generative Programming*, Addison-Wesley, May 2000.

[CD00] Cheesman, J. and Daniels, J., *UML Components: A Simple Process for Specifying Component-Based Software*, Addison-Wesley, 2000.

[CDK98] Chidamber, S., Darcy, D. and Kemerer, C., "Managerial Use of Metrics for Object-Oriented Software: An Exploratory Analysis," *IEEE Transactions on Software Engineering*, 24(8), 629–639, 1998.

[CD92] Cooks, S. and Daniels, J., *Designing Object Systems*, Prentice Hall, 1992.

[CN93] Coad, P. and Nicola, E., *Object-Oriented Programming*, Prentice Hall, 1993.

[Col94] Coleman, D., Arnold, P., Bodoff. S., Dollin, C., Gilchrist, H., Hayes, F. and Jeremaes, P., *Object-oriented Development. The Fusion Method*, Prentice Hall, 1994.

[Cot95] Cotton, T., "Evolutionary Fusion: A Customer-oriented Incremental Life-Cycle for Fusion," in *Object-Oriented Development at Work: Fusion in the Real World*, eds R. Malan, R. Letsinger and D. Coleman, Prentice Hall, 1995.

[Cox86] Cox, B. J., *Object-oriented Programming: An Evolutionary Approach*, Addison-Wesley, Reading (Mass), 1986.

[CS95] Coplien, J. O. and Schmidt, D. C. (eds), *Pattern Languages of Program Design*, Addison-Wesley (Software Patterns Series), 1995.

[CY91] Coad, P. and Yourdon, E., *Object-Oriented Analysis*, 2nd edn, Yourdon Press/Prentice Hall, Englewood Cliffs, NJ, 1991.

[Dec96] Deck, M., "Cleanroom and Object-Oriented Software Engineering: A Unique Synergy," *Proceedings of the 8th Annual Software Technology Conference*, in Salt Lake City, USA, 1996.

[DW98] D'Souza, D. F. and Wills A. C., *Objects, Components and Frameworks with UML: The Catalysis Approach*, Addison-Wesley, 1998.

[Erd98] Erdogmus, H., *Representing Architectural Evolution, Institute for Information Technology*, National Research Council, Montreal Road, Ottawa, 1998.

[Fag76] Fagan, M. E., "Design and Code Inspections to Reduce Errors in Program Development," *IBM Systems Journal*, 15(3), 182–211, 1976.

[FHA00] Forsell, M., Halttunen, V. and Ahonen, J., "Use and Identification of Components in Component-Based Software Development Methods," in *Software Reuse. Advances in Software Reusability. 6th International Conference, ICSR-6*. Lecture Notes in Computer Science, 1844, Springer-Verlag, 2000.

[FHG97] Firesmith, D. G., Henderson-Sellers, B. and Graham, I. M., *OPEN Modeling Language Reference Manual*, SIGS Books, Cambridge University Press, 1997.

[Fle00] Flege, O., *Architecture Description Using the UML*. Report of the Fraunhofer Institute for Experimental Software Engineering, No. 092.00-E, 2000.

[Fowl97] Fowler, M., *Analysis Patterns: Reusable Object Models*, Addison-Wesley, 1997.

[Fowl99] Fowler, M., *Refactoring. Improving the Design of Existing Code*, Addison-Wesley, 1999.

[FP98] Fenton, N. and Pfleeger, S., *Software Metrics, A Practical and Rigorous Approach*. International Thompson Computer Press, 1996.

[GA095] Garlan, D., Allen, R. and Ockerbloom, J., "Architectural Mismatch or Why it's hard to Build Systems out of Existing Parts," in *Proceedings of the 17th International Conference on Software Engineering*, April 1995.

[GFD98] Griss, M., Favaro, J. and d'Allesandro, M., "Integrating Feature Modeling with the RSEB," in *Proceedings of the 5th International Conference on Software Reuse (ICSR-5)*, 1998

[GG93] Gilb, T. and Graham, D., *Software Inspection*, Addison-Wesley, 1993.

[GHJ95] Gamma, E., Helm, R., Johnson, R. and Vlissides, J., *Design Patterns – Elements of Reusable Object-Oriented Software*, Addison-Wesley, 1995.

[GHY97] Graham, L., Henderson-Sellers, B. and Younessi, H., *The OPEN Process Specification*, Addison-Wesley, 1997.

[GOW01] Graham, I. with O'Callaghan, A. and Wills, A. C., *Object-Oriented Methods: Principles and Practice*, Addison-Wesley, 2001.

[Gro94] HUM Working Group, *HOOD User Manual*, HOOD User Group, July 1994.

[GS94] Garlan, D. and Shaw, M., "An Introduction to Software Architecture," Technical Report CMU–CS–94–166, School of Computer Science, Carnegie Mellon University, Pittsburgh, PA, USA, January 1994.

[GW94] Selic, B., Gullekson, G. and Ward, P. T., *Real-Time Object-Oriented Modeling*, John Wiley & Sons, 1994.

[Hill] The Hillside Patterns Homepage.
http://www.hillside.net/patterns/patterns.html

[HHC98] Hoek, A., Hall, R. S., Carzaniga, A., Heimbigner, D. and Wolf, A. L., "Software Deployment," *Crosstalk The Journal of Defense Software Engineering*, February 1998.

[ISO/IEC FCD 9126] ISO/IEC FCD 9126–1.2, "Information Technology – Software Product. Quality-Part 1: Quality Model," ISO, Geneva, 1998.

[IS012207] ISO/IEC 12207 "Information Technology – Software Life-cycle Processes," ISO, Geneva, 1995.

[Jac92] Jacobson, I., *Object-Oriented Software Engineering*, Addison-Wesley, 1992.

[JBR99] Jacobson, I., Booch, G. and Rumbaugh, J., *The Unified Software Development Process*, Addison-Wesley, 1999.

[JGJ97] Jacobson, I., Griss, M. and Jonsson, P. *Software Reuse – Architecture, Process, and Organization for Business Success*, Addison-Wesley, 1997.

[JEJ94] Jacobson, L., Ericsson, M. and Jacobson, A., *The Object Advantage: Business Process Reengineering with Object Technology*, Addison-Wesley, 1994.

[KCNP90] Kang, K. C., Cohen, S. G., Novak, W. E. and Peterson, A. S., "Feature-oriented Domain Analysis (FODA) Feasibility Study," Technical Report CMU/SEI–90–TR–21, Software Engineering Institute (SEI), November 1990.

[Kit95] Kit, E., *Software Testing in the Real World*, Addison-Wesley, 1995.

[Kont95] Kontio, J., "OTSO: A Systematic Process for Reusable Software Component Selection," Technical Report UMIAC–TR–95–63, Institute for Advanced Computer Studies and Department of Computer Science, University of Maryland, December 1995.

[Kru00] Kruchten, P. B., *The Rational Unified Process. An Introduction*, Addison-Wesley, 2000.

[Kru95] Kruchten, P. B., "The 4+1 View Model of Architecture," *IEEE Software*, November 1995.

[LA99] Laitenberger, O. and Atkinson, C., "Generalizing Perspective-based Inspection to handle Object-oriented Development Artifacts," in *Proceedings of the 21st International Conference of Software Engineering*, 1999.

[Lai00] Laitenberger, O., *Cost-effective Detection of Software Defects through Perspective-based Inspection*, Fraunhofer IRB Verlag, 2000.

[LASK00] Laitenberger, O., Atkinson, C., Schlich, M. and El Emam, K., "An Experimental Comparison of Reading Techniques for Defect Detection in UML Design Documents," *Journal of Systems and Software*, 2(53) pp. 183–204, 2000.

[LB85] Lehman M. M. and Belady, L., *Program Evolution: Processes of Software Change*, Academic Press, 1985.

[LD00] Laitenberger, O. and DeBaud, J.-M., "An Encompassing Life Cycle Centric Survey of Software Inspection," *Journal of Systems and Software*, 1(50) pp. 5–31, 2000.

[LH93] Li, W. and Henry, S., "Object-oriented Metrics that Predict Maintainability," *Journal Systems and Software*, 23(2), 111–122, 1993.

[Lim98] Lim, W. C., *Managing Software Reuse. A Comprehensive Guide to Strategically Reengineering the Organization for Reusable Components*, Prentice Hall, 1998.

[LT97] Lee, R. C. and Tepfenhart, W. M., *UML and C++: A Practical Guide to Object-Oriented Development*, Prentice Hall, 1997.

[Mac95] Macaulay, L. A., *Requirements Engineering, Applied Computing*, Springer Verlag, 1995.

[Mad90] Madhavji, N., *The Prism Model of Change – Part l and Part 2*, McGill University – School of Computer Science, Montreal, 1990.

[Mao93] Mao, H., "The Box-structure Development Method," PhD Thesis, University of Tennessee, Knoxville, USA, 1993.

[Mar95] Marick, B., *The Craft of Software Testing*, Prentice Hall, 1995.

[Mar00] Marshall, C., *Enterprise Modeling with UML*, Addison-Wesley, 2000.

[McIl68] Mcillroy, M. D., "Mass Produced Software Components," in *Proceedings of the NATO Conference on Software Engineering*, Garmisch, Germany, October 1968.

[Mey00] Meyer, T., "Dynamic Semantics Negotiation in Distributed and Evolving Software Systems: Towards Automated Semantic-Directed System Configuration," PhD thesis, University of Essen, Germany, Logos Verlag Berlin, 2000.

[Mil88] Mills, H. D., "Stepwise Refinement and Verification in Box-structured Systems," *IEEE Computer*, June, 1988.

[MLH87] Mills, H. D., Linger, R. C. and Hevner, A. R., "Box-structured Information Systems," *IBM Systems Journal*, 26(4), pp. 395–413,1987.

[MN98] Maiden, N. A. and Ncube, N., "Acquiring COTS Software Selection Requirements," *IEEE Software*, March/April 1998.

[MRB97] Martin, R. C., Riehle, D. and Buschmann, F. (eds.), *Pattern Languages of Program Design 3*, Addison-Wesley (Software Patterns Series), 1997.

[Musa98] Musa, J., *Software Reliability Engineering*, McGraw-Hill, 1998.

[NQ98] Nesi, P. and Querci, T., "Effort Estimation and Prediction of Object-oriented Systems," *Journal of Systems and Software* 42, pp. 89–102, 1998.

[OMGO0a] OMG Unified Modeling Language Specification, Version 1.4, Object Management Group, 2000.

[OMG00b] "Software Process Engineering: The Unified Process Model (UPM)," Initial Submission, OMG document ad/2000-05-05, May 12, 2000.

[OP00] Ochs, M., Pfahl, D., Chrobok-Diening, G. and Nothhelfer-Kolb, B., "A COTS Acquisition Process: Definition and Application Experience," in *Proceedings of ESCOM-SCOPE 2000*, Munich, Germany, April 2000.

[Paech00] Paech, B., *Aufgabenorientierte Softwareentwicklung – Integrierte Gestaltung von Unternehmen, Arbeit und Software*, Springer Verlag, 2000.

[Paech98a] Paech, B., "The Four Levels of Use Case Description," in E. Dubois, A. Opdahl and K. Pohl (eds) *International Workshop REFSQ 98*, Bericht Facultés Universitaires Notre-Dame de la Paix, Namurs, 1998.

[Paech98b] Paech, B., *On the Role of Activity Diagrams in UML*, UML '98, LNCS1618, pp. 267–277, Springer Verlag, 1998.

[Port] The Portland Pattern Repository. http://c2.com/ppr/index.html

[RBP91] Rumbaugh, J., Blaha, M., Premerlani, W., Eddy, F. and Lorensen, W., *Object-Oriented Modelling and Design*, Prentice Hall, 1991.

[RJB99] Rumbaugh, J., Jacobson, I. and Booch, G., *The Unified Modeling Language Reference Manual*, Addison-Wesley, 1999.

[Rei94] Reichenberger, C., *Konzepte und Verfahren für die Software-Versionsverwaltung*, Universitätsverlag Rudolf Trauner, Linz 1994.

[Rob92] Robinson, P. J., *Hierarchical Object-Oriented Design*, Prentice Hall, 1992.

[RWL96] Reenskaug, T., Wold, P. and Lehne, O., *Working with Objects: The OOram Software Development Method*, Manning/Prentice Hall 1996.

[Sch00] Schmid, K., "Scoping Software Product Lines – An Analysis of an Emerging Technology," in *Proceedings of the 1st Software Product Line Conference* (P. Donohoe, ed.), pp. 513–532, August 2000.

[SGW94] Selic, B., Gullekson, G. and Ward, P. T., *Real-Time Object-Oriented Modeling*, John Wiley and Sons, 1994.

[SG96] Shaw, M. and Garlan, D., *Software Architecture: Perspectives on an Emerging Discipline*, Prentice Hall, 1996.

[Szy99] Szyperski, C., *Component Software: Beyond Object-oriented Programming*, Addison-Wesley 1999.

[Trac95] Tracz, W., *Confessions of a Used Program Salesman. Institutionalizing Software Reuse*, Addison-Wesley, 1995.

[VCK96] Vlissides, J. M., Coplien J. O. and Kerth, N. L. (eds.), *Pattern Languages of Program Design 2*, Addison-Wesley (Software Patterns Series), 1996.

[Web95] Webster, B. F., *Pitfalls of Object-Oriented Development*, M&T Books, 1995.

[Wei97] Weiss, D. M., *Defining Families: The Commonality Analysis*, Lucent Technologies Bell Laboratories, 1997

[WL99] Weiss, D. M. and Lai, C. T. R., *Software Product Line Engineering: A Family Based Software Engineering Process*, Addison-Wesley, 1999.

[WWW90] Wirfs-Brock, R., Wilkerson B. and Wiener, L., *Designing Object-oriented Software*, Prentice Hall, 1990.

[Zel97] Zeller, A., *Configuration Management with Version Sets*, Technical University of Braunschweig, 1997.

Index